Applied Mathematics: Principles and Techniques

Applied Mathematics: Principles and Techniques

Editor: Gregory Rago

NY RESEARCH
PRESS

New York

Published by NY Research Press
118-35 Queens Blvd., Suite 400,
Forest Hills, NY 11375, USA
www.nyresearchpress.com

Applied Mathematics: Principles and Techniques
Edited by Gregory Rago

International Standard Book Number: 978-1-63238-731-8 (Hardback)

Cataloging-in-Publication Data

Applied mathematics : principles and techniques / edited by Gregory Rago.
p. cm.
Includes bibliographical references and index.
ISBN 978-1-63238-731-8
1. Mathematics. I. Rago, Gregory.
QA8.4 .A67 2020
510--dc23

Contents

Preface

Over the recent decade, advancements and applications have progressed exponentially. This has led to the increased interest in this field and projects are being conducted to enhance knowledge. The main objective of this book is to present some of the critical challenges and provide insights into possible solutions. This book will answer the varied questions that arise in the field and also provide an increased scope for furthering studies.

The application of mathematical methods in different fields such as engineering, science, industry, business and computer science is known as applied mathematics. It combines mathematical science with specialized knowledge. Applied mathematics is broadly subdivided into three parts- applied analysis, approximation theory and applied probability. These categorizations are made complex due to the changes in mathematics and science over time. Numerical analysis, algebra, logic, decision theory, financial mathematics are some of the areas of mathematics which are widely applied to the domains of scientific computing, actuarial science, computer science and mathematical economics. This book discusses the fundamentals as well as modern approaches to the field of applied mathematics, and its various principles and techniques. Students, researchers, experts and all associated with applied mathematics will benefit alike from this book.

I hope that this book, with its visionary approach, will be a valuable addition and will promote interest among readers. Each of the authors has provided their extraordinary competence in their specific fields by providing different perspectives as they come from diverse nations and regions. I thank them for their contributions.

Editor

1

Some New Volterra-Fredholm-Type Nonlinear Discrete Inequalities with Two Variables Involving Iterated Sums and Their Applications

Run Xu

Department of Mathematics, Qufu Normal University, Qufu, Shandong 273165, China

Correspondence should be addressed to Run Xu; xurun_2005@163.com

Academic Editor: Samir H. Saker

Some generalized discrete Volterra-Fredholm-type inequalities were developed, which can be used as effective tools in the qualitative analysis of the solution to difference equations.

1. Introduction

In recent years, various forms of inequalities played increasingly important roles in the study of quantitative properties of solutions of differential and integral equations [1–15]. Discrete inequalities, especially the discrete Volterra-Fredholm-type inequalities, have been applied to study the discrete equations widely. For example, see [1–3, 9–11] and the references therein. In this paper, some new Volterra-Fredholm-type discrete inequalities involving four iterated infinite sums were established. Furthermore, to illustrate the usefulness of the established results, some examples were provided for the studying of their solutions on the boundedness, uniqueness, and continuous dependence.

We design the needed symbols as follows:

(a) N_0 denotes the set of nonnegative integers and Z denotes the set of integers, while R denotes the set of real numbers $R_+ = [0, \infty)$.

(b) Let $\Omega := ([m_0, M] \times [n_0, N]) \cap Z^2$, where $m_0, n_0 \in Z$, and $M, N \in Z \cup \{\infty\}$ are two constants.

(c) $K_i > 0$ $(i = 1, 2, 3, 4)$ are all constants, and $l_1, l_2 \in Z$ are two constants.

(d) If U is a lattice, then we denote the set of all R−valued functions on U by $\wp(U)$ and denote the set of all R_+−valued functions on U by $\wp_+(U)$.

(e) For a function $g \in \wp_+(U)$, we have $\sum_{s=m_0}^{m_1} g(s) = 0$ provided $m_0 > m_1$.

We need the following lemmas in the discussions of our main results.

Lemma 1 (see [4]). *Let $u(m,n) \in \wp_+(\Omega)$, $b(s,t,m,n) \in \wp_+(\Omega^2)$ be nondecreasing in the third variable; $k \geq 0$ is a constant. For $(m,n) \in \Omega$, if*

$$u(m,n) \leq k + \sum_{s=m_0}^{m-1} \sum_{t=n_0}^{n-1} b(s,t,m,n)\, u(s,t), \quad (1)$$

then

$$u(m,n) \leq k \exp\left\{\sum_{s=m_0}^{m-1} \sum_{t=n_0}^{n-1} b(s,t,m,n)\right\}. \quad (2)$$

Lemma 2 (see [4]). *Let $u(m,n), a(m,n), c(m,n) \in \wp_+(\Omega)$. If $a(m,n)$ is nondecreasing in the first variable, then, for $(m,n) \in \Omega$,*

$$u(m,n) \leq a(m,n) + \sum_{s=m_0}^{m-1} c(s,n)\, u(s,n), \quad (3)$$

then

$$u(m,n) \leq a(m,n) \prod_{s=m_0}^{m-1} [1 + c(s,n)]. \quad (4)$$

Lemma 3 (see [5]). *Let $a \geq 0$, $p \geq q \geq 0$, and $p \neq 0$; then, for any $K > 0$,*

$$a^{q/p} \leq \frac{q}{p} K^{(q-p)/p} a + \frac{p-q}{p} K^{q/p}. \tag{5}$$

2. Main Results

Theorem 4. *Suppose that $u(m,n), a(m,n), b_1(m,n), b_2(m,n) \in \wp_+(\Omega)$, $c_i(s,t,m,n)$, $d_i(s,t,m,n)$, $e_i(s,t,m,n)$, $f_j(s,t,m,n)$, $g_j(s,t,m,n)$, $w_j(s,t,m,n) \in \wp_+(\Omega^2)$, and p, q_i, r_i, h_j, v_j are nonnegative constants with $p \geq q_i > 0$, $p \geq r_i > 0$ ($i = 1,2,\ldots,l_1$), $p \geq h_j > 0$, $p \geq v_j > 0$ ($j = 1,2,\ldots,l_2$), and $c_i, d_i, e_i, f_j, g_j, w_j$ being nondecreasing in the last two variables, $b_1(m,n)$ and $b_2(m,n)$ are also nondecreasing. If*

$$\begin{aligned} u^p(m,n) &\leq a(m,n) + b_1(m,n) \\ &\cdot \sum_{i=1}^{l_1} \sum_{s=m_0}^{m-1} \sum_{t=n_0}^{n-1} \left[c_i(s,t,m,n) u^{q_i}(s,t) + d_i(s,t,m,n) \right. \\ &\cdot \left. u^{r_i}(s,t) + e_i(s,t,m,n)\right] + b_2(m,n) \\ &\cdot \sum_{j=1}^{l_2} \sum_{s=m_0}^{M-1} \sum_{t=n_0}^{N-1} \left[f_j(s,t,m,n) u^{h_j}(s,t) \right. \\ &+ \left. g_j(s,t,m,n) u^{v_j}(s,t) + w_j(s,t,m,n)\right], \end{aligned} \tag{6}$$

then, for $(m,n) \in \Omega$, we have

$$u(m,n) \leq \left\{a(m,n) + b(m,n) \frac{J(M,N)}{1-\lambda(M,N)} C(m,n)\right\}^{1/p}, \tag{7}$$

provided that $\lambda(M,N) < 1$, where

$$b(m,n) = \max\{b_1(m,n), b_2(m,n)\}, \tag{8}$$

$$C(m,n) = \exp\left\{\sum_{s=m_0}^{m-1} \sum_{t=n_0}^{n-1} B(s,t,m,n)\right\}, \tag{9}$$

$$B(s,t,m,n) = \sum_{i=1}^{l_1} \left[c_i(s,t,m,n) \frac{q_i}{p} K_1^{(q_i-p)/p} + d_i(s,t,m,n) \frac{r_i}{p} K_2^{(r_i-p)/p}\right] b(s,t), \tag{10}$$

$$\begin{aligned} J(m,n) &= \sum_{i=1}^{l_1} \sum_{s=m_0}^{m-1} \sum_{t=n_0}^{n-1} \left\{c_i(s,t,m,n) \left[\frac{q_i}{p} K_1^{(q_i-p)/p} a(s,t) + \frac{p-q_i}{p} K_1^{q_i/p}\right]\right. \\ &+ \left. d_i(s,t,m,n) \left[\frac{r_i}{p} K_2^{(r_i-p)/p} a(s,t) + \frac{p-r_i}{p} K_2^{r_i/p}\right] + e_i(s,t,m,n)\right\} \\ &+ \sum_{j=1}^{l_2} \sum_{s=m_0}^{M-1} \sum_{t=n_0}^{N-1} \left\{f_j(s,t,m,n) \left[\frac{h_j}{p} K_3^{(h_j-p)/p} a(s,t) + \frac{p-h_j}{p} K_3^{h_j/p}\right]\right. \\ &+ \left. g_j(s,t,m,n) \left[\frac{v_j}{p} K_4^{(v_j-p)/p} a(s,t) + \frac{p-v_j}{p} K_4^{v_j/p}\right] + w_j(s,t,m,n)\right\}, \end{aligned} \tag{11}$$

$$\lambda(m,n) = \sum_{j=1}^{l_2} \sum_{s=m_0}^{M-1} \sum_{t=n_0}^{N-1} \left[f_j(s,t,m,n) \frac{h_j}{p} K_3^{(h_j-p)/p} + g_j(s,t,m,n) \frac{v_j}{p} K_4^{(v_j-p)/p}\right] b(s,t) C(s,t). \tag{12}$$

Proof. Given $b(m,n) = \max\{b_1(m,n), b_2(m,n)\}$, for $(m,n) \in \Omega$, we have

$$\begin{aligned} u^p(m,n) &\leq a(m,n) + b(m,n) \\ &\cdot \sum_{i=1}^{l_1} \sum_{s=m_0}^{m-1} \sum_{t=n_0}^{n-1} \left[c_i(s,t,m,n) u^{q_i}(s,t) \right. \\ &+ \left. d_i(s,t,m,n) u^{r_i}(s,t) + e_i(s,t,m,n)\right] + b(m,n) \\ &\cdot \sum_{j=1}^{l_2} \sum_{s=m_0}^{M-1} \sum_{t=n_0}^{N-1} \left[f_j(s,t,m,n) u^{h_j}(s,t) \right. \\ &+ \left. g_j(s,t,m,n) u^{v_j}(s,t) + w_j(s,t,m,n)\right]. \end{aligned} \tag{13}$$

Define a function $z(m,n)$ by

$$\begin{aligned} z(m,n) &= \sum_{i=1}^{l_1}\sum_{s=m_0}^{m-1}\sum_{t=n_0}^{n-1}\left[c_i(s,t,m,n)u^{q_i}(s,t)+d_i(s,t,m,n)u^{r_i}(s,t)+e_i(s,t,m,n)\right] \\ &\quad+\sum_{j=1}^{l_2}\sum_{s=m_0}^{M-1}\sum_{t=n_0}^{N-1}\left[f_j(s,t,m,n)u^{h_j}(s,t)+g_j(s,t,m,n)u^{v_j}(s,t)+w_j(s,t,m,n)\right]. \end{aligned} \tag{14}$$

Then

$$u^p(m,n)\le a(m,n)+b(m,n)z(m,n), \tag{15}$$

or

$$u(m,n)\le\left(a(m,n)+b(m,n)z(m,n)\right)^{1/p}. \tag{16}$$

By using Lemma 3, for any $K_i>0$ $(i=1,2,3,4)$, we have

$$\begin{aligned} z(m,n) &\le \sum_{i=1}^{l_1}\sum_{s=m_0}^{m-1}\sum_{t=n_0}^{n-1}\left\{c_i(s,t,m,n)\left[\frac{q_i}{p}K_1^{(q_i-p)/p}\left(a(s,t)+b(s,t)z(s,t)\right)+\frac{p-q_i}{p}K_1^{q_i/p}\right]\right. \\ &\quad\left.+d_i(s,t,m,n)\left[\frac{r_i}{p}K_2^{(r_i-p)/p}\left(a(s,t)+b(s,t)z(s,t)\right)+\frac{p-r_i}{p}K_2^{r_i/p}\right]+e_i(s,t,m,n)\right\} \\ &\quad+\sum_{j=1}^{l_2}\sum_{s=m_0}^{M-1}\sum_{t=n_0}^{N-1}\left\{f_j(s,t,m,n)\left[\frac{h_j}{p}K_3^{(h_j-p)/p}\left(a(s,t)+b(s,t)z(s,t)\right)+\frac{p-h_j}{p}K_3^{h_j/p}\right]\right. \\ &\quad\left.+g_j(s,t,m,n)\left[\frac{v_j}{p}K_4^{(v_j-p)/p}\left(a(s,t)+b(s,t)z(s,t)\right)+\frac{p-v_j}{p}K_4^{v_j/p}\right]+w_j(s,t,m,n)\right\}=R(m,n) \\ &\quad+\sum_{i=1}^{l_1}\sum_{s=m_0}^{m-1}\sum_{t=n_0}^{n-1}\left[c_i(s,t,m,n)\frac{q_i}{p}K_1^{(q_i-p)/p}+d_i(s,t,m,n)\frac{r_i}{p}K_2^{(r_i-p)/p}\right]b(s,t)z(s,t), \end{aligned} \tag{17}$$

where

$$R(m,n)=J(m,n)+\sum_{j=1}^{l_2}\sum_{s=m_0}^{M-1}\sum_{t=n_0}^{N-1}\left[f_j(s,t,m,n)\frac{h_j}{p}K_3^{(h_j-p)/p}+g_j(s,t,m,n)\frac{v_j}{p}K_4^{(v_j-p)/p}\right]b(s,t)z(s,t), \tag{18}$$

and $J(m,n)$ is defined in (11). Then, using that $R(m,n)$ is nondecreasing in every variable, we get

$$\begin{aligned} z(m,n) &\le R(M,N)+\sum_{i=1}^{l_1}\sum_{s=m_0}^{m-1}\sum_{t=n_0}^{n-1}\left[c_i(s,t,m,n)\frac{q_i}{p}K_1^{(q_i-p)/p}+d_i(s,t,m,n)\frac{r_i}{p}K_2^{(r_i-p)/p}\right]b(s,t)z(s,t) \\ &= R(M,N)+\sum_{s=m_0}^{m-1}\sum_{t=n_0}^{n-1}B(s,t,m,n)z(s,t), \end{aligned} \tag{19}$$

where $B(s,t,m,n)$ is defined in (10).

Since $b(m,n)$ is nondecreasing and $c_i(s,t,m,n), d_i(s,t,m,n)$ are nondecreasing in the last two variables, then $B(s,t,m,n)$ is also nondecreasing in the last two variables, and, by Lemma 1 and (19), we get

$$z(m,n) \le R(M,N)\exp\left\{\sum_{s=m_0}^{m-1}\sum_{t=n_0}^{n-1} B(s,t,m,n)\right\} = R(M,N)\,C(m,n), \tag{20}$$

where $C(m,n)$ is defined in (9). Considering the definition of $R(m,n)$ and (20), we have

$$\begin{aligned} R(M,N) &= J(M,N) + \sum_{j=1}^{l_2}\sum_{s=m_0}^{M-1}\sum_{t=n_0}^{N-1}\left[f_j(s,t,M,N)\frac{h_j}{p}K_3^{(h_j-p)/p} + g_j(s,t,M,N)\frac{v_j}{p}K_4^{(v_j-p)/p}\right] b(s,t)\,z(s,t) \le J(M,N) + R(M,N) \\ &\quad\cdot \sum_{j=1}^{l_2}\sum_{s=m_0}^{M-1}\sum_{t=n_0}^{N-1}\left[f_j(s,t,M,N)\frac{h_j}{p}K_3^{(h_j-p)/p} + g_j(s,t,M,N)\frac{v_j}{p}K_4^{(v_j-p)/p}\right] b(s,t)\,C(s,t) \\ &= J(M,N) + R(M,N)\,\lambda(M,N), \end{aligned} \tag{21}$$

where $\lambda(m,n)$ is defined in (12). Then,

$$R(M,N) \le \frac{J(M,N)}{1-\lambda(M,N)}. \tag{22}$$

Combining (20) and (22), we deduce

$$z(m,n) \le \frac{J(M,N)}{1-\lambda(M,N)}C(m,n), \tag{23}$$

where $C(m,n)$, $\lambda(m,n)$ are defined in (9) and (12).

Then, combining (16) and (23), we obtain the desired result.

Corollary 5. *Let $r_{1i}(m,n), d_{1i}(m,n), c_{1i}(m,n), e_{1i}(m,n) \in \wp_+(\Omega)$, $(i = 1,2,\ldots,l_1)$, $f_{1j}(m,n)$, $g_{1j}(m,n)$, $w_{1j}(m,n)$, $r_{2j}(m,n) \in \wp_+(\Omega)$, $(j = 1,2,\ldots,l_2)$, $r_{1i}(m,n), r_{2j}(m,n)$, $b_1(m,n)$ and $b_2(m,n)$ be nondecreasing in every variable. $u(m,n), a(m,n), b_1(m,n), b_2(m,n), p, q_i, r_i, h_j, v_j$ are defined as in Theorem 4. If*

$$\begin{aligned} u^p(m,n) &\le a(m,n) + b_1(m,n)\sum_{i=1}^{l_1} r_{1i}(m,n) \cdot \sum_{s=m_0}^{m-1}\sum_{t=n_0}^{n-1}\left[c_{1i}(s,t)u^{q_i}(s,t) + d_{1i}(s,t)u^{r_i}(s,t) + e_{1i}(s,t)\right] + b_2(m,n)\sum_{j=1}^{l_2} r_{2j}(m,n) \\ &\quad\cdot \sum_{s=m_0}^{M-1}\sum_{t=n_0}^{N-1}\left[f_{1j}(s,t)u^{h_j}(s,t) + g_{1j}(s,t)u^{v_j}(s,t) + w_{1j}(s,t)\right], \end{aligned} \tag{24}$$

then, for $(m,n) \in \Omega$, we have

$$u(m,n) \le \left\{a(m,n) + b(m,n)\frac{J(M,N)}{1-\lambda(M,N)}C(m,n)\right\}^{1/p}, \tag{25}$$

provided that $\lambda(M,N) < 1$, where

$$\begin{aligned} b(m,n) &= \max\{b_1(m,n), b_2(m,n)\}, \\ C(m,n) &= \exp\left\{\sum_{s=m_0}^{m-1}\sum_{t=n_0}^{n-1} B(s,t,m,n)\right\}, \\ B(s,t,m,n) &= \sum_{i=1}^{l_1} r_{1i}(m,n)\left[c_{1i}(s,t)\frac{q_i}{p}K_1^{(q_i-p)/p} + d_{1i}(s,t)\frac{r_i}{p}K_2^{(r_i-p)/p}\right] b(s,t), \\ J(m,n) &= \sum_{i=1}^{l_1} r_{1i}(m,n)\sum_{s=m_0}^{m-1}\sum_{t=n_0}^{n-1}\left\{c_{1i}(s,t) \cdot \left[\frac{q_i}{p}K_1^{(q_i-p)/p}a(s,t) + \frac{p-q_i}{p}K_1^{q_i/p}\right] + d_{1i}(s,t) \cdot \left[\frac{r_i}{p}K_2^{(r_i-p)/p}a(s,t) + \frac{p-r_i}{p}K_2^{r_i/p}\right] + e_{1i}(s,t)\right\} \\ &\quad + \sum_{j=1}^{l_2} r_{2j}(m,n)\sum_{s=m_0}^{M-1}\sum_{t=n_0}^{N-1}\left\{f_{1j}(s,t) \cdot \left[\frac{h_j}{p}K_3^{(h_j-p)/p}a(s,t) + \frac{p-h_j}{p}K_3^{h_j/p}\right] + g_{1j}(s,t) \cdot \left[\frac{v_j}{p}K_4^{(v_j-p)/p}a(s,t) + \frac{p-v_j}{p}K_4^{v_j/p}\right] + w_{1j}(s,t)\right\}, \\ \lambda(m,n) &= \sum_{j=1}^{l_2} r_{2j}(m,n)\sum_{s=m_0}^{M-1}\sum_{t=n_0}^{N-1}\left[f_{1j}(s,t)\frac{h_j}{p}K_3^{(h_j-p)/p} + g_{1j}(s,t)\frac{v_j}{p}K_4^{(v_j-p)/p}\right] b(s,t)\,C(s,t). \end{aligned} \tag{26}$$

The proof of Corollary 5 can be completed by setting $c_i(s,t,m,n) = r_{1i}(m,n)c_{1i}(s,t)$, $d_i(s,t,m,n) = r_{1i}(m,n)d_{1i}(s,t)$, $e_i(s,t,m,n) = r_{1i}(m,n)e_{1i}(s,t)$, $f_j(s,t,m,n) = r_{2j}(m,$

$n)f_{1j}(s,t)$, $g_j(s,t,m,n) = r_{2j}(m,n)g_{1j}(s,t)$, $w_j(s,t,m,n) = r_{2j}(m,n)w_{1j}(s,t)$ in Theorem 4.

Letting $p = 1$, we get the following corollary.

Corollary 6. *Let $u(m,n), a(m,n), b_1(m,n), b_2(m,n), c_i(s,t,m,n), d_i(s,t,m,n), e_i(s,t,m,n)$, $f_j(s,t,m,n)$, $g_j(s,t,m,n), w_j(s,t,m,n)$ be defined as in Theorem 4. If*

$$\begin{aligned} u(m,n) &\le a(m,n) + b_1(m,n) \sum_{i=1}^{l_1} \sum_{s=m_0}^{m-1} \sum_{t=n_0}^{n-1} [c_i(s,t,m,n)u(s,t) + d_i(s,t,m,n)u(s,t) + e_i(s,t,m,n)] \\ &\quad + b_2(m,n) \sum_{j=1}^{l_2} \sum_{s=m_0}^{M-1} \sum_{t=n_0}^{N-1} [f_j(s,t,m,n)u(s,t) + g_j(s,t,m,n)u(s,t) + w_j(s,t,m,n)], \end{aligned} \tag{27}$$

then, for $(m,n) \in \Omega$, we have

$$u(m,n) \le a(m,n) + b(m,n)\frac{J(M,N)}{1-\lambda(M,N)}C(m,n), \tag{28}$$

provided that $\lambda(M,N) < 1$, where

$$\begin{aligned} b(m,n) &= \max\{b_1(m,n), b_2(m,n)\}, \\ C(m,n) &= \exp\left\{\sum_{s=m_0}^{m-1} \sum_{t=n_0}^{n-1} B(s,t,m,n)\right\}, \\ B(s,t,m,n) &= \sum_{i=1}^{l_1} [c_i(s,t,m,n) + d_i(s,t,m,n)]\, b(s,t), \\ J(m,n) &= \sum_{i=1}^{l_1} \sum_{s=m_0}^{m-1} \sum_{t=n_0}^{n-1} \{[c_i(s,t,m,n) + d_i(s,t,m,n)]\, a(s,t) + e_i(s,t,m,n)\} \\ &\quad + \sum_{j=1}^{l_2} \sum_{s=m_0}^{M-1} \sum_{t=n_0}^{N-1} \{[f_j(s,t,m,n) + g_j(s,t,m,n)]\, a(s,t) + w_j(s,t,m,n)\}, \\ \lambda(m,n) &= \sum_{j=1}^{l_2} \sum_{s=m_0}^{M-1} \sum_{t=n_0}^{N-1} [f_j(s,t,m,n) + g_j(s,t,m,n)]\, b(s,t)\, C(s,t). \end{aligned} \tag{29}$$

Theorem 7. *Let $\varphi(m,n) \in \wp_+(\Omega)$, $u(m,n), a(m,n)$, $b_1(m,n)$, $b_2(m,n)$, $c_i(s,t,m,n)$, $d_i(s,t,m,n)$, $e_i(s,t,m,n)$, $f_j(s,t,m,n)$, $g_j(s,t,m,n), w_j(s,t,m,n)$, p, q_i, r_i, h_j, v_j be defined as in Theorem 4. Assume that $a(m,n)$ is nondecreasing in the first variable. If*

$$\begin{aligned} u^p(m,n) &\le a(m,n) + \sum_{s=m_0}^{m-1} \varphi(s,n)u^p(s,n) + b_1(m,n) \\ &\quad \cdot \sum_{i=1}^{l_1} \sum_{s=m_0}^{m-1} \sum_{t=n_0}^{n-1} [c_i(s,t,m,n)u^{q_i}(s,t) \\ &\quad + d_i(s,t,m,n)u^{r_i}(s,t) + e_i(s,t,m,n)] + b_2(m,n) \\ &\quad \cdot \sum_{j=1}^{l_2} \sum_{s=m_0}^{M-1} \sum_{t=n_0}^{N-1} [f_j(s,t,m,n) \\ &\quad \cdot u^{h_j}(s,t) + g_j(s,t,m,n)u^{v_j}(s,t) + w_j(s,t,m,n)], \end{aligned} \tag{30}$$

then, for $(m,n) \in \Omega$, we have

$$\begin{aligned} u(m,n) &\le \left\{\left[a(m,n) + b(m,n)\frac{\widetilde{J}(M,N)}{1-\widetilde{\lambda}(M,N)}\widetilde{C}(m,n)\right] \right. \\ &\quad \left. \cdot \widetilde{\varphi}(m,n)\right\}^{1/p}, \end{aligned} \tag{31}$$

provided that $\widetilde{\lambda}(M,N) < 1$, where

$$\widetilde{\varphi}(m,n) = \prod_{s=m_0}^{m-1}\left[1+\varphi(s,n)\right], \tag{32}$$

$$b(m,n) = \max\{b_1(m,n), b_2(m,n)\}, \tag{33}$$

$$\begin{aligned}
&\widetilde{c}_i(s,t,m,n) = c_i(s,t,m,n)\left(\widetilde{\varphi}(s,t)\right)^{q_i/p},\\
&\widetilde{d}_i(s,t,m,n) = d_i(s,t,m,n)\left(\widetilde{\varphi}(s,t)\right)^{r_i/p}, \quad i=1,2,\ldots,l_1,\\
&\widetilde{f}_j(s,t,m,n) = f_j(s,t,m,n)\left(\widetilde{\varphi}(s,t)\right)^{h_j/p},\\
&\widetilde{g}_j(s,t,m,n) = g_j(s,t,m,n)\left(\widetilde{\varphi}(s,t)\right)^{v_j/p}, \quad j=1,2,\ldots,l_2,
\end{aligned} \tag{34}$$

$$\widetilde{C}(m,n) = \exp\left\{\sum_{s=m_0}^{m-1}\sum_{t=n_0}^{n-1}\widetilde{B}(s,t,m,n)\right\}, \tag{35}$$

$$\widetilde{B}(s,t,m,n) = \sum_{i=1}^{l_1}\left[\widetilde{c}_i(s,t,m,n)\frac{q_i}{p}K_1^{(q_i-p)/p} + \widetilde{d}_i(s,t,m,n)\frac{r_i}{p}K_2^{(r_i-p)/p}\right]b(s,t), \tag{36}$$

$$\begin{aligned}
\widetilde{J}(m,n) = &\sum_{i=1}^{l_1}\sum_{s=m_0}^{m-1}\sum_{t=n_0}^{n-1}\left\{\widetilde{c}_i(s,t,m,n)\left[\frac{q_i}{p}K_1^{(q_i-p)/p}a(s,t) + \frac{p-q_i}{p}K_1^{q_i/p}\right] + \widetilde{d}_i(s,t,m,n)\left[\frac{r_i}{p}K_2^{(r_i-p)/p}a(s,t) + \frac{p-r_i}{p}\right.\right.\\
&\left.\left.\cdot K_2^{r_i/p}\right] + e_i(s,t,m,n)\right\} + \sum_{j=1}^{l_2}\sum_{s=m_0}^{M-1}\sum_{t=n_0}^{N-1}\left\{\widetilde{f}_j(s,t,m,n)\left[\frac{h_j}{p}K_3^{(h_j-p)/p}a(s,t) + \frac{p-h_j}{p}K_3^{h_j/p}\right]\right.\\
&\left. + \widetilde{g}_j(s,t,m,n)\left[\frac{v_j}{p}K_4^{(v_j-p)/p}a(s,t) + \frac{p-v_j}{p}K_4^{v_j/p}\right] + w_j(s,t,m,n)\right\},
\end{aligned} \tag{37}$$

$$\widetilde{\lambda}(m,n) = \sum_{j=1}^{l_2}\sum_{s=m_0}^{M-1}\sum_{t=n_0}^{N-1}\left[\widetilde{f}_j(s,t,m,n)\frac{h_j}{p}K_3^{(h_j-p)/p} + \widetilde{g}_j(s,t,m,n)\frac{v_j}{p}K_4^{(v_j-p)/p}\right]b(s,t)\widetilde{C}(s,t). \tag{38}$$

Proof. Given $b(m,n) = \max\{b_1(m,n), b_2(m,n)\}$, for $(m,n) \in \Omega$, we have

$$\begin{aligned}
u^p(m,n) \le\ & a(m,n) + \sum_{s=m_0}^{m-1}\varphi(s,n)u^p(s,n) + b(m,n)\\
&\cdot\sum_{i=1}^{l_1}\sum_{s=m_0}^{m-1}\sum_{t=n_0}^{n-1}\left[c_i(s,t,m,n)u^{q_i}(s,t)\right.\\
&\left. + d_i(s,t,m,n)u^{r_i}(s,t) + e_i(s,t,m,n)\right] + b(m,n)\\
&\cdot\sum_{j=1}^{l_2}\sum_{s=m_0}^{M-1}\sum_{t=n_0}^{N-1}\left[f_j(s,t,m,n)\right.\\
&\left.\cdot u^{h_j}(s,t) + g_j(s,t,m,n)u^{v_j}(s,t) + w_j(s,t,m,n)\right].
\end{aligned} \tag{39}$$

Define function $\widetilde{z}(m,n)$ by

$$\begin{aligned}
\widetilde{z}(m,n) =\ & a(m,n) + b(m,n)\\
&\cdot\sum_{i=1}^{l_1}\sum_{s=m_0}^{m-1}\sum_{t=n_0}^{n-1}\left[c_i(s,t,m,n)u^{q_i}(s,t)\right.\\
&\left. + d_i(s,t,m,n)u^{r_i}(s,t) + e_i(s,t,m,n)\right] + b(m,n)\\
&\cdot\sum_{j=1}^{l_2}\sum_{s=m_0}^{M-1}\sum_{t=n_0}^{N-1}\left[f_j(s,t,m,n)u^{h_j}(s,t)\right.\\
&\left. + g_j(s,t,m,n)u^{v_j}(s,t) + w_j(s,t,m,n)\right].
\end{aligned} \tag{40}$$

Then,

$$u^p(m,n) \le \widetilde{z}(m,n) + \sum_{s=m_0}^{m-1}\varphi(s,n)u^p(s,n). \tag{41}$$

Clearly $z(m,n)$ is nondecreasing in the first variable. Then, by Lemma 2, we get

$$\begin{aligned}
u^p(m,n) &\le \widetilde{z}(m,n)\prod_{s=m_0}^{m-1}\left[1+\varphi(s,n)\right]\\
&= \widetilde{z}(m,n)\widetilde{\varphi}(m,n),
\end{aligned} \tag{42}$$

where $\widetilde{\varphi}(m,n)$ is defined in (32). Define function

$$v(m,n) = \sum_{i=1}^{l_1}\sum_{s=m_0}^{m-1}\sum_{t=n_0}^{n-1}\left[c_i(s,t,m,n)u^{q_i}(s,t) + d_i(s,t,m,n)u^{r_i}(s,t) + e_i(s,t,m,n)\right] + \sum_{j=1}^{l_2}\sum_{s=m_0}^{M-1}\sum_{t=n_0}^{N-1}\left[f_j(s,t,m,n)u^{h_j}(s,t) + g_j(s,t,m,n)u^{v_j}(s,t) + w_j(s,t,m,n)\right]. \quad (43)$$

From (40), we get

$$\widetilde{z}(m,n) = a(m,n) + b(m,n)v(m,n). \quad (44)$$

Then (42) becomes

$$u(m,n) \le \left\{\left[a(m,n) + b(m,n)v(m,n)\right]\widetilde{\varphi}(m,n)\right\}^{1/p}. \quad (45)$$

By (45) and Lemma 3, from (43), we have

$$\begin{aligned} v(m,n) \le {} & \sum_{i=1}^{l_1}\sum_{s=m_0}^{m-1}\sum_{t=n_0}^{n-1}\Big\{c_i(s,t,m,n)(\widetilde{\varphi}(s,t))^{q_i/p}\left[\frac{q_i}{p}K_1^{(q_i-p)/p}(a(s,t)+b(s,t)v(s,t)) + \frac{p-q_i}{p}K_1^{q_i/p}\right] \\ & + d_i(s,t,m,n)(\widetilde{\varphi}(s,t))^{r_i/p}\left[\frac{r_i}{p}K_2^{(r_i-p)/p}(a(s,t)+b(s,t)v(s,t)) + \frac{p-r_i}{p}K_2^{r_i/p}\right] + e_i(s,t,m,n)\Big\} \\ & + \sum_{j=1}^{l_2}\sum_{s=m_0}^{M-1}\sum_{t=n_0}^{N-1}\Big\{f_j(s,t,m,n)(\widetilde{\varphi}(s,t))^{h_j/p}\left[\frac{h_j}{p}K_3^{(h_j-p)/p}(a(s,t)+b(s,t)v(s,t)) + \frac{p-h_j}{p}K_3^{h_j/p}\right] \\ & + g_j(s,t,m,n)(\widetilde{\varphi}(s,t))^{v_j/p}\left[\frac{v_j}{p}K_4^{(v_j-p)/p}(a(s,t)+b(s,t)v(s,t)) + \frac{p-v_j}{p}K_4^{v_j/p}\right] + w_j(s,t,m,n)\Big\} = \widetilde{R}(m,n) \\ & + \sum_{i=1}^{l_1}\sum_{s=m_0}^{m-1}\sum_{t=n_0}^{n-1}\left[\widetilde{c}_i(s,t,m,n)\frac{q_i}{p}K_1^{(q_i-p)/p} + \widetilde{d}_i(s,t,m,n)\frac{r_i}{p}K_2^{(r_i-p)/p}\right]b(s,t)v(s,t), \end{aligned} \quad (46)$$

where

$$\widetilde{R}(m,n) = \widetilde{J}(m,n) + \sum_{j=1}^{l_2}\sum_{s=m_0}^{M-1}\sum_{t=n_0}^{N-1}\left[\widetilde{f}_j(s,t,m,n)\frac{h_j}{p}K_3^{(h_j-p)/p} + \widetilde{g}_j(s,t,m,n)\frac{v_j}{p}K_4^{(v_j-p)/p}\right]b(s,t)v(s,t), \quad (47)$$

$\widetilde{c}_i, \widetilde{d}_i, \widetilde{f}_j, \widetilde{g}_j$ and $\widetilde{J}(m,n)$ are defined in (34) and (37), respectively.

Similar to the process of (17)–(23), we deduce that

$$v(m,n) \le \frac{\widetilde{J}(M,N)}{1-\widetilde{\lambda}(M,N)}\widetilde{C}(m,n), \quad (48)$$

where $\widetilde{C}(m,n), \widetilde{\lambda}(m,n)$ are defined in (35) and (38).

Combining (45) and (48), we get the desired result.

Theorem 8. *Let $u(m,n), a(m,n), b_1(m,n), b_2(m,n), c_i(s,t,m,n), d_i(s,t,m,n), e_i(s,t,m,n), f_j(s,t,m,n), g_j(s,t,m,n), w_j(s,t,m,n), p, q_i, r_i, h_j, v_j$ be defined as in Theorem 4. $H_j, L_j : \Omega \times R_+ \to R_+$ $(j = 1,2,\ldots,l_2)$ satisfies $0 \le H_j(m,n,u) - H_j(m,n,v) \le L_j(m,n,v)(u-v)$ for $u \ge v \ge 0$. If*

$$\begin{aligned} u^p(m,n) \le {} & a(m,n) + b_1(m,n) \cdot \sum_{i=1}^{l_1}\sum_{s=m_0}^{m-1}\sum_{t=n_0}^{n-1}\left[c_i(s,t,m,n)u^{q_i}(s,t) + d_i(s,t,m,n)u^{r_i}(s,t) + e_i(s,t,m,n)\right] + b_2(m,n) \\ & \cdot \sum_{j=1}^{l_2}\sum_{s=m_0}^{M-1}\sum_{t=n_0}^{N-1}\left[f_j(s,t,m,n)H_j\left(s,t,u^{h_j}(s,t)\right) + g_j(s,t,m,n)H_j\left(s,t,u^{v_j}(s,t)\right) + w_j(s,t,m,n)\right], \end{aligned} \quad (49)$$

then, for $(m,n) \in \Omega$, we have

$$u(m,n) \le \left\{a(m,n) + b(m,n)\frac{\overline{J}(M,N)}{1-\overline{\lambda}(M,N)}\overline{C}(m,n)\right\}^{1/p}, \quad (50)$$

provided that $\overline{\lambda}(M,N) < 1$, *where*

$$b(m,n) = \max\{b_1(m,n), b_2(m,n)\}, \tag{51}$$

$$\overline{C}(m,n) = \exp\left\{\sum_{s=m_0}^{m-1}\sum_{t=n_0}^{n-1}\overline{B}(s,t,m,n)\right\}, \tag{52}$$

$$\overline{B}(s,t,m,n) = \sum_{i=1}^{l_1}\left[c_i(s,t,m,n)\frac{q_i}{p}K_1^{(q_i-p)/p} + d_i(s,t,m,n)\frac{r_i}{p}K_2^{(r_i-p)/p}\right]b(s,t), \tag{53}$$

$$\begin{aligned}\overline{J}(m,n) = {} & \sum_{i=1}^{l_1}\sum_{s=m_0}^{m-1}\sum_{t=n_0}^{n-1}\left\{c_i(s,t,m,n)\left[\frac{q_i}{p}K_1^{(q_i-p)/p}a(s,t) + \frac{p-q_i}{p}K_1^{q_i/p}\right]\right.\\ & \left. + d_i(s,t,m,n)\left[\frac{r_i}{p}K_2^{(r_i-p)/p}a(s,t) + \frac{p-r_i}{p}K_2^{r_i/p}\right] + e_i(s,t,m,n)\right\}\\ & + \sum_{j=1}^{l_2}\sum_{s=m_0}^{M-1}\sum_{t=n_0}^{N-1}\left\{f_j(s,t,m,n)H_j\left[s,t,\frac{h_j}{p}K_3^{(h_j-p)/p}a(s,t) + \frac{p-h_j}{p}K_3^{h_j/p}\right]\right.\\ & \left. + g_j(s,t,m,n)H_j\left[s,t,\frac{v_j}{p}K_4^{(v_j-p)/p}a(s,t) + \frac{p-v_j}{p}K_4^{v_j/p}\right] + w_j(s,t,m,n)\right\},\end{aligned} \tag{54}$$

$$\overline{f}_j(s,t,m,n) = f_j(s,t,m,n)L_j\left(s,t,\frac{h_j}{p}K_3^{(h_j-p)/p}a(s,t) + \frac{p-h_j}{p}K_3^{h_j/p}\right), \quad j = 1,2,\ldots,l_2, \tag{55}$$

$$\overline{g}_j(s,t,m,n) = g_j(s,t,m,n)L_j\left(s,t,\frac{v_j}{p}K_4^{(v_j-p)/p}a(s,t) + \frac{p-v_j}{p}K_4^{v_j/p}\right), \quad j = 1,2,\ldots,l_2, \tag{56}$$

$$\overline{\lambda}(m,n) = \sum_{j=1}^{l_2}\sum_{s=m_0}^{M-1}\sum_{t=n_0}^{N-1}\left[\overline{f}_j(s,t,m,n)\frac{h_j}{p}K_3^{(h_j-p)/p} + \overline{g}_j(s,t,m,n)\frac{v_j}{p}K_4^{(v_j-p)/p}\right]b(s,t)\overline{C}(s,t). \tag{57}$$

Proof. Given $b(m,n) = \max\{b_1(m,n), b_2(m,n)\}$, for $(m,n) \in \Omega$, we have

$$\begin{aligned} u^p(m,n) \le {} & a(m,n) + b(m,n)\\ & \cdot \sum_{i=1}^{l_1}\sum_{s=m_0}^{m-1}\sum_{t=n_0}^{n-1}\left[c_i(s,t,m,n)u^{q_i}(s,t)\right.\\ & \left. + d_i(s,t,m,n)u^{r_i}(s,t) + e_i(s,t,m,n)\right] + b(m,n)\\ & \cdot \sum_{j=1}^{l_2}\sum_{s=m_0}^{M-1}\sum_{t=n_0}^{N-1}\left[f_j(s,t,m,n)H_j\left(s,t,u^{h_j}(s,t)\right)\right.\\ & \left. + g_j(s,t,m,n)H_j\left(s,t,u^{v_j}(s,t)\right) + w_j(s,t,m,n)\right].\end{aligned} \tag{58}$$

Define function $\overline{v}(m,n)$ by

$$\begin{aligned}\overline{v}(m,n) = {} & \sum_{i=1}^{l_1}\sum_{s=m_0}^{m-1}\sum_{t=n_0}^{n-1}\left[c_i(s,t,m,n)u^{q_i}(s,t) + d_i(s,t,m,n)u^{r_i}(s,t) + e_i(s,t,m,n)\right]\\ & + \sum_{j=1}^{l_2}\sum_{s=m_0}^{M-1}\sum_{t=n_0}^{N-1}\left[f_j(s,t,m,n)H_j\left(s,t,u^{h_j}(s,t)\right) + g_j(s,t,m,n)H_j\left(s,t,u^{v_j}(s,t)\right) + w_j(s,t,m,n)\right].\end{aligned} \tag{59}$$

Then

$$u^p(m,n) \le a(m,n) + b(m,n)\overline{v}(m,n), \tag{60}$$

or

$$u(m,n) \le \left(a(m,n) + b(m,n)\overline{v}(m,n)\right)^{1/p}. \tag{61}$$

By Lemma 3, we have

$$
\begin{aligned}
\bar{v}(m,n) \le & \sum_{i=1}^{l_1}\sum_{s=m_0}^{m-1}\sum_{t=n_0}^{n-1}\{c_i(s,t,m,n)(a(s,t)+b(s,t)\bar{v}(s,t))^{q_i/p} \\
& + d_i(s,t,m,n)(a(s,t)+b(s,t)\bar{v}(s,t))^{r_i/p} + e_i(s,t,m,n)\} \\
& + \sum_{j=1}^{l_2}\sum_{s=m_0}^{M-1}\sum_{t=n_0}^{N-1}\{f_j(s,t,m,n)H_j\left(s,t,(a(s,t)+b(s,t)\bar{v}(s,t))^{h_j/p}\right) \\
& + g_j(s,t,m,n)H_j\left(s,t,(a(s,t)+b(s,t)\bar{v}(s,t))^{v_j/p}\right) + w_j(s,t,m,n)\} \\
\le & \sum_{i=1}^{l_1}\sum_{s=m_0}^{m-1}\sum_{t=n_0}^{n-1}\left\{c_i(s,t,m,n)\left[\frac{q_i}{p}K_1^{(q_i-p)/p}(a(s,t)+b(s,t)\bar{v}(s,t)) + \frac{p-q_i}{p}K_1^{q_i/p}\right]\right. \\
& \left. + d_i(s,t,m,n)\left[\frac{r_i}{p}K_2^{(r_i-p)/p}(a(s,t)+b(s,t)\bar{v}(s,t)) + \frac{p-r_i}{p}K_2^{r_i/p}\right] + e_i(s,t,m,n)\right\} \\
& + \sum_{j=1}^{l_2}\sum_{s=m_0}^{M-1}\sum_{t=n_0}^{N-1}\left\{f_j(s,t,m,n)\left[H_j\left(s,t,\frac{h_j}{p}K_3^{(h_j-p)/p}(a(s,t)+b(s,t)\bar{v}(s,t)) + \frac{p-h_j}{p}K_3^{h_j/p}\right)\right.\right. \\
& - H_j\left(s,t,\frac{h_j}{p}K_3^{(h_j-p)/p}a(s,t) + \frac{p-h_j}{p}K_3^{h_j/p}\right) + \left.H_j\left(s,t,\frac{h_j}{p}K_3^{(h_j-p)/p}a(s,t) + \frac{p-h_j}{p}K_3^{h_j/p}\right)\right] + g_j(s,t,m,n) \\
& \cdot\left[H_j\left(s,t,\frac{v_j}{p}K_4^{(v_j-p)/p}(a(s,t)+b(s,t)\bar{v}(s,t)) + \frac{p-v_j}{p}K_4^{v_j/p}\right) - H_j\left(s,t,\frac{v_j}{p}K_4^{(v_j-p)/p}a(s,t) + \frac{p-v_j}{p}K_4^{v_j/p}\right)\right. \\
& \left.\left. + H_j\left(s,t,\frac{v_j}{p}K_4^{(v_j-p)/p}a(s,t) + \frac{p-v_j}{p}K_4^{v_j/p}\right)\right] + w_j(s,t,m,n)\right\} \\
\le & \sum_{i=1}^{l_1}\sum_{s=m_0}^{m-1}\sum_{t=n_0}^{n-1}\left\{c_i(s,t,m,n)\left[\frac{q_i}{p}K_1^{(q_i-p)/p}(a(s,t)+b(s,t)\bar{v}(s,t)) + \frac{p-q_i}{p}K_1^{q_i/p}\right]\right. \\
& \left. + d_i(s,t,m,n)\left[\frac{r_i}{p}K_2^{(r_i-p)/p}(a(s,t)+b(s,t)\bar{v}(s,t)) + \frac{p-r_i}{p}K_2^{r_i/p}\right] + e_i(s,t,m,n)\right\} \\
& + \sum_{j=1}^{l_2}\sum_{s=m_0}^{M-1}\sum_{t=n_0}^{N-1}\left\{f_j(s,t,m,n)\left[L_j\left(s,t,\frac{h_j}{p}K_3^{(h_j-p)/p}a(s,t) + \frac{p-h_j}{p}K_3^{h_j/p}\right)\right.\right. \\
& \cdot\frac{h_j}{p}K_3^{(h_j-p)/p}b(s,t)\bar{v}(s,t) + \left.H_j\left(s,t,\frac{h_j}{p}K_3^{(h_j-p)/p}a(s,t) + \frac{p-h_j}{p}K_3^{h_j/p}\right)\right] + g_j(s,t,m,n)\left[L_j\left(s,t,\frac{v_j}{p}K_4^{(v_j-p)/p}a(s,t)\right.\right. \\
& \left.\left. + \frac{p-v_j}{p}K_4^{v_j/p}\right)\frac{v_j}{p}K_4^{(v_j-p)/p}b(s,t)\bar{v}(s,t) + H_j\left(s,t,\frac{v_j}{p}K_4^{(v_j-p)/p}a(s,t) + \frac{p-v_j}{p}K_4^{v_j/p}\right)\right] + \left.w_j(s,t,m,n)\right\} = \bar{R}(m,n) \\
& + \sum_{i=1}^{l_1}\sum_{s=m_0}^{m-1}\sum_{t=n_0}^{n-1}\left[c_i(s,t,m,n)\frac{q_i}{p}K_1^{(q_i-p)/p} + d_i(s,t,m,n)\frac{r_i}{p}K_2^{(r_i-p)/p}\right]b(s,t)\bar{v}(s,t),
\end{aligned}
\tag{62}
$$

where

$$\overline{R}(m,n)=\overline{J}(m,n)+\sum_{j=1}^{l_2}\sum_{s=m_0}^{M-1}\sum_{t=n_0}^{N-1}\left[\overline{f}_j(s,t,m,n)\frac{h_j}{p}K_3^{(h_j-p)/p}+\overline{g}_j(s,t,m,n)\frac{v_j}{p}K_4^{(v_j-p)/p}\right]b(s,t)\overline{v}(s,t), \tag{63}$$

and $\overline{J}(m,n), \overline{f}_j(s,t,m,n), \overline{g}_j(s,t,m,n)$ are defined in (54)–(56).

Similar to the process of (17)–(23), we get

$$\overline{v}(m,n)\le\frac{\overline{J}(M,N)}{1-\overline{\lambda}(M,N)}\overline{C}(m,n), \tag{64}$$

where $\overline{C}(m,n), \overline{\lambda}(m,n)$ are defined in (52) and (57).

Combining (61) and (64), we get the desired result.

Theorem 9. *Let $\varphi(m,n)\in\wp_+(\Omega), u(m,n), a(m,n), b_1(m,n), b_2(m,n), c_i(s,t,m,n), d_i(s,t,m,n), e_i(s,t,m,n), f_j(s,t,m,n), g_j(s,t,m,n), w_j(s,t,m,n), p, q_i, r_i, h_j, v_j$ be defined as in Theorem 4. Assume that $a(m,n)$ is nondecreasing in the first variable. H_j, L_j $(j=1,2,\ldots,l_2)$ are defined as in Theorem 7. If*

$$\begin{aligned}u^p(m,n)&\le a(m,n)+\sum_{s=m_0}^{m-1}\varphi(s,n)u^p(s,n)\\&\quad+b_1(m,n)\sum_{i=1}^{l_1}\sum_{s=m_0}^{m-1}\sum_{t=n_0}^{n-1}\left[c_i(s,t,m,n)u^{q_i}(s,t)+d_i(s,t,m,n)u^{r_i}(s,t)+e_i(s,t,m,n)\right]\\&\quad+b_2(m,n)\sum_{j=1}^{l_2}\sum_{s=m_0}^{M-1}\sum_{t=n_0}^{N-1}\left[f_j(s,t,m,n)H_j\left(s,t,u^{h_j}(s,t)\right)+g_j(s,t,m,n)H_j\left(s,t,u^{v_j}(s,t)\right)+w_j(s,t,m,n)\right],\end{aligned} \tag{65}$$

then, for $(m,n)\in\Omega$, we have

$$u(m,n)\le\left\{a(m,n)+b(m,n)\frac{\widehat{J}(M,N)}{1-\widehat{\lambda}(M,N)}\widehat{C}(m,n)\right\}^{1/p}, \tag{66}$$

provided that $\widehat{\lambda}(M,N)<1$, where

$$\begin{aligned}\widehat{\varphi}(m,n)&=\prod_{s=m_0}^{m-1}\left[1+\varphi(s,n)\right],\\b(m,n)&=\max\left\{b_1(m,n),b_2(m,n)\right\},\\\widehat{C}(m,n)&=\exp\left\{\sum_{s=m_0}^{m-1}\sum_{t=n_0}^{n-1}\widehat{B}(s,t,m,n)\right\},\\\widehat{B}(s,t,m,n)&=\sum_{i=1}^{l_1}\left[\widehat{c}_i(s,t,m,n)\frac{q_i}{p}K_1^{(q_i-p)/p}+\widehat{d}_i(s,t,m,n)\frac{r_i}{p}K_2^{(r_i-p)/p}\right]b(s,t),\\\widehat{J}(m,n)&=\sum_{i=1}^{l_1}\sum_{s=m_0}^{m-1}\sum_{t=n_0}^{n-1}\left\{\widehat{c}_i(s,t,m,n)\left[\frac{q_i}{p}K_1^{(q_i-p)/p}a(s,t)+\frac{p-q_i}{p}K_1^{q_i/p}\right]\right.\\&\quad\left.+\widehat{d}_i(s,t,m,n)\left[\frac{r_i}{p}K_2^{(r_i-p)/p}a(s,t)+\frac{p-r_i}{p}K_2^{r_i/p}\right]+e_i(s,t,m,n)\right\}\\&\quad+\sum_{j=1}^{l_2}\sum_{s=m_0}^{M-1}\sum_{t=n_0}^{N-1}\left\{f_j(s,t,m,n)H_j\left[s,t,(\widehat{\varphi}(s,t))^{h_j/p}\left(\frac{h_j}{p}K_3^{(h_j-p)/p}a(s,t)+\frac{p-h_j}{p}K_3^{h_j/p}\right)\right]\right.\\&\quad\left.+g_j(s,t,m,n)H_j\left[s,t,(\widehat{\varphi}(s,t))^{v_j/p}\left(\frac{v_j}{p}K_4^{(v_j-p)/p}a(s,t)+\frac{p-v_j}{p}K_4^{v_j/p}\right)\right]+w_j(s,t,m,n)\right\},\end{aligned}$$

$$\widehat{\lambda}(m,n)=\sum_{j=1}^{l_2}\sum_{s=m_0}^{M-1}\sum_{t=n_0}^{N-1}\left[\widehat{f}_j(s,t,m,n)\frac{h_j}{p}K_3^{(h_j-p)/p}+\widehat{g}_j(s,t,m,n)\frac{v_j}{p}K_4^{(v_j-p)/p}\right]b(s,t)\widehat{C}(s,t),$$

$$\widetilde{c}_i(s,t,m,n)=c_i(s,t,m,n)(\widehat{\varphi}(s,t))^{q_i/p},$$

$$\widetilde{d}_i(s,t,m,n)=d_i(s,t,m,n)(\widehat{\varphi}(s,t))^{r_i/p},\quad i=1,2,\ldots,l_1,$$

$$\widehat{f}_j(s,t,m,n)=f_j(s,t,m,n)(\widehat{\varphi}(s,t))^{h_j/p}L_j\left[s,t,(\widehat{\varphi}(s,t))^{h_j/p}\left(\frac{h_j}{p}K_3^{(h_j-p)/p}a(s,t)+\frac{p-h_j}{p}K_3^{h_j/p}\right)\right],$$

$$\widehat{g}_j(s,t,m,n)=g_j(s,t,m,n)(\widehat{\varphi}(s,t))^{v_j/p}L_j\left[s,t,(\widehat{\varphi}(s,t))^{v_j/p}\left(\frac{v_j}{p}K_4^{(v_j-p)/p}a(s,t)+\frac{p-v_j}{p}K_4^{v_j/p}\right)\right],$$

$$j=1,2,\ldots,l_2. \tag{67}$$

The proof for Theorem 9 is similar to the combination of Theorems 7 and 8, and we omit the details here.

3. Applications

In this section, we will present some applications for the established results to study boundedness, uniqueness, and continuous dependence of solutions of certain difference equations.

Consider the following Volterra-Fredholm sum-difference equations:

$$\begin{aligned}u^p(m,n)&=a(m,n)+\sum_{s=m_0}^{m-1}\sum_{t=n_0}^{n-1}[C(s,t,m,n,u(s,t))\\&\quad+D(s,t,m,n,u(s,t))+E(s,t,m,n)]\\&\quad+\sum_{s=m_0}^{M-1}\sum_{t=n_0}^{N-1}[F(s,t,m,n,u(s,t))\\&\quad+G(s,t,m,n,u(s,t))+W(s,t,m,n)],\end{aligned}\tag{68}$$

where $u(m,n),a(m,n)\in\wp(\Omega)$, $p\geq 1$ is an odd number, $C,D,F,G:\Omega^2\times R\to R$, $E,W\in\wp(\Omega^2)$.

Theorem 10. *Assume that functions C,D,E,F,G,W in equation (68) satisfy the following conditions:*

$$\begin{aligned}|C(s,t,m,n,u_1)|&\leq c_1(s,t,m,n)|u_1^q|,\\|D(s,t,m,n,u_1)|&\leq d_1(s,t,m,n)|u_1^r|,\\|E(s,t,m,n)|&\leq e_1(s,t,m,n),\\|F(s,t,m,n,u_1)|&\leq f_1(s,t,m,n)|u_1^h|,\\|G(s,t,m,n,u_1)|&\leq g_1(s,t,m,n)|u_1^v|,\\|W(s,t,m,n)|&\leq w_1(s,t,m,n)\end{aligned}\tag{69}$$

for $(m,n)\in\Omega$, $u_1\in R$, where q,r,h,v are nonnegative constants satisfying $p\geq q>0$, $p\geq r>0$, $p\geq h>0$, $p\geq v>0$, $c_1,d_1,e_1,f_1,g_1,w_1\in\wp_+(\Omega^2)$ which are nondecreasing in the last two variables; then one has

$$|u(m,n)|\leq\left\{|a(m,n)|+\frac{J_1(M,N)}{1-\lambda_1(M,N)}C_1(m,n)\right\}^{1/p},\tag{70}$$

provided that $\lambda_1(M,N)<1$, where

$$C_1(m,n)=\exp\left\{\sum_{s=m_0}^{m-1}\sum_{t=n_0}^{n-1}B_1(s,t,m,n)\right\},$$

$$B_1(s,t,m,n)=c_1(s,t,m,n)\frac{q}{p}K_1^{(q-p)/p}+d_1(s,t,m,n)\cdot\frac{r}{p}K_2^{(r-p)/p},$$

$$\begin{aligned}J_1(m,n)&=\sum_{s=m_0}^{m-1}\sum_{t=n_0}^{n-1}\left\{c_1(s,t,m,n)\cdot\left[\frac{q}{p}K_1^{(q-p)/p}|a(s,t)|+\frac{p-q}{p}K_1^{q/p}\right]\right.\\&\quad+d_1(s,t,m,n)\left[\frac{r}{p}K_2^{(r-p)/p}|a(s,t)|+\frac{p-r}{p}K_2^{r/p}\right]\\&\quad\left.+e_1(s,t,m,n)\right\}+\sum_{s=m_0}^{M-1}\sum_{t=n_0}^{N-1}\left\{f_1(s,t,m,n)\cdot\left[\frac{h}{p}K_3^{(h-p)/p}|a(s,t)|+\frac{p-h}{p}K_3^{h/p}\right]\right.\\&\quad+g_1(s,t,m,n)\left[\frac{v}{p}K_4^{(v-p)/p}|a(s,t)|+\frac{p-v}{p}K_4^{v/p}\right]\\&\quad\left.+w_1(s,t,m,n)\right\},\end{aligned}$$

$$\lambda_1(m,n) = \sum_{s=m_0}^{M-1}\sum_{t=n_0}^{N-1}\left[f_1(s,t,m,n)\frac{h}{p}K_3^{(h-p)/p} + g_1(s,t,m,n)\frac{v}{p}K_4^{(v-p)/p}\right]C_1(s,t). \tag{71}$$

Proof. Using conditions (69) to (68), we have

$$\begin{aligned}|u^p(m,n)| &\le |a(m,n)| + \sum_{s=m_0}^{m-1}\sum_{t=n_0}^{n-1}[|C(s,t,m,n,u(s,t))| + |D(s,t,m,n,u(s,t))| + |E(s,t,m,n)|] \\ &+ \sum_{s=m_0}^{M-1}\sum_{t=n_0}^{N-1}[|F(s,t,m,n,u(s,t))| + |G(s,t,m,n,u(s,t))| + |W(s,t,m,n)|] \le |a(m,n)| \\ &+ \sum_{s=m_0}^{m-1}\sum_{t=n_0}^{n-1}[c_1(s,t,m,n)|u^q(s,t)| + d_1(s,t,m,n)|u^r(s,t)| + e_1(s,t,m,n)] \\ &+ \sum_{s=m_0}^{M-1}\sum_{t=n_0}^{N-1}\left[f_1(s,t,m,n)|u^h(s,t)| + g_1(s,t,m,n)|u^v(s,t)| + w_1(s,t,m,n)\right].\end{aligned} \tag{72}$$

Then a suitable application of Theorem 4 (with $l_1 = l_2 = 1$) to (72) yields the desired result.

The following theorem deals with the uniqueness of the solutions of (68).

Theorem 11. *Supposing that*

$$\begin{aligned}&|C(s,t,m,n,u_1) - C(s,t,m,n,u_2)| \le c_1(s,t,m,n)\left|u_1^p - u_2^p\right|, \\ &|F(s,t,m,n,u_1) - F(s,t,m,n,u_2)| \le f_1(s,t,m,n)\left|u_1^p - u_2^p\right|, \\ &|D(s,t,m,n,u_1) - D(s,t,m,n,u_2)| \le d_1(s,t,m,n)\left|u_1^p - u_2^p\right|, \\ &|G(s,t,m,n,u_1) - G(s,t,m,n,u_2)| \le g_1(s,t,m,n)\left|u_1^p - u_2^p\right|\end{aligned} \tag{73}$$

hold for $u_1, u_2 \in R$, where $c_1, d_1, f_1, g_1 \in \wp_+(\Omega^2)$ are nondecreasing in the last two variables,

$$\lambda(M,N) = \sum_{s=m_0}^{M-1}\sum_{t=n_0}^{N-1}[f_1(s,t,M,N) + g_1(s,t,M,N)]C(s,t) < 1, \tag{74}$$

$$B(s,t,m,n) = c_1(s,t,m,n) + d_1(s,t,m,n),$$

$$C(s,t) = \exp\left\{\sum_{s=m_0}^{m-1}\sum_{t=n_0}^{n-1}B(s,t,m,n)\right\},$$

then (68) has at most one solution.

Proof. Assume that $u(m,n), \bar{u}(m,n)$ are two solutions of (68). Then

$$\begin{aligned}&|u^p(m,n) - \bar{u}^p(m,n)| \\ &\le \sum_{s=m_0}^{m-1}\sum_{t=n_0}^{n-1}[|C(s,t,m,n,u(s,t)) - C(s,t,m,n,\bar{u}(s,t))| + |D(s,t,m,n,u(s,t)) - D(s,t,m,n,\bar{u}(s,t))|] \\ &+ \sum_{s=m_0}^{M-1}\sum_{t=n_0}^{N-1}[|F(s,t,m,n,u(s,t)) - F(s,t,m,n,\bar{u}(s,t))| + |G(s,t,m,n,u(s,t)) - G(s,t,m,n,\bar{u}(s,t))|] \\ &\le \sum_{s=m_0}^{m-1}\sum_{t=n_0}^{n-1}[c_1(s,t,m,n) + d_1(s,t,m,n)]|u^p(s,t) - \bar{u}^p(s,t)| + \sum_{s=m_0}^{M-1}\sum_{t=n_0}^{N-1}[f_1(s,t,m,n) + g_1(s,t,m,n)] \\ &\cdot |u^p(s,t) - \bar{u}^p(s,t)|.\end{aligned} \tag{75}$$

Treat $|u^p(m,n) - \bar{u}^p(m,n)|$ as one variable, and a suitable application of Corollary 6 yields $|u^p(m,n) - \bar{u}^p(m,n)| \le 0$, which implies that $u^p(m,n) \equiv \bar{u}^p(m,n)$. Since p is an odd number, then we have $u^p(m,n) = \bar{u}^p(m,n)$, and the proof is complete.

Finally we study the continuous dependence of the solutions of (68) on functions a, C, D, E, F, G, W. For this, we consider the following variation of (68):

$$\begin{aligned}\bar{u}^p(m,n) &= \bar{a}(m,n) + \sum_{s=m_0}^{m-1}\sum_{t=n_0}^{n-1}\left[\widetilde{C}(s,t,m,n,\bar{u}(s,t)) + \widetilde{D}(s,t,m,n,\bar{u}(s,t)) + \widetilde{E}(s,t,m,n)\right] \\ &+ \sum_{s=m_0}^{M-1}\sum_{t=n_0}^{N-1}\left[\widetilde{F}(s,t,m,n,\bar{u}(s,t)) + \widetilde{G}(s,t,m,n,\bar{u}(s,t)) + \widetilde{W}(s,t,m,n)\right],\end{aligned} \tag{76}$$

where $\widetilde{C}, \widetilde{D}, \widetilde{F}, \widetilde{G} : \Omega^2 \times R \to R$, $\widetilde{E}, \widetilde{W} \in \wp(\Omega^2)$ and $p \geq 1$ is an odd number.

Theorem 12. *Consider (68) and (76). If*

$$\begin{aligned} &\left|C(s,t,m,n,u_1(s,t)) - C(s,t,m,n,u_2(s,t))\right| \\ &\quad \leq c_1(s,t,m,n)\left|u_1^p - u_2^p\right|, \\ &\left|D(s,t,m,n,u_1(s,t)) - D(s,t,m,n,u_2(s,t))\right| \\ &\quad \leq d_1(s,t,m,n)\left|u_1^p - u_2^p\right|, \\ &\left|F(s,t,m,n,u_1(s,t)) - F(s,t,m,n,u_2(s,t))\right| \\ &\quad \leq f_1(s,t,m,n)\left|u_1^p - u_2^p\right|, \\ &\left|G(s,t,m,n,u_1(s,t)) - G(s,t,m,n,u_2(s,t))\right| \\ &\quad \leq g_1(s,t,m,n)\left|u_1^p - u_2^p\right|, \end{aligned} \tag{77}$$

hold for $u_1, u_2 \in R$, where $c_1, d_1, f_1, g_1 \in \wp_+(\Omega^2)$, and are nondecreasing in the last two variables, furthermore, for all solution $\widetilde{u}$ of (76), the following conditions hold for $(m,n) \in \Omega$:

$$\begin{aligned} &\left|a(m,n) - \widetilde{a}(m,n)\right| \leq \frac{\varepsilon}{4}, \\ &\sum_{s=m_0}^{m-1}\sum_{t=n_0}^{n-1}\left|E(s,t,m,n) - \widetilde{E}(s,t,m,n)\right| \leq \frac{\varepsilon}{8}, \\ &\sum_{s=m_0}^{M-1}\sum_{t=n_0}^{N-1}\left|W(s,t,m,n) - \widetilde{W}(s,t,m,n)\right| \leq \frac{\varepsilon}{8}, \\ &\sum_{s=m_0}^{m-1}\sum_{t=n_0}^{n-1}\left|F(s,t,m,n,\widetilde{u}) - \widetilde{F}(s,t,m,n,\widetilde{u})\right| \leq \frac{\varepsilon}{8}, \\ &\sum_{s=m_0}^{M-1}\sum_{t=n_0}^{N-1}\left|C(s,t,m,n,\widetilde{u}) - \widetilde{C}(s,t,m,n,\widetilde{u})\right| \leq \frac{\varepsilon}{8}, \\ &\sum_{s=m_0}^{m-1}\sum_{t=n_0}^{n-1}\left|D(s,t,m,n,\widetilde{u}) - \widetilde{D}(s,t,m,n,\widetilde{u})\right| \leq \frac{\varepsilon}{8}, \\ &\sum_{s=m_0}^{M-1}\sum_{t=n_0}^{N-1}\left|G(s,t,m,n,\widetilde{u}) - \widetilde{G}(s,t,m,n,\widetilde{u})\right| \leq \frac{\varepsilon}{8}, \end{aligned} \tag{78}$$

where $\varepsilon > 0$ is an arbitrary constant. Then

$$\begin{aligned} &\left|u^p(m,n) - \widetilde{u}^p(m,n)\right| \\ &\quad \leq \varepsilon\left[1 + \frac{J_2(M,N)}{1 - \lambda_2(M,N)}C_2(m,n)\right], \end{aligned} \tag{79}$$

where $\lambda_2(M,N) < 1$, and

$$\begin{aligned} &C_2(m,n) = \exp\left\{\sum_{s=m_0}^{m-1}\sum_{t=n_0}^{n-1}B_2(s,t,m,n)\right\}, \\ &B_2(s,t,m,n) = \left[c_1(s,t,m,n) + d_1(s,t,m,n)\right], \\ &\lambda_2(m,n) \\ &\quad = \sum_{s=m_0}^{M-1}\sum_{t=n_0}^{N-1}\left[f_1(s,t,m,n) + g_1(s,t,m,n)\right]C_2(s,t), \\ &J_2(m,n) \\ &\quad = \sum_{s=m_0}^{m-1}\sum_{t=n_0}^{n-1}\left[c_1(s,t,m,n) + d_1(s,t,m,n)\right] \\ &\quad\quad + \sum_{s=m_0}^{M-1}\sum_{t=n_0}^{N-1}\left[f_1(s,t,m,n) + g_1(s,t,m,n)\right] \end{aligned} \tag{80}$$

for $(m,n) \in \Omega$. That is, u^p depends continuously on the functions a, C, D, E, F, G, W.

Proof. Let $u(m,n)$ and $\widetilde{u}(m,n)$ be solutions of (68) and (76), respectively. Then $u(m,n)$ satisfies (68) and $\widetilde{u}(m,n)$ satisfies (76). Hence

$$\begin{aligned} &\left|u^p(m,n) - \widetilde{u}^p(m,n)\right| \leq \left|a(m,n) - \widetilde{a}(m,n)\right| \\ &\quad + \sum_{s=m_0}^{m-1}\sum_{t=n_0}^{n-1}\left[\left|C(s,t,m,n,u(s,t)) - \widetilde{C}(s,t,m,n,\widetilde{u}(s,t))\right|\right. \\ &\quad + \left|D(s,t,m,n,u(s,t)) - \widetilde{D}(s,t,m,n,\widetilde{u}(s,t))\right| \\ &\quad + \left.\left|E(s,t,m,n) - \widetilde{E}(s,t,m,n)\right|\right] \\ &\quad + \sum_{s=m_0}^{M-1}\sum_{t=n_0}^{N-1}\left[\left|F(s,t,m,n,u(s,t)) - \widetilde{F}(s,t,m,n,\widetilde{u}(s,t))\right|\right. \\ &\quad + \left|G(s,t,m,n,u(s,t)) - \widetilde{G}(s,t,m,n,\widetilde{u}(s,t))\right| \\ &\quad + \left.\left|W(s,t,m,n) - \widetilde{W}(s,t,m,n)\right|\right] \leq \left|a(m,n) - \widetilde{a}(m,\right. \\ &\left.n)\right| \\ &\quad + \sum_{s=m_0}^{m-1}\sum_{t=n_0}^{n-1}\left[\left|C(s,t,m,n,u(s,t)) - C(s,t,m,n,\widetilde{u}(s,t))\right|\right. \\ &\quad + \left|C(s,t,m,n,\widetilde{u}(s,t)) - \widetilde{C}(s,t,m,n,\widetilde{u}(s,t))\right| \\ &\quad + \left|D(s,t,m,n,u(s,t)) - D(s,t,m,n,\widetilde{u}(s,t))\right| \\ &\quad + \left|D(s,t,m,n,\widetilde{u}(s,t)) - \widetilde{D}(s,t,m,n,\widetilde{u}(s,t))\right| \\ &\quad + \left.\left|E(s,t,m,n) - \widetilde{E}(s,t,m,n)\right|\right] \\ &\quad + \sum_{s=m_0}^{M-1}\sum_{t=n_0}^{N-1}\left[\left|F(s,t,m,n,u(s,t)) - F(s,t,m,n,\widetilde{u}(s,t))\right|\right. \\ &\quad + \left|F(s,t,m,n,\widetilde{u}(s,t)) - \widetilde{F}(s,t,m,n,\widetilde{u}(s,t))\right| \end{aligned}$$

$$+ \left|G(s,t,m,n,u(s,t)) - G(s,t,m,n,\tilde{u}(s,t))\right|$$

$$+ \left|G(s,t,m,n,\tilde{u}(s,t)) - \widetilde{G}(s,t,m,n,\tilde{u}(s,t))\right|$$

$$+ \left|W(s,t,m,n) - \widetilde{W}(s,t,m,n)\right|\Big] \leq \varepsilon$$

$$+ \sum_{s=m_0}^{m-1} \sum_{t=n_0}^{n-1} \left[c_1(s,t,m,n) + d_1(s,t,m,n)\right]\left[u^p - \tilde{u}^p\right]$$

$$+ \sum_{s=m_0}^{M-1} \sum_{t=n_0}^{N-1} \left[f_1(s,t,m,n) + g_1(s,t,m,n)\right]\left|u^p - \tilde{u}^p\right|. \quad (81)$$

Treat $|u^p(m,n) - \tilde{u}^p(m,n)|$ as one variable, and a suitable application of Corollary 6 (with $l_1 = l_2 = 1$) yields the desired result (79). Hence u^p depends continuously on a, C, D, E, F, G, W.

4. Conclusions

The author carried out some new Volterra-Fredholm-type discrete inequalities involving four iterated infinite sums and their corresponding applications. The results are more effective to qualitative analysis of solutions for sum-difference equations, such as the boundedness, uniqueness, and continuous dependence on solutions.

Conflicts of Interest

The author declares that there are no conflicts of interest.

Acknowledgments

Run Xu received a grant from National Science Foundation of China (11671227).

References

[1] B. Zheng and B. Fu, "Some Volterra-Fredholm type nonlinear discrete inequalities involving four iterated infinite sums," *Advances in Difference Equations*, vol. 2012, no. 228, 2012.

[2] B. Zheng and Q. Feng, "Some new Volterra-Fredholm-type discrete inequalities and their applications in the theory of difference equations," *Abstract and Applied Analysis*, vol. 2011, Article ID 584951, 24 pages, 2011.

[3] B. Zheng, "Qualitative and quantitative analysis for solutions to a class of Volterra-Fredholm type difference equation," *Advances in Difference Equations*, vol. 2011, article no. 30, 2011.

[4] B. G. Pachpatte, "On some fundamental integral inequalities and their discrete analogues," *Journal of Inequalities in Pure and Applied Mathematics*, vol. 2, no. 2, article 15, 2001.

[5] F. Jiang and F. Meng, "Explicit bounds on some new nonlinear integral inequalities with delay," *Journal of Computational and Applied Mathematics*, vol. 205, no. 1, pp. 479–486, 2007.

[6] R. Xu and Y. Zhang, "Generalized Gronwall fractional summation inequalities and their applications," *Journal of Inequalities and Applications*, vol. 2015, no. 42, 2015.

[7] F. W. Meng and D. Ji, "On some new nonlinear discrete inequalities and their applications," *Journal of Computational and Applied Mathematics*, vol. 208, no. 2, pp. 425–433, 2007.

[8] T. Wang and R. Xu, "Some integral inequalities in two independent variables on time scales," *Journal of Mathematical Inequalities*, vol. 6, no. 1, pp. 107–118, 2012.

[9] Q. Feng, F. Meng, and B. Fu, "Some new generalized Volterra-Fredholm type finite difference inequalities involving four iterated sums," *Applied Mathematics and Computation*, vol. 219, no. 15, pp. 8247–8258, 2013.

[10] Q.-H. Ma and J. Pecarić, "Estimates on solutions of some new nonlinear retarded Volterra-Fredholm type integral inequalities," *Nonlinear Analysis*, vol. 69, no. 2, pp. 393–407, 2008.

[11] Q.-H. Ma, "Some new nonlinear Volterra-Fredholm-type discrete inequalities and their applications," *Journal of Computational and Applied Mathematics*, vol. 216, no. 2, pp. 451–466, 2008.

[12] R. Xu, F. Meng, and C. Song, "On some integral inequalities on time scales and their applications," *Journal of Inequalities and Applications*, vol. 2010, article 464976, 2010.

[13] L. Wan and R. Xu, "Some generalized integral inequalities and their applications," *Journal of Mathematical Inequalities*, vol. 7, no. 3, pp. 495–511, 2013.

[14] R. Xu and F. Meng, "Some new weakly singular integral inequalities and their applications to fractional differential equations," *Journal of Inequalities and Applications*, vol. 2016, no. 78, 2016.

[15] J. Huang and W.-S. Wang, "Some Volterra-Fredholm type nonlinear inequalities involving four iterated infinite integral and application," *Journal of Mathematical Inequalities*, vol. 10, no. 4, pp. 1105–1118, 2016.

2

Convergence Analysis on Unstructured Meshes of a DDFV Method for Flow Problems with Full Neumann Boundary Conditions

A. Kinfack Jeutsa,[1] A. Njifenjou,[2] and J. Nganhou[3]

[1]*Higher Technical Teachers' Training College, The University of Buea, P.O. Box 249, Buea Road, Kumba, Cameroon*
[2]*Faculty of Industrial Engineering, The University of Douala, P.O. Box 2701, Douala, Cameroon*
[3]*National Advanced School of Engineering, The University of Yaounde I, P.O. Box 8390, Yaounde, Cameroon*

Correspondence should be addressed to A. Kinfack Jeutsa; jeutsa2001@yahoo.fr

Academic Editor: Pedro J. Coelho

A Discrete Duality Finite Volume (DDFV) method to solve on unstructured meshes the flow problems in anisotropic nonhomogeneous porous media with full Neumann boundary conditions is proposed in the present work. We start with the derivation of the discrete problem. A result of existence and uniqueness of a solution for that problem is given thanks to the properties of its associated matrix combined with adequate assumptions on data. Their theoretical properties, namely, stability and error estimates (in discrete energy norms and L^2-norm), are investigated. Numerical test is provided.

1. Introduction and the Model Problem

Efficient schemes are required for addressing flow problems in geologically complex media. The most important criteria of efficiency are mass conservation in grid blocks, accurate approximation of Darcy velocity, capability for dealing with anisotropic flow on unstructured grids and diverse heterogeneities (relevant to absolute permeability, porosity, etc.), and robust and easy implementation. Schemes well known in the literature for meeting many of the previous criteria are the following: Mixed Finite Element methods (see, e.g., [1, 2]), Control-Volume Finite Element methods (see, e.g., [3, 4]), Mimetic Finite Difference methods (see, e.g., [5, 6] and references therein), Cell-Centered Finite Volume methods (see, e.g., [7–11] and certain references therein; see also [12–14]), Multipoint Flux Approximation (see, e.g., [15–18] and some contributions to convergence analysis of MPFA O-scheme like [19]) and Discrete Duality Finite Volume methods (DDFV methods for short).

The DDFV methods come in two formulations. The first formulation is based on interface flux computations for primary and dual meshes, accounting with the interface flux continuity (see, e.g., [20, 21]) and the second formulation of DDFV is based on pressure gradient reconstructions over a diamond grid (see [22–24]). Note that this second formulation attracted the attention of some mathematicians as Andreianov, Boyer, and Hubert who have greatly contributed to its mathematical development. Indeed, key ideas involved in the pressure gradient reconstruction approach have been generalized by these authors to nonlinear operators of Leray-Lions type. Motivated by the possibility of increasing the order of convergence of the pressure gradient reconstruction method for nonlinear operators, Boyer and Hubert have proposed in [25] the so-called modified DDFV.

Beyond flow problems, we find the applications of DDFV methods in many areas: the numerical modeling of the surface erosion occurring at a fluid/soil interface undergoing a flow process in [26], the discretisation of partial differential equation appearing in image processing in [27], the assessment of nuclear waste repository safety in the context of simulating flow, transport in porous media in [28], and so on.

For presenting our analysis of DDFV method, let us consider the 2D diffusion problem consisting of finding a function φ which satisfies the following partial differential equation associated with nonhomogeneous Neumann boundary conditions:

$$-\mathrm{div}\,(K\,\mathrm{grad}\,\varphi) = f \quad \text{in } \Omega, \tag{1}$$

$$[-K\,\mathrm{grad}\,\varphi]\cdot\eta = g \quad \text{on } \Gamma, \tag{2}$$

where Ω is a given open polygonal domain, Γ is its boundary, f and g are two given functions (defined, resp., in Ω and Γ, at least almost everywhere in Lebesgue measure sense), and η is the unit normal vector to Γ outward to Ω. The permeability $K(x)$, with $x = (x_1, x_2)^t \in \Omega$, may be a full matrix depending on solely spatial variables and obeying the following conditions.

(i) The primary mesh is such that the discontinuity of K lies on mesh interfaces.

(ii) Symmetry is

$$K_{ij}(x) = K_{ji}(x) \quad \text{a.e. in } \Omega,\ \forall 1 \le i, j \le 2. \tag{3}$$

(iii) Uniform ellipticity and boundedness are

$$\exists \gamma_{\min}, \gamma_{\max} \in \mathbb{R}_+^* \quad \text{such that } \forall \xi \in \mathbb{R}^2,\ \xi \neq 0,$$
$$\gamma_{\min} \le \frac{\xi^T K(x)\,\xi}{|\xi|^2} \le \gamma_{\max} \quad \text{a.e. in } \Omega, \tag{4}$$

where $|\cdot|$ denotes the Euclidian norm in $\mathbb{R}^2$ and where $K_{ij}(\cdot)$ are the components of K satisfying

$$K_{ij} \in \mathrm{PWC}(\Omega), \tag{5}$$

where PWC(Ω) denotes the subspace of $L^\infty(\Omega)$ made of piecewise constant functions defined in Ω.

One also assume that f and g satisfy the following conditions.

(i) Compatibility condition is

$$f \in L^2(\Omega), \quad g \in L^2(\Gamma),$$
$$\text{such that } \int_\Omega f(x)\,dx - \int_\Gamma g(x)\,d\gamma(x) = 0. \tag{6}$$

(ii) Null average condition is

$$\int_\Omega \varphi(x)\,dx = 0. \tag{7}$$

(iii) Let us emphasize that the novelty of this work is that Neumann boundary conditions are imposed on the whole boundary, which causes additional difficulties in the analysis.

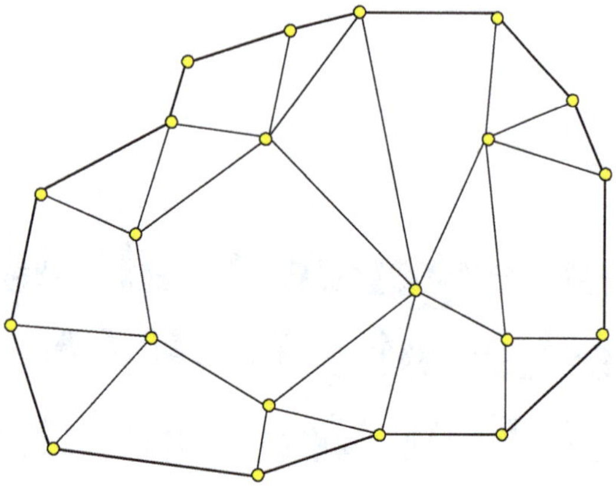

Figure 1: Example of a primary unstructured conforming mesh.

2. A Finite Volume Formulation of the Model Problem

We briefly present finite volume formulation of the model problem (1)-(2) on unstructured meshes into the same spirit as [20, 25, 29–31]. We assume that

the diffusion matrix K is a piecewise constant function over Ω.

This assumption is currently used at least in industrial problems (e.g., reservoir simulation problems). Indeed, a subsurface area is made up of a collection of various geologic formations that may be characterized at intermediate scales by averaged full permeability tensors over grid blocks of the primary grid: for more details on this topic, see [32, 33].

2.1. Formulation of the Discrete Problem. First of all, notice that the method under consideration is analyzed in this work for general polygonal domains covered with unstructured matching primary meshes $\mathcal{P}$ made up of arbitrary convex polygons (see Figure 1). Let us consider some definitions needed in what follows.

Definition 1. A mesh defined on Ω is compatible with the discontinuities of the permeability tensor K if these discontinuities are located along the mesh interfaces.

Definition 2. One defines an edge-point as any point located over an edge and different from the extremities of that edge.

Definition 3. Two edge-points I and J are named "neighboring edge-points" if they share the same vertex V in the sense that I and J belong to two different edges that intersect in V.

Let us recall our main objectives in this work:

(i) Compute at the cell-points (to be defined later) and at the interior vertices from the mesh $\mathcal{P}$ the values of the unknown function φ as a solution (expected unique) of a linear system.

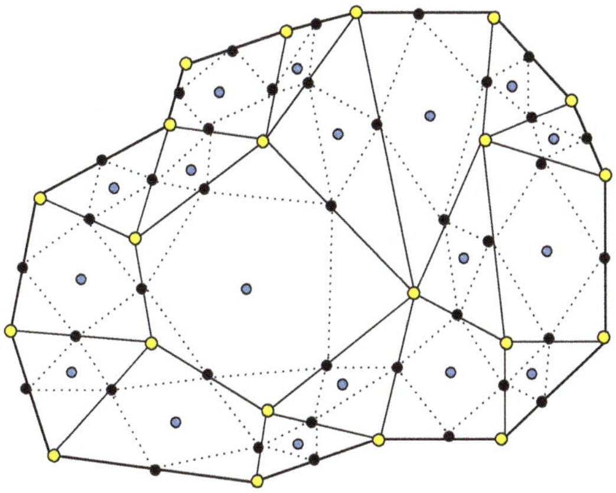

FIGURE 2: A primary mesh (full lines) and the associated auxiliary mesh (dotted lines), including edge-points and cell-points, respectively, in black and blue colors.

(ii) Analyze the stability and the convergence of this solution in some discrete energy norm similar to the one introduced in [31].

In the context of unstructured primary meshes (including square primary meshes with cell-points chosen different from cell-centers), the definition of a discrete energy norm requires that the cell-points lie inside certain perimeters. In this connection, the main steps for defining the cell-points are as follows:

(a) Choose arbitrarily a unique point (different from a vertex) on each edge of the mesh $\mathscr{P}$. This process generates a finite family of edge-points.

(b) Join every pair of neighboring edge-points by a dotted straight line.

By this way, we generate an auxiliary mesh denoted $\mathscr{A}$.

(c) Fix arbitrarily a unique point inside each intersection of a primary cell and an auxiliary cell. These points define a finite family of cell-points.

Figure 2 illustrates the location of the edge-points and cell-points within both primary and auxiliary meshes.

Remark 4. Note that, in the 3D framework, for a given primary mesh the corresponding auxiliary mesh is generated very easily as follows. Any primary cell C involves a finite number of faces and to each face is assigned one and only one face point lying necessarily on the boundary of C. Therefore, one could associate with these face points the smallest convex polygon containing all of them. This is the process leading to generation of the auxiliary mesh associated with the primary mesh under consideration.

We should emphasize that we need to take cell-points inside auxiliary cells (see Figure 2) in view of achieving the following goal which is to define a vector space for the piecewise constant solution (named "weak approximate" solution in the sequel) and to equip it with a discrete energy norm.

Remark 5. Note that there exists a DDFV approach based on pressure gradient reconstructions that can address nonlinear elliptic problems: see [25, 34] for more details. Nevertheless in these works, the choice of edge-points depends strongly on that of cell-points as each edge-point coincides (by Definition 2) with the intersection of a primary edge and the straight line joining two cell-points located in both sides of that edge.

From the boundary-value problem theory (see, e.g., [35]), system (1)-(2) possesses a unique solution in $H^1(\Omega)$ under assumptions (3)–(7). We have assumed that the diffusion coefficient K is a piecewise constant function over Ω. The discontinuities of K naturally divide Ω into a finite number of convex subdomains $(\Omega_s)_{s\in S}$. We now make the additional assumption that the restriction over Ω_s of the exact solution to system (1)-(2), denoted by $\varphi_{|\Omega_s}$, satisfies

$$\varphi_{|\Omega_s} \in C^2\left(\overline{\Omega}_s\right) \quad \forall s \in S. \tag{8}$$

Let us now focus on a finite volume formulation of problem (1)-(2) in terms of a linear system which should be derived from the elimination of auxiliary unknowns, namely, edge-point pressures, in flux balance equations over primary cells and also dual cells (to be introduced later). This linear system should involve the real numbers $\{u_P\}_{P\in\mathscr{P}}$ and $\{u_{D^*}\}_{D^*\in\mathscr{D}}$ as discrete unknowns which are expected to be reasonable approximations of $\{\varphi_P\}_{P\in\mathscr{P}}$ (cell-point pressures) and $\{\varphi_{D^*}\}_{D^*\in\mathscr{D}}$ (interior vertex pressures), respectively, where $\varphi_P = \varphi(x_1^P, x_2^P)$ and $\varphi_{D^*} = \varphi(x_1^{D^*}, x_2^{D^*})$ and where $\mathscr{D}$ stands for dual mesh. We now give a description of the procedure leading to the linear discrete system.

Let C_P be a primary cell, where P is the corresponding cell-point. We integrate the two sides of the mass balance equation (1) in C_P. Applying Ostrogradsky's theorem to the integral in the left-hand side leads to computing the flux across the boundary of C_P. Thanks to a suitable quadrature formula this computation yields a relation involving edge-point pressures. Due to the flux continuity over the mesh interfaces, the edge-point pressures can be eliminated from the previous relation.

As an illustration of this technique for computing the fluxes across primary cell boundaries, we consider the edge $[A^*B^*]$ associated with the primary cell C_P (see Figure 3).

Let K^P be the diffusion tensor of the primary cell C_P. Denoting $\xi^P_{[A^*B^*]}$ as the unit normal vector to $[A^*B^*]$ exterior to C_P, the flux expression over the edge $[A^*B^*]$ viewed as part of the boundary of C_P is given by

$$\begin{aligned} Q^P_{[A^*B^*]} &= \left[\frac{a_h\left(K^P\right)\widehat{a}_h\left(K^L\right)h_{A^*B^*}}{a_h\left(K^P\right)h_{IL} + \widehat{a}_h\left(K^L\right)h_{PI}}\right]\left[\varphi_P - \varphi_L\right] \\ &+ T^P_{I(A^*,B^*)} \\ &+ \left[\frac{\widehat{a}_h\left(K^L\right)b_h\left(K^P\right)h_{PI} + a_h\left(K^P\right)\widehat{b}_h\left(K^L\right)h_{IL}}{a_h\left(K^P\right)h_{IL} + \widehat{a}_h\left(K^L\right)h_{PI}}\right] \\ &\cdot \left[\varphi_{B^*} - \varphi_{A^*}\right], \end{aligned} \tag{9}$$

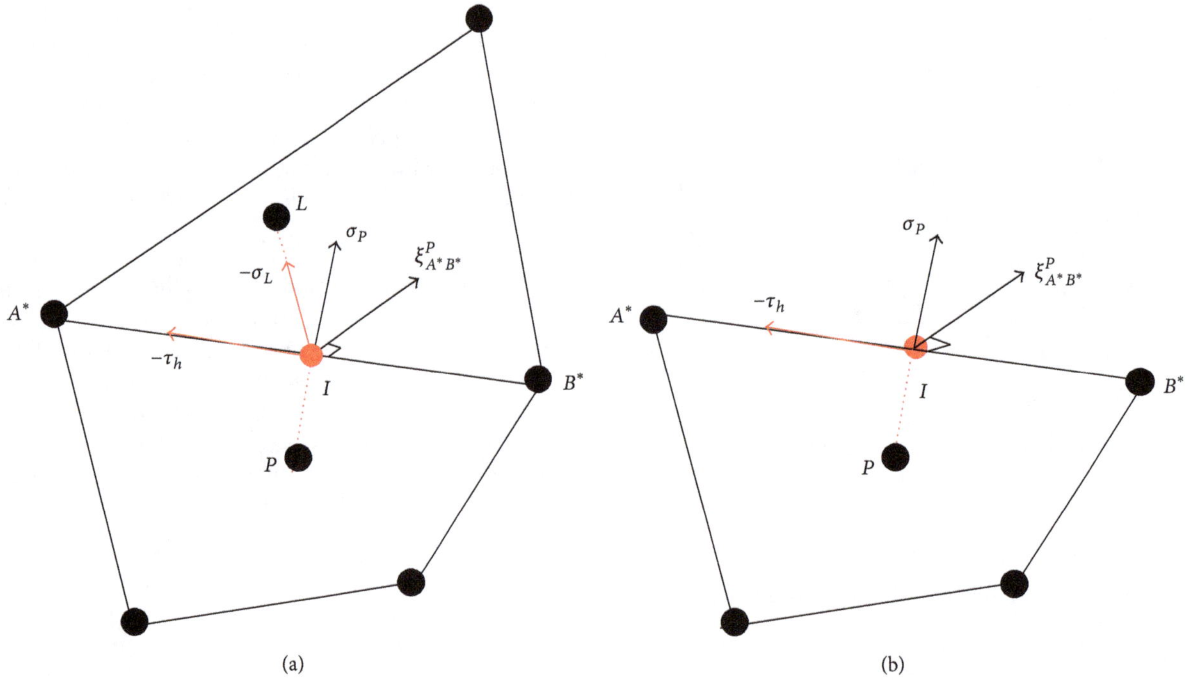

FIGURE 3: Two molecules for a finite volume computation of the flux across the edge $[A^*B^*]$. (a) The edge $[A^*B^*]$ is lying inside the domain Ω. (b) The edge $[A^*B^*]$ is part of the boundary of Ω.

where $T^P_{I(A^*,B^*)}$ is the truncation error and where we introduce the following necessary ingredients:

$$\sigma_P = \frac{\overrightarrow{PI}}{\left|\overrightarrow{PI}\right|},$$

$$\tau_h = \frac{\overrightarrow{A^*B^*}}{\left|\overrightarrow{A^*B^*}\right|},$$

$$h = \max\left\{\text{size}\left(\mathscr{P}\right), \text{size}\left(\mathscr{D}\right), \text{size}\left(\mathscr{A}\right)\right\},$$

$$a_h\left(K^P\right) = \frac{1}{\cos\theta_h^{P,I}}\left(\xi^P_{[A^*B^*]}\right)^t K^P\left(\xi^P_{[A^*B^*]}\right), \tag{10}$$

$$b_h\left(K^P\right) = \frac{1}{\cos\theta_h^{P,I}}\left(\xi^{B^*}_{[PI]}\right)^t K^P\left(\xi^P_{[A^*B^*]}\right),$$

$$\widehat{a}_h\left(K^L\right) = \frac{1}{\cos\theta_h^{L,I}}\left(\xi^P_{[A^*B^*]}\right)^t K^L\left(\xi^P_{[A^*B^*]}\right),$$

$$\widehat{b}_h\left(K^P\right) = \frac{1}{\cos\theta_h^{L,I}}\left(\xi^{B^*}_{[IL]}\right)^t K^L\left(\xi^P_{[A^*B^*]}\right),$$

where $\theta_h^{P,I}$ (resp., $\theta_h^{L,I}$) is the angle defined by the vectors σ_P and $\xi^P_{[A^*B^*]}$ (resp., $-\sigma_P$ and $\xi^P_{[A^*B^*]}$) and where $\xi^{B^*}_{[PI]}$ (resp., $\xi^{B^*}_{[IL]}$) denotes the unit normal vector to (PI) (resp., to $[IL]$) exterior to the half-plane from $\mathbb{R}^2$ containing the point B^* and bordered by the straight line (PI) (resp., (IL)). Notice that $0 \le \theta_h^{P,I}, \theta_h^{L,I} < \pi/2$. Moreover if the primary mesh $(\mathscr{P}, \mathscr{E})$ is regular in the sense of Definition 6, we have $0 \le \theta_h^{L,I} < \pi/2 - \theta$ and therefore $0 < \cos(\pi/2 - \theta) < \cos(\theta_h^{L,I}) \le 1$, where $\theta \in]0, \pi/2[$ is a certain real number not depending on h.

In what follows, we will need the following notations. We denote by $\mathscr{E}$ the set of all edge-points (from the primary mesh of course), $\mathscr{E}^{\text{int}}$ the subset of $\mathscr{E}$ made up of internal edge-points, that is, edge-points lying on Ω, $\mathscr{E}^{\text{ext}}$ the subset of $\mathscr{E}$ made up of edge-points lying on edges included in the boundary of Ω, and $\mathscr{E}^P$ (for $P \in \mathscr{P}$) the subset of $\mathscr{E}$ made up of edge-points lying on the boundary of the primary cell C_P.

We now introduce the notion of regular primary mesh that should play a central role in the sequel.

Definition 6. The set $\{\mathscr{P}, \mathscr{E}\}$ defines a regular mesh system if the following conditions are fulfilled.

There exist $0 < \varpi \le 1$ and $0 < \theta \le \pi/2$, both of them mesh independent, such that

$$d_P \le \text{diam}\left(P\right) \le \frac{1}{\varpi} d_P \quad \forall P \in \mathscr{P}, \tag{11}$$

$$0 \le \theta_h^{P,I} < \frac{\pi}{2} - \theta \quad \forall P \in \mathscr{P}\ \forall I \in \mathscr{E}^P, \tag{12}$$

where $\mathscr{P}$ stands for the set of primary cells and where d_P is the distance between a cell-point or a vertex from P and an edge-point from P.

Proposition 7. *Assume that (i) the primary mesh $\mathscr{P}$ is compatible with the discontinuities of the permeability tensor*

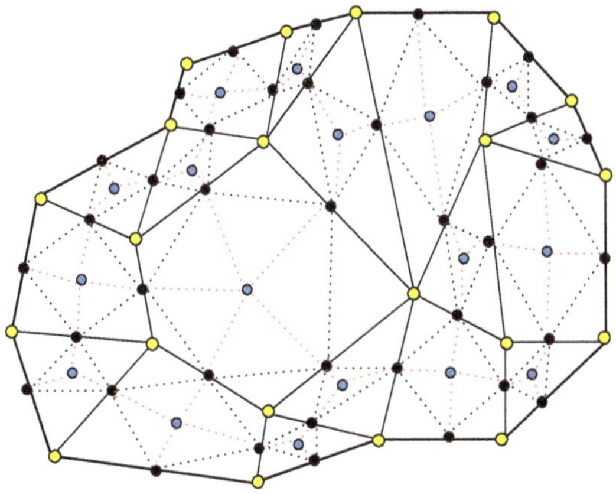

FIGURE 4: A primary mesh in full lines with the associated dual mesh in red discontinuous lines and the corresponding auxiliary mesh in black dotted lines.

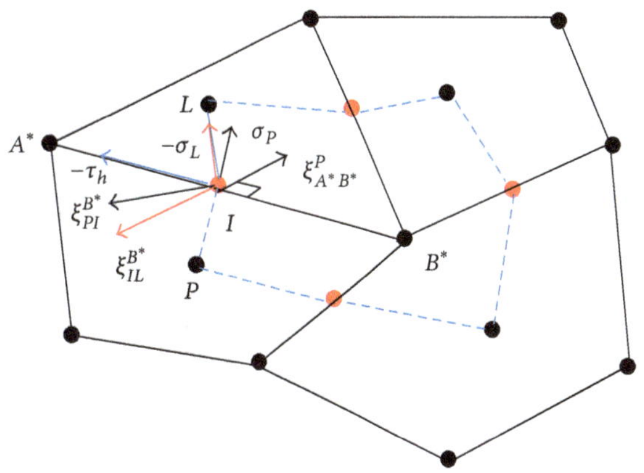

FIGURE 5: A dual cell (blue discontinuous line) with its four pseudoedges "centered" on red edge-points.

K (see Definition 1), (ii) the set $\{\mathcal{P}, \mathcal{E}\}$ is a regular mesh system, and (iii) relation (8) is honored.

Then, there exists a strictly positive number C mesh independent such that

$$T^P_{I(A^*,B^*)} \leq Ch^2. \tag{13}$$

Using the previous notations and thanks to the consistency of the flux approximation across cell edges (see Proposition 7), one reasonably can approximate the flux balance within any primary cell C_P as follows:

$$\sum_{I\in\mathcal{E}^P\cap\mathcal{E}^{\text{int}}} \left[\left(\frac{a_h(K^P)\,\widehat{a}_h(K^L)\,h_{A^*B^*}}{a_h(K^P)\,h_{IL} + \widehat{a}_h(K^L)\,h_{PI}} \right) [\varphi_P - \varphi_L] + \left(\frac{\widehat{a}_h(K^L)\,b_h(K^P)\,h_{PI} + a_h(K^P)\,\widehat{b}_h(K^L)\,h_{IL}}{a_h(K^P)\,h_{IL} + \widehat{a}_h(K^L)\,h_{PI}} \right) \cdot [\varphi_{B^*} - \varphi_{A^*}] \right] \approx \int_{C_P} f(x)\,dx - \int_{\Gamma_P\cap\Gamma} g(x)\,d\gamma(x) = 0 \quad \forall P \in \mathcal{P}, \tag{14}$$

where $L \in \mathcal{P}$, $A^* \in \mathcal{D}$, and $B^* \in \mathcal{D}$ are such that $\Gamma_P \cap \Gamma_L = [A^*B^*]$ and where $\mathcal{E}^P \cap \mathcal{E}^L = \{I\}$. It is clear that the number of discrete unknowns $\{\varphi_P\}_{P\in\mathcal{P}}$ and $\{\varphi_{D^*}\}_{D^*\in\mathcal{D}}$ is greater than the number of equations in system (14). It is then natural to complete this system with discrete equations obtained from mass flux balance over dual cells; see Figure 4 for the definition of the dual mesh.

In what follows, we will need to use the notion of pseudoedge associated with dual cells. Let us define now this notion that is illustrated in Figure 5.

Definition 8. Let P and L be two cell-points from the primary mesh (i.e., $P, L \in \mathcal{P}$) such that the corresponding primary cells C_P and C_L are adjacent, and consider $I \in \mathcal{E}^P \cap \mathcal{E}^L$ (recall that $\mathcal{E}^P$, for $E \in \mathcal{P}$, is the set of edge-points lying in the boundary of the primary cell C_E). The line $[PI] \cup [IL]$ defines a pseudoedge denoted by $[PIL]$ and "centered" on I, with extremities P and L.

We will say that a pseudoedge is associated with a dual cell C_{A^*} if it is lying in the boundary of C_{A^*}.

Remark 9. The boundary of each dual cell is a union of a finite number of pseudoedges (see Figure 4).

Let us now look for discrete flux balance equations over dual cells. Integrating the two sides of the balance equation (1) in a dual cell C_{B^*}, applying Ostrogradsky's theorem for the left-hand side, and exploiting Remark 9 lead to

$$\sum_{I\in\mathcal{E}^{B^*}} -\int_{[PIL]} K\,\text{grad}\,\varphi \cdot n_{B^*}\,d\gamma = \int_{C_{B^*}} f(x)\,dx, \tag{15}$$

where n_{B^*} stands for the outward unit normal vector to the boundary of C_{B^*} and where $[PIL]$ is a pseudoedge associated with the dual cell C_{B^*}. Recall that $\mathcal{E}^{B^*}$ is the set of edge-points lying in the boundary of the dual cell C_{B^*}.

Let us look for a flux approximation across the pseudoedge $[PIL]$ viewed as part of the boundary of C_{B^*}. Denoting by $Q^{B^*}_{[PIL]}$ the exact flux over $[PIL]$, it can be expressed by the relation

$$Q^{B^*}_{[PIL]} = -\int_{[PI]} \text{grad}\,\varphi \cdot \left(K^P \xi^{B^*}_{[PI]}\right) d\gamma - \int_{[IL]} \text{grad}\,\varphi \cdot \left(K^L \xi^{B^*}_{[IL]}\right) d\gamma. \tag{16}$$

By using the same process, the computation of the flux across $[PI]$ is

$$-\int_{[PI]} \operatorname{grad}\varphi \cdot \left(K^P \xi^{B^*}_{[PI]}\right) d\gamma = \left(\frac{c_h\left(K^P\right) \widehat{a}_h\left(K^L\right) h_{PI}}{a_h\left(K^P\right) h_{IL} + \widehat{a}_h\left(K^L\right) h_{PI}} \right) \left[\varphi_P - \varphi_L\right] + T^{B^*}_{[PI]}$$
$$+ h_{PI} \left(\frac{d_h\left(K^P\right)\left\{a_h\left(K^P\right) h_{IL} + \widehat{a}_h\left(K^L\right) h_{PI}\right\} + c_h\left(K^P\right) h_{IL}\left\{\widehat{b}_h\left(K^L\right) - b_h\left(K^P\right)\right\}}{h_{A^*B^*}\left[a_h\left(K^P\right) h_{IL} + \widehat{a}_h\left(K^L\right) h_{PI}\right]} \right) \cdot \left[\varphi_{B^*} - \varphi_{A^*}\right], \tag{17}$$

where

$$c_h\left(K^P\right) = \frac{1}{\cos\theta_h^{P,I}} \left(\xi^P_{[A^*B^*]}\right)^t K^P \left(\xi^{B^*}_{[PI]}\right),$$
$$d_h\left(K^P\right) = \frac{1}{\cos\theta_h^{P,I}} \left(\xi^{B^*}_{[PI]}\right)^t K^P \left(\xi^{B^*}_{[PI]}\right) \tag{18}$$

(recall that $\theta_h^{P,I}$ denotes the angle defined by the vectors σ_P and $\xi^P_{[A^*B^*]}$).

Similarly, the computation of the flux across $[IL]$ leads to

$$-\int_{[IL]} \operatorname{grad}\varphi \cdot \left(K^L \xi^{B^*}_{[IL]}\right) d\gamma = \left(\frac{\widehat{c}_h\left(K^L\right) a_h\left(K^P\right) h_{IL}}{a_h\left(K^P\right) h_{IL} + \widehat{a}_h\left(K^L\right) h_{PI}} \right) \left[\varphi_P - \varphi_L\right] + T^{B^*}_{[IL]}$$
$$+ h_{IL} \left(\frac{\widehat{d}_h\left(K^L\right)\left\{a_h\left(K^P\right) h_{IL} + \widehat{a}_h\left(K^L\right) h_{PI}\right\} + \widehat{c}_h\left(K^P\right) h_{PI}\left\{b_h\left(K^P\right) - \widehat{b}_h\left(K^L\right)\right\}}{h_{A^*B^*}\left[a_h\left(K^P\right) h_{IL} + \widehat{a}_h\left(K^L\right) h_{PI}\right]} \right) \cdot \left[\varphi_{B^*} - \varphi_{A^*}\right], \tag{19}$$

where

$$\widehat{c}_h\left(K^P\right) = \frac{1}{\cos\theta_h^{L,I}} \left(\xi^P_{[A^*B^*]}\right)^t K^L \left(\xi^{B^*}_{[IL]}\right),$$
$$\widehat{d}_h\left(K^L\right) = \frac{1}{\cos\theta_h^{L,I}} \left(\xi^{B^*}_{[IL]}\right)^t K^L \left(\xi^{B^*}_{[IL]}\right). \tag{20}$$

We should now formulate a global estimate for the truncation errors associated with the flux approximation over the pseudoedge$[PIL]$.

Proposition 10. *Under the same assumptions as those of Proposition 7, there exists a positive number C mesh independent such that*

$$\left|T^{B^*}_{[PI]}\right| + \left|T^{B^*}_{[IL]}\right| \le Ch^2. \tag{21}$$

The above inequality shows the consistency of the flux approximation across the pseudoedge $[PIL]$. We summarize what precedes as

$$Q^{B^*}_{[PIL]} \approx \left(\frac{c_h\left(K^P\right) \widehat{a}_h\left(K^L\right) h_{PI} + \widehat{c}_h\left(K^L\right) a_h\left(K^P\right) h_{IL}}{a_h\left(K^P\right) h_{IL} + \widehat{a}_h\left(K^L\right) h_{PI}} \right) \cdot \left[\varphi_P - \varphi_L\right] + \left(\frac{\omega_h(P,L,I)}{h_{A^*B^*}\left[a_h\left(K^P\right) h_{IL} + \widehat{a}_h\left(K^L\right) h_{PI}\right]} \right) \cdot \left[\varphi_{B^*} - \varphi_{A^*}\right], \tag{22}$$

where we have set

$$\omega_h(P,L,I) = \left[a_h\left(K^P\right) h_{IL} + \widehat{a}_h\left(K^L\right) h_{PI}\right] \cdot \left[d_h\left(K^P\right) h_{PI} + \widehat{d}_h\left(K^L\right) h_{IL}\right] + h_{PI} h_{IL}\left[\widehat{b}_h\left(K^L\right) - b_h\left(K^P\right)\right] \cdot \left[c_h\left(K^P\right) - \widehat{c}_h\left(K^L\right)\right]. \tag{23}$$

We deduce from (22) that an approximate flux balance equation over any dual cell C_{B^*} can be formulated as follows:

$$\sum_{I \in \mathscr{E}^P \cap \mathscr{E}^{\text{int}}} \left[\left(\frac{c_h\left(K^P\right) \widehat{a}_h\left(K^L\right) h_{PI} + \widehat{c}_h\left(K^L\right) a_h\left(K^P\right) h_{IL}}{a_h\left(K^P\right) h_{IL} + \widehat{a}_h\left(K^L\right) h_{PI}} \right) \cdot \left[\varphi_P - \varphi_L\right] + \left(\frac{\omega_h(P,L,I)}{h_{A^*B^*}\left[a_h\left(K^P\right) h_{IL} + \widehat{a}_h\left(K^L\right) h_{PI}\right]} \right) \cdot \left[\varphi_{B^*} - \varphi_{A^*}\right] \right]$$

$$+ \sum_{I\in\mathscr{E}^{B^*}\cap\mathscr{E}^{\text{ext}}} \left(\frac{h_{IE}\left[a\left(K^E\right)d_h\left(K^E\right) - c_h\left(K^E\right)b_h\left(K^E\right)\right]}{a_h\left(K^E\right)h_{B^*D^*}} \right)$$
$$\cdot\left[\varphi_{B^*} - \varphi_{D^*}\right] \approx \int_{C_{B^*}} f(x)\,dx - \int_{\Gamma_{B^*}\cap\Gamma} g\,d\gamma$$
$$+ \sum_{I\in\mathscr{E}^{B^*}\cap\mathscr{E}^{\text{ext}}} \frac{c_h\left(K^E\right)h_{IE}}{a\left(K^E\right)h_{B^*D^*}} \int_{[B^*D^*]} g\,d\gamma \quad \forall B^* \in \mathscr{D}, \tag{24}$$

where $P \in \mathscr{P}$, $L \in \mathscr{P}$, and $A^* \in \mathscr{D}$ are such that $\Gamma_P \cap \Gamma_L = [A^*B^*]$ and $[PL] \cap [A^*B^*] = \{I\}$ and where $E \in \mathscr{P}$, $D^* \in \mathscr{D}$ a boundary-vertex are such that $[B^*D^*] \subset \Gamma \cap \Gamma_E$ and $I \in [B^*D^*]$.

Systems (14) and (24) naturally suggest defining a finite volume formulation of the diffusion problem (1)-(2) as follows.

Find $\{\overline{\varphi}_P\}_{P\in\mathscr{P}}$ and $\{\overline{\varphi}_{D^*}\}_{D^*\in\mathscr{D}}$ such that

$$\sum_{I\in\mathscr{E}^P\cap\mathscr{E}^{\text{int}}} \left[\left(\frac{a_h\left(K^P\right)\widehat{a}_h\left(K^L\right)h_{A^*B^*}}{a_h\left(K^P\right)h_{IL} + \widehat{a}_h\left(K^L\right)h_{PI}} \right)\left[\overline{\varphi}_P - \overline{\varphi}_L\right] \right.$$
$$+ \left(\frac{\widehat{a}_h\left(K^L\right)b_h\left(K^P\right)h_{PI} + a_h\left(K^P\right)\widehat{b}_h\left(K^L\right)h_{IL}}{a_h\left(K^P\right)h_{IL} + \widehat{a}_h\left(K^L\right)h_{PI}} \right)$$
$$\left. \cdot\left[\overline{\varphi}_{B^*} - \overline{\varphi}_{A^*}\right] \right] = \int_{C_P} f(x)\,dx - \int_{\Gamma_P\cap\Gamma} g\,d\gamma = 0 \quad \forall P \in \mathscr{P}, \tag{25}$$

$$\sum_{I\in\mathscr{E}^{B^*}\cap\mathscr{E}^{\text{int}}} \left[\left(\frac{c_h\left(K^P\right)\widehat{a}_h\left(K^L\right)h_{PI} + \widehat{c}_h\left(K^L\right)a_h\left(K^P\right)h_{IL}}{a_h\left(K^P\right)h_{IL} + \widehat{a}_h\left(K^L\right)h_{PI}} \right) \right.$$
$$\cdot\left[\overline{\varphi}_P - \overline{\varphi}_L\right] + \left(\frac{\omega_h(P,L,I)}{h_{A^*B^*}\left[a_h\left(K^P\right)h_{IL} + \widehat{a}_h\left(K^L\right)h_{PI}\right]} \right)$$
$$\left. \cdot\left[\overline{\varphi}_{B^*} - \overline{\varphi}_{A^*}\right] \right]$$
$$+ \sum_{I\in\mathscr{E}^{B^*}\cap\mathscr{E}^{\text{ext}}} \left(\frac{h_{IE}\left[a\left(K^E\right)d_h\left(K^E\right) - c_h\left(K^E\right)b_h\left(K^E\right)\right]}{a_h\left(K^E\right)h_{B^*D^*}} \right)$$
$$\cdot\left[\overline{\varphi}_{B^*} - \overline{\varphi}_{D^*}\right] = \int_{C_{B^*}} f(x)\,dx - \int_{C_{B^*}\cap\Gamma} g\,d\gamma$$
$$+ \sum_{I\in\mathscr{E}^{B^*}\cap\mathscr{E}^{\text{ext}}} \frac{c_h\left(K^E\right)h_{IE}}{a\left(K^E\right)h_{B^*D^*}} \int_{[B^*D^*]} g\,d\gamma \quad \forall B^* \in \mathscr{D}. \tag{26}$$

Remark 11. It is useful to note that

$$\sum_{B^*\in\mathscr{D}} \sum_{I\in\mathscr{E}^{B^*}\cap\mathscr{E}^{\text{ext}}} \frac{c_h\left(K^E\right)h_{IE}}{a\left(K^E\right)h_{B^*D^*}} \int_{[B^*D^*]} g\,d\gamma = 0. \tag{27}$$

2.2. Existence of Solutions to System (25)-(26) Conditions for Uniqueness

(i) Matrix Properties of the DDFV Problem (25)-(26). It is easily seen that the symmetry of the matrix associated with the linear system (25)-(26) essentially follows from the symmetry of the diffusion coefficient K (see assumption (3)). We assume that all the cell-points and all the vertices (with respect to the primary mesh) are numbered. Therefore one can identify $\mathscr{P}$ and $\mathscr{D}$ with two disjoint subsets of the set of positive integers. To fix ideas, let us set

$$\mathscr{P} \equiv \{1,\ldots,n\},$$
$$\mathscr{D} \equiv \{n+1,\ldots,n+m\}. \tag{28}$$

Then Card($\mathscr{P}$) denotes the total number of cell-points and Card($\mathscr{D}$) stands for the total number of vertices. On the other hand, define the subvectors Φ_{cc} and Φ_{vc} by

$$\Phi_{cc} = \{\overline{\varphi}_P\}_{P\in\mathscr{P}},$$
$$\Phi_{vc} = \{\overline{\varphi}_{D^*}\}_{D^*\in\mathscr{D}} \tag{29}$$

and set

$$\mathbb{M} = \begin{bmatrix} \mathbb{A} & \mathbb{B} \\ \mathbb{B}^t & \mathbb{C} \end{bmatrix}, \tag{30}$$

where

$$\mathbb{M}\begin{bmatrix} \Phi_{cc} \\ \Phi_{vc} \end{bmatrix} = \begin{bmatrix} F_{cc} \\ F_{vc} \end{bmatrix}, \tag{31}$$

where F_{cc} is a subvector with Card($\mathscr{P}$) components defined by the right-hand side of (25) and F_{vc} is a subvector with Card($\mathscr{D}$) components defined by the right-hand side of (26).

Remark 12. The matrix $\mathbb{M}$ satisfies the following properties:

(1) $1 \le i,j \le m$ $\mathbb{M}_{ij} = \mathbb{M}_{ji}$;

(2) $1 \le i,j \le m$ $\sum_{j=1}^{m}\mathbb{M}_{ij} = 0$;

(3) $1 \le i,j \le m$ $\sum_{i=1}^{m}\mathbb{M}_{ij} = 0$.

Let us introduce two vectors of $\mathbb{R}^{n+m}$ named $\mathscr{F}_{cc}$ and $\mathscr{F}_{vc}$ and defined by

$$\left(\mathscr{F}_{cc}\right)_i = \begin{cases} 1 & \text{if } 1 \le i \le n \\ 0 & \text{if } n+1 \le i \le n+m, \end{cases}$$
$$\left(\mathscr{F}_{vc}\right)_i = \begin{cases} 0 & \text{if } 1 \le i \le n \\ 1 & \text{if } n+1 \le i \le n+m. \end{cases} \tag{32}$$

Proposition 13 (characterization of Kernel space of $\mathbb{M}$).

(i) *The matrix $\mathbb{M}$ is singular.*

(ii) *Moreover, let* $\ker(\mathbb{M})$ *denote the subset of* $\mathbb{R}^{n+m}$ *defined as follows:*
$\ker(\mathbb{M}) = \{V \in \mathbb{R}^{n+m},\ \mathbb{M}V = 0\}$ *(named Kernel space of $\mathbb{M}$ in the sequel); then we have*

$$\ker(\mathbb{M}) = \left\langle \mathscr{F}_{cc}, \mathscr{F}_{vc} \right\rangle. \tag{33}$$

Sketch for the Proof

(i) The singularity of $\mathbb{M}$ is an immediate consequence of Remark 12.

(ii) Define the space $\mathbb{R}^{\mathscr{P},\mathscr{D}}$ by

$$\mathbb{R}^{\mathscr{P},\mathscr{D}} = \{V = [V_{cc}, V_{vc}]; V_{cc} = \{V_P\}_{P\in\mathscr{P}} \subset \mathbb{R},\ V_{vc} = \{V_{D^*}\}_{D^*\in\mathscr{D}} \subset \mathbb{R}\} \qquad (34)$$

and endow it with the seminorm $|\cdot|_{\mathbb{R}^{\mathscr{P},\mathscr{D}}}$ defined as

$$|V|_{\mathbb{R}^{\mathscr{P},\mathscr{D}}} = \left\{ \sum_{\substack{(P,L)\in\mathscr{P}^2,\ (A^*,B^*)\in\mathscr{D}^2 \\ \text{s.t. } \Gamma_P\cap\Gamma_L=[A^*B^*]}} \left[(V_P - V_L)^2 + (V_{B^*} - V_{A^*})^2\right] + \sum_{(A^*,B^*)\in\mathscr{D}^2 \text{ s.t. } [A^*B^*]\in\mathscr{E}^{\text{ext}}} (V_{B^*} - V_{A^*})^2 \right\}^{1/2} \quad \forall V \in \mathbb{R}^{\mathscr{P},\mathscr{D}}. \qquad (35)$$

Then, find the above identification of the Kernel space of $\mathbb{M}$ using the following Lemma.

Lemma 14. *The matrix* $\mathbb{M}$ *is positive; that is,*

$$[V_{cc}, V_{vc}]\,\mathbb{M}\begin{bmatrix} V_{cc} \\ V_{vc} \end{bmatrix} \geq 0 \quad \forall V \in \mathbb{R}^{\mathscr{P},\mathscr{D}}. \qquad (36)$$

Proof. Let $V = [V_{cc}, V_{vc}] \in \mathbb{R}^{\mathscr{P},\mathscr{D}}$, where $V_{cc} = \{V_P\}_{P\in\mathscr{P}}$ and $V_{vc} = \{V_{A^*}\}_{A^*\in\mathscr{D}}$.

It follows from what precedes that

$$\begin{aligned}
[V_{cc}, V_{vc}]\,\mathbb{M}\begin{bmatrix} V_{cc} \\ V_{vc} \end{bmatrix} &= \sum_{P\in\mathscr{P}} \sum_{I\in\mathscr{E}^P\cap\mathscr{E}^{\text{int}}} \left[\left(\frac{a_h(K^P)\hat{a}_h(K^L)h_{A^*B^*}}{a_h(K^P)h_{IL} + \hat{a}_h(K^L)h_{PI}}\right)(V_P - V_L)V_P \right.\\
&\quad \left. + \left(\frac{\hat{a}_h(K^L)b_h(K^P)h_{PI} + a_h(K^P)\hat{b}_h(K^L)h_{IL}}{a_h(K^P)h_{IL} + \hat{a}_h(K^L)h_{PI}}\right)(V_{B^*} - V_{A^*})V_P\right] \\
&\quad + \sum_{B^*\in\mathscr{D}} \sum_{I\in\mathscr{E}^{B^*}\cap\mathscr{E}^{\text{int}}} \left[\left(\frac{c_h(K^P)\hat{a}_h(K^L)h_{PI} + \hat{c}_h(K^L)a_h(K^P)h_{IL}}{a_h(K^P)h_{IL} + \hat{a}_h(K^L)h_{PI}}\right)(V_P - V_L)V_{B^*} \right.\\
&\quad \left. + \left(\frac{\omega_h(P,L,I)}{h_{A^*B^*}\left[a_h(K^P)h_{IL} + \hat{a}_h(K^L)h_{PI}\right]}\right)(V_{B^*} - V_{A^*})V_{B^*}\right] \\
&\quad + \sum_{B^*\in\mathscr{D}} \sum_{I\in\mathscr{E}^{B^*}\cap\mathscr{E}^{\text{ext}}} \left(\frac{h_{IE}\left[a(K^E)d_h(K^E) - c_h(K^E)b_h(K^E)\right]}{a_h(K^E)h_{B^*D^*}}\right)(V_{B^*} - V_{D^*})V_{B^*}.
\end{aligned} \qquad (37)$$

Define

$$\begin{aligned}
K_{11}^{PL} &= \frac{a_h(K^P)\hat{a}_h(K^L)h_{A^*B^*}}{a_h(K^P)h_{IL} + \hat{a}_h(K^L)h_{PI}}, \\
K_{22}^{PL} &= \frac{\omega_h(P,L,I)}{h_{A^*B^*}\left[a_h(K^P)h_{IL} + \hat{a}_h(K^L)h_{PI}\right]}, \\
\Pi_{P(I,A^*,B^*)} &= \frac{\left[a(K^P)d_h(K^P) - c_h(K^P)b_h(K^P)\right]h_{IP}}{a_h(K^P)h_{A^*B^*}},
\end{aligned} \qquad (38)$$

where $P, L \in \mathscr{P}$ are two adjacent primary cells sharing the edge $[A^*B^*]$ as interface containing the edge-point I. Then, relation (37) becomes

$$\begin{aligned}
[V_{cc}, V_{vc}]\,\mathbb{M}\begin{bmatrix} V_{cc} \\ V_{vc} \end{bmatrix} &= \sum_{\substack{(P,L)\in\mathscr{P}^2,(A^*,B^*)\in\mathscr{D}^2 \\ \text{s.t. } \Gamma_P\cap\Gamma_L=[A^*B^*]}} \left[K_{11}^{PL}(V_P - V_L)^2 + K_{22}^{PL}(V_{B^*} - V_{A^*})^2 + 2K_{12}^{PL}(V_P - V_L)(V_{B^*} - V_{A^*})\right] \\
&\quad + \sum_{\substack{P\in\mathscr{P}^{\text{ext}},(A^*,B^*)\in\mathscr{D}^2\cap\Gamma^2 \\ \text{s.t. } [A^*B^*]\in\mathscr{E}^P}} \Pi_{P(I,A^*,B^*)}(V_{B^*} - V_{A^*})^2,
\end{aligned} \qquad (39)$$

where $\mathscr{P}^{\text{ext}}$ denotes the subset of $\mathscr{P}$ made of primary cells adjacent to the domain boundary Γ. Let us prove that the homogenized symmetric matrix K^{PL} is positive definite; that is,

$$K_{11}^{PL}K_{22}^{PL} - \left(K_{12}^{PL}\right)^2 > 0. \quad (40)$$

Setting

$$\Delta^{PL} = K_{11}^{PL}K_{22}^{PL} - \left(K_{12}^{PL}\right)^2, \quad (41)$$

it is easy to check that

$$\begin{aligned}\Delta^{PL} &= N_1\left[a_h\left(K^P\right)d_h\left(K^P\right) - \left(b_h\left(K^P\right)\right)^2\right] \\ &\quad + N_2\left[\widehat{a}_h\left(K^L\right)\widehat{d}_h\left(K^L\right) - \left(\widehat{b}_h\left(K^L\right)\right)^2\right],\end{aligned} \quad (42)$$

where we have set

$$\begin{aligned}N_1 &= \left(\frac{\widehat{a}_h\left(K^L\right)h_{IL}}{a_h\left(K^P\right)h_{IL} + \widehat{a}_h\left(K^L\right)h_{PI}}\right)^2 \\ &\quad + \frac{a_h\left(K^P\right)\widehat{a}_h\left(K^L\right)h_{PI}h_{IL}}{\left(a_h\left(K^P\right)h_{IL} + \widehat{a}_h\left(K^L\right)h_{PI}\right)^2}, \\ N_2 &= \left(\frac{a_h\left(K^P\right)h_{IL}}{a_h\left(K^P\right)h_{IL} + \widehat{a}_h\left(K^L\right)h_{PI}}\right)^2 \\ &\quad + \frac{a_h\left(K^P\right)\widehat{a}_h\left(K^L\right)h_{PI}h_{IL}}{\left(a_h\left(K^P\right)h_{IL} + \widehat{a}_h\left(K^L\right)h_{PI}\right)^2}\end{aligned} \quad (43)$$

which are strictly positive numbers.

Since the diffusion matrix K is symmetric and positive definite (see assumptions (3)-(4)), Cauchy-Schwarz's inequality for the inner product associated with K ensures that

$$\begin{aligned}&a_h\left(K^P\right)d_h\left(K^P\right) - \left(b_h\left(K^P\right)\right)^2 > 0, \\ &\widehat{a}_h\left(K^L\right)\widehat{d}_h\left(K^L\right) - \left(\widehat{b}_h\left(K^L\right)\right)^2 > 0\end{aligned} \quad (44)$$

as either $\xi_{[A^*B^*]}^P$ and $\xi_{[PI]}^{B^*}$ or $\xi_{[A^*B^*]}^P$ and $\xi_{[IL]}^{B^*}$ are not collinear. Therefore, $\Delta^{PL} > 0$ and thus K^{PL} is a symmetric and positive definite matrix.

It follows from what precedes that the matrix K^{PL} possesses strictly positive eigenvalues. Let $\lambda_{\min}^{PL}$ be its least eigenvalue. So we have

$$\begin{aligned}&\left[V_{cc}, V_{vc}\right]\mathbb{M}\begin{bmatrix}V_{cc}\\ V_{vc}\end{bmatrix} \\ &\geq \sum_{\substack{(P,L)\in\mathcal{P}^2,(A^*,B^*)\in\mathcal{D}^2 \\ \text{s.t. } \Gamma_P\cap\Gamma_L=[A^*B^*]}} \lambda_{\min}^{PL}\left[\left(V_P - V_L\right)^2 \right. \\ &\quad \left. + \left(V_{B^*} - V_{A^*}\right)^2\right] \\ &\quad + \sum_{\substack{P\in\mathcal{P}^{\text{ext}},(A^*,B^*)\in\mathcal{D}^2\cap\Gamma^2 \\ \text{s.t. } [A^*B^*]\in\mathcal{E}^P}} \Pi_{P(I,A^*,B^*)}\left(V_{B^*} - V_{A^*}\right)^2.\end{aligned} \quad (45)$$

Remarking that there exist two real numbers δ and η strictly positive depending exclusively on the geological structure of the domain such that

$$\delta \leq K_{11}^P K_{22}^P - \left(K_{12}^P\right)^2 \leq \eta \quad \forall P \in \mathcal{P}, \quad (46)$$

the following result is easily seen.

Lemma 15.

$$\Pi_{P(I,A^*,B^*)} = \left[K_{11}^P K_{22}^P - \left(K_{12}^P\right)^2\right]\frac{h_{PI}}{a_h\left(K^P\right)h_{A^*B^*}}. \quad (47)$$

Moreover

$$\frac{\delta\varpi\sin\theta}{\gamma_{max}} < \Pi_{P(I,A^*,B^*)} < \frac{\eta}{\varpi\gamma_{min}}. \quad (48)$$

Thanks to inequality (45) and to Lemma 15 the proof of Lemma 14 ends. □

Lemma 16. *The matrix $\mathbb{M}$ satisfies the relation*

$$\gamma\left(|V|_{\mathbb{R}^{\mathcal{P},\mathcal{D}}}\right)^2 \leq \left[V_{cc}, V_{vc}\right]\mathbb{M}\begin{bmatrix}V_{cc}\\ V_{vc}\end{bmatrix} \quad \forall V \in \mathbb{R}^{\mathcal{P},\mathcal{D}}, \quad (49)$$

where γ is a strictly positive number mesh independent and where $|\cdot|_{\mathbb{R}^{\mathcal{P},\mathcal{D}}}$ is a seminorm defined on $\mathbb{R}^{\mathcal{P},\mathcal{D}}$ by relation (35).

Proof. It is conducted with the same arguments as those developed in the previous proof, except that one should go much farther by proving that there exists $\rho > 0$ mesh independent such that

$$\rho \leq \lambda_{\min}^{PL} \quad \forall P, L \in \mathcal{P}. \quad (50)$$

The eigenvalues λ of the symmetric positive definite matrix K^{PL} satisfy the so-called characteristic equation associated with K^{PL}; that is,

$$\lambda^2 - \left[K_{11}^{PL} + K_{22}^{PL}\right]\lambda + \left[K_{11}^{PL}K_{22}^{PL} - \left(K_{12}^{PL}\right)^2\right] = 0. \quad (51)$$

The least eigenvalue of K^{PL} denoted by

$$\lambda_{\min}^{PL} = \frac{\left[K_{11}^{PL} + K_{22}^{PL}\right] - \sqrt{\Delta}}{2}, \quad (52)$$

where $\Delta = [K_{11}^{PL} + K_{22}^{PL}]^2 - 4[K_{11}^{PL}K_{22}^{PL} - (K_{12}^{PL})^2]$, is a strictly positive number. One can easily deduce that

$$\lambda_{\min}^{PL} \geq \frac{\det\left(K^{PL}\right)}{\left[K_{11}^{PL} + K_{22}^{PL}\right] + \det\left(K^{PL}\right) + 1}, \quad (53)$$

where $\det(K^{PL}) = K_{11}^{PL}K_{22}^{PL} - (K_{12}^{PL})^2$ is a strictly positive number. We should bound the quantities K_{11}^{PL}, K_{22}^{PL}, and $\det(K^{PL})$ by mesh independent strictly positive numbers. Let us start first with $\det(K^{PL})$. We consider a change of coordinates by moving from the initial Cartesian coordinates to a local one, namely, $(J, \overrightarrow{C^*D^*}/|\overrightarrow{C^*D^*}|, \xi_{[C^*D^*]}^{\perp})$, where J

is the edge-point on the interface $[C^*D^*]$ between the cells C_G and C_H and where $\xi^{\perp}_{[C^*D^*]}$ is a vector orthogonal to $[C^*D^*]$ and oriented such that the basis change matrix M is a rotation. Denoting the permeability tensor of the cell C_G by $K^G = \{K^G_{ij}\}$ in the initial Cartesian coordinates and by $\widetilde{K}^G = \{\widetilde{K}^G_{ij}\}$ in the local coordinates we have

$$\widetilde{K}^G = M^{-1}K^G M. \tag{54}$$

Similarly, we get for the cell C_H

$$\widetilde{K}^H = M^{-1}K^H M. \tag{55}$$

Then it is easy to check that

$$a_h\left(K^G\right)d_h\left(K^G\right) - \left(b_h\left(K^G\right)\right)^2 = \widetilde{K}^G_{11}\widetilde{K}^G_{22} - \left[\widetilde{K}^G_{12}\right]^2; \tag{56}$$

that is,

$$\begin{aligned} a_h\left(K^G\right)d_h\left(K^G\right) - \left(b_h\left(K^G\right)\right)^2 &= \det\left(\widetilde{K}^G\right) \\ &= \det\left(K^G\right), \end{aligned} \tag{57}$$

where $\det(\cdot)$ denotes the determinant. Similarly, we have for the cell C_H

$$\begin{aligned} \widehat{a}_h\left(K^H\right)\widehat{d}_h\left(K^H\right) - \left(\widehat{b}_h\left(K^H\right)\right)^2 &= \det\left(\widetilde{K}^H\right) \\ &= \det\left(K^H\right). \end{aligned} \tag{58}$$

It follows from what precedes that

$$\det\left(K^{PL}\right) = N_1 \det\left(K^G\right) + N_2 \det\left(K^H\right), \tag{59}$$

where N_1 and N_2 are given, respectively, by relations (43).

On one hand, we can deduce from (4) and Definition 6 that

$$N_i \geq \frac{1}{2}\left(\frac{\varpi\gamma_{\min}\sin\theta}{\gamma_{\max}}\right)^2 \quad \forall i = 1, 2. \tag{60}$$

On the other hand, we remark that

$$\det\left(K^P\right) \geq \min\left\{\det\left(K^s\right),\ s \in S\right\} \quad \forall P \in \mathscr{P}, \tag{61}$$

where the set S (introduced in (8)) depends exclusively on the lithologic structure of the medium Ω. Then we deduce from (59), (60), and (61) that

$$\begin{aligned} &\det\left(K^P\right) \\ &\quad \geq \frac{1}{2}\left(\frac{\varpi\gamma_{\min}\sin\theta}{\gamma_{\max}}\right)^2 \left[\min\left\{\det\left(K^s\right),\ s \in S\right\}\right] \\ &\qquad \forall P \in \mathscr{P}. \end{aligned} \tag{62}$$

Remarking that

$$\det\left(K^{PL}\right) \leq K^{PL}_{11}K^{PL}_{22} + \left(K^{PL}_{12}\right)^2 \tag{63}$$

and exploiting again (4) and Definition 6 lead to the following inequality:

$$\det\left(K^{PL}\right) \leq \frac{2\left(\gamma_{\max}\right)^4 \max\left\{2, \varpi\right\}}{\left(\varpi\right)^3\left(\gamma_{\min}\right)^2\left(\sin\theta\right)^4}. \tag{64}$$

Thanks again to (4) and Definition 6 one can easily check that

$$K^{PL}_{11} + K^{PL}_{22} \leq \left(\frac{\gamma_{\max}}{\sqrt{2}\varpi\sin\theta}\right)^2\left[1 + \frac{2}{\varpi} + \frac{4\gamma_{\max}}{\varpi\sin\theta}\right]. \tag{65}$$

Lemma 16 follows from the combination of (53), (62), (64), and (65).

Proposition 17 (discrete compatibility condition). *The right-hand side of the discrete system (25)-(26) satisfies the following discrete compatibility condition:*

$$\begin{aligned} &\sum_{P\in\mathscr{P}}\left[\int_{C_P} f(x)\,dx - \int_{\Gamma_P\cap\Gamma} g\,d\gamma\right] = 0, \\ &\sum_{B^*\in\mathscr{D}}\left[\int_{C_{B^*}} f(x)\,dx - \int_{\Gamma_{B^*}\cap\Gamma} g\,d\gamma\right] \\ &\quad + \sum_{B^*\in\mathscr{D}}\ \sum_{I\in\mathscr{E}^{B^*}\cap\mathscr{E}^{ext}} \frac{c_h\left(K^E\right)h_{IE}}{a\left(K^E\right)h_{B^*D^*}}\int_{[B^*D^*]} g\,d\gamma = 0. \end{aligned} \tag{66}$$

Proof. First of all, note that the double summation in the previous proposition is equal to zero thanks to Remark 11. Let us consider two vectors of $\mathbb{R}^{n+m}$ called $\mathscr{F}_{cc}$ and $\mathscr{F}_{vc}$ and defined by

$$\begin{aligned} \left(\mathscr{F}_{cc}\right)_i &= \begin{cases} 1 & \text{if } 1 \leq i \leq n \\ 0 & \text{if } n+1 \leq i \leq n+m, \end{cases} \\ \left(\mathscr{F}_{vc}\right)_i &= \begin{cases} 0 & \text{if } 1 \leq i \leq n \\ 1 & \text{if } n+1 \leq i \leq n+m. \end{cases} \end{aligned} \tag{67}$$

According to the matrix form of the discrete system (see relation (31)), we have

$$\left(\mathscr{F}_{cc}\right)^t \mathbb{M}\begin{bmatrix}\Phi_{cc} \\ \Phi_{vc}\end{bmatrix} = \left(\mathscr{F}_{cc}\right)^t\begin{bmatrix}F_{cc} \\ F_{vc}\end{bmatrix}. \tag{68}$$

It follows from Remark 12 that

$$\left(\mathscr{F}_{cc}\right)^t \mathbb{M}\begin{bmatrix}\Phi_{cc} \\ \Phi_{vc}\end{bmatrix} = \left[\Phi_{cc}\ \ \Phi_{vc}\right]^t \mathbb{M}\mathscr{F}_{cc} = 0. \tag{69}$$

Therefore

$$0 = \left(\mathscr{F}_{cc}\right)^t\begin{bmatrix}F_{cc} \\ F_{vc}\end{bmatrix} = \sum_{P\in\mathscr{P}}\left[\int_{C_P} f(x)\,dx - \int_{\Gamma_P\cap\Gamma} g\,d\gamma\right]. \tag{70}$$

Similarly we have

$$\begin{aligned} 0 = \left(\mathscr{F}_{vc}\right)^t \mathbb{M}\begin{bmatrix}\Phi_{cc} \\ \Phi_{vc}\end{bmatrix} &= \left(\mathscr{F}_{vc}\right)^t\begin{bmatrix}F_{cc} \\ F_{vc}\end{bmatrix} \\ &= \sum_{B^*\in\mathscr{D}}\left[\int_{C_{B^*}} f(x)\,dx - \int_{C_{B^*}\cap\Gamma} g\,d\gamma\right]. \end{aligned} \tag{71}$$

This ends the proof of Proposition 17.

An immediate consequence of the previous result is the following existence result for a solution to the discrete problem (25)-(26).

Proposition 18 (existence result). *(1) The linear system (25)-(26) possesses an infinite number of solutions.*

(2) More precisely, if s_h is a solution to the discrete system (25)-(26), then $s_h + \ker(\mathbb{M})$ is the set of all solutions for this system.

The question to know how to get the physical solution of the model problem has a natural answer; that is, one should follow the same way as the continuous problem analysis. Since the dimension of $\ker(\mathbb{M})$ is 2, required are two (linearly independent) discrete versions of the null average condition (7). This naturally leads to the following result.

Proposition 19 (uniqueness result). *Under the following (null average) conditions,*

$$\begin{aligned}&\text{(i) } \sum_{P\in\mathcal{P}} mes\,(C_P)\,\overline{\varphi}_P = 0,\\ &\text{(ii) } \sum_{D^*\in\mathcal{D}} mes\,(C_{D^*})\,\overline{\varphi}_{D^*} = 0,\end{aligned} \tag{72}$$

the discrete problem (25)-(26) possesses a unique solution.

Proof. Let us consider the space

$$\mathbb{R}^{\mathcal{P},\mathcal{D}} = \left\{V = [V_{cc}, V_{vc}]\,; V_{cc} = \{V_P\}_{P\in\mathcal{P}} \subset \mathbb{R},\ V_{vc} = \{V_{D^*}\}_{D^*\in\mathcal{D}} \subset \mathbb{R}\right\}. \tag{73}$$

One knows from Lemma 14 that

$$V^t \mathbb{M} V \geq 0 \quad \forall V \in \mathbb{R}^{\mathcal{P},\mathcal{D}}. \tag{74}$$

Then it follows that, for all V in the subspace of $\mathbb{R}^{\mathcal{P},\mathcal{D}}$ made up of $V = [\{V_P\}_{P\in\mathcal{P}}, \{V_{D^*}\}_{D^*\in\mathcal{D}}]$ such that conditions (72) are fulfilled, one has $V^t\mathbb{M}V = 0 \Leftrightarrow V = 0$.

3. Stability and Error Estimates

We deal here with a theoretical analysis of the solution for the discrete system (25), (26), and (72). Recall that the existence and uniqueness of that solution (under explicit conditions) is proven in the previous section. We assume in what follows that the primary mesh is regular in the sense of Definition 6.

3.1. Preliminaries and a Stability Result. Let us consider the auxiliary mesh $\mathcal{A}$ introduced in the previous section (see Figure 4). Note that each mesh cell of $\mathcal{A}$ contains either one cell-point or one corner point and only one which should be lying inside or on the boundary of Ω. In the sequel, a *node* is a cell-point or a corner point with respect to the primary mesh.

Definition 20. An auxiliary mesh cell is degenerate if the corresponding node is lying on the boundary of Ω.

In the sequel, we will say sometimes "auxiliary cell" instead of "auxiliary mesh cell." We denote by $\mathbf{E}(\mathcal{A})$ the subspace of $\mathbb{R}^{\mathcal{P},\mathcal{D}}$ made of functions v which satisfies conditions (72). This space is obviously nonempty as there is the null function. Moreover, the solution of the discrete system (25), (26), and (72) could clearly be identified with one (and only one) function from $\mathbf{E}(\mathcal{A})$. In the sequel we denote by $\overline{\varphi}_h$ such a function named "cellwise constant (approximate) solution" of the diffusion problem (1)-(2). Let us endow $\mathbb{R}^{\mathcal{P},\mathcal{D}}$ with the following discrete seminorm.

For all $v \equiv [\{V_P\}_{P\in\mathcal{P}}, \{V_{D^*}\}_{D^*\in\mathcal{D}}] \in \mathbb{R}^{\mathcal{P},\mathcal{D}}$, define

$$|v|_{\mathbb{R}^{\mathcal{P},\mathcal{D}}} = \left\{ \sum_{\substack{(P,L)\in\mathcal{P}^2,(A^*,B^*)\in\mathcal{D}^2\\ \text{s.t. } \Gamma_P\cap\Gamma_L=[A^*B^*]}} \left[(V_P - V_L)^2 + (V_{B^*} - V_{A^*})^2\right] + \sum_{(A^*,B^*)\in\mathcal{D}^2 \text{ s.t. } [A^*B^*]\in\mathcal{E}^{\text{ext}}} (V_{B^*} - V_{A^*})^2 \right\}^{1/2} \quad \forall V \in \mathbb{R}^{\mathcal{P},\mathcal{D}}, \tag{75}$$

where "s.t." means "such that." As this seminorm is actually a discrete energy norm for the space $\mathbf{E}(\mathcal{A})$, we denote it by $\|\cdot\|_{\mathcal{A}}$ as far as functions from $\mathbf{E}(\mathcal{A})$ are concerned. A norm on the space $\mathbb{R}^{\mathcal{P},\mathcal{D}}$ is defined by the mapping

$$\|\cdot\|_{\mathbb{R}^{\mathcal{P},\mathcal{D}}} = \left\{|\cdot|^2_{\mathbb{R}^{\mathcal{P},\mathcal{D}}} + \|\cdot\|^2_{L^2(\Omega)}\right\}^{1/2}. \tag{76}$$

We focus here on the proof of the stability of the solution to the system of (25), (26), and (72). For this purpose, we need to introduce as in [31] a preliminary result, namely, a discrete version of Poincaré inequality which is based upon the following ingredients. Consider the following spaces

$$\begin{aligned}\mathbb{R}^{\mathcal{P}} &= \left\{v = \{v_P\}_{P\in\mathcal{P}}\,;\ v_P \in \mathbb{R}\ \forall P \in \mathcal{P}\right\},\\ \mathbb{R}^{\mathcal{D}} &= \left\{v = \{v_{D^*}\}_{D^*\in\mathcal{D}}\,;\ v_{D^*} \in \mathbb{R}\ \forall D^* \in \mathcal{D}\right\}\end{aligned} \tag{77}$$

and the linear operators $\Pi_{\mathcal{P}}$ and $\Pi_{\mathcal{D}}$ defined as follows:

$$\begin{aligned}\mathbf{v} \in \mathbf{E}(\mathcal{A}) &\longmapsto \Pi_{\mathcal{P}}(\mathbf{v}) \in \mathbf{E}(\mathcal{P}),\\ \mathbf{v} \in \mathbf{E}(\mathcal{A}) &\longmapsto \Pi_{\mathcal{D}}(\mathbf{v}) \in \mathbf{E}(\mathcal{D})\end{aligned} \tag{78}$$

with $\Pi_{\mathcal{P}}(\mathbf{v})=\{v_P\}_{P\in\mathcal{P}}$ and $\Pi_{\mathcal{D}}(\mathbf{v}) = \{v_{D^*}\}_{D^*\in\mathcal{D}}$, where $\mathbf{E}(\mathcal{P})$ and $\mathbf{E}(\mathcal{D})$ are, respectively, subspaces of $\mathbb{R}^{\mathcal{P}}$ and $\mathbb{R}^{\mathcal{D}}$ made of functions satisfying conditions (72)-(i) and (72)-(ii), respectively.

Let us equip the function spaces $\mathbf{E}(\mathscr{P})$ and $\mathbf{E}(\mathscr{D})$ with the following discrete energy norms:

$$\|v\|_{\mathscr{P}} = \left\{ \sum_{\substack{(P,L)\in\mathscr{P}^2,(A^*,B^*)\in\mathscr{D}^2 \\ \text{s.t. } \Gamma_P\cap\Gamma_L=[A^*B^*]}} \left[v_P - v_L\right]^2 \right\}^{1/2},$$

$$\|v\|_{\mathscr{D}} = \left\{ \sum_{\substack{(P,L)\in\mathscr{P}^2,(A^*,B^*)\in\mathscr{D}^2 \\ \text{s.t. } \Gamma_P\cap\Gamma_L=[A^*B^*]}} \left[\left(v_{B^*} - v_{A^*}\right)^2\right] + \sum_{(A^*,B^*)\in\mathscr{D}^2 \text{ s.t. } [A^*B^*]\in\mathscr{E}^{\text{ext}}} \left(V_{B^*} - V_{A^*}\right)^2 \right\}^{1/2}. \quad (79)$$

These mappings are only seminorms, respectively, for $\mathbb{R}^{\mathscr{P}}$ and $\mathbb{R}^{\mathscr{D}}$ that one can transform into norms for these spaces as follows:

$$\|\cdot\|_{\mathbb{R}^{\mathscr{P}}} = \left\{|\cdot|^2_{\mathbb{R}^{\mathscr{P}}} + \|\cdot\|^2_{L^2(\Omega)}\right\}^{1/2},$$
$$\|\cdot\|_{\mathbb{R}^{\mathscr{D}}} = \left\{|\cdot|^2_{\mathbb{R}^{\mathscr{D}}} + \|\cdot\|^2_{L^2(\Omega)}\right\}^{1/2}, \quad (80)$$

where we have set $|\cdot|_{\mathbb{R}^{\mathscr{P}}} = \|\cdot\|_{\mathscr{P}}$ and $|\cdot|_{\mathbb{R}^{\mathscr{D}}} = \|\cdot\|_{\mathscr{D}}$. The following results are key ingredients for proving the stability of the discrete solution in the sense of the discrete energy norm defined previously on the space $\mathbf{E}(\mathscr{A})$.

Lemma 21 (discrete versions of Poincaré inequality). *We have the following inequalities:*

$$\|\cdot\|^2_{L^2(\Omega)} \le C\|v\|^2_{\mathscr{P}} + \frac{2}{mes(\Omega)}\left(\int_\Omega v(x)\,dx\right)^2 \quad \forall v \in \mathbb{R}^{\mathscr{P}}, \quad (81)$$

$$\|\cdot\|^2_{L^2(\Omega)} \le C\|v\|^2_{\mathscr{D}} + \frac{2}{mes(\Omega)}\left(\int_\Omega v(x)\,dx\right)^2 \quad \forall v \in \mathbb{R}^{\mathscr{D}}, \quad (82)$$

where C represents diverse strictly positive numbers mesh independent.

Note that the right-hand side of inequalities (81) and (82) is, respectively, norms for $\mathbb{R}^{\mathscr{P}}$ and $\mathbb{R}^{\mathscr{D}}$. Since Ω is bounded, the previous lemma permits seeing that these norms are equivalent to standard ones, namely, (80). For proving the preceding lemma one can use the same arguments as in [36].

Lemma 22 (a key-result). *Let $(\overline{\mathscr{P}}, \overline{\mathscr{E}})$ be a regular mesh system (defined on) in the sense of Definition 6 and denote by $\mathscr{D}$ the corresponding dual mesh. For $v \in E(\overline{\mathscr{P}})$, denote by v_P the value of v in the control volume P. Let $\overline{\gamma}(v)$ be a discrete trace function defined a.e. (for the $(d-1)$-Lebesgue measure) by $\overline{\gamma}(v) = v_P$ on $\Gamma_P \cap \Gamma$, for all $P \in \overline{\mathscr{P}}^{ext}$, where $\overline{\mathscr{P}}^{ext}$ denotes the set of mesh elements adjacent to the domain boundary Γ. Then*

$$\left\|\overline{\gamma}(v)\right\|_{L^2(\Gamma)} \le C\left(\|v\|_{\overline{\mathscr{P}}} + \|\cdot\|_{L^2(\Omega)}\right) \quad \forall v \in E\left(\overline{\mathscr{P}}\right). \quad (83)$$

Similarly, we have

$$\left\|\overline{\gamma}(v)\right\|_{L^2(\Gamma)} \le C\left(\|v\|_{\overline{\mathscr{D}}} + \|\cdot\|_{L^2(\Omega)}\right) \quad \forall v \in E\left(\overline{\mathscr{D}}\right), \quad (84)$$

where C stands for diverse positive numbers mesh independent.

Proof. See, for instance, [36]. □

Remark 23. It is more than useful to note that

$$\|v\|^2_{\mathscr{A}} = \left\|v^{\mathscr{P}}\right\|^2_{\mathscr{P}} + \left\|v^{\mathscr{D}}\right\|^2_{\mathscr{D}} \quad \forall v \in \mathbf{E}(\mathscr{A}). \quad (85)$$

Let us give now one of the main results of this section.

Proposition 24 (stability result). *The cellwise constant approximate solution $\overline{\varphi}_h$ of the diffusion problem (1)-(2) satisfies the following inequality:*

$$\left\|\overline{\varphi}_h\right\|_{\mathscr{A}} \le C\left(\|f\|_{L^2(\Omega)} + \|g\|_{L^2(\Gamma)}\right), \quad (86)$$

where C is a strictly positive real number not depending on the spatial discretization.

Proof. Multiplying (25) by $\overline{\varphi}_P$ and (26) by $\overline{\varphi}_{D^*}$ and summing over $P \in \mathscr{P}$ and $D^* \in \mathscr{D}$, respectively, lead to

$$\begin{bmatrix}\Phi_{cc} & \Phi_{vc}\end{bmatrix}\begin{bmatrix}\mathbb{A} & \mathbb{B}\\ \mathbb{B}^t & \mathbb{C}\end{bmatrix}\begin{bmatrix}\Phi_{cc}\\ \Phi_{vc}\end{bmatrix} = \begin{bmatrix}\Phi_{cc} & \Phi_{vc}\end{bmatrix}\begin{bmatrix}F_{cc}\\ F_{vc}\end{bmatrix}. \quad (87)$$

Recall that

$$F_{cc} = \left\{\int_{C_P} f(x)\,dx - \int_{\Gamma_P\cap\Gamma} g\,d\gamma\right\}_{P\in\mathscr{P}},$$

$$F_{vc} = \left\{\int_{C_{B^*}} f(x)\,dx - \int_{C_{B^*}\cap\Gamma} g\,d\gamma + \sum_{\substack{A^*\in\mathscr{N}(B^*),I\in\mathscr{E}^{B^*}\cap\mathscr{E}^{\text{ext}} \\ \text{with } \Gamma_{C_{A^*}}\cap\Gamma_{C_{B^*}}=[PI]}} \frac{\mathcal{C}_h\left(K^E\right)h_{IE}}{a\left(K^E\right)h_{B^*D^*}} \cdot \int_{[B^*D^*]} g\,d\gamma\right\}_{B^*\in\mathscr{D}}. \quad (88)$$

Let us set

$$\text{LHS} = \begin{bmatrix}\Phi_{cc} & \Phi_{vc}\end{bmatrix}\begin{bmatrix}\mathbb{A} & \mathbb{B}\\ \mathbb{B}^t & \mathbb{C}\end{bmatrix}\begin{bmatrix}\Phi_{cc}\\ \Phi_{vc}\end{bmatrix},$$
$$\text{RHS} = \begin{bmatrix}\Phi_{cc} & \Phi_{vc}\end{bmatrix}\begin{bmatrix}F_{cc}\\ F_{vc}\end{bmatrix}. \quad (89)$$

We know from Lemma 16 that there exists a strictly positive number ρ mesh independent such that

$$\rho \left\|\overline{\varphi}_h\right\|_{\mathscr{A}}^2 \leq \text{LHS}. \quad (90)$$

On the other hand, according to Remark 11 we have

$$\begin{aligned}\text{RHS} &= \sum_{P\in\mathscr{P}} \int_{C_P} f\overline{\varphi}_P dx - \sum_{P\in\mathscr{P}} \int_{\Gamma_P\cap\Gamma} g\overline{\varphi}_P d\gamma \\ &+ \sum_{B^*\in\mathscr{D}} \int_{C_{B^*}} f(x)\,dx - \sum_{B^*\in\mathscr{D}} \int_{\Gamma_{B^*}\cap\Gamma} g\overline{\varphi}_{B^*} d\gamma \\ &+ \sum_{B^*\in\mathscr{D}} \sum_{\substack{A^*\in\mathscr{N}(B^*)\text{ s.t.},\\ \exists I\in\mathscr{E}^{B^*}\cap\mathscr{E}^{A^*}\cap\mathscr{E}^{\text{ext}},\exists L\in\mathscr{P},\\ \text{with } \Gamma_{A^*}\cap\Gamma_{B^*}=[LI]}} \overline{\varphi}_{B^*}\frac{c_h(K^L)h_{LI}}{a(K^L)h_{A^*B^*}} \\ &\cdot \int_{[A^*B^*]} g\,d\gamma.\end{aligned} \quad (91)$$

Cauchy-Schwarz's inequality leads to

$$\begin{aligned}\text{RHS} &\leq \|f\|_{L^2(\Omega)} \left\|\overline{\varphi}_h^{\mathscr{P}}\right\|_{L^2(\Omega)} + \|g\|_{L^2(\Gamma)} \left\|\overline{\gamma}_{\mathscr{P}}\left(\overline{\varphi}_h^{\mathscr{P}}\right)\right\|_{L^2(\Gamma)} \\ &+ \|f\|_{L^2(\Omega)} \left\|\overline{\varphi}_h^{\mathscr{D}}\right\|_{L^2(\Omega)} + \|g\|_{L^2(\Gamma)} \left\|\overline{\gamma}_{\mathscr{D}}\left(\overline{\varphi}_h^{\mathscr{D}}\right)\right\|_{L^2(\Gamma)} \\ &+ \sum_{\substack{A^*,B^*\in\mathscr{D}^{\text{ext}}\text{ s.t.},\\ \exists I\in\mathscr{E}^{B^*}\cap\mathscr{E}^{A^*},\exists L\in\mathscr{P},\\ \text{with } \Gamma_{A^*}\cap\Gamma_{B^*}=[LI]}} \left|\overline{\varphi}_{B^*}-\overline{\varphi}_{A^*}\right| \frac{c_h(K^L)h_{LI}}{a(K^L)h_{A^*B^*}} \\ &\cdot \int_{[A^*B^*]} g\,d\gamma,\end{aligned} \quad (92)$$

where $\overline{\gamma}_{\mathscr{P}}$ and $\overline{\gamma}_{\mathscr{D}}$ are, respectively, discrete trace operators associated with grids $\mathscr{P}$ and $\mathscr{D}$. Thanks to relation (4), Definition 6, and Cauchy-Schwarz inequality we get

$$\begin{aligned}\text{RHS} &\leq \|f\|_{L^2(\Omega)} \left[\left\|\overline{\varphi}_h^{\mathscr{P}}\right\|_{L^2(\Omega)} + \left\|\overline{\varphi}_h^{\mathscr{D}}\right\|_{L^2(\Omega)}\right] + \|g\|_{L^2(\Gamma)} \\ &\cdot \left[\left\|\overline{\gamma}_{\mathscr{P}}\left(\overline{\varphi}_h^{\mathscr{P}}\right)\right\|_{L^2(\Gamma)} + \left\|\overline{\gamma}_{\mathscr{D}}\left(\overline{\varphi}_h^{\mathscr{D}}\right)\right\|_{L^2(\Gamma)}\right] \\ &+ \frac{1}{\varpi\gamma_{\min}\sin\theta} \sum_{\substack{A^*,B^*\in\mathscr{D}^{\text{ext}}\text{ s.t.},\\ \mathscr{E}^{A^*}\cap\mathscr{E}^{B^*}\neq\emptyset}} \left|\overline{\varphi}_{B^*}-\overline{\varphi}_{A^*}\right| \\ &\cdot \int_{[A^*B^*]} |g|\,d\gamma.\end{aligned} \quad (93)$$

Accounting with the fact that $\overline{\varphi}_h$ satisfies null average condition on the grids $\mathscr{P}$ and $\mathscr{D}$, Lemma 21 ensures that

$$\left\|\overline{\varphi}_h^{\mathscr{P}}\right\|_{L^2(\Omega)} + \left\|\overline{\varphi}_h^{\mathscr{D}}\right\|_{L^2(\Omega)} \leq C\left[\left\|\overline{\varphi}_h^{\mathscr{P}}\right\|_{\mathscr{P}} + \left\|\overline{\varphi}_h^{\mathscr{D}}\right\|_{\mathscr{D}}\right], \quad (94)$$

whereas Lemma 22 guarantees that

$$\begin{aligned}&\left\|\overline{\gamma}_{\mathscr{P}}\left(\overline{\varphi}_h^{\mathscr{P}}\right)\right\|_{L^2(\Gamma)} + \left\|\overline{\gamma}_{\mathscr{D}}\left(\overline{\varphi}_h^{\mathscr{D}}\right)\right\|_{L^2(\Gamma)} \\ &\quad\leq C\left[\left\|\overline{\varphi}_h^{\mathscr{P}}\right\|_{\mathscr{P}} + \left\|\overline{\varphi}_h^{\mathscr{D}}\right\|_{\mathscr{D}}\right].\end{aligned} \quad (95)$$

A double application of Cauchy-Schwarz's inequality leads to

$$\begin{aligned}&\sum_{\substack{A^*,B^*\in\mathscr{D}^{\text{ext}}\text{ s.t.},\\ \mathscr{E}^{A^*}\cap\mathscr{E}^{B^*}\neq\emptyset}} \left|\overline{\varphi}_{B^*}-\overline{\varphi}_{A^*}\right| \int_{[A^*B^*]} |g|\,d\gamma \\ &\quad\leq [\text{meas}(\Gamma)]^{1/2} \left\|\overline{\varphi}_h^{\mathscr{D}}\right\|_{\mathscr{D}} \|g\|_{L^2(\Gamma)}.\end{aligned} \quad (96)$$

Combining relations (93), (94), (95), and (96) on the one hand and using the inequality of Cauchy-Schwarz and Remark 23 on the other hand yield

$$\text{RHS} \leq C\left[\|f\|_{L^2(\Omega)} + \|g\|_{L^2(\Gamma)}\right] \left\|\overline{\varphi}_h\right\|_{\mathscr{A}}. \quad (97)$$

Recall that C represents diverse strictly positive numbers mesh independent. The stability result follows from relations (90) and (97). The proof of Proposition 24 ends.

3.2. Error Estimates. Following the original technique exposed in [28], we investigate in this subsection the error estimates for the finite volume approximate solution to the model problem. Recall that φ denotes the exact solution to the model problem while $\overline{\varphi}_h \in \mathbf{E}(\mathscr{A})$ is the cellwise constant approximate solution obtained from a DDFV formulation of this model problem. View also $\overline{\varphi}_h$ as a representation in terms of function of the vector $\Phi = [\Phi_{cc}\ \Phi_{vc}]^t$, where $\Phi_{cc} = \{\overline{\varphi}_P\}_{P\in\mathscr{P}}$ and $\Phi_{vc} = \{\overline{\varphi}_{D^*}\}_{D^*\in\mathscr{D}}$. Recall that the vector Φ is defined above as the unique solution of the DDFV discrete system, satisfying null average conditions on grids $\mathscr{P}$ and $\mathscr{D}$ (see (25)-(26) and (72)). Let us set

$$\begin{aligned}E_P &= \varphi(x_P) - \overline{\varphi}_h(x_P) \equiv \varphi_P - \overline{\varphi}_P \quad \forall P\in\mathscr{P}, \\ E_{D^*} &= \varphi(x_{D^*}) - \overline{\varphi}_h(x_{D^*}) \equiv \varphi_{D^*} - \overline{\varphi}_{D^*} \quad \forall D^*\in\mathscr{D}.\end{aligned} \quad (98)$$

Note that the components of the so-called error-vector $[\{E_P\}_{P\in\mathscr{P}}, \{E_{D^*}\}_{D^*\in\mathscr{D}}]^t \in \mathbb{R}^{\mathscr{P},\mathscr{D}}$ can be viewed as values in auxiliary cells of the so-called error-function E_h defined a.e. in Ω. More precisely let us recall that any auxiliary cell A is associated with a unique cell-point or a unique vertex. Let E_A be the error corresponding to the auxiliary cell A; that is,

$$E_A = \begin{cases}\varphi(x_P) - \overline{\varphi}_h(x_P) & \text{if } P \text{ is a cell point associated with } A \\ \varphi(x_{D^*}) - \overline{\varphi}_h(x_{D^*}) & \text{if } D^* \text{ is a cell point associated with } A.\end{cases} \quad (99)$$

Since the auxiliary mesh define a partition of Ω, one can set

$$E_h = \sum_{A\in\mathscr{A}} E_A\left(\chi_A(x)\right), \tag{100}$$

where χ_A is the characteristic function of any cell A.

Our first purpose is to show that the error-vector is a solution to a square linear system of the same type as the discrete system (25)-(26) in the sense that both of them are associated with the same matrix $\mathbb{M}$ defined above. Note that unfortunately $E_h \in \mathbb{R}^{\mathscr{P},\mathscr{D}}$ is not in $\mathbf{E}(\mathscr{A})$ as it does meet the null average conditions (72) on grids $\mathscr{P}$ and $\mathscr{D}$. So the error estimates, which are our main purpose in this section, should be investigated in the sense of the norm defined by (76) on the space $\mathbb{R}^{\mathscr{P},\mathscr{D}}$. In this connection we first should prove that

$$\left|E_h\right|_{\mathbb{R}^{\mathscr{P},\mathscr{D}}} \le Ch \tag{101}$$

and in the next step we should prove that

$$\left\|E_h\right\|_{L^2(\Omega)} \le Ch. \tag{102}$$

(i) Let us start with proving that the components of the error-vector are solution to a square linear system of the same type as the discrete system (25)-(26).

Accounting with truncation errors, one can see from the DDFV flux computations that the exact flux across the edge $[A^*B^*]$ viewed as part of the boundary of the primary cell C_P is given by

$$\begin{aligned} Q^P_{[A^*B^*]} &= \left[\frac{a_h\left(K^P\right)\widehat{a}_h\left(K^L\right)h_{A^*B^*}}{a_h\left(K^P\right)h_{IL}+\widehat{a}_h\left(K^L\right)h_{PI}}\right]\left[\varphi_P-\varphi_L\right] \\ &+\left[\frac{\widehat{a}_h\left(K^L\right)b_h\left(K^P\right)h_{PI}+a_h\left(K^P\right)\widehat{b}_h\left(K^L\right)h_{IL}}{a_h\left(K^P\right)h_{IL}+\widehat{a}_h\left(K^L\right)h_{PI}}\right] \\ &\cdot\left[\varphi_{B^*}-\varphi_{A^*}\right]+R^P_{I(A^*,B^*,L)}, \end{aligned} \tag{103}$$

where the truncation error $R^P_{I(A^*,B^*,L)}$ is given by

$$\begin{aligned} R^P_{I(A^*,B^*,L)} &= \left[1+\frac{a_h\left(K^P\right)h_{IL}}{a_h\left(K^P\right)h_{IL}+\widehat{a}_h\left(K^L\right)h_{PI}}\right]T^P_{I(A^*,B^*)} \\ &+\left[\frac{a_h\left(K^P\right)h_{IL}}{a_h\left(K^P\right)h_{IL}+\widehat{a}_h\left(K^L\right)h_{PI}}\right]T^L_{I(A^*,B^*)} \end{aligned} \tag{104}$$

and meets the following inequality (see Proposition 7):

$$\left|R^P_{I(A^*,B^*,L)}\right| \le Ch^2. \tag{105}$$

On the other hand, the exact fluxes across the pseudoedge $[PIL]$ for $I\in\Omega$ and across the edge $[PI]$ for $I\in\Gamma$, both of them acting as part of the boundary of the dual cell C_{B^*}, are given by the following relations (accounting with truncation errors):

$$\begin{aligned} Q^{B^*}_{[PIL]} &= \left(\frac{c_h\left(K^P\right)\widehat{a}_h\left(K^L\right)h_{PI}+\widehat{c}_h\left(K^L\right)a_h\left(K^P\right)h_{IL}}{a_h\left(K^P\right)h_{IL}+\widehat{a}_h\left(K^L\right)h_{PI}}\right) \\ &\cdot\left[\varphi_P-\varphi_L\right] \\ &+\left(\frac{\omega_h\left(P,L,I\right)}{h_{A^*B^*}\left[a_h\left(K^P\right)h_{IL}+\widehat{a}_h\left(K^L\right)h_{PI}\right]}\right) \\ &\cdot\left[\varphi_{B^*}-\varphi_{A^*}\right]+R^{B^*}_{I(P,L,A^*)}, \\ Q^{B^*}_{[PI]} &= \left[\frac{\left(a_h\left(K^P\right)d_h\left(K^P\right)-\left(c_h\left(K^P\right)\right)^2\right)h_{PI}}{a_h\left(K^P\right)h_{A^*B^*}}\right] \\ &\cdot\left[\varphi_{B^*}-\varphi_{A^*}\right]-\frac{c_h\left(K^P\right)h_{PI}}{a\left(K^P\right)h_{A^*B^*}}\int_{[A^*B^*]}g\,d\gamma \\ &+R^{B^*}_{I(P,A^*)}, \end{aligned} \tag{106}$$

where the truncation errors $R^{B^*}_{I(P,L,A^*)}$ and $R^{B^*}_{I(P,A^*)}$ meet the following inequalities (according to Propositions 7 and 10):

$$\begin{aligned} \left|R^{B^*}_{I(P,L,A^*)}\right| &\le Ch^2, \\ \left|R^{B^*}_{I(P,A^*)}\right| &\le Ch^2. \end{aligned} \tag{107}$$

The system of equations satisfied by the *exact nodal* potentials $\{\varphi_P\}_{P\in\mathscr{P}}$ and $\{\varphi_{D^*}\}_{D^*\in\mathscr{D}}$ reads as

$$\begin{aligned} &\sum_{\substack{L\in\mathscr{N}(P)\text{ s.t.}\\ \exists I\in\mathscr{E},\exists A^*,B^*\in\mathscr{D},\\ \text{with } I\in\mathscr{E}^P\cap\mathscr{E}^L,\Gamma_P\cap\Gamma_L=[A^*B^*]}}\left[\left(\frac{a_h\left(K^P\right)\widehat{a}_h\left(K^L\right)h_{A^*B^*}}{a_h\left(K^P\right)h_{IL}+\widehat{a}_h\left(K^L\right)h_{PI}}\right)\left[\varphi_P-\varphi_L\right]+\left(\frac{\widehat{a}_h\left(K^L\right)b_h\left(K^P\right)h_{PI}+a_h\left(K^P\right)\widehat{b}_h\left(K^L\right)h_{IL}}{a_h\left(K^P\right)h_{IL}+\widehat{a}_h\left(K^L\right)h_{PI}}\right)\right. \\ &\left.\cdot\left[\varphi_{B^*}-\varphi_{A^*}\right]\right]=\int_{C_P}f(x)\,dx-\int_{\Gamma_P\cap\Gamma}g\,d\gamma-\sum_{\substack{L\in\mathscr{N}(P)\text{ s.t.}\\ \exists I\in\mathscr{E},\exists A^*,B^*\in\mathscr{D},\\ \text{with } I\in\mathscr{E}^P\cap\mathscr{E}^L,\Gamma_P\cap\Gamma_L=[A^*B^*]}}R^P_{I(A^*,B^*,L)}\quad\forall P\in\mathscr{P}, \end{aligned}$$

$$\sum_{\substack{A^* \in \mathcal{N}(B^*) \text{ s.t.} \\ \exists I \in \mathcal{E}, \exists P, L \in \mathcal{P}, \\ \text{with } I \in \mathcal{E}^{A^*} \cap \mathcal{E}^{B^*}, \Gamma_{A^*} \cap \Gamma_{B^*} = [PIL]}} \left[\left(\frac{c_h(K^P) \widehat{a}_h(K^L) h_{PI} + \widehat{c}_h(K^L) a_h(K^P) h_{IL}}{a_h(K^P) h_{IL} + \widehat{a}_h(K^L) h_{PI}} \right) [\varphi_P - \varphi_L] \right.$$

$$\left. + \left(\frac{\omega_h(P, L, I)}{h_{A^*B^*} [a_h(K^P) h_{IL} + \widehat{a}_h(K^L) h_{PI}]} \right) [\varphi_{B^*} - \varphi_{A^*}] \right]$$

$$+ \sum_{\substack{A^* \in \mathcal{N}(B^*) \text{ s.t.} \\ \exists I \in \mathcal{E}, \exists P \in \mathcal{P} \\ \text{with } I \in \mathcal{E}^{A^*} \cap \mathcal{E}^{B^*} \cap \mathcal{E}^{\text{ext}}, \Gamma_{A^*} \cap \Gamma_{B^*} = [PI]}} \left(\frac{\left(a(K^P) d_h(K^P) - (c_h(K^P))^2 \right) h_{PI}}{a_h(K^P) h_{A^*B^*}} \right) [\varphi_{B^*} - \varphi_{D^*}] = \int_{C_{B^*}} f(x)\, dx - \int_{C_{B^*} \cap \Gamma} g\, d\gamma$$

$$- \sum_{\substack{A^* \in \mathcal{N}(B^*) \text{ s.t.} \\ \exists I \in \mathcal{E}, \exists P, L \in \mathcal{P} \\ \text{with } I \in \mathcal{E}^{A^*} \cap \mathcal{E}^{B^*}, \Gamma_{A^*} \cap \Gamma_{B^*} = [PIL]}} R^{B^*}_{I(P,L,A^*)} - \sum_{\substack{A^* \in \mathcal{N}(B^*) \text{ s.t.} \\ \exists I \in \mathcal{E}, \exists P \in \mathcal{P} \\ \text{with } I \in \mathcal{E}^{A^*} \cap \mathcal{E}^{B^*} \cap \mathcal{E}^{\text{ext}}, \Gamma_{A^*} \cap \Gamma_{B^*} = [PI]}} R^{B^*}_{I(P,A^*)}$$

$$+ \sum_{\substack{A^* \in \mathcal{N}(B^*) \text{ s.t.} \\ \exists I \in \mathcal{E}, \exists P \in \mathcal{P} \\ \text{with } I \in \mathcal{E}^{A^*} \cap \mathcal{E}^{B^*} \cap \mathcal{E}^{\text{ext}}, \Gamma_{A^*} \cap \Gamma_{B^*} = [PI]}} \frac{c_h(K^P) h_{PI}}{a(K^P) h_{A^*B^*}} \int_{[A^*B^*]} g\, d\gamma \quad \forall B^* \in \mathcal{D}, \tag{108}$$

where $\omega_h(P, L, I)$ is defined in (23). Recall that $\mathcal{N}(\cdot)$ is the set of neighboring cells of a given (primary or dual) cell.

Due to the conservation of our flux approximation schemes, the truncation errors naturally obey to the following relation:

$$R^a_{I(\cdot,\cdot,b)} + R^b_{I(\cdot,\cdot,a)} = 0 \quad \forall I \in \mathcal{E}^{\text{int}}, \ \{I\} = \mathcal{E}^a \cap \mathcal{E}^b, \tag{109}$$

$$R^a_{I(\cdot,b)} + R^b_{I(\cdot,a)} = 0 \quad \forall I \in \mathcal{E}^{\text{ext}}, \ \{I\} = \mathcal{E}^a \cap \mathcal{E}^b, \tag{110}$$

where a and b are two cell-points or corner points associated with primary or dual adjacent cells. Note that, in the previous equality, the truncation errors are written in a very formal way. This is with a view to involving both primary and dual adjacent cells. Let us introduce now the set of diamond cells called in what follows the diamond mesh and denoted as $\mathcal{M}$ (the concept of diamond cell has been introduced for a different usage in earlier works of some authors: see, e.g., [22, 25, 34]). Each diamond cell is associated with one edge-point and vice versa (see Figure 6). A diamond cell is declared degenerate if the corresponding edge-point is lying on the boundary of Ω. An example of a degenerate diamond cell is provided in Figure 6.

The following assumption plays a key-role in our proof of the convergence of the piecewise constant (approximate) solution $\overline{\varphi}_h$.

$$\exists \nu \in \mathbb{R}^*_+ \text{ such that } \nu h^2 \leq \text{meas}(M) \quad \forall P \in \mathcal{M}, \tag{111}$$

where meas$(\cdot)$ is the Lebesgue measure in any spatial dimension.

Recall that we have set $E_P = \varphi_P - \overline{\varphi}_P$, for $P \in \mathcal{P}$, and $E_{D^*} = \varphi_{D^*} - \overline{\varphi}_{D^*}$, for $D^* \in \mathcal{D}$. An adequate linear combination of equations from system (108)-(109) with those from system (25)-(26) shows that the quantities $\{E_P = \varphi_P - \overline{\varphi}_P\}_{P \in \mathcal{P}}$ and $\{E_{D^*} = \varphi_{D^*} - \overline{\varphi}_{D^*}\}_{D^* \in \mathcal{D}}$ satisfy the following equations:

$$\sum_{\substack{L \in \mathcal{N}(P) \text{ s.t.} \\ \exists I \in \mathcal{E}, \exists A^*, B^* \in \mathcal{D}, \\ \text{with } I \in \mathcal{E}^P \cap \mathcal{E}^L, \Gamma_P \cap \Gamma_L = [A^*B^*]}} \left[\left(\frac{a_h(K^P) \widehat{a}_h(K^L) h_{A^*B^*}}{a_h(K^P) h_{IL} + \widehat{a}_h(K^L) h_{PI}} \right) [E_P - E_L] + \left(\frac{\widehat{a}_h(K^L) b_h(K^P) h_{PI} + a_h(K^P) \widehat{b}_h(K^L) h_{IL}}{a_h(K^P) h_{IL} + \widehat{a}_h(K^L) h_{PI}} \right) \right. \tag{112}$$

$$\left. \cdot [E_{B^*} - E_{A^*}] \right] = \sum_{\substack{L \in \mathcal{N}(P) \text{ s.t.} \\ \exists I \in \mathcal{E}, \exists A^*, B^* \in \mathcal{D}, \\ \text{with } I \in \mathcal{E}^P \cap \mathcal{E}^L, \Gamma_P \cap \Gamma_L = [A^*B^*]}} R^P_{I(A^*,B^*,L)} \quad \forall P \in \mathcal{P},$$

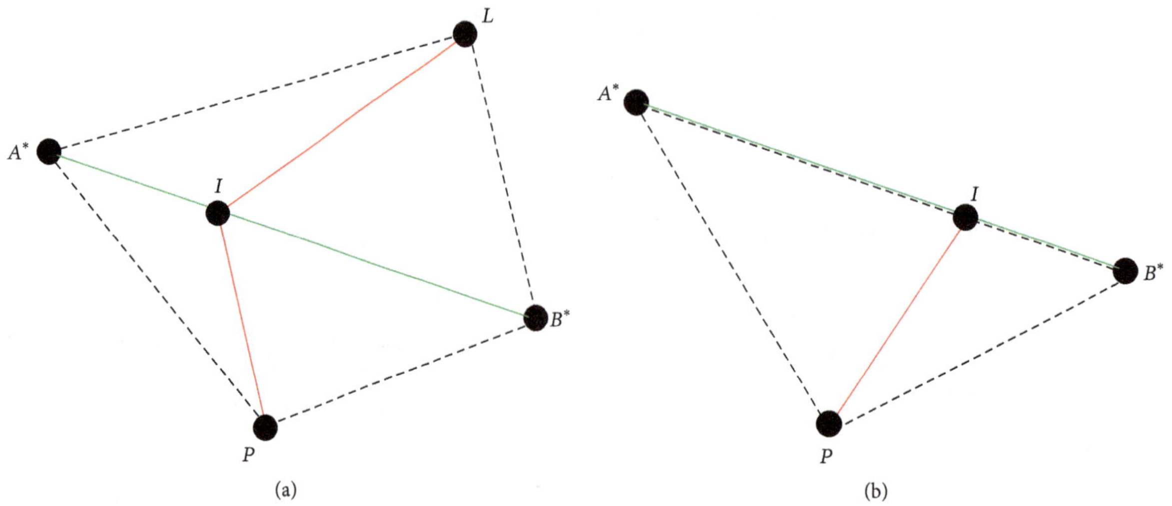

FIGURE 6: Examples of diamond cells. (a) A normal diamond cell. (b) A degenerate diamond cell.

$$\sum_{\substack{A^*\in\mathcal{N}(B^*)\text{ s.t.}\\ \exists I\in\mathcal{E},\exists P,L\in\mathcal{P},\\ \text{with } I\in\mathcal{E}^{A^*}\cap\mathcal{E}^{B^*},\Gamma_{A^*}\cap\Gamma_{B^*}=[PIL]}} \left[\left(\frac{c_h(K^P)\hat{a}_h(K^L)h_{PI}+\hat{c}_h(K^L)a_h(K^P)h_{IL}}{a_h(K^P)h_{IL}+\hat{a}_h(K^L)h_{PI}}\right)[E_P-E_L]\right.$$

$$\left.+\left(\frac{\omega_h(P,L,I)}{h_{A^*B^*}[a_h(K^P)h_{IL}+\hat{a}_h(K^L)h_{PI}]}\right)[E_{B^*}-E_{A^*}]\right]$$

$$+\sum_{\substack{A^*\in\mathcal{N}(B^*)\text{ s.t.}\\ \exists I\in\mathcal{E},\exists P\in\mathcal{P}\\ \text{with } I\in\mathcal{E}^{A^*}\cap\mathcal{E}^{B^*}\cap\mathcal{E}^{\text{ext}},\Gamma_{A^*}\cap\Gamma_{B^*}=[PI]}} \left[\frac{\left(a_h(K^P)d_h(K^P)-\left(c_h(K^P)^2\right)\right)h_{PI}}{a_h(K^P)h_{A^*B^*}}\right][E_{B^*}-E_{A^*}] \quad (113)$$

$$=-\sum_{\substack{A^*\in\mathcal{N}(B^*)\text{ s.t.}\\ \exists I\in\mathcal{E},\exists P,L\in\mathcal{P}\\ \text{with } I\in\mathcal{E}^{A^*}\cap\mathcal{E}^{B^*},\Gamma_{A^*}\cap\Gamma_{B^*}=[PIL]}} R^{B^*}_{I(P,L,A^*)}-\sum_{\substack{A^*\in\mathcal{N}(B^*)\text{ s.t.}\\ \exists I\in\mathcal{E},\exists P\in\mathcal{P}\\ \text{with } I\in\mathcal{E}^{A^*}\cap\mathcal{E}^{B^*}\cap\mathcal{E}^{\text{ext}},\Gamma_{A^*}\cap\Gamma_{B^*}=[PI]}} R^{B^*}_{I(P,A^*)} \quad \forall B^*\in\mathcal{D}.$$

(ii) Let us now prove that $|E_h|_{\mathbb{R}^{\mathcal{P},\mathcal{D}}}\le Ch$, where $C>0$ is a mesh independent positive number.

Multiplying (112) by E_P and (113) by E_{B^*} and summing over $P\in\mathcal{P}$ and $B^*\in\mathcal{D}$ yield (thanks to Lemma 16)

$$\gamma|E_h|^2_{\mathbb{R}^{\mathcal{P},\mathcal{D}}}$$

$$\le\sum_{P\in\mathcal{P}}-E_P\left(\sum_{\substack{L\in\mathcal{N}(P)\text{ s.t.}\\ \exists I\in\mathcal{E}^P\cap\mathcal{E}^L,\exists A^*,B^*\in\mathcal{D},\\ \text{with }\Gamma_P\cap\Gamma_L=[A^*B^*]}} R^P_{I(A^*,B^*,L)}\right)$$

$$+\sum_{B^*\in\mathcal{D}}\left\{-E_{B^*}\left(\sum_{\substack{A^*\in\mathcal{N}(B^*)\text{ s.t.}\\ \exists I\in\mathcal{E}^{A^*}\cap\mathcal{E}^{B^*},\exists P,L\in\mathcal{P}\\ \text{with }\Gamma_{A^*}\cap\Gamma_{B^*}=[PIL]}} R^{B^*}_{I(P,L,A^*)}\right)\right.$$

$$\left.-E_{B^*}\left(\sum_{\substack{A^*\in\mathcal{N}(B^*)\text{ s.t.}\\ \exists I\in\mathcal{E}^{A^*}\cap\mathcal{E}^{B^*}\cap\mathcal{E}^{\text{ext}},\exists P\in\mathcal{P}\\ \text{with }\Gamma_{A^*}\cap\Gamma_{B^*}=[PI]}} R^{B^*}_{I(P,A^*)}\right)\right\}, \quad (114)$$

where γ is some strictly positive real number mesh independent.

Thanks to (109)-(110) (consequence of the conservation of the flux approximation scheme) we have

$$\gamma \left|E_h\right|^2_{\mathbb{R}^{\mathcal{P},\mathcal{D}}} \leq \sum_{\substack{I\in\mathcal{E} \text{ s.t.} \\ \exists P,L\in\mathcal{P} \text{ with } I\in\mathcal{E}^P\cap\mathcal{E}^L, \\ \exists A^*,B^*\in\mathcal{D} \text{ with } \Gamma_P\cap\Gamma_L=[A^*B^*]}} \left|E_P - E_L\right| \left|R^P_{I(A^*,B^*,L)}\right| + \sum_{\substack{I\in\mathcal{E} \text{ s.t.} \\ \exists A^*,B^*\in\mathcal{D} \text{ with } I\in\mathcal{E}^{A^*}\cap\mathcal{E}^{B^*}, \\ \exists P,L\in\mathcal{P} \text{ with } \Gamma_{A^*}\cap\Gamma_{B^*}=[PIL]}} \left|E_{B^*} - E_{A^*}\right| \left|R^{B^*}_{I(P,L,A^*)}\right| + \sum_{\substack{I\in\mathcal{E} \text{ s.t.} \\ \exists A^*,B^*\in\mathcal{D} \text{ with } I\in\mathcal{E}^{A^*}\cap\mathcal{E}^{B^*}\cap\Gamma, \\ \exists P\in\mathcal{P} \text{ with } \Gamma_{A^*}\cap\Gamma_{B^*}=[PI]}} \left|E_{B^*} - E_{A^*}\right| \left|R^{B^*}_{I(P,A^*)}\right|. \quad (115)$$

Therefore

$$\gamma \left|E_h\right|^2_{\mathbb{R}^{\mathcal{P},\mathcal{D}}} \leq \sum_{I(P,L,A^*,B^*)\in\mathcal{E}^{\text{int}}} R^{\max}_{I(P,L,A^*,B^*)} \left(\left|E_P - E_L\right| + \left|E_{B^*} - E_{A^*}\right|\right) + \sum_{I(P,A^*,B^*)\in\mathcal{E}^{\text{ext}}} R^{\max}_{I(P,A^*,B^*)} \left|E_{B^*} - E_{A^*}\right|, \quad (116)$$

where we have set

$$R^{\max}_{I(P,L,A^*,B^*)} = \max\left\{\left|R^P_{I(A^*,B^*,L)}\right|, \left|R^{B^*}_{I(P,L,A^*)}\right|\right\}, \qquad R_{I(P,A^*,B^*)} = \left|R^{B^*}_{I(P,A^*)}\right|. \quad (117)$$

It is important to mention that, due to (105) and (107), we have

$$0 \leq R^{\max}_{I(P,L,A^*,B^*)} \leq Ch^2, \qquad 0 \leq R_{I(P,A^*,B^*)} \leq Ch^2. \quad (118)$$

Define $S_{I(P,L,A^*,B^*)}$ = 2D Lebesgue measure of the diamond cell defined by the points P, L, A^*, B^* and associated with $I \in \mathcal{E}^{\text{int}}$ and $S_{I(P,A^*,B^*)}$ = 2D Lebesgue measure of the diamond cell defined by the points P, A^*, B^* and associated with $I \in \mathcal{E}^{\text{ext}}$.

Therefore, thanks to Cauchy-Schwarz's inequality, it follows from (116) that

$$\gamma \left|E_h\right|^2_{\mathbb{R}^{\mathcal{P},\mathcal{D}}} \leq \left(\sum_{I(P,L,A^*,B^*)\in\mathcal{E}^{\text{int}}} S_{I(P,L,A^*,B^*)} + \sum_{I(P,A^*,B^*)\in\mathcal{E}^{\text{int}}} S_{I(P,A^*,B^*)}\right)^{1/2} \cdot \left(\sum_{I(P,L,A^*,B^*)\in\mathcal{E}^{\text{int}}} \frac{2\left(R^{\max}_{I(P,L,A^*,B^*)}\right)^2}{S_{I(P,L,A^*,B^*)}} \cdot \left[\left(E_P - E_L\right)^2 + \left(E_{B^*} - E_{A^*}\right)^2\right]\right)^{1/2} + \left(\sum_{I(P,A^*,B^*)\in\mathcal{E}^{\text{ext}}} \frac{2\left(R_{I(P,A^*,B^*)}\right)^2}{S_{I(P,A^*,B^*)}} \left(E_{B^*} - E_{A^*}\right)^2\right)^{1/2}. \quad (119)$$

One concludes with the help of assumption (111) and relation (118) that

$$\left|E_h\right|_{\mathbb{R}^{\mathcal{P},\mathcal{D}}} \leq Ch, \quad (120)$$

where C is a strictly positive constant that is mesh independent.

(iii) Let us now prove that $\|E_h\|_{L^2(\Omega)} \leq Ch$.

We start with setting

$$\mathcal{A}(\mathcal{P}) = \left\{E \in \mathcal{A}; \exists P \in \mathcal{P} \text{ with } x_P \in E\right\}, \qquad \mathcal{A}(\mathcal{D}) = \left\{E \in \mathcal{A}; \exists D^* \in \mathcal{D} \text{ with } x_{D^*} \in E\right\}. \quad (121)$$

Recall A is the auxiliary grid and x_P is a cell-point while x_{D^*} is a vertex (with respect to the primary grid introduced in a preceding section).

Define

$$\Omega_{\mathcal{P}} = \bigcup_{E\in\mathcal{A}(\mathcal{P})} E, \qquad \Omega_{\mathcal{D}} = \bigcup_{E\in\mathcal{A}(\mathcal{D})} E. \quad (122)$$

Note that $\Omega_{\mathcal{P}}$ and $\Omega_{\mathcal{D}}$ are a partition of Ω in the sense that

$$\text{(i) } \Omega_{\mathcal{P}} \cap \Omega_{\mathcal{D}} = \emptyset, \qquad \text{(ii) } \overline{\Omega} = \overline{\Omega}_{\mathcal{P}} \cup \overline{\Omega}_{\mathcal{D}}. \quad (123)$$

Two main steps are necessary in our technique to get the estimates of $\|\varphi - \overline{\varphi}\|_{L^2(\Omega)}$; before exposing these steps, we develop some preliminaries.

$$\int_\Omega \left|\varphi(x) - \overline{\varphi}_h(x)\right|^2 dx = \int_{\Omega_{\mathcal{P}}} \left|\varphi(x) - \overline{\varphi}_h(x)\right|^2 dx + \int_{\Omega_{\mathcal{D}}} \left|\varphi(x) - \overline{\varphi}_h(x)\right|^2 dx \leq \int_\Omega \left|\varphi(x) - \Pi_{\mathcal{P}}\overline{\varphi}_h(x)\right|^2 dx + \int_\Omega \left|\varphi(x) - \Pi_{\mathcal{D}}\overline{\varphi}_h(x)\right|^2 dx, \quad (124)$$

where $\Pi_{\mathscr{P}}$ and $\Pi_{\mathscr{D}}$ are operators from the function space $\mathbf{E}(\mathscr{A})$ into the function spaces $\mathscr{A}(\mathscr{P})$ and $\mathscr{A}(\mathscr{D})$, respectively (see (78) for the definition of $\Pi_{\mathscr{P}}$ and $\Pi_{\mathscr{D}}$).

Let $\lambda_{\mathscr{P}}$ and $\lambda_{\mathscr{D}}$ be two real numbers and $\varphi_{\text{aux}}^{\mathscr{P}}$ and $\varphi_{\text{aux}}^{\mathscr{D}}$ two real functions defined almost everywhere in Ω such that

$$\begin{aligned} &\text{(i)}\ \varphi_{\text{aux}}^{\mathscr{P}} = \varphi(x) - \lambda_{\mathscr{P}},\\ &\text{(ii)}\ \sum_{P\in\mathscr{P}} \operatorname{meas}(C_P)\,\varphi_{\text{aux}}^{\mathscr{P}}(x_P) = 0, \end{aligned} \tag{125}$$

$$\begin{aligned} &\text{(i)}\ \varphi_{\text{aux}}^{\mathscr{D}} = \varphi(x) - \lambda_{\mathscr{D}},\\ &\text{(ii)}\ \sum_{D^*\in\mathscr{D}} \operatorname{meas}(C_{D^*})\,\varphi_{\text{aux}}^{\mathscr{D}}(x_{D^*}) = 0. \end{aligned} \tag{126}$$

Define

$$\begin{aligned} \mathscr{E}_{\text{aux}}^{\mathscr{P}} &= \left\{\varphi_{\text{aux}}^{\mathscr{P}}(x_P) - \overline{\varphi}_P\right\}_{P\in\mathscr{P}},\\ \mathscr{E}_{\text{aux}}^{\mathscr{D}} &= \left\{\varphi_{\text{aux}}^{\mathscr{D}}(x_{D^*}) - \overline{\varphi}_{D^*}\right\}_{D^*\in\mathscr{D}}. \end{aligned} \tag{127}$$

It is then clear that the vector

$$\mathscr{E}_{\text{aux}} = \left[\mathscr{E}_{\text{aux}}^{\mathscr{P}}, \mathscr{E}_{\text{aux}}^{\mathscr{D}}\right]^t \in \mathbb{R}^{\mathscr{P},\mathscr{D}} \tag{128}$$

is identifiable with a function denoted again by $\mathscr{E}_{\text{aux}}$ which lies in the space $\mathbf{E}(\mathscr{A})$ and thus in the space $\mathbb{R}^{\mathscr{P},\mathscr{D}}$.

Proposition 25. *The function $\mathscr{E}_{aux}$ meets the following trivial properties:*

$$\left|E_h\right|_{\mathbb{R}^{\mathscr{P},\mathscr{D}}} = \left|\mathscr{E}_{aux}\right|_{\mathbb{R}^{\mathscr{P},\mathscr{D}}}, \tag{129}$$

$$\left|\mathscr{E}_{aux}\right|^2_{\mathbb{R}^{\mathscr{P},\mathscr{D}}} = \left|\mathscr{E}_{aux}^{\mathscr{P}}\right|^2_{\mathbb{R}^{\mathscr{P}}} + \left|\mathscr{E}_{aux}^{\mathscr{D}}\right|^2_{\mathbb{R}^{\mathscr{D}}}, \tag{130}$$

$$\left\|\mathscr{E}_{aux}^{\mathscr{P}}\right\|_{L^2(\Omega)} \le C\left|\mathscr{E}_{aux}^{\mathscr{P}}\right|_{\mathbb{R}^{\mathscr{P}}}, \tag{131}$$

$$\left\|\mathscr{E}_{aux}^{\mathscr{D}}\right\|_{L^2(\Omega)} \le C\left|\mathscr{E}_{aux}^{\mathscr{D}}\right|_{\mathbb{R}^{\mathscr{D}}}, \tag{132}$$

where C stands for diverse positive numbers that are mesh independent.

Let us look for the estimates of $\int_\Omega |\varphi(x) - \Pi_{\mathscr{P}}\overline{\varphi}_h(x)|^2 dx$:

$$\begin{aligned} &\int_\Omega \left|\varphi(x) - \Pi_{\mathscr{P}}\overline{\varphi}_h(x)\right|^2 dx\\ &\quad\le 3\int_\Omega \left|\varphi(x) - \varphi_{\text{aux}}^{\mathscr{P}}(x)\right|^2 dx\\ &\quad+ 3\int_\Omega \left|\varphi_{\text{aux}}^{\mathscr{P}}(x) - \Pi_{\mathscr{P}}\varphi_{\text{aux}}^{\mathscr{P}}(x)\right|^2 dx\\ &\quad+ 3\int_\Omega \left|\Pi_{\mathscr{P}}\varphi_{\text{aux}}^{\mathscr{P}}(x) - \Pi_{\mathscr{P}}\overline{\varphi}_h(x)\right|^2 dx. \end{aligned} \tag{133}$$

(i) Since the primary cells C_P (with $P \in \mathscr{P}$) are convex and the function $\varphi \in C^2(\overline{P})$ the Taylor-Lagrange expansion applies and gives rise to what follows:

$$\lambda_{\mathscr{P}} = \varphi(x) - \left[\varphi_{\text{aux}}^{\mathscr{P}}(x_P) + (x - x_P)^t \operatorname{grad}\varphi_{\text{aux}}^{\mathscr{P}}(\xi_P)\right] \quad \text{in } P,\ \forall P \in \mathscr{P}. \tag{134}$$

Since φ meets the null average condition (7) and $\varphi_{\text{aux}}^{\mathscr{P}}$ aux honors condition (125)-(ii), integrating the two sides of (134) in a primary cell P and summing on $P \in \mathscr{P}$ yield

$$\int_\Omega \left|\varphi(x) - \varphi_{\text{aux}}^{\mathscr{P}}(x)\right|^2 dx \le Ch^2 \tag{135}$$

when accounting with the fact that $|\operatorname{grad}\varphi_{\text{aux}}^{\mathscr{P}}(x)| \le C$ a.e. in Ω, where C represents diverse positive numbers that are mesh independent.

(ii) Thanks again to Taylor-Lagrange it is easily seen that

$$\int_\Omega \left|\varphi_{\text{aux}}^{\mathscr{P}}(x) - \Pi_{\mathscr{P}}\varphi_{\text{aux}}^{\mathscr{P}}(x)\right|^2 dx \le Ch^2. \tag{136}$$

(iii) At last, we have

$$\begin{aligned} &\int_\Omega \left|\Pi_{\mathscr{P}}\varphi_{\text{aux}}^{\mathscr{P}}(x) - \Pi_{\mathscr{P}}\overline{\varphi}_h(x)\right|^2 dx\\ &\quad= \sum_{P\in\mathscr{P}} \operatorname{meas}(C_P)\left|\varphi_{\text{aux}}^{\mathscr{P}}(x_P) - \overline{\varphi}_h(x_P)\right|^2\\ &\quad= \left\|\mathscr{E}_{\text{aux}}^{\mathscr{P}}\right\|^2_{L^2(\Omega)} \le C\left|\mathscr{E}_{\text{aux}}^{\mathscr{P}}\right|^2_{\mathbb{R}^{\mathscr{P}}} \quad (\text{due to (131)})\\ &\quad\le C\left|\mathscr{E}_{\text{aux}}^{\mathscr{P}}\right|^2_{\mathbb{R}^{\mathscr{P},\mathscr{D}}} \quad (\text{according to (130)})\\ &\quad\le C\left|E_h\right|^2_{\mathbb{R}^{\mathscr{P},\mathscr{D}}} \quad (\text{according to (129)})\\ &\quad\le Ch^2 \quad (\text{by virtue of (120)}). \end{aligned} \tag{137}$$

Summarizing what precedes we have proven the following.

Lemma 26. *One has*

$$\int_\Omega \left|\varphi(x) - \Pi_{\mathscr{P}}\overline{\varphi}_h(x)\right|^2 dx \le Ch^2. \tag{138}$$

Remarking that any dual cell C_{D^*} (with $D^* \in \mathscr{D}$) is actually a finite union of convex homogeneous polygons sharing D^* as a common vertex, the arguments that have led to the preceding lemma apply and give rise to the following.

Lemma 27. *One has*

$$\int_\Omega \left|\varphi(x) - \Pi_{\mathscr{D}}\overline{\varphi}_h(x)\right|^2 dx \le Ch^2. \tag{139}$$

Let us summarize the previous developments in the following proposition in which $\mathscr{N}_0$ denotes the set of nodes (i.e., cell-points and vertices) with respect to the mesh system $(\mathscr{P}, \mathscr{E}, \mathscr{A}, \mathscr{D})$ defined on Ω.

Proposition 28 (error estimate result). *Assume that the primary mesh is regular in the sense of Definition 6 and that the discontinuities in Ω of the piecewise constant permeability tensor K generate a finite number of convex subdomains $\{\Omega_s\}_{s\in S}$ over which the exact solution φ of (1)-(2) meets the following property:*

$$\varphi_{|\Omega_s} \in C^2\left(\overline{\Omega}_s\right) \quad \forall s \in S. \tag{140}$$

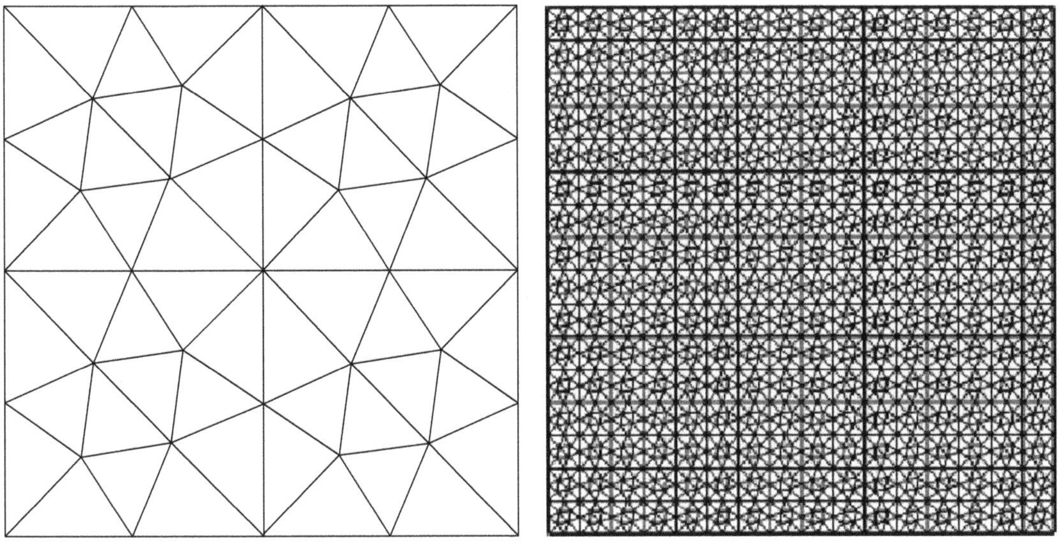

(a) (b)

FIGURE 7: Triangular mesh with acute angles. (a) Coarse mesh and (b) fine mesh.

Under conditions (3), (4), (6), (11), and (111), the error-function E_h associated with the error-vector from $\mathbb{R}^{\mathcal{P},\mathcal{D}}$ with components $\{E_N = \varphi_N - \overline{\varphi}_N\}_{N\in\mathcal{N}_0}$ satisfies the following estimate:

$$|E_h|^2_{\mathbb{R}^{\mathcal{P},\mathcal{D}}} + \|E_h\|^2_{L^2(\Omega)} \leq Ch^2, \tag{141}$$

where C represents mesh independent real number.

4. Numerical Test

The triangular coarse and fine meshes (see Figure 7) are also from FVCA5 Benchmark (see, e.g., [37]). Let us consider a diffusion problem formulated as (1)-(2), where the permeability tensor K is defined by the following relation:

$$K = \begin{pmatrix} 1 & 0 \\ 0 & 10^5 \end{pmatrix}. \tag{142}$$

The exact solution is taken to be $u(x, y) = \sin(2\pi x)e^{-2\pi y\sqrt{1/10^5}}$. Furthermore, we shall ensure the uniqueness of the solution to (1)-(2) by enforcing the condition

$$\int_\Omega \varphi(x)\,dx = 0. \tag{143}$$

Notations

(i) nunkw: number of unknowns.

(ii) nnmat: number of nonzero terms in the matrix.

(iii) sumflux: the discrete flux balance; that is, sumflux = flux0 + flux1 + fluy0 + fluy1, where flux0, flux1, fluy0, and fluy1 are, respectively, the outward numerical fluxes through the boundaries $x = 0$, $x = 1$, $y = 0$, and $y = 1$ (e.g., flux0 is an approximation of $\int_{x=0} K\nabla u \cdot \eta\, ds$), and sum$f = \sum_{K\in\mathcal{T}} |K| f(x_K)$, where x_K denotes some point of the control volume K; note that the residual sumflux is a measure of the global conservation of the scheme.

(iv) $u$min: value of the minimum of the approximate solution.

(v) $u$max: value of the maximum of the approximate solution.

(vi) ener1, ener2: approximations of the energy following the two expressions:$E_1 = \int_\Omega K\nabla u \cdot \nabla u\, dx, E_1 = \int_\Omega K\nabla u \cdot \eta u\, dx$.

Let us denote by u the exact solution, by $\mathcal{T}$ the mesh, and by $u_{\mathcal{T}} = (u_K)_{K\in\mathcal{T}}$ the piecewise constant approximate solution.

(vii) erl2: relative discrete L^2-norm of the error; that is, for instance,

$$\text{erl2} = \left(\frac{\sum_{P\in\mathcal{P}} \text{meas}(C_P)\left(\varphi(x_P) - \overline{\varphi}_h(x_P)\right)^2}{\sum_{P\in\mathcal{P}} \text{meas}(C_P)\left(\varphi(x_P)\right)^2} \right)^{1/2}. \tag{144}$$

(viii) ergrad: discrete L^2-norm of the error on the gradient; that is, for instance,

$$\text{ergrad} = \left(\sum_{\sigma\in\xi^{\text{int}}_{\mathcal{P}}} \frac{m_\sigma}{d_{KL}} \left|\overline{\varphi}_h(x_P) - \overline{\varphi}_h(x_L)\right|^2 \right. \\ \left. + \left(\sum_{\sigma^*\in\xi^{\text{int}}_{\mathcal{D}}} \frac{m_{\sigma^*}}{d_{A^*B^*}} \left|\overline{\varphi}_h(x_{A^*}) - \overline{\varphi}_h(x_{B^*})\right|^2 \right) \right)^{1/2}. \tag{145}$$

TABLE 1: Numerical results.

(a)

i	nunkw	nnmat	sumflux	erl2	ergrad	Ratiol2	ratiograd
1	21	68	$-1.16E-10$	$4.7E-01$	$1.62E-01$	$0E00$	$0E00$
2	109	944	$-6.03E-11$	$1.09E-01$	$2.01E02$	$1.78E00$	$-8.66E00$
3	385	3576	$3.50E-10$	$2.12E-02$	$3.80E01$	$2.60E00$	$2.64E00$
4	1441	13880	$-1.02E-10$	$4.82E-03$	$9.70E00$	$2.24E00$	$2.07E00$
5	5569	54648	$-6.55E-11$	$1.18E-03$	$2.45E00$	$2.09E00$	$2.03E00$

(b)

i	erflx0	erflx1	erfly0	erfly1	erflm	$u$min	$u$max
1	$1.43E-16$	$1.43E-16$	$5.13E02$	$5.11E01$	$4.14E-01$	$-1.79E00$	$1.79E00$
2	$2.62E-13$	$6.65E-14$	$4.1E02$	$9.22E03$	$4.59E02$	$-9.42E-01$	$9.23E-01$
3	$1.91E-13$	$1.37E-13$	$1.23E04$	$4.61E03$	$8.56E01$	$-9.88E-01$	$9.88E-01$
4	$-0E00$	$0E00$	$6.14E03$	$2.78E01$	$-9.97E-01$	$-9.98E-01$	$9.97E-01$
5	$-0E00$	$0E00$	$7.23E02$	$7.37E00$	$-9.99E-01$	$-9.99E-01$	$9.99E-01$

(c)

i	flux0	flux1	fluy0	fluy1	ener1	ener2	eren	enerdisc
1	$6.22E00$	$-6.22E00$	$-5.82E-11$	$-5.82E-11$	$4.15E01$	$1.12E00$	$7.3E-01$	2.9905
2	$6.22E00$	$-6.22E00$	$4.66E-10$	$-5.24E-10$	$3.03E01$	$2.46E01$	$1.89E-01$	1.2247
3	$6.22E00$	$-6.22E00$	$-1.75E-10$	$5.24E-10$	$3.62E01$	$3.30E01$	$8.86E-02$	0.3368
4	$6.22E00$	$-6.22E00$	$-1.31E-10$	$2.91E-11$	$3.80E01$	$3.69E01$	$2.83E-02$	0.1046
5	$6.22E00$	$-6.22E00$	$1.24E-10$	$-1.89E-10$	$3.85E01$	$3.82E01$	$7.86E-03$	0.0389

(d)

ocvl2	ocvgradl2	ocvenerdisc
$2.04E00$	$1.98E00$	$1.42E00$

(ix) ratiol2: for $i \geq 2$,

$$\text{ratiol2}(i) = -2\frac{\ln(\text{erl2}(i)) - \ln(\text{erl2}(i-1))}{\ln(\text{nunkw}(i)) - \ln(\text{nunkw}(i-1))}. \quad (146)$$

(x) ratiograd: for $i \geq 2$, same formula as above with ergrad instead of erl2.

(xi) erflx0, erflx1, erfly0, erfly1: relative error between flux0, flux1, fluy0, fluy1 and the corresponding flux of the exact solution:

$$\text{erflx0} = \left| \frac{\text{flux0} + \int_{x=0} K\nabla u \cdot \eta \, ds}{\int_{x=0} K\nabla u \cdot \eta \, ds} \right|. \quad (147)$$

(xii) ocvl2: order of convergence of the method for L^2-norm of the solution as defined by erl2 with respect to the mesh size:

$$\text{ocvl2} = \frac{\ln(\text{erl2}(i\max)) - \ln(\text{erl2}(i\max-1))}{\ln(h(i\max)) - \ln(h(i\max-1))}, \quad (148)$$

where h is the maximum of the diameter of the control volume.

(xiii) ocvenerdisc: order of convergence of the method for the norm defined by Remark 23.

(xiv) ocvgradl2: order of convergence of the method in L^2-norm of the gradient as defined by ergrad with respect to the mesh size, same formula as above with ergrad instead of erl2.

Comments about the Numerical Test Results. The results of numerical computations of pressure and its gradient show a convergence of order two in L^2-norm and a convergence of order 1.4 in discrete energy norm (see, e.g., Table 1). Moreover the similarity of curves erl2 and ergrad (Figure 8) confirm the closeness of their order of convergence. There is no discordance with the theoretical result where a linear convergence is announced (see Proposition 28). Indeed, the presence of discontinuities in the permeability tensor coefficients prevents the exact solution from being regular enough in the whole domain. More precisely in presence of such discontinuities, the exact pressure is in $H^1(\Omega)$ and never in $H^2(\Omega)$ no matter how regular may be the data f, g and the

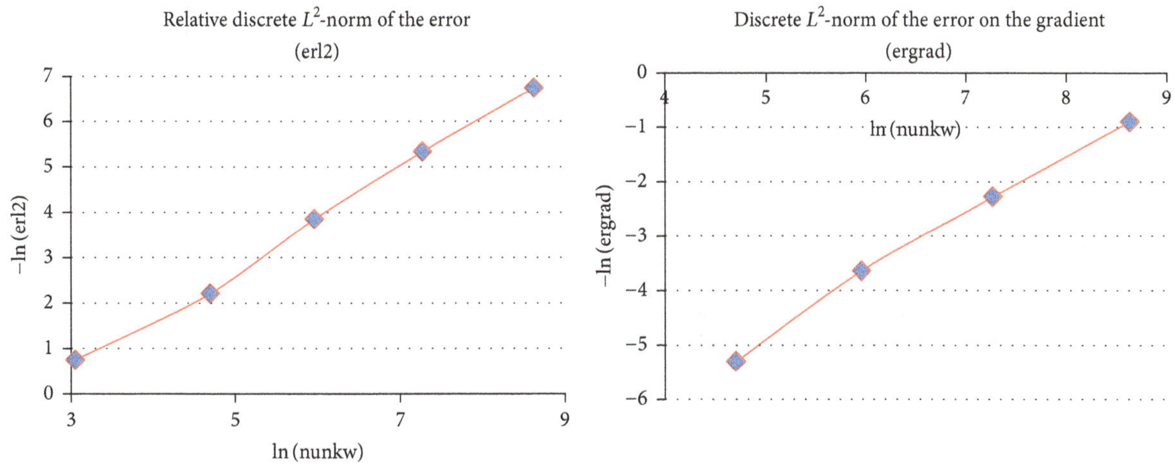

FIGURE 8: Relative discrete L^2-norm of the error (erl2) and of the error on the gradient (ergrad).

domain boundary. So a much slower convergence should take place.

Competing Interests

The authors declare that they have no competing interests.

References

[1] F. Brezzi, J. Douglas, and L. D. Marini, "Two families of mixed finite elements for second order elliptic problems," *Numerische Mathematik*, vol. 47, no. 2, pp. 217–235, 1985.

[2] P. A. Raviart and J. M. Thomas, "A mixed finite element method for 2nd order elliptic problems," in *Mathematical Aspects of the Finite Element Method*, A. Dold and B. Eckmann, Eds., vol. 606 of *Lecture Notes in Mathematics*, pp. 292–315, Springer, Berlin, Germany, 1977.

[3] S.-H. Chou, D. Y. Kwak, and K. Y. Kim, "A general framework for constructing and analyzing mixed finite volume methods on quadrilateral grids: the overlapping covolume case," *SIAM Journal on Numerical Analysis*, vol. 39, no. 4, pp. 1170–1196, 2001.

[4] T. F. Russell, M. F. Wheeler, and I. Yotov, "Superconvergence for control-volume mixed finite element methods on rectangular grids," *SIAM Journal on Numerical Analysis*, vol. 45, no. 1, pp. 223–235, 2007.

[5] M. Berndt, K. Lipnikov, M. Shashkov, M. F. Wheeler, and I. Yotov, "Superconvergence of the velocity in mimetic finite difference methods on quadrilaterals," *SIAM Journal on Numerical Analysis*, vol. 43, no. 4, pp. 1728–1749, 2005.

[6] F. Brezzi, K. Lipnikov, and M. Shashkov, "Convergence of mimetic finite difference method for diffusion problems on polyhedral meshes with curved faces," *Mathematical Models & Methods in Applied Sciences*, vol. 16, no. 2, pp. 275–297, 2006.

[7] R. Eymard, T. Gallouet, and R. Herbin, "Finite volume methods," in *Handbook of Numerical Analysis, VII*, P. G. Ciarlet and J.-L. Lions, Eds., pp. 713–1020, North-Holland, Amsterdam, The Netherlands, 2000.

[8] R. Eymard, T. Gallouët, and R. Herbin, "A cell-centered finite-volume approximation for anisotropic diffusion operators on unstructured meshes in any space dimension," *IMA Journal of Numerical Analysis*, vol. 26, no. 2, pp. 326–353, 2006.

[9] R. Eymard, T. Gallouet, and R. Herbin, "A new finite volume scheme for anisotropic diffusion problems on general grids: convergence analysis," *Comptes Rendus Mathematique, Académie des Sciences de Paris*, vol. 344, no. 6, pp. 403–406, 2007.

[10] R. Eymard, T. Gallouet, and R. Herbin, "Discretization schemes for linear diffusion operators on general non-conforming meshes," in *Proceedings of the FVCA5*, R. Eymard and J. M. Herard, Eds., pp. 375–382, Aussoie, France, 2008.

[11] R. Eymard, T. Gallouet, and R. Herbin, "Discretization of heterogeneous and anisotropic diffusion problems on general nonconforming meshes SUSHI: a scheme using stabilization and hybrid interfaces," *IMA Journal of Numerical Analysis*, vol. 30, no. 4, pp. 1009–1043, 2010.

[12] T. Arbogast, M. F. Wheeler, and I. Yotov, "Mixed finite elements for elliptic problems with tensor coefficients as cell-centered finite differences," *SIAM Journal on Numerical Analysis*, vol. 34, no. 2, pp. 828–852, 1997.

[13] R. D. Lazarov and P. S. Vassilevski, "Numerical methods for convection-diffusion problems on general grids," in *Proceedings of the International Conference on Approximation Theory*, B. Bojanov, Ed., pp. 258–283, Sofia, Bulgaria, 2002.

[14] I. D. Mishev, "Nonconforming finite volume methods," *Computational Geosciences*, vol. 6, no. 3-4, pp. 253–268, 2002.

[15] I. Aavatsmark, T. Barkve, O. Boe, and T. Mannseth, "Discretization on non-orthogonal curvilinear grids for multi-phase flow," in *Proceedings of the 4th European Conference on the Mathematics of Oil Recovery (ECMOR '94)*, Roros, Norway, 1994.

[16] M. G. Rogers and C. F. Rogers, "A flux continuous scheme for the full tensor pressure equation," in *Proceedings of the 4th European Conference on the Mathematics of Oil Recovery (ECMOR '94)*, Roros, Norway, June 1994.

[17] I. Aavatsmark, "Multi-point flux approximation methods for quadrilateral grids," in *Proceedings of the 9th International Forum on Reservoir Simulation*, Abu Dhabi, United Arab Emirates, 2007.

[18] Q.-Y. Chen, J. Wan, Y. Yang, and R. T. Mifflin, "Enriched multi-point flux approximation for general grids," *Journal of Computational Physics*, vol. 227, no. 3, pp. 1701–1721, 2008.

[19] R. A. Klausen, F. A. Radu, and G. T. Eigestad, "Convergence of MPFA on triangulations and for Richards' equation," *International Journal for Numerical Methods in Fluids*, vol. 58, no. 12, pp. 1327–1351, 2008.

[20] F. Hermeline, "Approximation of diffusion operators with discontinuous tensor coefficients on distorted meshes," *Computer Methods in Applied Mechanics and Engineering*, vol. 192, no. 16–18, pp. 1939–1959, 2003.

[21] A. Njifenjou and I. Moukouop-Nguena, "Traitement des anisotropies de perméabilité en simulation d'écoulement en milieu poreux par les volumes finis," in *Proceedings of the International Conference on "Systèmes Informatiques pour la Gestion de l'Environnement"*, M. Tchuente, Ed., pp. 245–259, Douala, Cameroon, 2001.

[22] K. Domelevo and P. Omnes, "A finite volume method for the Laplace equation on almost arbitrary two-dimensional grids," *ESAIM Mathematical Modelling and Numerical Analysis*, vol. 39, no. 6, pp. 1203–1249, 2005.

[23] P. Omnes, "Error estimates for a finite volume method for the Laplace equation in dimension one through discrete Green functions," *International Journal on Finite*, vol. 6, no. 1, pp. 24–41, 2009.

[24] P. Omnes, "On the second-order convergence of a function reconstructed from finite volume approximations of the Laplace equation on Delaunay-Voronoi meshes," *ESAIM: Mathematical Modelling and Numerical Analysis*, vol. 45, no. 4, pp. 627–650, 2011.

[25] F. Boyer and F. Hubert, "Finite volume method for 2D linear and nonlinear elliptic problems with discontinuities," *SIAM Journal on Numerical Analysis*, vol. 46, no. 6, pp. 3032–3070, 2008.

[26] J. Lakhlili, C. Galusinski, and F. Golay, "Discrete duality finite volume applied to soil erosion," in *22ème Congrès Français de Mécanique*, Lyon, France, August 2015.

[27] N. Hartung and F. Hubert, "An efficient implementation of a 3D CeVeFE DDFV scheme on cartesian grids and an application in image processing," in *Finite Volumes for Complex Applications VII-Elliptic, Parabolic and Hyperbolic Problems*, J. Fuhrmann, M. Ohlberger, and C. Rohde, Eds., vol. 78 of *Springer Proceedings in Mathematics & Statistics*, pp. 637–645, Springer, New York, NY, USA, 2014.

[28] P. M. Berthe, C. Japhet, and P. Omnes, "Space-time domain decomposition with finite volumes for Porous media applications," in *Domain Decomposition Methods in Science and Engineering XXI*, vol. 98 of *Lecture Notes in Computational Science and Engineering*, pp. 567–576, Springer, 2014.

[29] F. Hermeline, "Approximation of 2-D and 3-D Diffusion operators with variable full tensor coefficients on arbitrary meshes," *Computer Methods in Applied Mechanics and Engineering*, vol. 196, no. 21–24, pp. 2497–2526, 2007.

[30] A. Njifenjou, H. Donfack, and I. Moukouop-Nguena, "Analysis on general meshes of a discrete duality finite volume method for subsurface flow problems," *Computational Geosciences*, vol. 17, no. 2, pp. 391–415, 2013.

[31] A. Njifenjou and A. J. Kinfack, "Convergence analysis of an MPFA method for flow problems in anisotropic heterogeneous porous media," *International Journal on Finite Volumes*, vol. 5, no. 1, pp. 17–56, 2008.

[32] L. J. Durlofsky, *Upscaling and Gridding of Geologically Complex Systems*, Department of Petroleum Engineering, Stanford University Chevron Texaco E & P Technology Company, 2005.

[33] P. Renard, "Averaging methods for permeability fields, development, protection, management and sequestration of subsurface," in *Fluid Flow and Transport in Porous and Fractured Media*, Cargèse Summer School, Cargèse, France, 2005.

[34] B. Andreianov, F. Boyer, and F. Hubert, "Discrete duality finite volume schemes for Leray-LIOns-type elliptic problems on general 2D meshes," *Numerical Methods for Partial Differential Equations*, vol. 23, no. 1, pp. 145–195, 2007.

[35] H. Brézis, *Analyse Fonctionnelle, Théorie et Applications*, Masson, 1983.

[36] A. H. Le and P. Omnes, *Discrete Poincaré Inequalities for Arbitrary Meshes in the Discrete Duality Finite Volume Context*, Electronic Transactions on Numerical Analysis, Kent State University Library, 2013.

[37] R. Herbin and F. Hubert, "Benchmark on discretization schemes for anisotropic diffusion problems on general grids," in *ISTE. Finite Volumes for Complex Applications V*, pp. 659–692, John Wiley & Sons, Paris, France, 2008.

3

Adomian Decomposition Method with Modified Bernstein Polynomials for Solving Ordinary and Partial Differential Equations

Ahmed Farooq Qasim and Ekhlass S. AL-Rawi

College of Computer Sciences and Mathematics, University of Mosul, Iraq

Correspondence should be addressed to Ahmed Farooq Qasim; ahmednumerical@yahoo.com

Academic Editor: Jafar Biazar

In this paper, we used Bernstein polynomials to modify the Adomian decomposition method which can be used to solve linear and nonlinear equations. This scheme is tested for four examples from ordinary and partial differential equations; furthermore, the obtained results demonstrate reliability and activity of the proposed technique. This strategy gives a precise and productive system in comparison with other traditional techniques and the arrangements methodology is extremely straightforward and few emphasis prompts high exact solution. The numerical outcomes showed that the acquired estimated solutions were in appropriate concurrence with the correct solution.

1. Introduction

Adomian decomposition technique was established by George Adomian and has as of late turned into an extremely recognized strategy in connected sciences. The technique does not require any diminutiveness presumptions or linearization to solve the ordinary and partial differential equations and this produces the strategy extremely effective among alternate strategies. Recently, many iteration techniques have been used for solving nonlinear equations from ordinary, partial, and fractional equations [1], like variational iteration method and differential transform method [2], homotopy perturbation, and analysis methods [3]. Numerous works have been tested in various different regions, for example, warmth or mass exchange, incompressible fluid, nonlinear optics and gas elements wonders [4, 5], fractional Maxwell fluid [6, 7], and the Oldroyd-B fluid model [8].

The approximation used polynomials extremely important in scientific experiments where many rely on topics such as the study of statistics different population and the temperatures and others on the approximation theory. In addition, many experiments rely mainly on the approximate measurements and observations to be studied and processed by the appropriate scientific methods in order to reach the results expected from the study.

The Adomian decomposition technique is improved via Chebyshev polynomials in [9, 10], with Legendre polynomials [11] and with Laguerre polynomials [12].

This paper is organized as follows. In Section 2, the basic ideas of the modified Bernstein polynomials are described. Section 3 is devoted to solving a nonlinear differential equations using Adomian decomposition method based on modified Bernstein polynomials, the results and comparisons of the numerical solutions are presented in Section 4, and concluding remarks are given in Section 5.

2. The Modified Bernstein Polynomials

Polynomials are the mathematical technique as these can be characterized, figured, separated, and incorporated effortlessly. The Bernstein premise polynomials are trying to inexact the capacities. Bernstein polynomials are the better guess to a capacity with a couple of terms. These polynomials are utilized as a part of the fields of connected arithmetic and material science and PC helped geometric outlines and are likewise joined with different techniques like Galerkin and collocation technique to solve some differential and integral equations [13].

Definition 1 (Bernstein basis polynomials). The Bernstein basis polynomials of degree m over the interval $[0,1]$ are defined by

$$B_{i,m}(x) = \binom{m}{i} x^i (1-x)^{m-i} \tag{1}$$

where the binomial coefficient is

$$\binom{m}{i} = \frac{m!}{i!(m-i)!} \tag{2}$$

For example, when m=5, then the Bernstein terms are

$$\begin{aligned} B_{0,5}(x) &= (1-x)^5 \\ B_{1,5}(x) &= 5x(1-x)^4 \\ B_{2,5}(x) &= 10x^2(1-x)^3 \\ B_{3,5}(x) &= 10x^3(1-x)^2 \\ B_{4,5}(x) &= 5x^4(1-x) \\ B_{5,5}(x) &= x^5 \end{aligned} \tag{3}$$

Definition 2 (Bernstein polynomials). A linear combination of Bernstein basis polynomials

$$B_m(x) = \sum_{i=0}^{m} B_{i,m}(x)\beta_i \tag{4}$$

is called the Bernstein polynomials of degree m, where β_i are the Bernstein coefficients.

Definition 3. Let f be a real valued function defined and bounded on $[0,1]$; let $B_m(f)$ be the polynomial on $[0,1]$, defined by

$$B_m(f) = \sum_{i=0}^{m} \binom{m}{i} x^i (1-x)^{m-i} f\left(\frac{i}{m}\right) \tag{5}$$

where $B_m(f)$ is the m-th Bernstein polynomials for $f(x)$.

For each function $f : [0,1] \longrightarrow R$, we have

$$\lim_{m\to\infty} B_m^f(x) = f(x) \tag{6}$$

Example. If $f(x) = e^x$, $x \in [0,1]$ then the Bernstein expanded for the function $f(x)$ when m=5 is

$$\begin{aligned} B_m(f) &= f(0)(1-x)^5 + f\left(\frac{1}{5}\right)5x(1-x)^4 \\ &+ f\left(\frac{2}{5}\right)10x^2(1-x)^3 \\ &+ f\left(\frac{3}{5}\right)10x^3(1-x)^2 \\ &+ f\left(\frac{4}{5}\right)5x^4(1-x) + f(1)x^5 \end{aligned}$$

$$\begin{aligned} B_m(f) &= e^0(1-x)^5 + 5e^{1/5}x(1-x)^4 \\ &+ 10e^{2/5}x^2(1-x)^3 + 10e^{3/5}x^3(1-x)^2 \\ &+ 5e^{4/5}x^4(1-x) + e^1x^5 \end{aligned} \tag{7}$$

In (1986) [14] Lorentz, prove that if the $2k$-th order derivative $f^{2k}(x)$ is bounded in the interval (0,1) then for each $x \in [0,1]$

$$B_m^f(x) = f(x) + \sum_{a=2}^{2k-1} \frac{f^{(a)}(x)}{a!m^a} T_{m,a}(x) + O\left(\frac{1}{m^k}\right) \tag{8}$$

where

$$T_{m,a}(x) = \sum_k (k-mx)^a \binom{m}{k} x^k (1-x)^{m-k} \tag{9}$$

Remark (see [15]). Notice that $T_{m,a}(x)$ is the a-th central moment of a random variable with a binomial appropriation with parameters m and x. Clearly, $T_{m,0} = 1$, $T_{m,1} = 0$. It is well known that the sequence $\{T_{m,a}(x)\}$ satisfies the following recurrence:

$$T_{m,a+1}(x) = x(1-x)\left(T'_{m,a}(x) + maT_{m,a-1}(x)\right) \tag{10}$$

If we apply (8) to $k = 1; 2; 3$, then we obtain

$$\begin{aligned} B_m^f(x) &= f(x) + O\left(\frac{1}{m}\right) \\ B_m^f(x) &= f(x) + \frac{x(1-x)f''(x)}{2m} + O\left(\frac{1}{m^2}\right) \\ B_m^f(x) &= f(x) + \frac{x(1-x)f''(x)}{2m} \\ &+ \frac{x(1-x)\left(4(1-2x)f^{(3)}(x) + 3x(1-x)f^{(4)}(x)\right)}{24m^2} \\ &+ O\left(\frac{1}{m^3}\right) \end{aligned} \tag{11}$$

and higher level approximations can be computed.

3. ADM Based on Modified Bernstein Polynomials

Let us consider the following equation:

$$Lu + Nu + Ru = g(x) \tag{12}$$

where L is an invertible linear term, N represents the nonlinear term, and R is the remaining linear part; from (12) we have

$$Lu = g(x) - Nu - Ru. \tag{13}$$

Now, applying the inverse factorL^{-1} to both sides of (13) then via the initial conditions we find

$$u = f(x) - L^{-1}Nu - L^{-1}Ru, \tag{14}$$

where $L^{-1} = \int_0^x (.)\, ds$ and $f(x)$ are the terms having from integrating the rest of the term $g(x)$ and from utilizing the given initial or boundary conditions. The ADM assumes that $N(u)$ (nonlinear term) can be decomposed by an infinite series of polynomials which is expressed in form

$$N(u) = \sum_{n=0}^{\infty} A_n (u_o, u_1, \ldots, u_n) \tag{15}$$

where A_n are the Adomian's polynomials [16] defined as

$$A_n = \frac{1}{n!}\frac{d^n}{d\lambda^n}\left[N\left(\sum_{i=0}^{\infty}\lambda^i u_i\right)\right]_{\lambda=0}, \quad n = 0, 1, 2, \ldots \tag{16}$$

We expand the function $g(x)$ by Bernstein series

$$g(x) = \sum_{i=0}^{m} a_i B_i(x) \tag{17}$$

where $B_i(x)$ is the Bernstein polynomials.

Now, using (14) and (17) we have

$$\begin{aligned} u_0 &= L^{-1}(a_0 B_0(x) + a_1 B_1(x) + a_2 B_2(x) \\ &\quad + \cdots .a_m B_m(x)) + \theta(x), \\ u_1 &= -L^{-1}(Ru_0) - L^{-1}(Nu_0), \\ u_2 &= -L^{-1}(Ru_1) - L^{-1}(Nu_1), \\ &\vdots \end{aligned} \tag{18}$$

and so on. These formulas are easy to compute by using Maple 13 software.

In this paper, we improve the function $g(x)$ using modified Bernstein series

$$\begin{aligned} g(x) &= \sum_{i=0}^{m}\binom{m}{i} x^i (1-x)^{m-i} f\left(\frac{i}{m}\right) \\ &\quad - \sum_{a=2}^{2k-1}\frac{f^{(a)}(x)}{a!m^a} T_{m,a}(x) \end{aligned} \tag{19}$$

And we can approach the derivatives using the Bernstein polynomials

$$\frac{d}{dx}B_{i,m}(x) = m\left(B_{i-1,m-1}(x) - B_{i,m-1}(x)\right), \tag{20}$$

Then (19) becomes

$$\begin{aligned} g(x) &= \sum_{i=0}^{m}\binom{m}{i} x^i (1-x)^{m-i} f\left(\frac{i}{m}\right) \\ &\quad - \sum_{a=2}^{2k-1}\frac{\left(d^{(a)}/dx^{(a)}\right) B_{i,m}(x)}{a!m^a} T_{m,a}(x) \end{aligned} \tag{21}$$

Now, using (18) and (21) we have

$$\begin{aligned} u_0 &= L^{-1}(B_{i,m}(x)) + \theta(x), \\ u_1 &= -L^{-1}(Ru_0) - L^{-1}(Nu_0), \\ u_2 &= -L^{-1}(Ru_1) - L^{-1}(Nu_1), \\ &\vdots \end{aligned} \tag{22}$$

The above equation is governing equation of ADM using modified Bernstein polynomials. The obtained approximate solution, $\omega_V(x) = \sum_{i=0}^{V} u_i$, by (22) has a comparison with the classic approximation solution and the correct solution.

4. Numerical Results

In this section, we solve ordinary and partial differential equations by ADM based on Bernstein polynomials and we compare with ADM based on classical Bernstein polynomial.

Example 1. Consider the ordinary equation

$$\begin{aligned} \frac{d^2y}{dt^2} + t\frac{dy}{dt} + t^2y^3 &= \left(2+6t^2\right)e^{t^2} + t^2e^{3t^2}, \\ y(0) &= 1, \\ \frac{dy}{dt}(0) &= 0, \end{aligned} \tag{23}$$

with the exact solution $(t) = e^{t^2}$. Using (12) we have

$$Ly + Ny + Ry = g(x) \tag{24}$$

where $L = d^2/dt^2, Ry = t(d/dt), Ny = t^2y^3$, and $g(t) = (2 + 6t^2)e^{t^2} + t^2e^{3t^2}$.

The Adomian polynomials for representing the nonlinear term Ny are

$$\begin{aligned} A_0 &= t^2 y_0^3, \\ A_1 &= t^2\left(3y_0^2 y_1\right), \\ A_2 &= t^2\left(3y_0^2 y_2 + 3y_0 y_1^2\right), \\ &\vdots \end{aligned} \tag{25}$$

Now $L^{-1} = \int_0^t \int_0^t (.)dt\, dt$; then using (5) the classical Bernstein polynomials of $g(t)$ when v=m=6 are

$$\begin{aligned} g_b(t) &= 2 + 1.547324t + 9.290164t^2 + 7.83289t^3 \\ &\quad + 9.751887t^4 + 7.659668t^5 + 3.749864t^6 \end{aligned} \tag{26}$$

And modified Bernstein polynomials (21) of $g(t)$ with k=2 are

$$\begin{aligned} g_{mb}(t) &= 2 - 0.001037t + 6.922082t^2 + 1.997441t^3 \\ &\quad + 6.737662t^4 + 11.051121t^5 + 13.124523t^6 \end{aligned} \tag{27}$$

TABLE 1: Comparison of absolute errors using y_6 when m=v=6 and k=2.

t	$\lvert y_{exact} - y_b \rvert$	$\lvert y_{exact} - y_{mb} \rvert$
0	0	0
0.01	2.580000 E-7	2.000000 E-9
0.02	2.068000 E-6	2.900000 E-8
0.03	6.992000 E-6	1.420000 E-7
0.04	1.660300 E-5	4.450000 E-7
0.05	3.249600 E-5	1.074000 E-6
0.06	5.629000 E-5	2.207000 E-6
0.07	8.962800 E-5	4.057000 E-6
0.08	1.341850 E-4	6.872000 E-6
0.09	1.916690 E-4	1.093000 E-5
0.1	2.638350 E-4	1.654900 E-5
MSE	1.369383957 E-8	4.632475500 E-11

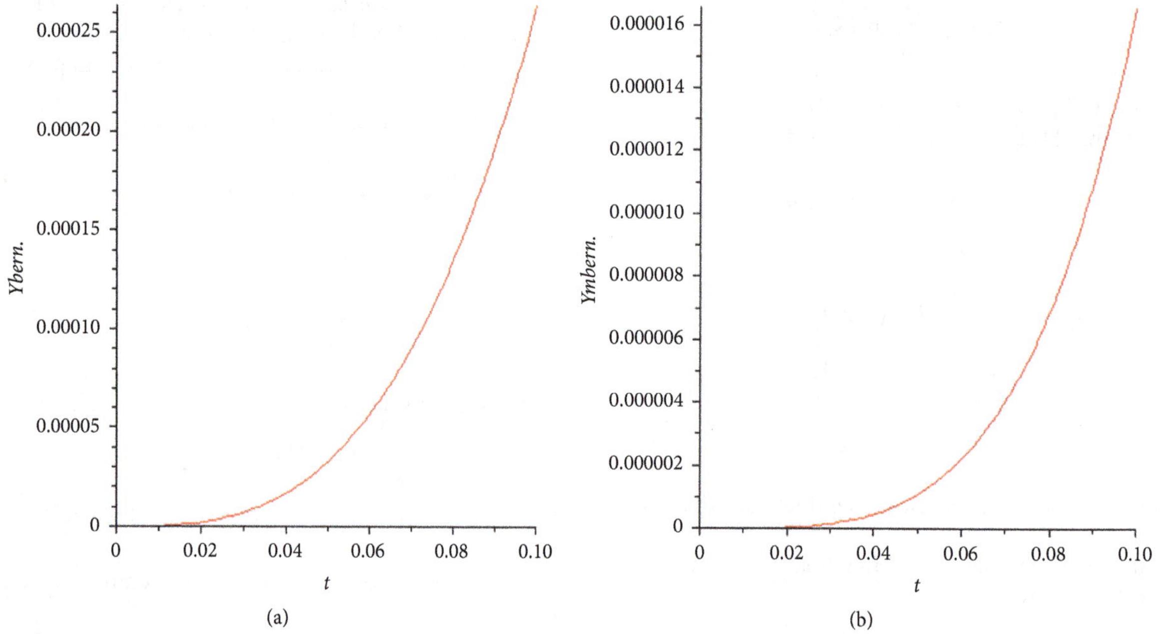

FIGURE 1: The absolute error between ADM with modified Bernstein polynomials and the exact solution when m=v=6 and k=2.

By (22), we have

$$y_0 = L^{-1}(g_{mb}(t)) + y(0) + \frac{dy}{dt}(0)t = 1 + t^2 - 0.000173t^3 + 0.57684t^4 + 0.099872t^5 + 0.224589t^6 + 0.263122t^7 + 0.234367t^8,$$

$$y_1 = -L^{-1}\left(t\frac{d}{dt}y_0\right) - L^{-1}(A_0) = -0.25t^4 + 0.000026t^5 - 0.176912t^6 - 0.011877t^7 + \cdots,$$

$$y_2 = -L^{-1}\left(t\frac{d}{dt}y_1\right) - L^{-1}(A_1) = 0.033333t^6 - 0.000003t^7 + 0.032348t^8 + 0.011536t^9 + \cdots,$$

$$\vdots \tag{28}$$

And we obtain

$$y_{mb}(t) = \sum_{i=0}^{6} y_i = 1 + t^2 - 0.000173t^3 + 0.32684t^4 + \cdots. \tag{29}$$

The absolute error of $y_{mb}(t)$ and $y_b(t)$ is presented in Table 1 and Figure 1.

Figure 1 presents the absolute error of ADM with Bernstein polynomial in (a) and ADM with modified Bernstein

TABLE 2: Comparison of absolute errors using y_{10} when m=16, v=4, and k=2.

t	$\lvert y_{exact} - y_b \rvert$	$\lvert y_{exact} - y_{mb} \rvert$
0	0	0
0.1	2.061126 E-6	3.074560 E-7
0.2	1.145239 E-5	1.058636 E-5
0.3	1.315732 E-5	5.114716 E-5
0.4	2.165603 E-4	1.331415 E-4
0.5	8.646500 E-4	2.420463 E-4
0.6	2.311555 E-3	3.299021 E-4
0.7	4.905204 E-3	3.231831 E-4
0.8	8.829753 E-3	1.540876 E-4
0.9	1.391114 E-2	1.870564 E-4
1	1.946226 E-2	6.088701 E-4
MSE	6.804632 E-5	7.217798344 E-8

polynomial in (b) at m=v=6 and k=2. The absolute errors generated using the ADM with Bernstein polynomial are of 10^{-4} while the errors yielded from ADM with modified Bernstein polynomial are of 10^{-5}.

Example 2. Consider the ordinary equation

$$\frac{d^2y}{dt^2} + y\frac{dy}{dt} = t\sin\left(2t^2\right) - 4t^2\sin\left(t^2\right) + 2\cos\left(t^2\right), \quad 0 \le t \le 1,$$
$$y(0) = 0, \qquad (30)$$
$$\frac{dy}{dt}(0) = 0,$$

with the exact solution $(t) = \sin(t^2)$.

Here, $L = d^2/dt^2$, $Ny = y(dy/dt)$, and $g(t) = t\sin(2t^2) - 4t^2\sin(t^2) + 2\cos(t^2)$.

The Adomian polynomials for represent the nonlinear term Nu are

$$A_0 = y_0\frac{d}{dt}y_0,$$
$$A_1 = y_1\frac{d}{dt}y_0 + y_0\frac{d}{dt}y_1, \qquad (31)$$
$$A_2 = y_2\frac{d}{dt}y_0 + y_1\frac{d}{dt}y_1 + y_0\frac{d}{dt}y_1,$$
$$\vdots$$

Then using (5) the classical Bernstein polynomials of $g(t)$ when v=4 and m=16 is

$$g_b(t) = 20.00659171t + 0.2233190t^2 + 0.098085t^3 - 3.39540t^4 + \cdots \qquad (32)$$

And modified Bernstein polynomials (21) of g(t) with k=2 are

$$g_{mb}(t) = 2 - 0.007366t + 0.2188855t^2 + 1.389751t^3 - 4.50213t^4 + \cdots \qquad (33)$$

By (22), we have

$$y_{mb}(t) = \sum_{i=0}^{4} y_i = t^2 - 0.001228t^3 + 0.018241t^4 - 0.030513t^5 + \cdots, \qquad (34)$$

The absolute error of $y_{mb}(t)$ and $y_b(t)$ is presented in Table 2 and Figure 2.

Figure 2 presents the absolute error of ADM with Bernstein polynomial in (a) and ADM with modified Bernstein polynomial in (b) at m=v=10 and k=3. The absolute errors generated using the ADM with Bernstein polynomial are of 10^{-2} while the errors yielded from ADM with modified Bernstein polynomial are of 10^{-4}.

Example 3. Consider the ordinary equation

$$\frac{dy}{dt} - ty + y^2 = e^{t^2}, \qquad (35)$$
$$y(0) = 1,$$

with the exact solution $(t) = e^{t^2/2}$.

Here $L = d/dt$, $Ru = -ty$, $Ny = y^2$, and $g(t) = e^{t^2}$.

Then using (5) the classical Bernstein polynomials of $g(t)$ when v=8 and m=12 is

$$g_b(t) = 1 + 0.083623t + 0.939177t^2 + 0.197729t^3 + 0.331572t^4 + \cdots, \qquad (36)$$

TABLE 3: Comparison of absolute errors using y_{10} when m=12, v=8, and k=3.

t	$\lvert y_{exact} - y_b \rvert$	$\lvert y_{exact} - y_{mb} \rvert$
0	0	0
0.01	4.134000 E-6	1.750000 E-7
0.02	1.635300 E-5	6.400000 E-7
0.03	3.639700 E-5	1.314000 E-6
0.04	6.402400 E-5	2.123000 E-6
0.05	9.900900 E-5	2.999000 E-6
0.06	1.411460 E-4	3.883000 E-6
0.07	1.902410 E-4	4.720000 E-6
0.08	2.461180 E-4	5.463000 E-6
0.09	3.086100 E-4	6.069000 E-6
0.1	3.775690 E-4	6.501000 E-6
MSE	3.699974901 E-8	1.619641710 E-11

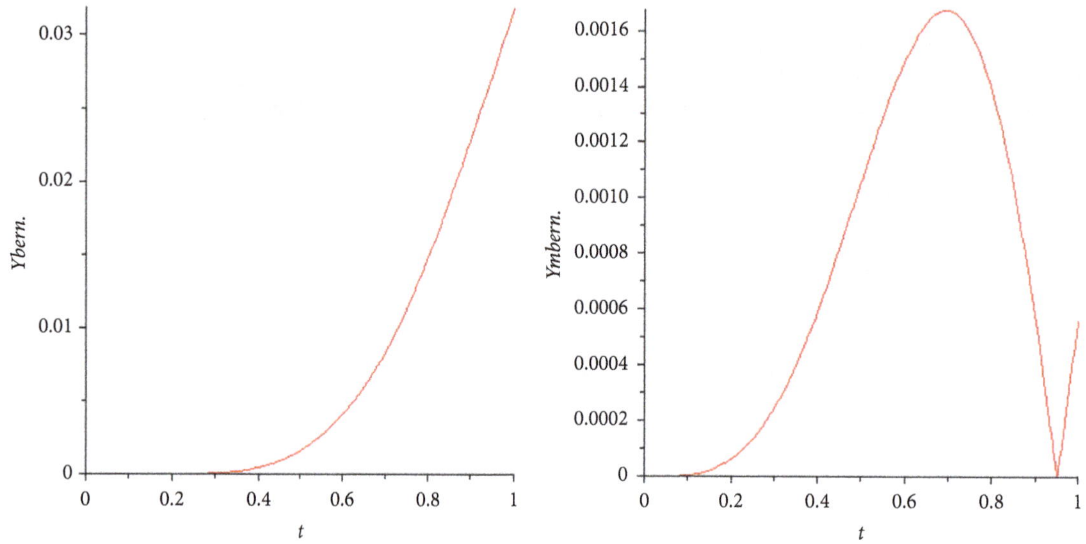

(a) (b)

FIGURE 2: The absolute error between ADM with modified Bernstein polynomials and the exact solution when m=v=10 and k=3.

and modified Bernstein polynomials (21) of $g(t)$ with k=3 are

$$g_{mb}(t) = 1 + 0.003794t + 0.957094t^2 + 0.103509t^3 + 0.410289t^4 + \cdots . \tag{37}$$

By (22), we have

$$u_{mb}(t) = \sum_{i=0}^{8} u_i = 1 + 0.501897t^2 - 0.015567t^3 + 0.159135t^4 + \cdots , \tag{38}$$

The absolute error of $u_{mb}(t)$ and $u_b(t)$ is presented in Table 3 and Figure 3.

Figure 3 presents the absolute error of ADM with Bernstein polynomial in (a) and ADM with modified Bernstein polynomial in (b) at m=12, v=8, and k=3. The absolute errors generated using the ADM with Bernstein polynomial are of 10^{-4} while the errors yielded from ADM with modified Bernstein polynomial are of 10^{-6}.

Example 4. Consider the partial differential equation

$$\frac{\partial y}{\partial t} + y\frac{\partial y}{\partial x} - v\frac{\partial^2 y}{\partial x^2} = x\left(2t\cos\left(t^2\right) + \sin^2\left(t^2\right)\right), \quad y(x,0) = 0, \tag{39}$$

Using (12) we have

$$Ly + Ny + Ry = g(x,t) \tag{40}$$

where $L = d/dt$, $Ry = -v(\partial^2 y/\partial x^2)$, $Ny = y(\partial y/\partial x)$, and $g(x,t) = x(2t\cos(t^2) + \sin^2(t^2))$.

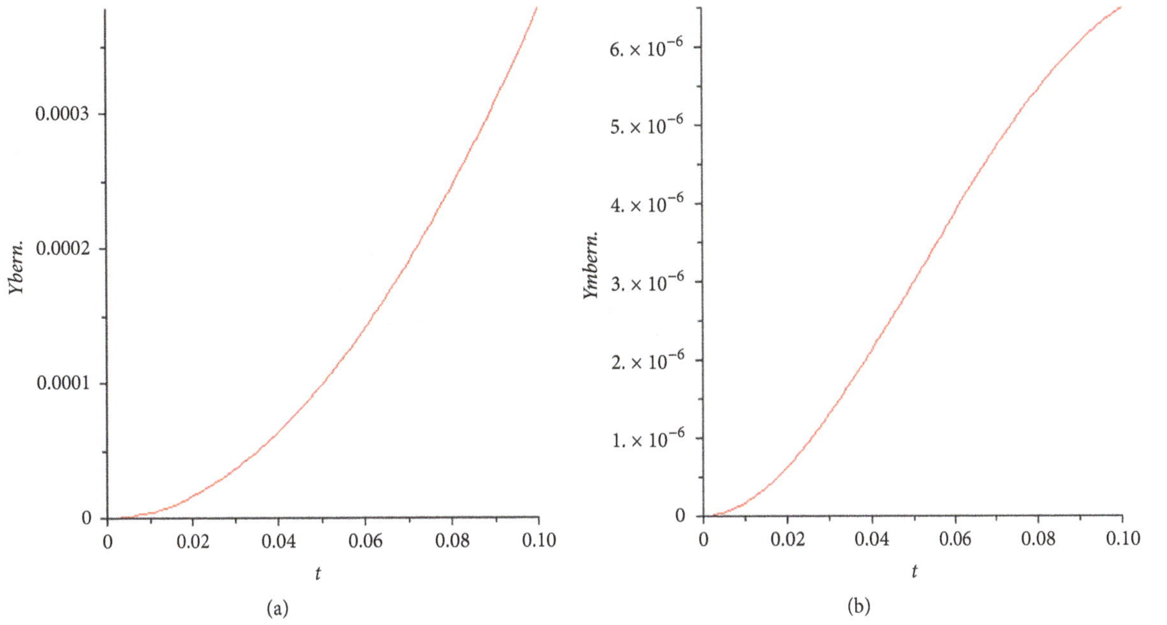

FIGURE 3: The absolute error between ADM with modified Bernstein polynomials and the exact solution when m=12, v=8, and k=3.

The Adomian polynomials for Ny are

$$A_0 = y_o \frac{\partial y_o}{\partial x},$$
$$A_1 = y_o \frac{\partial y_1}{\partial x} + y_1 \frac{\partial y_o}{\partial x}, \quad (41)$$
$$A_2 = y_o \frac{\partial y_2}{\partial x} + y_2 \frac{\partial y_o}{\partial x} + y_1 \frac{\partial y_1}{\partial x},$$
$$\vdots$$

Then using (5) the classical Bernstein polynomials of $g(x,t)$ when v=m=6 is

$$g_b(x,t) = 2.003859xt + 0.103475xt^2 + 0.149954xt^3 - 0.27935xt^4 + \cdots, \quad (42)$$

and modified Bernstein polynomials (21) of $g(x,t)$ with k=2 are

$$g_{mb}(x,t) = 1.986611xt + 0.045744xt^2 + 0.504281xt^3 - 0.25375xt^4 + \cdots. \quad (43)$$

By (22) with $v = 1$, we have

$$y_0 = L^{-1}(g_{mb}(x,t)) + y(x,0) = 0.993306xt^2 + 0.015248xt^3 + 0.12607xt^4 - 0.050750xt^5 \ldots,$$
$$y_1 = -L^{-1}\left(-v\frac{\partial^2 y_0}{\partial x^2}\right) - L^{-1}(A_0) = -0.197331xt^5 - 0.005049xt^6 - 0.035812xt^7 + 0.012122xt^8 + \cdots,$$
$$y_2 = -L^{-1}\left(-v\frac{\partial^2 y_1}{\partial x^2}\right) - L^{-1}(A_1) = 0.049003xt^8 + 0.001783xt^9 + 0.012105xt^{10} - 0.003795xt^{11} + \cdots,$$
$$\vdots \quad (44)$$

And we obtain

$$y_{mb}(x,t) = \sum_{i=0}^{6} y_i = 0.993306xt^2 + 0.015248xt^3 + 0.126070xt^4 - 0.248081xt^5 - 0.083749xt^6 + \cdots, \quad (45)$$

The absolute error of $y_{mb}(x,t)$ and $y_b(x,t)$ is presented in Table 4 and Figure 4 with the exact solution $y(x,t) = x\sin(t^2)$.

Also Figure 4 presents the absolute error of ADM with Bernstein polynomial in (a) and ADM with modified Bernstein polynomial in (c) at m=v=6 and k=2. The absolute errors generated using the ADM with Bernstein polynomial are of 10^{-3} while the errors yielded from ADM with modified Bernstein polynomial are of 10^{-4}.

5. Conclusions

In this paper, we show that utilizing modified Bernstein polynomials is smartly thought to modify the performance

Table 4: Comparison of absolute errors using y_6 when m=v=6, k=2, and x=0.1.

t	$\lvert y_{exact} - y_b \rvert$	$\lvert y_{exact} - y_{mb} \rvert$
0	0	0
0.1	5.512102 E-6	4.149093 E-6
0.2	3.401160 E-5	1.849504 E-6
0.3	8.851342 E-5	2.838128 E-5
0.4	1.404426 E-5	9.121566 E-5
0.5	1.256008 E-4	1.620426 E-4
0.6	4.250973 E-5	1.880417 E-4
0.7	4.360788 E-4	1.086168 E-4
0.8	1.050197 E-3	1.020235 E-4
0.9	1.718854 E-3	3.752218 E-4
1	2.024768 E-3	4.863297 E-4
MSE	8.393551358 E-7	4.702782622 E-8

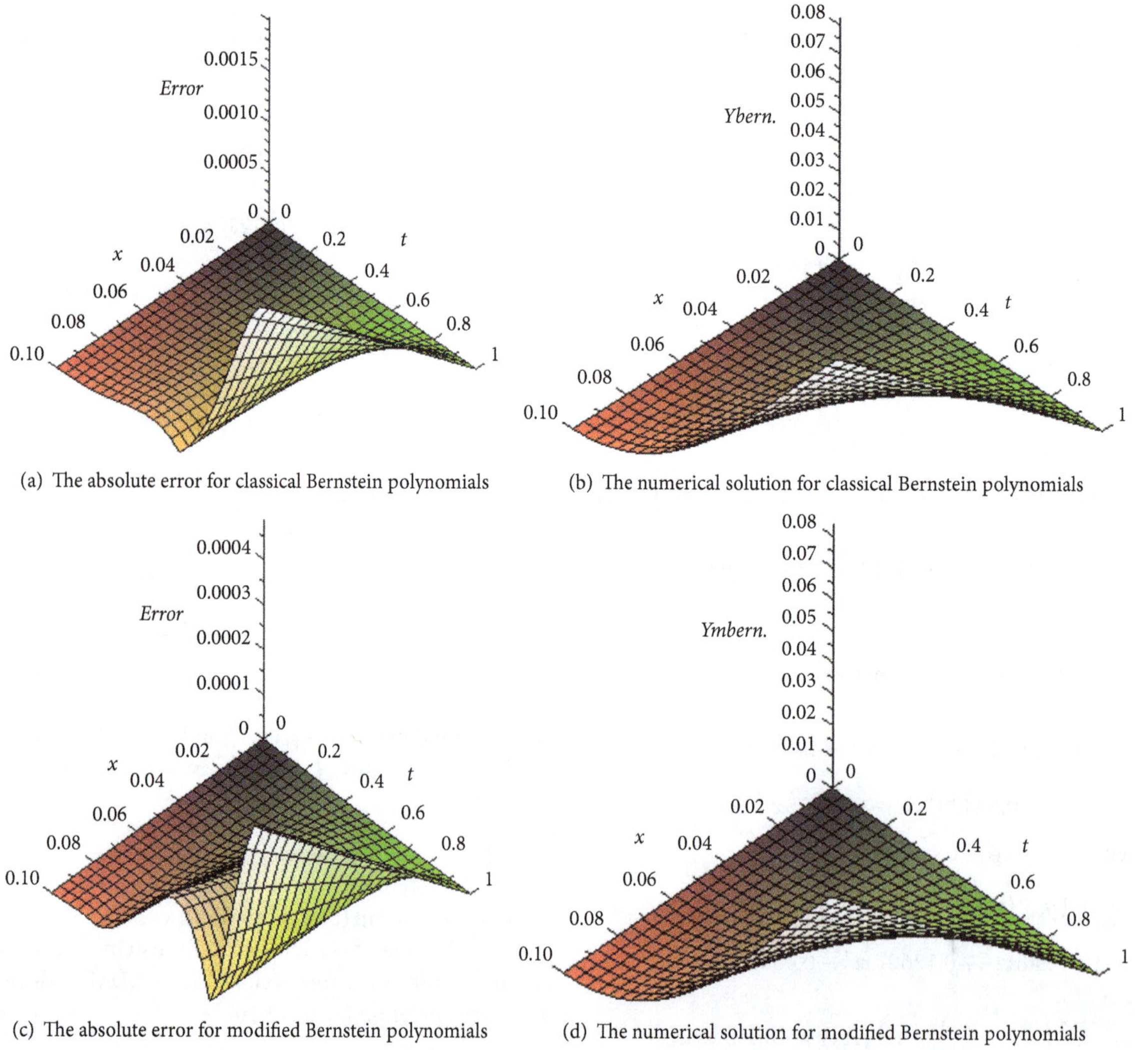

(a) The absolute error for classical Bernstein polynomials

(b) The numerical solution for classical Bernstein polynomials

(c) The absolute error for modified Bernstein polynomials

(d) The numerical solution for modified Bernstein polynomials

Figure 4: The absolute error between ADM with classical and modified Bernstein polynomials and the exact solution using y_6 when m=v=6, k=2, and x=0.1.

of the Adomian decomposition technique. The fundamental preferred standpoint of this strategy is that it can be used specifically for all sort of differential and integral equations.

We utilize modified Bernstein extensions of the nonlinear term to get more exact outcomes. Figures empower us to consider the difference between utilizing two strategies graphically. Tables are additionally given to demonstrate the variety of the outright mistakes for bigger estimation, to be specific for bigger m. We observed from the numerical outcomes in Tables 1–4 and Figures 1–4 that the ADM with modification Bernstein polynomials gives more exact and robust numerical solution than the classical Bernstein polynomials. Every one of the calculations was done with the guide of Maple 13 programming.

Conflicts of Interest

The authors declare that there are no conflicts of interest regarding the publication of this paper.

Acknowledgments

The research is supported by College of Computer Sciences and Mathematics, University of Mosul, Republic of Iraq, under Project no. 11251357.

References

[1] S. Muhammad and I. Muhammad, "Some multi-step iterative methods for solving nonlinear equations," *Open Journal of Mathematical Sciences*, vol. 1, no. 1, pp. 25–33, 2017.

[2] Y. Yang and L. Hua, "Variational iteration transform method for fractional differential equations with local fractional derivative," *Abstract and Applied Analysis*, vol. 2014, Article ID 760957, 9 pages, 2014.

[3] U. R. Hamood, S. S. Muhammad, and A. Ayesha, "Combination of homotopy perturbation method (HPM) and double sumudu transform to solve fractional KDV equations," *Open Journal of Mathematical Sciences*, vol. 2, no. 1, pp. 29–38, 2018.

[4] A. M. Wazwaz, "Construction of solitary wave solutions and rational solutions for the KdV equation by Adomian decomposition method," *Chaos, Solitons & Fractals*, vol. 12, no. 12, pp. 2283-2293, 2001.

[5] N. Bildik and A. Konuralp, "The use of variational iteration method, differential transform method and Adomian decomposition method for solving different types of nonlinear partial differential equations," *International Journal of Nonlinear Sciences and Numerical Simulation*, vol. 7, no. 1, pp. 65–70, 2006.

[6] T. Madeeha, N. N. Muhammad, S. Rabia, V. Dumitru, I. Muhammad, and S. Naeem, "On unsteady flow of a viscoelastic fluid through rotating cylinders," *Open Physics*, vol. 1, no. 1, pp. 1–15, 2017.

[7] A. Sannia, Q. Haitao, A. Muhammad, J. Maria, and I. Muhammad, "Exact solutions of fractional Maxwell fluid between two cylinders," *Open Journal of Mathematical Sciences*, vol. 1, no. 1, pp. 52–61, 2017.

[8] Q. Haitao, F. Nida, W. Hassan, and S. Junaid, "Analytical solution for the flow of a generalized Oldroyd-B fluid in a circular cylinder," *Open Journal of Mathematical Sciences*, vol. 1, no. 1, pp. 85–96, 2017.

[9] M. M. Hosseini, "Adomian decomposition method with Chebyshev polynomials," *Applied Mathematics and Computation*, vol. 175, no. 2, pp. 1685–1693, 2006.

[10] N. Bildik and Deniz, "Modified Adomian decomposition method for solving Riccati differential equations," *Review of the Air Force Academy*, vol. 3, no. 30, pp. 21–26, 2015.

[11] Y. Liu, "Adomian decomposition method with orthogonal polynomials: Legendre polynomials," *Mathematical and Computer Modelling*, vol. 49, no. 5-6, pp. 1268–1273, 2009.

[12] Y. Mahmoudi et al., "Adomian decomposition method with Laguerre polynomials for solving ordinary differential equation," *Journal of Basic and Applied Scientific Research*, vol. 2, no. 12, pp. 12236–12241, 2012.

[13] D. Rani and V. Mishra, "Approximate solution of boundary value problem with bernstein polynomial laplace decomposition method," *International Journal of Pure and Applied Mathematics*, vol. 114, no. 4, pp. 823–833, 2017.

[14] G. G. Lorentz, *Bernstein Polynomials*, Chelsea Publishing Series, 1986.

[15] J. Cicho and Z. Goi, "On Bernoulli sums and Bernstein polynomials," *Discrete Mathematics and Theoretical Computer Science*, pp. 1–12, 2009.

[16] G. Adomian, *Nonlinear Stochastic Operator Equations*, Academic Press, San Diego, CA, USA, 1986.

4

Modelling of Rabies Transmission Dynamics using Optimal Control Analysis

Joshua Kiddy K. Asamoah,[1] Francis T. Oduro,[1] Ebenezer Bonyah,[2] and Baba Seidu[3]

[1] *Department of Mathematics, Kwame Nkrumah University of Science and Technology, Kumasi, Ghana*
[2] *Department of Statistics and Mathematics, Kumasi Technical University, Kumasi, Ghana*
[3] *Department of Mathematics, University for Development Studies, Navrongo, Ghana*

Correspondence should be addressed to Joshua Kiddy K. Asamoah; jasamoah@aims.edu.gh

Academic Editor: Zhen Jin

We examine an optimal way of eradicating rabies transmission from dogs into the human population, using preexposure prophylaxis (vaccination) and postexposure prophylaxis (treatment) due to public education. We obtain the disease-free equilibrium, the endemic equilibrium, the stability, and the sensitivity analysis of the optimal control model. Using the Latin hypercube sampling (LHS), the forward-backward sweep scheme and the fourth-order Range-Kutta numerical method predict that the global alliance for rabies control's aim of working to eliminate deaths from canine rabies by 2030 is attainable through mass vaccination of susceptible dogs and continuous use of pre- and postexposure prophylaxis in humans.

1. Introduction

Rabies is an infection that mostly affects the brain of an infected animal or individual, caused by viruses belonging to the genus *Lyssavirus* of the family Rhabdoviridae and order Mononegavirales [1, 2]. This disease has become a global threat and it is also estimated that rabies occurs in more than 150 countries and territories [2]. Raccoons, skunks, bats, and foxes are the main animals that transmit the virus in the United States [2]. In Asia, Africa, and Latin America, it is known that dogs are the main source of transmission of the rabies virus into the human population [2]. When the rabies virus enters the human body or that of an animal, the infection (virus) moves rapidly along the neural pathways to the central nervous system; from there the virus continues to spread to other organs and causes injury by interrupting various nerves [2]. The symptoms of rabies are quite similar to those of encephalitis (see [3]). Due to movement of dogs in homes or the surroundings, the risk of not being infected by a rabid dog can never be guaranteed. Rabies is a major health problem in many populations dense with dogs, especially in areas where there are less or no preventive measures (vaccination and treatment) for dogs and humans. Treatment after exposure to the rabies virus is known as postexposure prophylaxis (PEP) and vaccination before exposure to the infection is known as preexposure prophylaxis.

The study of optimal control analysis in maximizing or minimizing a said target was introduced by Pontryagin and his collaborators around 1950. They developed the key idea of introducing the adjoint function to a differential equation, by forming an objective functional [4], and since then there has been a considerable study of infectious disease using optimal control analysis (see [4–12]).

Research published by Aubert [13], on the advancement of the expense of wildlife rabies in France, incorporated various variables. They follow immunization of domestic animals, the reinforcement of epidemiological reconnaissance system and the bolster given to indicative research laboratories, the costs connected with outbreaks of rabies, the clinical perception of those mammals which had bitten humans, the preventive immunization, and postexposure treatment of people. A significant percentage (72%) of the cost was the preventive immunization of local animals. In France, as in other European nations in which the red fox (Vulpes) is the species most affected, two primary procedures for controlling rabies were assessed in [13] at the repository level to be specific: fox termination and the oral immunization of foxes. The consolidated costs and advantages of both systems were looked at and included either the expenses of fox separation or the cost of oral immunization. The total yearly costs of both techniques stayed practically identical until the fourth year, after which the oral immunization methodology turned out to be more cost effective. This estimate was made in 1988 and readjusted in 1993 and affirmed by ex-postinvestigation five years later. Accordingly, it was presumed that fox termination brought about a transient diminishment in the event of the infection while oral immunization turned out to be equipped for wiping out rabies even in circumstances in which fox population was growing. Anderson and May [14] formulated a mathematical model based on each time step dynamic which was calculated independently in every cell. Later, Bohrer et al. [15] published a paper on the viability of different rabies spatial immunization designs in a simulated host population.

The research presented by Bohrer [15] stated that, in desert environments, where host population size varies over time, nonuniform spreading of oral rabies vaccination may, under certain circumstances, be more effective than the commonly used uniform spread. The viability of a nonarbitrary spread of the immunization depends, to some extent, on the dispersal behavior of the carriers. The outcomes likewise exhibit that, in a warm domain in a few high-density regions encompassed by populations with densities below the critical threshold for the spread of the disease, the rabies infection can persist.

Levin et al. [16] also presented a model for the immune responses to rabies virus in bats. Coyne et al. [17] proposed an SEIR model, which was also used in a study predicting the local dynamics of rabies among raccoons in the United States. Childs et al. [18] also researched rabies epidemics in raccoons with a seasonal birth pulse, using optimal control of an SEIRS model which describes the population dynamics. Hampson et al. [19] also noted that rabies epidemic cycles have a period of 3–6 years in dog populations in Africa, so they built a susceptible, exposed, infectious, and vaccinate model with an intervention response variable, which showed significant synchrony.

Carroll et al. [20] also used compartmental models to describe rabies epidemiology in dog populations and explored three control methods: vaccination, vaccination pulse fertility control, and culling. An ordinary differential equation model was used to characterize the transmission dynamics of rabies between humans and dogs by [21, 22]. The work by Zinsstag et al. [23] further extended the existing models on rabies transmission between dogs to include dog-to-human transmission and concluded that human postexposure prophylaxis (PEP) with a dog vaccination campaign was the more cost effective in controlling the disease in the long run. Furthermore, Ding et al. [24] formulated an epidemic model for rabies in raccoons with discrete time and spatial features. Their goal was to analyze the strategies for optimal distribution of vaccine baits to minimize the spread of the disease and the cost of carrying out the control. Smith and Cheeseman [25] show that culling could be more effective than vaccination, given the same efficacy of control, but Tchuenche and Bauch suggest that culling could be counterproductive, for some parameter values (see [26]).

The work in [27, 28] also presented a mathematical model of rabies transmission in dogs and from the dog population to the human population in China. Their study did not consider the use optimal control analysis to the study of the rabies virus in dogs and from the dog population to the human population. Furthermore, the insightful work of Wiraningsih et al. [29] studied the stability analysis of a rabies model with vaccination effect and culling in dogs, where they introduced postexposure prophylaxis to a rabies transmission model, but the paper did not consider the noneffectiveness of the pre- and postprophylaxis on the susceptible humans and exposed humans and that of the dog population and the use of optimal control analysis. Therefore, motivated by the research predictions of the global alliance of rabies control [30] and the work mention above, we seek to adjust the model presented in [27–29], by formulating an optimal control model, so as to ascertain an optimal way of controlling rabies transmission in dogs and from the dog population to the human population taking into account the noneffectiveness (failure) of vaccination and treatment.

The paper is petition as follows. Section 2 contains the model formulation, mathematical assumptions, the mathematical flowchart, and the model equations. Section 3 contains the model analysis, invariant region, equilibrium points, basic reproduction number $\mathscr{R}_0$, and the stability analysis of the equilibria. In Section 4 we present the parameter values leading to numerical values of the basic reproduction number $\mathscr{R}_0$, the herd immunity threshold and sensitivity analysis using Latin hypercube sampling (LHS), and some numerical plots. Section 5 contains the objective functional and the optimality system of the model. Finally, Sections 6 and 7 contain discussion and conclusion, respectively.

2. Model Formulation

We present two subpopulation transmission models of rabies virus in dogs and that of the human population (see Figure 1), based on the work presented in [27–29]. The dog population has a total of four compartments. The compartments represent the susceptible dogs, $S_D(t)$, exposed dogs, E_D, infected dogs, $I_D(t)$, and partially immune dogs, $R_D(t)$. Thus, the total dog population is $N_D(t) = S_D(t) + E_D(t) + I_D(t) + R_D(t)$. The human population also has four compartments representing susceptible humans, $S_H(t)$, exposed humans, $E_H(t)$, infected humans, $I_H(t)$, and partially immune humans, $R_H(t)$. Thus, the total human population is $N_H(t) = S_H(t) + E_H(t) + I_H(t) + R_H(t)$. It is assumed that there is no human to human transmission of the rabies virus in the human submodel (see [29]). In the dog submodel, it is assumed that there is a direct transmission of the rabies virus from one dog to the other and from the infected dog compartment to the susceptible human population. It is further assumed that the susceptible dog population, $S_D(t)$, is increased by recruitment at a rate A_D and B_H is the birth or immigration rate into the susceptible human population, $S_H(t)$. It is assumed that the transmission

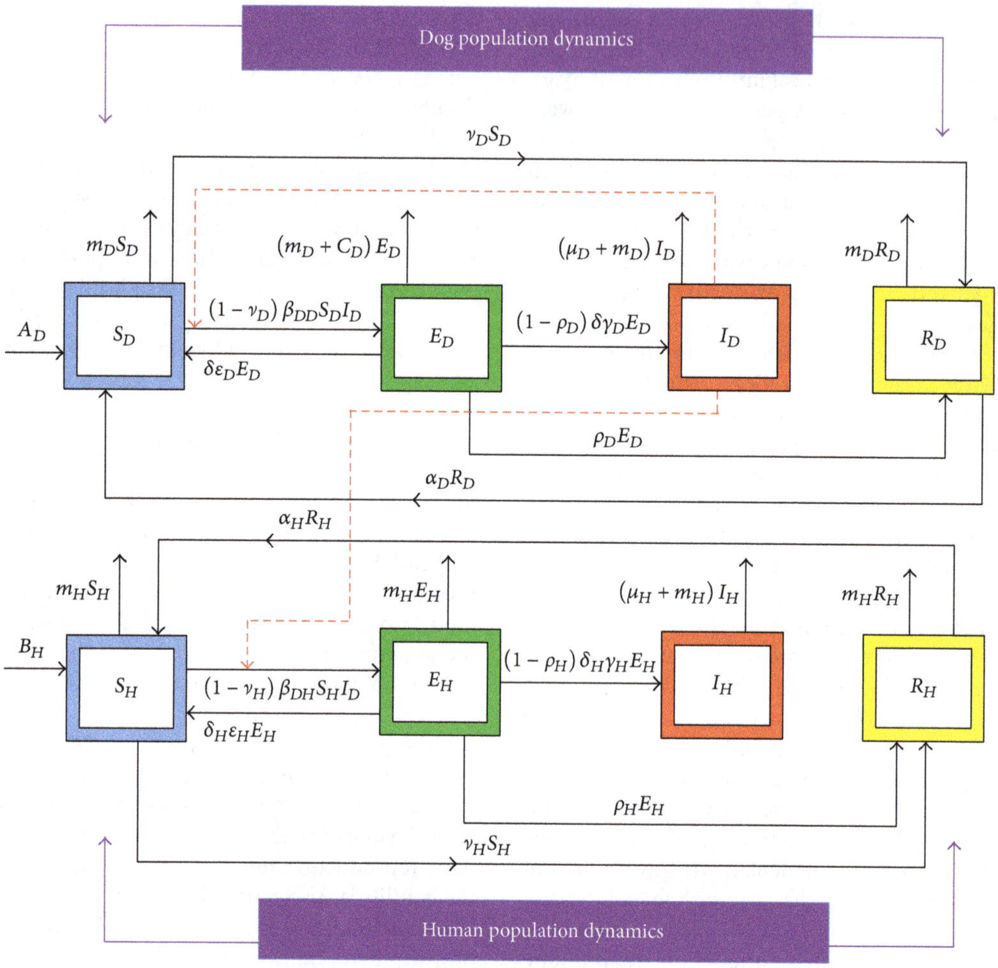

FIGURE 1: Optimal control model of rabies transmission dynamics.

and contact rate of the rabid dog into the dog compartment is β_{DD}. Suppose that ν_D represents the control strategy due to public education and vaccination in the dogs compartment; then the transmission dynamics become $(1 - \nu_D)\beta_{DD}S_D I_D$, where $(1 - \nu_D)$ is the noneffectiveness (failure) of the vaccine. It is also assumed that the contact rate of infectious dogs to the human population is β_{DH}. Similarly, administrating vaccination to the susceptible humans the progression rate of the susceptible humans to the exposed stage becomes $(1 - \nu_H)\beta_{DH}S_H I_D$, where ν_H is the preexposure prophylaxis (vaccination), $(1 - \nu_H)$ represents the failure of the preexposure prophylaxis in the human compartment. Furthermore, administrating postexposure prophylaxis (treatment) to affected humans at the rate ρ_H decreases the progression rate of the rabies virus, at the exposed class to the infectious class as $(1 - \rho_H)\delta_H\gamma_H E_H$, where $(1 - \rho_H)$ is the failure rate of the postexposure prophylaxis and $\delta_H\gamma_H$ represents the rate at which exposed humans progress to the infected compartment [27]. The rate of losing immunity in both compartments is represented by α_D and α_H, respectively.

The exposed humans without clinical rabies that move back to the susceptible population are denoted by the rate $\delta_H\varepsilon_H$. The natural death rate of dogs is m_D, and m_H denotes the mortality rate of humans (natural death rate), μ_D represents the death rate associated with rabies infection in dogs, and μ_H represents the disease induce death in humans. The rate at which exposed dogs die due to culling is C_D, and $\delta\varepsilon_D$ represents the rate at which exposed dogs without clinical rabies move back to the susceptible dog compartment. Subsequently, using the idea presented in [29], we assumed that the exposed dogs are treated or quarantined by their owners at the rate ρ_D; this implies that $(1 - \rho_D)\delta\gamma_D E_D$ is the progression rate of the exposed dogs to the infectious compartment, where $(1 - \rho_D)$ is the failure of the treatment or quarantined strategy, and $\delta\gamma_D E_D$ denotes those exposed dogs that develop clinical rabies [27]. Figure 1 shows the mathematical dynamics of the rabies virus in both compartments.

From Figure 1 transmission flowchart and assumptions give the disease pathways as

$$\frac{dS_D}{dt} = A_D - (1 - \nu_D)\beta_{DD}S_D I_D - (m_D + \nu_D)S_D + \delta\varepsilon_D E_D + \alpha_D R_D,$$

$$\frac{dE_D}{dt} = (1 - \nu_D)\beta_{DD}S_DI_D - ((1 - \rho_D)\delta\gamma_D + m_D + \rho_D + \delta\varepsilon_D + C_D)E_D,$$
$$\frac{dI_D}{dt} = (1 - \rho_D)\delta\gamma_D E_D - (m_D + \mu_D)I_D,$$
$$\frac{dR_D}{dt} = \nu_D S_D + \rho_D E_D - (m_D + \alpha_D)R_D,$$
$$\frac{dS_H}{dt} = B_H - (1 - \nu_H)\beta_{DH}S_HI_D - (m_H + \nu_H)S_H + \delta_H\varepsilon_H E_H + \alpha_H R_H,$$
$$\frac{dE_H}{dt} = (1 - \nu_H)\beta_{DH}S_HI_D - ((1 - \rho_H)\delta_H\gamma_H + m_H + \rho_H + \delta_H\varepsilon_H)E_H,$$
$$\frac{dI_H}{dt} = (1 - \rho_H)\delta_H\gamma_H E_H - (m_H + \mu_H)I_H,$$
$$\frac{dR_H}{dt} = \nu_H S_H + \rho_H E_H - (m_H + \alpha_H)R_H,$$
$$\text{with } S_D(0) > 0,\ E_D(0) \geq 0,\ I_D(0) \geq 0,\ R_D(0) \geq 0,\ S_H(0) > 0,\ E_H(0) \geq 0,\ I_H(0) > 0,\ R_H(0) > 0. \quad (1)$$

3. Model Analysis

Model system (1) will be studied in a biological feasible region as outlined below. Model system (1) is basically divided into two regions; thus $\Omega = \Omega_D \times \Omega_H$.

Lemma 1. *The solution set* $\{S_D, E_D, I_D, R_D, S_H, E_H, I_H, R_H\} \in R^8_+$ *of model system (1) is contained in the feasible region* Ω.

Proof. Suppose $\{S_D, E_D, I_D, R_D, S_H, E_H, I_H, R_H\} \in R^8_+$ for all $t > 0$. We want to show that the region Ω is positively invariant, so that it becomes sufficient to look at the dynamics of model system (1), given that

$$N_D(t) = S_D(t) + E_D(t) + I_D(t) + R_D(t), \quad (2)$$
$$N_H(t) = S_H(t) + E_H(t) + I_H(t) + R_H(t), \quad (3)$$

where $N_D(t)$ is the total population of dogs at any time (t) and $N_H(t)$ is total population of humans at any time (t).

Equation (2) gives

$$\frac{dN_D}{dt} = A_D - (S_D + E_D + I_D + R_D)m_D - \mu_D I_D - C_D E_D, \quad (4)$$

which yields

$$\frac{dN_D}{dt} = A_D - m_D N_D - \mu_D I_D - C_D E_D. \quad (5)$$

Similarly (3) gives

$$\frac{dN_H}{dt} = B_H - m_H N_H - \mu_H I_H. \quad (6)$$

Now, assuming that there are no disease induced death rate and culling effect in the dogs' compartment, it implies that (5) and (6) become

$$\frac{dN_D}{dt} = A_D - m_D N_D, \qquad \frac{dN_H}{dt} = B_D - m_H N_H. \quad (7)$$

Suppose $dN_D/dt \leq 0, dN_H/dt \leq 0, N_D \leq A_D/m_D$, and $N_H \leq B_H/m_H$, and then imposing the theorem proposed in [32] on differential inequality results in $0 \leq N_D \leq A_D/m_D$ and $0 \leq N_H \leq B_H/m_H$. Therefore (7) becomes

$$\frac{dN_D}{dt} \leq A_D - m_D N_D, \quad (8)$$
$$\frac{dN_H}{dt} \leq B_D - m_H N_H. \quad (9)$$

Solve (8) and (9) using the integrating factor (IF) method. Thus $dy/dt + p(t)y = Q$, $IF = e^{\int p(t)dt}$. After some algebraic manipulation the feasible solution of the dogs' population in model system (1) is in the region

$$\Omega_D = \left\{(S_D, E_D, I_D, R_D) \in \mathbb{R}^4_+,\ N_D \leq \frac{A_D}{m_D}\right\}. \quad (10)$$

Similarly the human population follows suit, and from (9) this implies that the feasible solution of the human population of model system (1) is in the region

$$\Omega_H = \left\{(S_H, E_H, I_H, R_H) \in \mathbb{R}^4_+,\ N_H \leq \frac{B_H}{m_H}\right\}. \quad (11)$$

Therefore, the feasible solutions are contained in Ω. Thus $\Omega = \Omega_D \times \Omega_H$. From the standard comparison theorem used on differential inequality in [33], it implies that

$$\begin{aligned} N_D(t) &\le N_D(0)\, e^{-(m_D)t} + \frac{A_D}{m_D}\left(1 - e^{-(m_D)t}\right), \\ N_H(t) &\le N_H(0)\, e^{-(m_H)t} + \frac{B_H}{m_H}\left(1 - e^{-(m_H)t}\right). \end{aligned} \tag{12}$$

Hence, the total dog population size $N_D(t) \to A_D/m_D$ as $t \to \infty$. Similarly, the total human population size $N_H(t) \to B_H/m_H$ as $t \to \infty$. This means that the infected state variables (E_D, I_D, E_H, I_H) of the two populations tend to zero as time goes to infinity. Therefore, the region Ω is pulling (attracting) all the solutions in $\mathbb{R}^8_+$. This gives the feasible solution set of model system (1) as

$$\left(\begin{array}{c} S_D \\ E_D \\ I_D \\ R_D \\ S_H \\ E_H \\ I_H \\ R_H \end{array} \right) \in \mathbb{R}^8_+ \mid \left(\begin{array}{c} S_D > 0 \\ E_D \ge 0 \\ I_D \ge 0 \\ I_D \ge 0 \\ R_D \ge 0 \\ S_H > 0 \\ E_H \ge 0 \\ I_H \ge 0 \\ R_H \ge 0 \\ N_D \le \dfrac{A_D}{m_D} \\ N_H \le \dfrac{B_H}{m_H} \end{array} \right). \tag{13}$$

□

Hence, (1) is mathematically well posed and epidemiologically meaningful.

3.1. Disease-Free Equilibrium $\mathscr{E}_0$. Suppose there is no infection of rabies in both compartments; then $(E_D = 0,\ I_D = 0,\ E_H = 0,\ I_H = 0)$. Incorporating this into (1) leads to

$$\begin{aligned} A_D - (m_D + \nu_D) S_D + \alpha_D R_D &= 0, \\ \nu_D S_D - (m_D + \alpha_D) R_D &= 0, \\ B_H - (m_H + \nu_H) S_H + \alpha_H R_H &= 0, \\ \nu_H S_H - (m_H + \alpha_H) R_H &= 0. \end{aligned} \tag{14}$$

After some algebraic manipulation of (14), the disease-free equilibrium point becomes $\mathscr{E}_0 = (S^0_D, E^0_D, I^0_D, R^0_D, S^0_H, E^0_H, I^0_H, R^0_H)$ with

$$\begin{aligned} \mathscr{E}_0 = \Bigg(& \frac{A_D (m_D + \alpha_D)}{m_D (m_D + \alpha_D + \nu_D)}, 0, 0, \\ & \frac{A_D \nu_D}{m_D (m_D + \alpha_D + \nu_D)}, \frac{B_H (m_H + \alpha_H)}{m_H (m_H + \alpha_H + \nu_H)}, 0, 0, \\ & \frac{B_H \nu_H}{m_H (m_H + \alpha_H + \nu_H)} \Bigg). \end{aligned} \tag{15}$$

3.2. Basic Reproduction Number $\mathscr{R}_0$. Here, the basic reproduction number $(\mathscr{R}_0)$ measures the average number of new infections produced by one infected dog in a completely susceptible (dog and human) population (see also [34]). Now taking E_D, I_D, E_H, and I_H as our infected compartments gives

$$\begin{aligned} f_1 &= (1 - \nu_D)\, \beta_{DD} S_D I_D \\ &\quad - \left((1 - \rho_D)\, \delta\gamma_D + m_D + \rho_D + \delta\varepsilon_D + C_D\right) E_D, \\ f_2 &= (1 - \rho_D)\, \delta\gamma_D E_D - (m_D + \mu_D)\, I_D, \\ f_3 &= (1 - \nu_H)\, \beta_{DH} S_H I_D \\ &\quad - \left((1 - \rho_H)\, \delta_H \gamma_H + m_H + \rho_H + \delta_H \varepsilon_H\right) E_H, \\ f_4 &= (1 - \rho_H)\, \delta_H \gamma_H E_H - (m_H + \mu_H)\, I_H, \end{aligned} \tag{16}$$

where $f_1 = dE_D/dt$, $f_2 = dI_D/dt$, $f_3 = dE_H/dt$, and $f_4 = dI_H/dt$.

Now, using the next generation matrix operator $G = FV^{-1}$ and the Jacobian matrix

$$J = \begin{pmatrix} \dfrac{\partial f_1}{\partial E_D} & \dfrac{\partial f_1}{\partial I_D} & \dfrac{\partial f_1}{\partial E_H} & \dfrac{\partial f_1}{\partial I_H} \\ \dfrac{\partial f_2}{\partial E_D} & \dfrac{\partial f_2}{\partial I_D} & \dfrac{\partial f_2}{\partial E_H} & \dfrac{\partial f_2}{\partial I_H} \\ \dfrac{\partial f_3}{\partial E_D} & \dfrac{\partial f_3}{\partial I_D} & \dfrac{\partial f_3}{\partial E_H} & \dfrac{\partial f_3}{\partial I_H} \\ \dfrac{\partial f_4}{\partial E_D} & \dfrac{\partial f_4}{\partial I_D} & \dfrac{\partial f_4}{\partial E_H} & \dfrac{\partial f_4}{\partial I_H} \end{pmatrix}, \tag{17}$$

as described in [34], results in

$$\begin{aligned} J &= \begin{pmatrix} -\left((1 - \rho_D)\, \delta\gamma_D + m_D + \rho_D + \delta\varepsilon_D + C_D\right) & (1 - \nu_D)\, \beta_{DD} S_D & 0 & 0 \\ (1 - \rho_D)\, \delta\gamma_D & -(m_D + \mu_D) & 0 & 0 \\ 0 & (1 - \nu_H)\, \beta_{DH} S_H & -\left((1 - \rho_H)\, \delta_H \gamma_H + m_H + \rho_H + \delta_H \varepsilon_H\right) & 0 \\ 0 & 0 & (1 - \rho_H)\, \delta_H \gamma_H & -(m_H + \mu_H) \end{pmatrix}. \end{aligned} \tag{18}$$

Using the fact that $J = F - V$ gives F and V evaluated at $\mathscr{E}_0$ as

$$F(\mathscr{E}_0) = \begin{pmatrix} 0 & \dfrac{(1-\nu_D)\beta_{DD}A_D(m_D+\alpha_D)}{m_D(m_D+\nu_D+\alpha_D)} & 0 & 0 \\ 0 & 0 & 0 & 0 \\ 0 & \dfrac{(1-\nu_H)\beta_{DH}(m_H+\alpha_H)B_H}{m_H(m_H+\nu_H+\alpha_H)} & 0 & 0 \\ 0 & 0 & 0 & 0 \end{pmatrix},$$

$$V(\mathscr{E}_0) = \begin{pmatrix} ((1-\rho_D)\delta\gamma_D + m_D + \rho_D + \delta\varepsilon_D + C_D) & 0 & 0 & 0 \\ -(1-\rho_D)\delta\gamma_D & (m_D+\mu_D) & 0 & 0 \\ 0 & 0 & ((1-\rho_H)\delta_H\gamma_H + m_H + \rho_H + \delta_H\varepsilon_H) & 0 \\ 0 & 0 & -(1-\rho_H)\delta_H\gamma_H & (m_H+\mu_H) \end{pmatrix}, \tag{19}$$

where the element in matrix F constitutes the new infection terms, while that of matrix V constitutes the new transfer of infection terms from one compartment to another. Now, splitting matrix V into four 2×2 submatrices and finding its corresponding inverses result in $G = FV^{-1}$, given by

$$G = \begin{pmatrix} \dfrac{(1-\rho_D)(1-\nu_D)\delta\gamma_D\beta_{DD}A_D(m_D+\alpha_D)}{((1-\rho_D)\delta\gamma_D + m_D + \rho_D + \delta\varepsilon_D + C_D)(m_D+\mu_D)m_D(m_D+\nu_D+\alpha_D)} & \dfrac{(1-\nu_D)\beta_{DD}A_D(m_D+\alpha_D)}{m_D(m_D+\nu_D+\alpha_D)} & 0 & 0 \\ 0 & 0 & 0 & 0 \\ \dfrac{(1-\rho_D)\delta\gamma_D(1-\nu_H)\beta_{DH}B_H(m_H+\alpha_H)}{((1-\rho_D)\delta\gamma_D + m_D + \rho_D + \delta\varepsilon_D + C_D)(m_D+\mu_D)m_H(m_H+\nu_H+\alpha_H)} & \dfrac{(1-\nu_H)\beta_{DH}(m_H+\alpha_H)B_H}{(m_D+\nu_D)m_H(m_H+\nu_H+\alpha_H)} & 0 & 0 \\ 0 & 0 & 0 & 0 \end{pmatrix} \tag{20}$$

Letting

$$\begin{aligned} a &= \frac{(1-\rho_D)(1-\nu_D)\delta\gamma_D\beta_{DD}A_D(m_D+\alpha_D)}{((1-\rho_D)\delta\gamma_D + m_D + \rho_D + \delta\varepsilon_D + C_D)(m_D+\mu_D)m_D(m_D+\nu_D+\alpha_D)}, \\ b &= \frac{(1-\nu_D)\beta_{DD}A_D(m_D+\alpha_D)}{m_D(m_D+\nu_D+\alpha_D)}, \\ c &= \frac{(1-\rho_D)(1-\nu_H)\delta\gamma_D\beta_{DH}(m_H+\alpha_H)}{((1-\rho_D)\delta\gamma_D + m_D + \rho_D + \delta\varepsilon_D + C_D)(m_D+\mu_D)m_H(m_H+\nu_H+\alpha_H)}, \\ d &= \frac{(1-\nu_H)\beta_{DH}(m_H+\alpha_H)B_H}{(m_D+\nu_D)m_H(m_H+\nu_H+\alpha_H)} \end{aligned} \tag{21}$$

implies

$$G = \begin{pmatrix} a & b & 0 & 0 \\ 0 & 0 & 0 & 0 \\ c & d & 0 & 0 \\ 0 & 0 & 0 & 0 \end{pmatrix}. \quad (22)$$

Finding the matrix determinant of (22) and denoting it by D give the expression $D = |G - \mathbb{I}\lambda|$, where $\mathbb{I}$ is the identity matrix of a 4×4 matrix; thus

$$D = \begin{vmatrix} a-\lambda & b & 0 & 0 \\ 0 & -\lambda & 0 & 0 \\ c & d & -\lambda & 0 \\ 0 & 0 & 0 & -\lambda \end{vmatrix} = 0. \quad (23)$$

This gives a characteristic equation of the form $\lambda^3(a - \lambda) = 0$; solving the characteristic polynomial results in the following eigenvalues: $\lambda_i = [0, 0, 0, a]$. The basic reproduction number $\mathscr{R}_0$ is the spectral radius (largest eigenvalue) $\rho(FV^{-1})$, also defined as the dominant eigenvalue of FV^{-1}.

Therefore,

$$\mathscr{R}_0 = \frac{(1-\rho_D)(1-\nu_D)\delta\gamma_D\beta_{DD}A_D(m_D+\alpha_D)}{((1-\rho_D)\delta\gamma_D + m_D + \rho_D + \delta\varepsilon_D + C_D)(m_D+\mu_D)m_D(m_D+\nu_D+\alpha_D)}. \quad (24)$$

Remark 2. $\mathscr{R}_0$ contains the secondary infection produced by the infectious compartment of dogs (in the presence of preexposure prophylaxis (vaccination), postexposure prophylaxis (treatment/quarantine), and culling of exposed dogs). When $\mathscr{R}_0 < 1$, the infection gradually leaves the dog compartment, but when $\mathscr{R}_0 > 1$, the rabies virus remains in the dog compartments for a longer time, thereby increasing the rate at which the susceptible dogs and humans get infected by a rabid dog.

3.3. *Endemic Equilibrium* $\mathscr{E}_1$. The endemic equilibrium is given as

$$\begin{aligned}
S_D^* &= \frac{A_D(m_D+\alpha_D)}{m_D(m_D+\nu_D+\alpha_D)\mathscr{R}_0}, \\
E_D^* &= \frac{(m_D+\mu_D)}{(1-\rho_D)\delta\gamma_D} I_D^*, \\
I_D^* &= \frac{[(1-\rho_D)\delta\gamma_D + m_D + \rho_D + \delta\gamma_D](m_D+\mu_D)m_D(m_D+\nu_D+\alpha_D)(\mathscr{R}_0 - 1)}{(m_D+\alpha_D)(1-\nu_D)\beta_{DD}[(1-\rho_D)\delta\gamma_D + m_D + C_D] + m_D(1-\nu_D)\beta_{DD}\rho_D}, \\
R_D^* &= \frac{A_D\nu_D(1-\nu_D)\beta_{DD}(1-\rho_D)\delta\gamma_D(m_D+\alpha_D) + (1-\nu_D)\beta_{DD}\rho_D(m_D+\mu_D)I_D^*}{m_D(m_D+\nu_D+\alpha_D)\mathscr{R}_0(1-\nu_D)\beta_{DD}(1-\rho_D)\delta\gamma_D(m_D+\alpha_D)}, \\
S_H^* &= \frac{B_H(m_H+\alpha_H) + [\delta_H\varepsilon_H + \alpha_H\rho_H]E_H^*}{[(1-\nu_H)(m_H+\alpha_H)\beta_{DH}I_D^* + m_H(m_H+\alpha_H+\nu_H)]}, \qquad (25) \\
E_H^* &= \frac{(1-\nu_H)B_H(m_H+\alpha_H)\beta_{DH}I_D^*}{(m_H+\alpha_H)[(1-\nu_H)\beta_{DH}I_D^*((1-\rho_H)\delta_H\gamma_H + m_H + \rho_H) + (m_H+\nu_H)((1-\rho_H)\delta_H\gamma_H + m_H + \rho_H + \delta_H\varepsilon_H)] - (1-\nu_H)\beta_{DH}I_D^*\alpha_H\rho_H}, \\
I_H^* &= \frac{(1-\rho_H)\delta_H\gamma_H}{m_H+\mu_H} E_H^*, \\
R_H^* &= \frac{B_H\nu_H(m_H+\nu_H) + [(\nu_H\delta_H\varepsilon_H + \nu_H\alpha_H\rho_H) + \rho_H(1-\nu_H)(m_H+\alpha_H)\beta_{DH}I_D^* + \rho_H m_H(m_H+\alpha_H+\nu_H)]E_H^*}{[(1-\nu_H)(m_H+\alpha_H)^2\beta_{DH}I_D^* + (m_H+\alpha_H)m_H(m_H+\alpha_H+\nu_H)]}.
\end{aligned}$$

Note that if $\mathscr{R}_0 = 1$, it results in the disease-free equilibrium; if $\mathscr{R}_0 > 1$, then there exists a unique endemic equilibrium; if $\mathscr{R}_0 < 1$, then there exist two endemic equilibriums.

3.4. Stability Analysis of $\mathscr{E}_0$. Linearizing (1) at $\mathscr{E}_0$ and subtracting eigenvalue λ along the main diagonal yield

$$\mathbb{J}(\mathscr{E}_0) = \begin{pmatrix} b_1-\lambda & b_7 & a_1 & \alpha_D & 0 & 0 & 0 & 0 \\ 0 & a_2-\lambda & a_3 & 0 & 0 & 0 & 0 & 0 \\ 0 & b_{10} & b_2-\lambda & 0 & 0 & 0 & 0 & 0 \\ \nu_D & \rho_D & 0 & b_3-\lambda & 0 & 0 & 0 & 0 \\ 0 & 0 & a_4 & 0 & b_4-\lambda & b_5 & 0 & \alpha_H \\ 0 & 0 & a_5 & 0 & 0 & a_6-\lambda & 0 & 0 \\ 0 & 0 & 0 & 0 & 0 & b_9 & b_6-\lambda & 0 \\ 0 & 0 & 0 & 0 & \nu_H & \rho_H & 0 & b_8-\lambda \end{pmatrix}, \tag{26}$$

where

$$\begin{aligned} a_1 &= \frac{-(1-\nu_D)\beta_{DD}(m_D+\alpha_D)A_D}{m_D(m_D+\nu_D+\alpha_D)}, \\ a_2 &= -((1-\rho_D)\delta\gamma_D+m_D+\rho_D+\delta\varepsilon_D+C_D), \\ a_3 &= \frac{(1-\nu_D)\beta_{DD}(m_D+\alpha_D)A_D}{m_D(m_D+\nu_D+\alpha_D)}, \\ a_4 &= \frac{-(1-\nu_H)\beta_{DH}(m_H+\alpha_H)}{m_H(m_H+\nu_H+\alpha_H)}, \\ a_5 &= \frac{(1-\nu_H)\beta_{DH}(m_H+\alpha_H)}{m_H(m_H+\nu_H+\alpha_H)}, \\ a_6 &= -((1-\rho_H)\delta_H\gamma_H+m_H+\rho_H+\delta_H\varepsilon_H), \\ b_1 &= -(m_D+\nu_D), \\ b_2 &= -(m_D+\mu_D), \\ b_3 &= -(m_D+\alpha_D), \\ b_4 &= -(\nu_H+m_H), \\ b_5 &= \delta_H\varepsilon_H, \\ b_6 &= -(m_H+\mu_H), \\ b_7 &= \delta\varepsilon_D, \\ b_8 &= -(m_H+\alpha_H), \\ b_9 &= (1-\rho_H)\delta_H\gamma_H, \\ b_{10} &= (1-\rho_D)\delta\gamma_D. \end{aligned} \tag{27}$$

Simplifying matrix $\mathbb{J}(\mathscr{E}_0)$ gives

$$\begin{aligned} &(b_6-\lambda)(a_6-\lambda)(b_4-\lambda)(b_8-\lambda) \\ &\quad\cdot\left[\lambda^4+a_{11}\lambda^3+a_{12}\lambda^2+a_{13}\lambda+a_{14}\right]=0, \end{aligned} \tag{28}$$

where

$$\begin{aligned} a_{11} &= (-b_2-a_2-b_1-b_3), \\ a_{12} &= \nu_D\alpha_D+a_2b_3+a_2b_1+b_2b_3+b_2b_1+b_3b_1+b_2a_2 \\ &\quad-(1-\rho_D)\delta\gamma_Da_3, \\ a_{13} &= -a_2\nu_D\alpha_D-b_2\nu_D\alpha_D+a_3(1-\rho_D)\delta\gamma_Db_2+a_3(1 \\ &\quad-\rho_D)\delta\gamma_Db_1-a_2b_2b_3-b_2a_2b_1-a_2b_3b_1-b_2b_3b_1, \\ a_{14} &= (b_1b_2b_3a_2+(1-\rho_D)\delta\gamma_Da_3b_3b_1 \\ &\quad+(1-\rho_D)\delta\gamma_Da_3\nu_D+\nu_D\alpha_Db_2a_2). \end{aligned} \tag{29}$$

From (28) the four characteristic factors that are negative are

$$\begin{aligned} \lambda_1 &= b_6, \\ \lambda_2 &= a_6, \\ \lambda_3 &= b_4, \\ \lambda_4 &= b_8, \end{aligned} \tag{30}$$

where $a_6 = -((1-\rho_H)\delta_H\gamma_H+m_H+\rho_H+\delta_H+\delta_H\varepsilon_H)$, $b_6 = -(m_H+\mu_H)$, $b_4 = -(\nu_H+m_H)$, and $b_8 = -(m_H+\alpha_H)$. The other four characteristic factors can be obtained using the Routh-Hurwitz criterion. Routh-Hurwitz stability criterion is a test to ascertain the nature of the eigenvalues. If the roots of the polynomial are all positive, then the polynomial has a negative real part [35, 36]. The remaining four characteristic eigenvalues are obtained as follows:

$$\lambda^4+a_{11}\lambda^3+a_{12}\lambda^2+a_{13}\lambda+a_{14}=0. \tag{31}$$

Hence, simplifying the coefficient of the above characteristic polynomial in (31) yields

$$\begin{aligned} a_{11} &= ((1-\rho_D)\delta\gamma_D+m_D+\rho_D+\delta\varepsilon_D+C_D) \\ &\quad+(m_D+\mu_D)+(m_D+\alpha_D)+(m_D+\nu_D), \end{aligned}$$

$$
\begin{aligned}
a_{12} &= \nu_D\alpha_D + ((1-\rho_D)\,\delta\gamma_D + m_D + \rho_D + \delta\varepsilon_D + C_D) \\
&\quad \cdot [(m_D+\alpha_D) + (m_D+\nu_D)] + (m_D+\mu_D) \\
&\quad \cdot [(m_D+\alpha_D) + (m_D+\nu_D)] + (m_D+\alpha_D) \\
&\quad \cdot (m_D+\nu_D) + ((1-\rho_D)\,\delta\gamma_D + m_D + \rho_D + C_D) \\
&\quad \cdot (m_D+\mu_D)\,(1-\mathscr{R}_0), \\
a_{13} &= ((1-\rho_D)\,\delta\gamma_D + m_D + \rho_D + \delta\varepsilon_D + C_D)\,\nu_D\alpha_D \\
&\quad + (m_D+\mu_D)\,\nu_D\alpha_D + (m_D+\mu_D)\,(m_D+\alpha_D) \\
&\quad \cdot (m_D+\nu_D) \\
&\quad + ((1-\rho_D)\,\delta\gamma_D + m_D + \rho_D + \delta\varepsilon_D + C_D) \\
&\quad \cdot (m_D+\alpha_D)\,(m_D+\mu_D)\left[1 - \frac{\mathscr{R}_0\,(m_D+\mu_D)}{m_D+\alpha_D}\right] \\
&\quad + ((1-\rho_D)\,\delta\gamma_D + m_D + \rho_D + \delta\varepsilon_D + C_D) \\
&\quad \cdot (m_D+\alpha_D)\,(m_D+\nu_D)\left[1 - \frac{\mathscr{R}_0}{m+\alpha}\right], \\
a_{14} &= \nu_D\alpha_D\,(m_D+\nu_D) \\
&\quad \cdot ((1-\rho_D)\,\delta\gamma_D + m_D + \rho_D + \delta\varepsilon_D + C_D) \\
&\quad + \frac{(1-\rho_D)\,\delta\gamma_D\,(1-\nu_D)\,\beta_{DD}\,(m_D+\alpha_D)\,A_D\nu_D}{m_D\,(m_D+m_D+\nu_D+\alpha_D)} \\
&\quad + (m_D+\nu_D)\,(m_D+\mu_D)\,(m_D+\alpha_D) \\
&\quad \cdot ((1-\rho_D)\,\delta_D\gamma_D + m_D + \rho_D + \delta\varepsilon_D + C_D) \\
&\quad \cdot (1-\mathscr{R}_0).
\end{aligned} \tag{32}
$$

Therefore, from the Routh-Hurwitz criterion of order four, it implies that the conditions, $a_{11} > 0$, $a_{12} > 0$, $a_{13} > 0$, $a_{14} > 0$, and $a_{11}a_{12}a_{13} > a_{13}^2 + a_{11}^2 a_{14}$, are satisfied if $\mathscr{R}_0 < 1$. Hence, the disease-free equilibrium $\mathscr{E}_0$ is locally asymptotically stable when $\mathscr{R}_0 < 1$ (see [37]).

3.4.1. Global Stability of $\mathscr{E}_0$

Theorem 3. *The disease-free equilibrium $\mathscr{E}_0$ of model (1) is globally asymptotically stable if $\mathscr{R}_0 \le 1$ and unstable if $\mathscr{R}_0 > 1$.*

Proof. Let $\mathscr{V}$ be a Lyapunov function with positive constants $\mathscr{K}_1$, $\mathscr{K}_2$, $\mathscr{K}_3$, and $\mathscr{K}_4$ such that

$$
\begin{aligned}
\mathscr{V} &= \left(S_D - S_D^0 - S_D^0 \ln \frac{S_D}{S_D^0}\right) + \mathscr{K}_1 E_D + \mathscr{K}_2 I_D \\
&\quad + \left(R_D - R_D^0 - R_D^0 \ln \frac{R_D}{R_D^0}\right) \\
&\quad + \left(S_H - S_H^0 - S_H^0 \ln \frac{S_H}{S_H^0}\right) + \mathscr{K}_3 E_H + \mathscr{K}_4 I_H \\
&\quad + \left(R_H - R_H^0 - R_H^0 \ln \frac{R_H}{R_H^0}\right).
\end{aligned} \tag{33}
$$

Taken the derivative of the Lyapunov function with respect to time gives

$$
\begin{aligned}
\frac{d\mathscr{V}}{dt} &= \left(1 - \frac{S_D^0}{S_D}\right)\frac{dS_D}{dt} + \mathscr{K}_1 \frac{dE_D}{dt} + \mathscr{K}_2 \frac{dI_D}{dt} \\
&\quad + \left(1 - \frac{R_D^0}{R_D}\right)\frac{dR_D}{dt} + \left(1 - \frac{S_H^0}{S_H}\right)\frac{dS_H}{dt} \\
&\quad + \mathscr{K}_3 \frac{dE_H}{dt} + \mathscr{K}_4 \frac{dI_H}{dt} + \left(1 - \frac{R_H^0}{R_H}\right)\frac{dR_H}{dt}.
\end{aligned} \tag{34}
$$

Plugging (1) into (34) results in

$$
\begin{aligned}
\frac{d\mathscr{V}}{dt} &= \left(1 - \frac{S_D^0}{S_D}\right)[A_D - (1-\nu_D)\,\beta_{DD}S_D I_D \\
&\quad - (m_D+\nu_D)\,S_D + \delta\varepsilon_D E_D + \alpha R_D] \\
&\quad + \mathscr{K}_1 [(1-\nu_D)\,\beta_{DD} S_D I_D \\
&\quad - ((1-\rho_D)\,\delta\gamma_D + m_D + \rho_D + \delta\varepsilon_D + C_D)\,E_D] \\
&\quad + \mathscr{K}_2 [(1-\rho_D)\,\delta\gamma_D E_D - (m_D+\mu_D)\,I_D] + \Bigg(1 \\
&\quad - \frac{R_D^0}{R_D}\Bigg)[\nu_D S_D + \rho_D E_D - (m_D+\alpha_D)\,R_D] + \Bigg(1 \\
&\quad - \frac{S_H^0}{S_H}\Bigg)[B_H - (1-\nu_H)\,\beta_{DH} S_H I_D - (m_H+\nu_H)\,S_H \\
&\quad + \delta_H\varepsilon_H E_H + \alpha_H R_H] + \mathscr{K}_3 [(1-\nu_H)\,\beta_{DH} S_H I_D \\
&\quad - ((1-\rho_H)\,\delta_H\gamma_H + m_H + \rho_H + \delta_H\varepsilon_H)\,E_H] \\
&\quad + \mathscr{K}_4 [(1-\rho_H)\,\delta_H\gamma_H E_H - (m_H+\mu_H)\,I_H] + \Bigg(1 \\
&\quad - \frac{R_H^0}{R_H}\Bigg)[\nu_H S_H + \rho_H E_H - (m_H+\alpha_H)\,R_H].
\end{aligned} \tag{35}
$$

Now, after forming the Lyapunov function $\mathscr{V}$ on the space of the eight state variables, thus $(S_D, E_D, I_D, R_D, S_H, E_H, I_H, R_H)$, and introducing the idea from [37], it is clear that if $E_D(t)$, $I_D(t)$, $E_H(t)$, and $I_H(t)$ at the disease-free equilibrium are globally stable (thus, $E_D = 0$, $I_D = 0$, $E_H = 0$, and $I_H = 0$), then $S_D(t) \to A_D(m_D+\alpha_D)/m_D(m_D+\alpha_D+\nu_D)$, $R_D(t) \to A_D\nu_D/m_D(m_D+\alpha_D+\nu_D)$, $S_H(t) \to B_H(m_H + \alpha_H)/m_H(m_H+\alpha_H+nu_H)$, and $R_H(t) \to B_H\nu_H/m_H(m_H + \alpha_H + H + \nu_H)$ as $t \to \infty$.

Therefore, it can be assumed that

$$S_D \leq S_D^0 = \frac{A_D(m_D+\alpha_D)}{m_D(m_D+\alpha_D+\nu_D)},$$
$$R_D \leq R_D^0 = \frac{A_D\nu_D}{m_D(m_D+\alpha_D+\nu_D)},$$
$$S_H \leq S_H^0 = \frac{B_H(m_H+\alpha_H)}{m_H(m_H+\alpha_H+nu_H)},$$
$$R_H \leq R_H^0 = \frac{B_H\nu_H}{m_H(m_H+\alpha_H+H+\nu_H)}, \quad (36)$$

(see [38]) and replacing it into (35) yields

$$\begin{aligned}\frac{d\mathcal{V}}{dt} \leq{}& \mathcal{K}_1\left[\frac{(1-\nu_D)\beta_{DD}A_D(m_D+\alpha_D)}{m_D(m_D+\alpha_D+\nu_D)}I_D\right.\\ &\left.-((1-\rho_D)\delta\gamma_D+m_D+\rho_D+\delta\varepsilon_D+C_D)E_D\right]\\ &+\mathcal{K}_2[(1-\rho_D)\delta\gamma_DE_D-(m_D+\mu_D)I_D]\\ &+\mathcal{K}_3\left[\frac{(1-\nu_H)\beta_{DH}B_H(mH+\alpha_H)}{m_H(m_H+\alpha_H+\nu_H)}I_d\right.\\ &\left.-((1-\rho_H)\delta_H\gamma_H+m_H+\rho_H+\delta_H\varepsilon_H)E_H\right]\\ &+\mathcal{K}_4[(1-\rho_H)\delta_H\gamma_HE_H-(m_H+\mu_H)I_H],\end{aligned} \quad (37)$$

This implies that

$$\begin{aligned}\frac{d\mathcal{V}}{dt} \leq{}& \left[\frac{\mathcal{K}_1(1-\nu_D)\beta_{DD}A_D(m_D+\alpha_D)}{m_D(m_D+\alpha_D+\nu_D)}\right.\\ &-\mathcal{K}_2(m_D+\mu_D)\\ &\left.+\frac{\mathcal{K}_3(1-\nu_H)\beta_{DH}B_H(m_H+\alpha_H)}{m_H(m_H+\alpha_H+\nu_H)}\right]I_D\\ &+[\mathcal{K}_2(1-\rho_D)\delta\gamma_D\\ &-\mathcal{K}_1((1-\rho_D)\delta\gamma_D+m_D+\rho_D+\delta\varepsilon_D+C_D)]E_D\\ &+[\mathcal{K}_4(1-\rho_H)\delta_H\gamma_H\\ &-\mathcal{K}_3((1-\rho_H)\delta_H\gamma_H+m_H+\delta_H\varepsilon_H)]E_H\\ &-\mathcal{K}_4(m_H+\mu_H).\end{aligned} \quad (38)$$

Equating the coefficient of I_D, E_D, I_H, and E_H in (38) to zero gives

$$\begin{aligned}&\mathcal{K}_4=\mathcal{K}_3=0,\\ &\mathcal{K}_2=((1-\rho_D)\delta\gamma_D+m_D+\rho_D+\delta\varepsilon_D+C_D),\\ &\mathcal{K}_1=(1-\rho_D)\delta\gamma_D,\end{aligned} \quad (39)$$

and we obtain

$$\begin{aligned}\frac{d\mathcal{V}}{dt} \leq{}& ((1-\rho_D)\delta\gamma_D+m_D+\rho_D+\delta\varepsilon_D+C_D)\\ &\cdot(m_D+\mu_D)(\mathcal{R}_0-1)I_D,\\ \leq{}& 0, \quad \text{if } \mathcal{R}_0 \leq 1.\end{aligned} \quad (40)$$

Additionally $d\mathcal{V}/dt = 0$ if and only if $I_D = 0$. Therefore, for $E_D = I_D = E_H = I_H = 0$ it shows that $S_D(t) \to A_D(m_D+\alpha_D)/m_D(m_D+\alpha_D+\nu_D)$, $R_D(t) \to A_D\nu_D/m_D(m_D+\alpha_D+\nu_D)$, $S_H(t) \to B_H(m_H+\alpha_H)/m_H(m_H+\alpha_H+\nu_H)$, and $R_H(t) \to B_H\nu_H/m_H(m_H+\alpha_H+\nu_H)$ as $t \to \infty$. Hence, the largest compact invariant set in $\{(S_D, E_D, I_D, R_D, S_H, E_H, I_H, R_H) \in \Omega : d\mathcal{V}/dt \leq 0\}$ is the singleton set $\{\mathcal{E}_0\}$. Therefore, from La Salle's invariance principle, we conclude that $\mathcal{E}_0$ is globally asymptotically stable in Ω if $\mathcal{R}_0 \leq 1$ (see also [38, 39]).

3.5. Global Stability of Endemic Equilibrium $\mathcal{E}_1$

Theorem 4. *The endemic equilibrium $\mathcal{E}_1$ of model (1) is globally asymptotically stable whenever $\mathcal{R}_0 > 1$.*

Proof. Suppose $\mathcal{R}_0 > 1$; then the existence of the endemic equilibrium point is assured. Using the common quadratic Lyapunov function

$$V(x_1, x_2, \dots, x_n) = \sum_{i=1}^{n}\frac{c_i}{2}(x_i - x_i^*)^2, \quad (41)$$

as illustrated in [40], we consider a Lyapunov function with the following candidate:

$$\begin{aligned}\mathcal{V}(S_D, E_D, I_D, R_D, S_H, E_H, I_H, R_H) ={}& \frac{1}{2}[(S_D-S_D^*)\\ &+(E_D-E_D^*)+(I_D-I_D^*)+(R_D-R_D^*)]^2\\ &+\frac{1}{2}[(S_H-S_H^*)+(E_H-E_H^*)+(I_H-I_H^*)\\ &+(R_H-R_H^*)]^2.\end{aligned} \quad (42)$$

Now, differentiating (42) along the solution curve of (1) gives

$$\begin{aligned}\frac{d\mathcal{V}}{dt} ={}& [(S_D-S_D^*)+(E_D-E_D^*)+(I_D-I_D^*)\\ &+(R_D-R_D^*)]\frac{d(S_D+E_D+I_D+R_D)}{dt}\\ &+[(S_H-S_H^*)+(E_H-E_H^*)+(I_H-I_H^*)\\ &+(R_H-R_H^*)]\frac{d(S_H+E_H+I_H+R_H)}{dt}.\end{aligned} \quad (43)$$

From (1) it implies that $d(S_D + E_D + I_D + R_D)/dt = A_D - m_D(S_D + E_D + I_D + R_D) - C_D E_D - \mu_D I_D$ and $d(S_H + E_H + I_H + R_H)/dt = B - m(S_H + E_H + I_H + R_H) - \mu_H I_H$, which when plugged into (43) gives

$$\begin{aligned}\frac{d\mathcal{V}}{dt} &= [(S_D - S_D^*) + (E_D - E_D^*) + (I_D - I_D^*) \\ &\quad + (R_D - R_D^*)] (A_D - m_D (S_D + E_D + I_D + R_D) \\ &\quad - C_D E_D - \mu_D I_D) + [(S_H - S_H^*) + (E_H - E_H^*) \\ &\quad + (I_H - I_H^*) + (R_H - R_H^*)] (B_H \\ &\quad - m (S_H + E_H + I_H + R_H) - \mu_H I_H).\end{aligned} \tag{44}$$

Now assuming

$$\begin{aligned}A_D &= m_D (S_D^* + E_D^* + I_D^* + R_D^*) + C_D E_D^* + \mu_D I_D^*, \\ B_H &= m_H (S_H^* + E_H^* + I_H^* + R_H^*) + \mu_H I_H^*\end{aligned} \tag{45}$$

and substituting it into (44), we have

$$\begin{aligned}\frac{d\mathcal{V}}{dt} &= [(S_D - S_D^*) + (E_D - E_D^*) + (I_D - I_D^*) + (R_D \\ &\quad - R_D^*)] [m_D (S_D^* + E_D^* + I_D^* + R_D^*) + C_D E_D^* + \mu_D I_D^* \\ &\quad - m_D (S_D + E_D + I_D + R_D) - C_D E_D - \mu_D I_D] \\ &\quad + [(S_H - S_H^*) + (E_H - E_H^*) + (I_H - I_H^*) + (R_H \\ &\quad - R_H^*)] [m_H (S_H^* + E_H^* + I_H^* + R_H^*) + \mu_H I_H^* \\ &\quad - m (S_H + E_H + I_H + R_H) - \mu_H I_H], \\ \frac{d\mathcal{V}}{dt} &= [(S_D - S_D^*) + (E_D - E_D^*) + (I_D - I_D^*) + (R_D \\ &\quad - R_D^*)] [(-m_D (S_D - S_D^*) - m_D (E_D - E_D^*) \\ &\quad - m_D (I_D - I_D^*) - m_D (R_D - R_D^*) - C_D (E_D - E_D^*) \\ &\quad - \mu_D (I_D - I_D^*))] + [(S_H - S_H^*) + (E_H - E_H^*) \\ &\quad + (I_H - I_H^*) + (R_H - R_H^*)] [(-m_H (S_H - S_H^*) \\ &\quad - m_H (E_H - E_H^*) - m_H (I_H - I_H^*) \\ &\quad - m_H (R_H - R_H^*) - \mu_H (I_H - I_H^*))].\end{aligned} \tag{46}$$

This also implies that

$$\begin{aligned}\frac{d\mathcal{V}}{dt} &= -m_D (S_D - S_D^*)^2 - (C_D + m_D) (E_D - E_D^*)^2 \\ &\quad - (m_D + \mu_D) (I_D - I_D^*)^2 - m_D (R_D - R_D^*)^2 - (2m_D \\ &\quad + C_D) (S_D - S_D^*) (E_D - E_D^*) - (2m_D + \mu_D) (S_D \\ &\quad - S_D^*) (I_D - I_D^*) - (2m_D + \mu_D + C_D) (E_D - E_D^*) (I_D \\ &\quad - I_D^*) - 2m_D (R_D - R_D^*) (I_D - I_D^*) - (2m_D + \mu_D \\ &\quad + C_D) (R_D - R_D^*) (I_D - I_D^*) - m_H (S_H - S_H^*)^2 \\ &\quad - m_H (E_H - E_H^*)^2 - (m_H - \mu_H) (I_H - I_H^*)^2 \\ &\quad - m_H (R_H - R_H^*)^2 - 2m_H (S_H - S_H^*) (E_H - E_H^*) \\ &\quad - (2m_H - \mu_H) (S_H - S_H^*) (I_H - I_H^*) - (2m_H + \mu_H) \\ &\quad \cdot (E_H - E_H^*) (I_H - I_H^*) \\ &\quad - m_H [(I_H - I_H^*) (R_H - R_H^*) \\ &\quad + (S_H - S_H^*) (R_H - R_H^*)].\end{aligned} \tag{47}$$

This shows that $d\mathcal{V}/dt$ is negative and $d\mathcal{V}/dt = 0$, if and only if $S_D = S_D^*$, $E_D = E_D^*$, $I_D = I_D^*$, $R_D = R_D^*$, $S_H = S_H^*$, $E_H = E_H^*$, $I_H = I_H^*$, $R_H = R_H^*$. Additionally every solution of (1) with the initial conditions approaches $\mathscr{E}_1$ as $t \to \infty$ (see [38, 39]); therefore, the largest compact invariant set in $\{(S_D, E_D, I_D, R_D, S_H, E_H, I_H, R_H) \in \Omega : d\mathcal{V}/dt \leq 0\}$ is the singleton set $\{\mathscr{E}_1\}$. Therefore, from Lasalle's invariant principle [41], it implies that the endemic equilibrium $\mathscr{E}_1$ is globally asymptotically stable in Ω whenever $\mathscr{R}_0 > 1$.

4. Numerical Analysis

Considering the parameter values in Table 1, we will ascertain the numerical importance of our analysis.

4.1. Different Scenarios of the Basic Reproduction Number $\mathscr{R}_0$. We shall denote $\mathscr{R}_0$ without pre- and postexposure prophylaxis (treatment) as $\mathscr{R}_0^*$ and $\mathscr{R}_0$ without preexposure prophylaxis and culling as $\mathscr{R}_0^{**}$ and the $\mathscr{R}_0$ without postexposure prophylaxis (treatment) and culling as $\mathscr{R}_0^{***}$. Therefore, using the parameter values in Table 1, $\mathscr{R}_0^*$, $\mathscr{R}_0^{**}$, and $\mathscr{R}_0^{***}$ are given as follows:

$$\begin{aligned}\mathscr{R}_0^* &= \frac{\beta_{DD} A_D \delta\gamma_D}{(\delta\gamma_D + m_D + \delta\varepsilon_D + C_D) \times (m_D + \mu_D) m_D}, \\ \mathscr{R}_0^* &= 3.027, \\ \mathscr{R}_0^{**} &\\ &= \frac{(1 - \rho_D) \delta\gamma_D \beta_{DD} A_D}{((1 - \rho_D) \delta\gamma_D + m_D + \rho_D + \delta\varepsilon_D) (m_D + \mu_D) m_D}, \\ \mathscr{R}_0^{**} &= 2.181, \\ \mathscr{R}_0^{***} &\\ &= \frac{(1 - \nu_D) \delta\gamma_D \beta_{DD} A_D (m_D + \alpha_D)}{(\delta\gamma_D + m_D + \delta\varepsilon_D) (m_D + \mu_D) m_D (m_D + \nu_D + \alpha_D)}, \\ \mathscr{R}_0^{***} &= 1.914.\end{aligned} \tag{48}$$

Therefore, from the above calculations it indicates that the best way in reducing or minimizing the rabies virus in the dogs compartment is to use more of preexposure prophylaxis (vaccination).

TABLE 1: Parameter values.

Parameter	Description	Standard value	Source
A_D	Recruitment rate of dogs	$3 \times 10^6 y^{-1}$	[27]
α_D	Loss of immunity in dogs	$1 y^{-1}$	[27]
C_D	Death rate of dogs due to culling	$0.3 y^{-1}$	Assumed
m_D	Natural death rate of dogs	$0.056 y^{-1}$	[27]
μ_D	Disease induced mortality in dogs	$1 y^{-1}$	[27]
ν_D	Preexposure prophylaxis for dogs	$0.25 y^{-1}$	Assumed
ρ_D	Postexposure prophylaxis for dogs	$0.2 y^{-1}$	[27]
β_{DD}	Transmission rate in dogs	$1.58 \times 10^{-7} y^{-1}$	[27]
γ_D	Latency period in dogs	$(2.37/6)\, y^{-1}$	[27]
$\delta\varepsilon_D$	Rate of no clinical rabies	$0.4 y^{-1}$	[27]
B_H	Birth rate (humans)	$0.0314 y^{-1}$	[31]
β_{DH}	Transmission rate (dog-humans)	$2.29 \times 10^{-12} y^{-1}$	[27]
α_H	Loss of immunity (humans)	$1 y^{-1}$	[27]
m_H	Natural death rate (humans)	$0.0074 y^{-1}$	[31]
μ_H	Disease induced mortality (humans)	$1 y^{-1}$	[27]
ν_H	Preexposure prophylaxis for humans	$0.54 y^{-1}$	Assumed
ρ_H	Postexposure prophylaxis for humans	$0.1 y^{-1}$	[27]
γ_H	Latency rate (humans)	$(1/6)\, y^{-1}$	[27]
$\gamma_H \varepsilon_H$	Rate of no clinical rabies (humans)	$2.4 y^{-1}$	[27]

4.1.1. Herd Immunity Threshold H_1. Therefore, from the above numerical values, we are motivated to know the number of humans or dogs that should be vaccinated when $\mathscr{R}_0^* = 3.027$.

$$H_1 := 1 - \frac{1}{\mathscr{R}_0^*} = 0.66. \tag{49}$$

This shows that if $\mathscr{R}_0^* = 3.027$, then 66% of individuals and dogs should receive vaccination.

4.2. Sensitivity Analysis. To determine parameters that contribute most to the rabies transmission, we used two sensitivity analysis approach: the normalised forward sensitivity index as presented in [37] and the Latin hypercube sampling as described in [42]. To determine the dependence of parameters in $\mathscr{R}_0$, using a sampling size, $n = 1000$, the partial rank correction coefficients (PRCC) value of the ten parameters in $\mathscr{R}_0$ are shown in Figure 2(a). The longer the bar in Figure 2(a) suggests that the statistical influence of those parameters to changes in $\mathscr{R}_0$ is high. Also, using the normalised forward sensitivity index gives the following values and the nature of their signs in Table 2, based on the parameter value given in Table 1. The plus sign or minus sign signifies that the influence is positive or negative, respectively [42],

$$\begin{aligned}
\Gamma_{\mathscr{R}_0}^{\beta_{DD}} &= \frac{\partial \mathscr{R}_0}{\partial \beta_D} \frac{\beta_{DD}}{\mathscr{R}_0} = 1, \\
\Gamma_{\mathscr{R}_0}^{A_D} &= \frac{\partial \mathscr{R}_0}{\partial A_D} \frac{A_D}{\mathscr{R}_0} = 1, \\
\Gamma_{\mathscr{R}_0}^{\mu_D} &= \frac{\partial \mathscr{R}_0}{\partial \mu_D} \frac{\mu_D}{\mathscr{R}_0} = \frac{-\mu_D}{(m_D + \mu_D)} = -0.95, \\
\Gamma_{\mathscr{R}_0}^{\delta\varepsilon_D} &= \frac{\partial \mathscr{R}_0}{\partial \delta\varepsilon_D} \frac{\delta\varepsilon_D}{\mathscr{R}_0} \\
&= \frac{\delta\varepsilon_D}{((1-\rho_D)\,\delta\gamma_D - \delta\varepsilon_D - C_D - m_D - \rho_D)} \\
&= -1.61, \\
\Gamma_{\mathscr{R}_0}^{C_D} &= \frac{\partial \mathscr{R}_0}{\partial C_D} \frac{C_D}{\mathscr{R}_0} \\
&= \frac{C_D}{((1-\rho_D)\,\delta\gamma_D - \delta\varepsilon_D - C_D - m_D - \rho_D)} \\
&= -0.45, \\
\Gamma_{\mathscr{R}_0}^{\alpha_D} &= \frac{\partial \mathscr{R}_0}{\partial \alpha_D} \frac{\alpha_D}{\mathscr{R}_0} = 0.28, \\
\Gamma_{\mathscr{R}_0}^{m_D} &= \frac{\partial \mathscr{R}_0}{\partial m_D} \frac{m_D}{\mathscr{R}_0} = -1.64, \\
\Gamma_{\mathscr{R}_0}^{\delta\gamma_D} &= \frac{\partial \mathscr{R}_0}{\partial \delta\gamma_D} \frac{\delta\gamma_D}{\mathscr{R}_0} = 1.33, \\
\Gamma_{\mathscr{R}_0}^{\rho_D} &= \frac{\partial \mathscr{R}_0}{\partial \rho_D} \frac{\rho_D}{\mathscr{R}_0} = -0.5, \\
\Gamma_{\mathscr{R}_0}^{\nu_D} &= \frac{\partial \mathscr{R}_0}{\partial \nu_D} \frac{\nu_D}{\mathscr{R}_0} = -0.52.
\end{aligned} \tag{50}$$

Therefore, from Table 2 it shows that an addition or a reduction in the values of β_{DD}, α_D, $\delta\gamma_D$, and A_D will have an increase or a decrease in the spread of the rabies virus. For

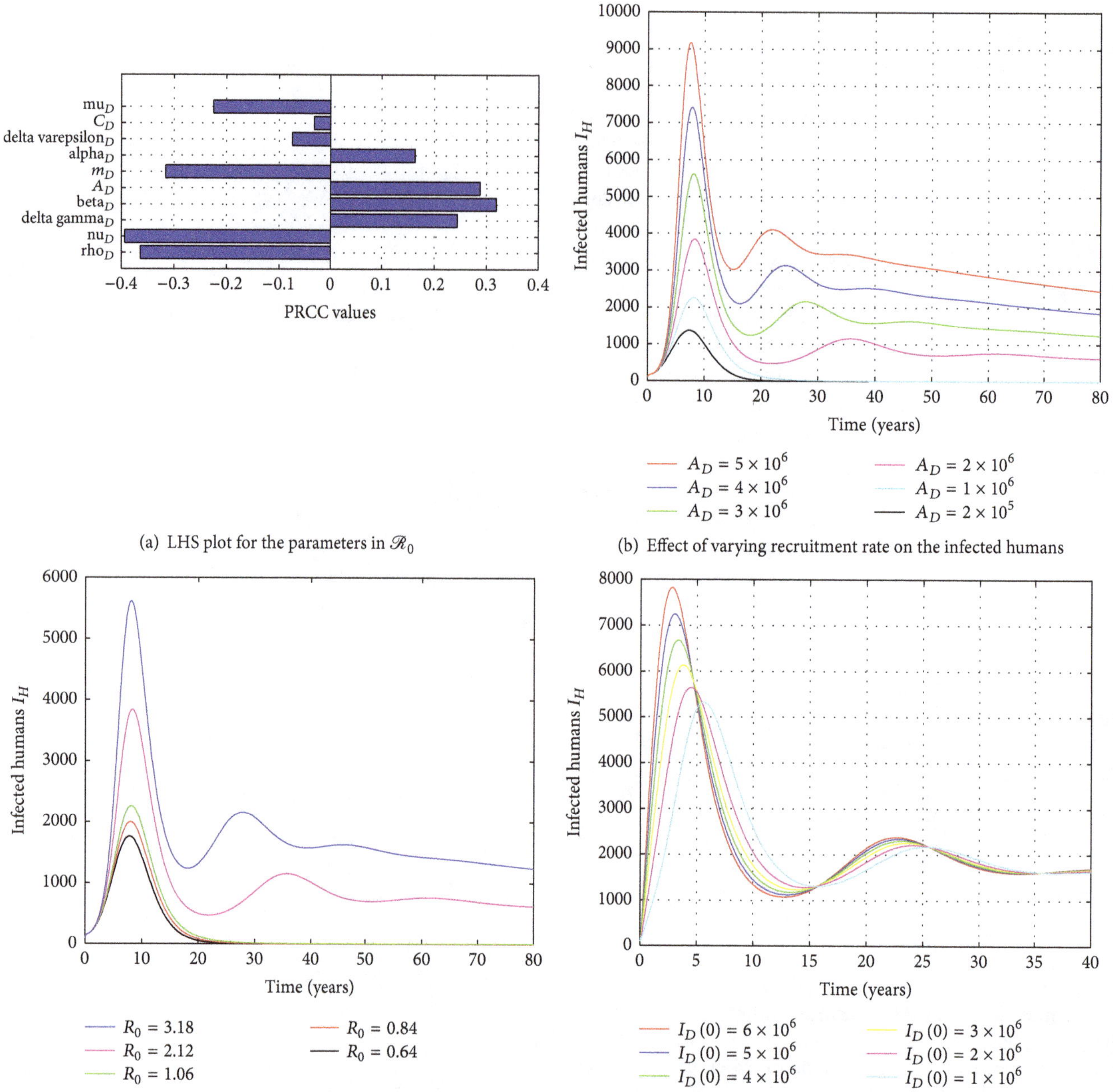

(a) LHS plot for the parameters in $\mathscr{R}_0$

(b) Effect of varying recruitment rate on the infected humans

(c) Effect of an increase in $\mathscr{R}_0$ on the infected humans

(d) Effect of varying the initial infected dog population size on the infected humans

FIGURE 2: The graphical representation of some parameters in $\mathscr{R}_0$ and the effect of varying some initial state values on the model.

example, $\Gamma_{\mathscr{R}_0}^{\beta_D} = 1$ indicates that increasing or reducing the transmission rate by 5% may increase or reduce the number of secondary infection by 5%. The negative sign in Table 2 will have a reduction in the basic reproduction number, $\mathscr{R}_0$, when the values of those parameters are increased, and a reduction in the values of ρ_D, ν_D, μ_D, m_D, and $\delta\varepsilon_D$ will lead to an increase in the number of secondary infections.

The Latin hypercube sampling (LHS) in Figure 2(a) shows that μ_D, C_D, α_D, and $\delta\gamma_D$ have a minimal influence on the rate at which the rabies virus is spread. The Latin hypercube sampling (LHS) plots for the ten parameters in $\mathscr{R}_0$ show that culling of exposed dogs does not actually minimize the spread of rabies as compared to vaccination of susceptible dogs. Figure 2(a) also shows that the most influential parameter in spreading the infection is β_{DD} followed by A_D. Figure 2(c) shows that an increase in the basic reproduction number will contribute to a high level of secondary infection in the human population. Similarly, Figure 2(a) shows that vaccination of dogs ν_D is the most effective way of controlling the rabies virus in the dog population as compared to the treatment/quarantine of exposed dogs, ρ_D. Figure 3(a) gives the contour nature of ν_D and ρ_D, which shows a more saturated effect on the basic reproduction number. Figure 3(b) shows that β_{DD} and α_D have a positive relation with the basic

TABLE 2: Sensitivity signs of $\mathscr{R}_0$ to the parameters in (24).

Parameter	Description	Sensitivity sign
β_{DD}	Transmission rate of dogs	+ve
A_D	Recruitment rate of dogs	+ve
μ_D	Disease induce death rate of dogs	−ve
$\delta\varepsilon_D$	Rate of no clinical rabies	−ve
C_D	Culling of exposed dogs	−ve
α_D	Loss of immunity in dogs	+ve
m_D	Natural death rate of dogs	−ve
$\delta\gamma_D$	Rate at which exposed dogs become infective (infective rate)	+ve
ρ_D	Postexposure prophylaxis (treatment/quarantined)	−ve
ν_D	Preexposure prophylaxis (vaccination)	−ve

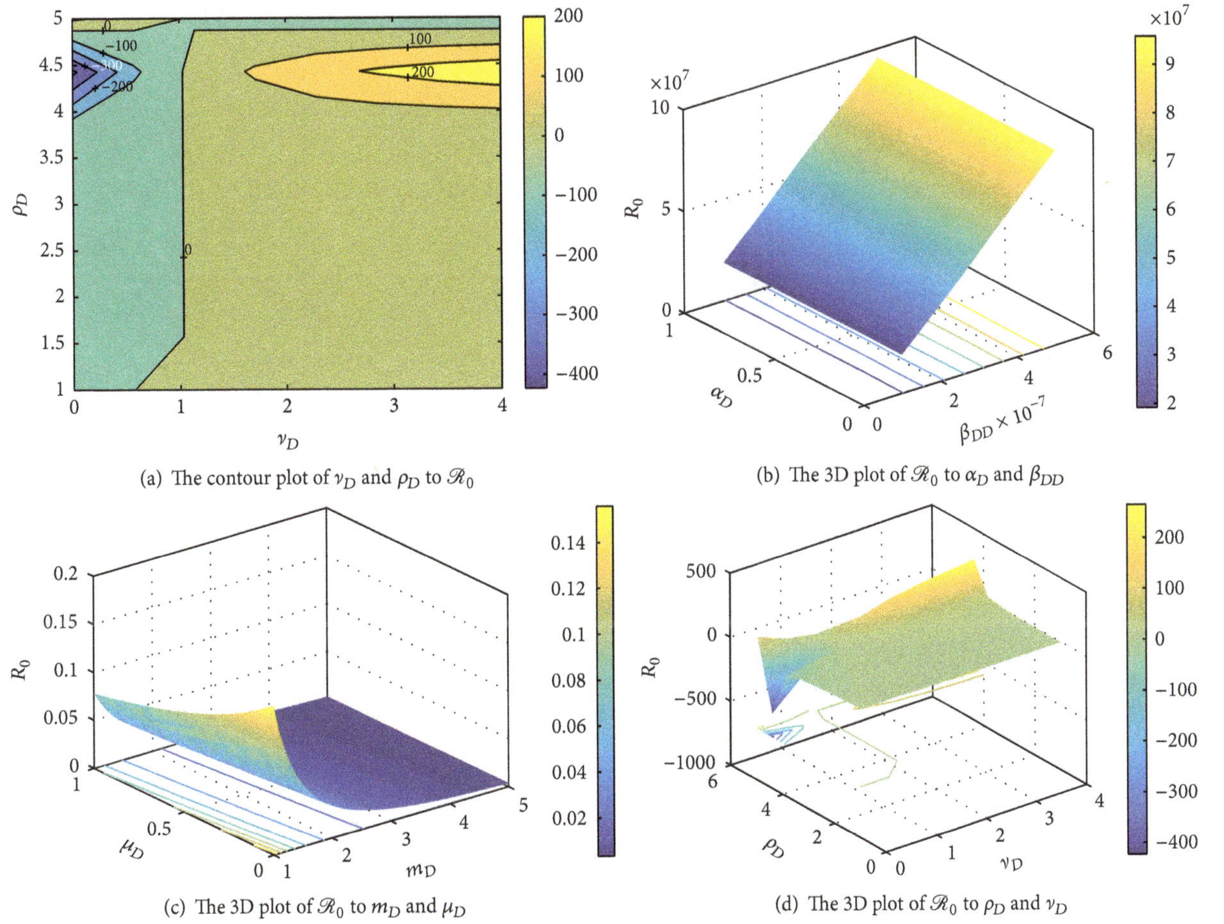

(a) The contour plot of ν_D and ρ_D to $\mathscr{R}_0$

(b) The 3D plot of $\mathscr{R}_0$ to α_D and β_{DD}

(c) The 3D plot of $\mathscr{R}_0$ to m_D and μ_D

(d) The 3D plot of $\mathscr{R}_0$ to ρ_D and ν_D

FIGURE 3: The graphical representation of some parameters in $\mathscr{R}_0$.

reproduction number $\mathscr{R}_0$. Therefore, an increase in β_{DD} and α_D will have a direct increase in the spread of the rabies virus. Figure 2(b) indicates that with a high number of recruitment of dogs into the susceptible dog's compartment will have a corresponding high increase in the number of infected humans. Figure 2(d) demonstrates that a high number of infected dogs in the compartment will lead to an increase in the number of infected humans. Figure 3(c) shows that a high increase in the number of disease induce death rate and natural death rate will have a negative reflection on $\mathscr{R}_0$; biologically, we would not recommend this approach in minimizing the spread of the disease, since an increase in both μ_D and m_D may result in a high rate of the disease in the human population, even though μ_D and m_D naturally reduce the number of susceptible and infected dogs in the population. Finally, Figure 3(d) shows the 3D plot of Figure 3(a).

5. Objective Functional

Given that $y(t) \in Y \in \mathbb{R}^n$ is a state variable of model system (1) and $u(t) \in U \in \mathbb{R}^n$ are the control variables at any time (t) with $t_{(0)} \leq t \leq t_{(f)}$, then an optimal control problem consists of finding a piecewise continuous control $u(t)$ and its corresponding state $y(t)$. This optimizes the cost functional $J[y(t), u(t)]$ using Pontryagin's maximum principle [43]. Therefore we set the following likelihood control strategies:

(1) $u_1 = \nu_D$ is the control effort aimed at increasing the immunity of susceptible dogs (preexposed prophylaxis).

(2) $u_2 = \rho_D$ is the control effort aimed at treating the exposed dogs (postexposed prophylaxis).

(3) $u_3 = \nu_H$ is the control effort aimed at increasing the immunity of susceptible humans (preexposure prophylaxis).

(4) $u_4 = \rho_H$ is the control effort aimed at treating the exposed humans (postexposed prophylaxis).

Our goal is to seek optimal controls such as ν_D^*, ρ_D^*, ν_H^*, and ρ_H^* that minimize the objective functional:

$$J = \min \int_{t_0}^{t_f} \left[A_1 E_D + A_2 E_H + A_3 I_D + A_4 I_H + \frac{B1}{2}\nu_D^2 + \frac{B2}{2}\rho_D^2 + \frac{B3}{2}\nu_H^2 + \frac{B4}{2}\rho_H^2 \right] dt. \quad (51)$$

Therefore, (51) is subject to

$$\begin{aligned}
\frac{dS_d}{dt} &= A_D - (1-\nu_D)\beta_{DD} S_D I_D - (m_D + \nu_D) S_D + \delta\varepsilon_D E_D + \alpha_D R_D, \\
\frac{dE_d}{dt} &= (1-\nu_D)\beta_{DD} S_D I_D - ((1-\rho_D)\delta\gamma_D + m_D + \rho_D + \delta\varepsilon_D + C_D) E_D, \\
\frac{dI_D}{dt} &= (1-\rho_D)\delta\gamma_D E_D - (m_D + \mu_D) I_D, \\
\frac{dR_D}{dt} &= \nu_D S_D + \rho_D E_D - (m_D + \alpha_D) R_D, \\
\frac{dS_H}{dt} &= B_H - (1-\nu_H)\beta_{DH} S_H I_d - (m_H + \nu_H) S_H + \delta_H \varepsilon_H E_H + \alpha_H R_H, \\
\frac{dE_H}{dt} &= (1-\nu_H)\beta_{DH} S_H I_d - ((1-\rho_H)\delta_H\gamma_H + m_H + \rho_H + \delta_H\varepsilon_H) E_H, \\
\frac{dI_H}{dt} &= (1-\rho_H)\delta_H\gamma_H E_H - (m_H + \mu_H) I_H, \\
\frac{dR_H}{dt} &= \nu_H S_H + \rho_H E_H - (m_H + \alpha_H) R_H,
\end{aligned} \quad (52)$$

$$S_D > 0,\ E_D \geq 0,\ I_D \geq 0,\ R_D \geq 0,\ S_H > 0,\ E_H \geq 0,\ I_H \geq 0,\ R_H \geq 0.$$

From (51) the quantities A_1 and A_2 denote the weight constants of the exposed classes and A_3 and A_4 are the weight of the infectious classes, respectively. $B1, B2, B3, B4$ are the weight constants for the dog and human controls. $B1\nu_D^2$, $B2\rho_D^2$, $B3\nu_H^2$, $B4\rho_H^2$ describe the cost associated with rabies vaccination and treatment. The square of the control variables shows the severity of the side effects of the vaccination and treatment. Employing Pontryagin's maximum principle, we form the Hamiltonian equation with state variables $S_D = S_D^*$, $E_D = E_D^*$, $I_D = I_D^*$, R_D^* and $S_H = S_H^*$, $E_H = E_H^*$, $I_H = I_H^*$, R_H^* as

$$\begin{aligned}
H &= A_1 E_D^* + A_2 E_H^* + A_3 I_D^* + A_4 I_H^* + \frac{B1}{2}\nu_D^2 + \frac{B2}{2} \\
&\cdot \rho_D^2 + \frac{B3}{2}\nu_H^2 + \frac{B4}{2}\rho_H^2 + \lambda_1 [A_D \\
&- (1-\nu_D)\beta_{DD} S_D^* I_D^* - (m_D + \nu_D) S_D^* + \delta\varepsilon E_D^*
\end{aligned}$$

$$+\alpha_D R_D^*] + \lambda_2 [(1-\nu_D)\beta_{DD} S_D^* I_D^*$$
$$- ((1-\rho_D)\delta\gamma_D + m_D + \rho_D + \delta\varepsilon_D + C_D) E_D^*]$$
$$+ \lambda_3 [(1-\rho_D)\delta\gamma_D E_D^* - (m_D+\mu_D) I_D^*]$$
$$+ \lambda_4 [\nu_D S_D^* + \rho_D E_D^* - (m_D + \alpha_D) R_D^*] + \lambda_5 [B_H$$
$$- (1-\nu_H)\beta_{DH} S_H^* I_D^* - (m_H + \nu_H) S_H^* + \delta_H \varepsilon_H E_H^*$$
$$+ \alpha_H R_H^*] + \lambda_6 [(1-\nu_H)\beta_{DH} S_H^* I_D^*$$
$$- ((1-\rho_H)\delta_H\gamma_H + m_H + \rho_H + \delta_H\varepsilon_H) E_H^*]$$
$$+ \lambda_7 [(1-\rho_H)\delta_H\gamma_H E_H^* - (m_H+\mu_H) I_H^*]$$
$$+ \lambda_8 [\nu_H S_H^* + \rho_H E_H^* - (m_H+\alpha_H) R_H^*]. \tag{53}$$

Considering the existence of adjoint functions λ_i, $i = 1, 2, \ldots, 8$, satisfying

$$\frac{d\lambda_1}{dt} = -\frac{\partial H}{\partial S_D^*}$$
$$= \lambda_1 ((1-\nu_D)\beta_{DD} I_D^* + m_D + \nu_D)$$
$$- \lambda_2 (1-\nu_D)\beta_{DD} I_D^* - \lambda_4 \nu_D,$$
$$\frac{d\lambda_2}{dt} = -\frac{\partial H}{\partial E_D^*}$$
$$= \lambda_2 ((1-\rho_D)\delta\gamma_D + m_D + \rho_D + \delta\varepsilon_D + C_D)$$
$$- \lambda_1 \delta\varepsilon_D - \lambda_3 (1-\rho_D)\delta\gamma_D - \lambda_4\rho_D - A_1,$$
$$\frac{d\lambda_3}{dt} = -\frac{\partial H}{\partial I_D^*}$$
$$= \lambda_3 (m_D+\mu_D) + \lambda_1 (1-\nu_D)\beta_{DD} S_D^*$$
$$+ \lambda_5 (1-\nu_H)\beta_{DH} S_H^* - \lambda_2 (1-\nu_D)\beta_D S_D^*$$
$$- \lambda_6 (1-\nu_H)\beta_{DH} S_H^* - A_3,$$
$$\frac{d\lambda_4}{dt} = -\frac{\partial H}{\partial R_D^*} = \lambda_4 (m_D+\alpha_D) - \lambda_1\alpha_D,$$
$$\frac{d\lambda_5}{dt} = -\frac{\partial H}{\partial S_H^*}$$
$$= \lambda_5 ((1-\nu_H)\beta_{dH} I_D^* + m_H + \nu_H)$$
$$- \lambda_6 (1-\nu_H)\beta_{DH} I_D^* - \lambda_8\nu_H,$$
$$\frac{d\lambda_6}{dt} = -\frac{\partial H}{\partial E_H^*}$$
$$= \lambda_6 ((1-\rho_H)\delta_H\gamma_H + m_H + \rho_H + \delta_H\varepsilon_H)$$
$$- \lambda_5\delta_H\varepsilon_H - \lambda_7 (1-\rho_H)\delta_H\gamma_H - \lambda_8\rho_H - A_2,$$
$$\frac{d\lambda_7}{dt} = -\frac{\partial H}{\partial I_H^*} = \lambda_7 (m_H+\mu_H) - A_4,$$
$$\frac{d\lambda_8}{dt} = -\frac{\partial H}{\partial R_H^*} = \lambda_8 (m_H+\alpha_H) - \lambda_5\alpha_H, \tag{54}$$

with transversality condition $\lambda_i(t_f) = 0$ for $i = 1, \ldots, 8$ for the control set u_i, hence we have

$$\frac{\partial H}{\partial u_i} = 0, \quad \text{where } i = 1,2,3,4,$$
$$\left.\frac{\partial H}{\partial \nu_D}\right|_{\nu_D=\nu_D^*} := B1\nu_D^* - \lambda_1 S_D^* + \lambda_4 S_D^* + \lambda_1\beta_{DD} S_D^* I_D^*$$
$$- \lambda_2\beta_D S_D^* I_D^* = 0,$$
$$\nu_D^* = \frac{(\lambda_1 S_D^* - \lambda_4 S_D^*) + (\lambda_2-\lambda_1)\beta_{DD} I_D^* S_D^*}{B1},$$
$$\left.\frac{\partial H}{\partial \rho_D}\right|_{\rho_D=\rho_D^*} := B2\rho_D^* - \lambda_1 E_D^* + \lambda_4 E_D^* + \lambda_2 E_D \delta\gamma_D E_D^*$$
$$- \lambda_3\delta\gamma_D E_D^* = 0,$$
$$\rho_D^* = \frac{(\lambda_2 E_D^* - \lambda_4 E_D^*) + (\lambda_3-\lambda_2)\delta\gamma_D E_D^*}{B2}, \tag{55}$$
$$\left.\frac{\partial H}{\partial \nu_H}\right|_{\nu_H=\nu_H^*} := B3\nu_H^* - \lambda_5 S_H + \lambda_8 S_H + \lambda_5\beta_{DH} S_H I_D$$
$$- \lambda_6\beta_{DH} S_H^* I_D^* = 0,$$
$$\nu_H^* = \frac{(\lambda_5 S_H^* - \lambda_8 S_H^*) + (\lambda_6-\lambda_5)\beta_{DH} S_H^* I_D^*}{B3},$$
$$\left.\frac{\partial H}{\partial \rho_H}\right|_{\rho_H=\rho_H^*} := B4\rho^* - \lambda_6 E_H^* + \lambda_8 E_H^* + \lambda_6\delta_H\gamma_H E_H^*$$
$$- \lambda_7\delta_H\gamma_H E_H^* = 0,$$
$$\rho_H^* = \frac{(\lambda_6 E_H^* - \lambda_8 E_H^*) + (\lambda_7-\lambda_6)\delta_H\gamma_H E_H^*}{B4}.$$

Now, using an appropriate variation argument and taking the bounds into account, the optimal control strategies are given as

$$\nu_D^* = \min\left\{\max\left(0, \frac{(\lambda_1-\lambda_4) S_D^* + (\lambda_2-\lambda_1)\beta_{DD} I_D^* S_D^*}{B1}\right), \nu_{D\max}\right\}, \tag{56}$$

$$\rho_D^* = \min\left\{\max\left(0, \frac{(\lambda_2-\lambda_4) E_D^* + (\lambda_3-\lambda_2)\delta\gamma_D E_D^*}{B2}\right), \rho_{D\max}\right\}, \tag{57}$$

$$\nu_H^* = \min\left\{\max\left(0, \frac{(\lambda_5 - \lambda_8) S_H^* + (\lambda_6 - \lambda_5) \beta_{DH} S_H^* I_D^*}{B3}\right), \nu_{H\max}\right\}, \tag{58}$$

$$\rho_H^* = \min\left\{\max\left(0, \frac{(\lambda_6 - \lambda_8) E_H^* + (\lambda_7 - \lambda_6) \delta_H \gamma_H E_H^*}{B4}\right), \rho_{H\max}\right\}. \tag{59}$$

Optimality System. Substituting the representation of the optimal vaccination and treatment control with corresponding adjoint function, we have the optimality system as

$$\begin{aligned}
\frac{dS_D}{dt} &= A_D - \left(1 - \min\left\{\max\left(0, \frac{(\lambda_1 - \lambda_4) S_D^* + (\lambda_2 - \lambda_1) \beta_{DD} I_D^* S_D^*}{B1}\right), \nu_{D\max}\right\}\right) \cdot \beta_{DD} S_D I_D - m_D S_D - \min\left\{\max\left(0, \frac{(\lambda_1 - \lambda_4) S_D^* + (\lambda_2 - \lambda_1) \beta_{DD} I_D^* S_D^*}{B1}\right), \nu_{D\max}\right\} S_D \\
&\quad + \delta \varepsilon_D E_D + \alpha_D R_D, \\
\frac{dE_D}{dt} &= \left(1 - \min\left\{\max\left(0, \frac{(\lambda_1 - \lambda_4) S_D^* + (\lambda_2 - \lambda_1) \beta_{DD} I_D^* S_D^*}{B1}\right), \nu_{D\max}\right\}\right) \cdot \beta_{DD} S_D I_D - \left(\left(1 - \min\left\{\max\left(0, \frac{(\lambda_2 - \lambda_4) E_D^* + (\lambda_3 - \lambda_2) \delta \gamma_D E_D^*}{B2}\right), \rho_{\max}\right\}\right) \delta \gamma_D + m_D + \delta \varepsilon_D + C_D\right) E_D \\
&\quad - \min\left\{\max\left(0, \frac{(\lambda_2 - \lambda_4) E_D^* + (\lambda_3 - \lambda_2) \delta \gamma_D E_D^*}{B2}\right), \rho_{D\max}\right\} E_D, \\
\frac{dI_D}{dt} &= \delta \gamma_D E_D - (m_D + \mu_D) I_D, \\
\frac{dR_D}{dt} &= \min\left\{\max\left(0, \frac{(\lambda_1 - \lambda_4) S_D^* + (\lambda_2 - \lambda_1) \beta_{DD} I_D^* S_D^*}{B1}\right), \nu_{D\max}\right\} S_D - (m_D + \alpha_D) R_D + \min\left\{\max\left(0, \frac{(\lambda_2 - \lambda_4) E_D^* + (\lambda_3 - \lambda_2) \delta \gamma_D E_D^*}{B2}\right), \rho_{D\max}\right\} E_D, \\
\frac{dS_H}{dt} &= B_H - \left(1 - \min\left\{\max\left(0, \frac{(\lambda_5 - \lambda_8) S_H^* + (\lambda_6 - \lambda_5) \beta_{DH} S_H^* I_D^*}{B3}\right), \nu_{H\max}\right\}\right) \cdot \beta_{DH} S_H I_D - m_H S_H - \min\left\{\max\left(0, \frac{(\lambda_5 - \lambda_8) S_H^* + (\lambda_6 - \lambda_5) \beta_{DH} S_H^* I_D^*}{B3}\right), \nu_{H\max}\right\} S_H \\
&\quad + \delta_H \varepsilon_H E_H + \alpha_H R_H, \\
\frac{dE_H}{dt} &= \left(1 - \min\left\{\max\left(0, \frac{(\lambda_6 - \lambda_8) E_H^* + (\lambda_7 - \lambda_6) \delta_H \gamma_H E_H^*}{B4}\right), \rho_{H\max}\right\}\right) \cdot \beta_{DH} S_H I_D - (\delta_H \gamma_H + m_H + \delta_H \varepsilon_H) E_H \\
&\quad - \min\left\{\max\left(0, \frac{(\lambda_6 - \lambda_8) E_H^* + (\lambda_7 - \lambda_6) \delta_H \gamma_H E_H^*}{B4}\right), \rho_{H\max}\right\} E_H, \\
\frac{dI_H}{dt} &= \delta_H \gamma_H E_H - (m_H + \mu_H) I_H, \\
\frac{dR_H}{dt} &= \min\left\{\max\left(0, \frac{(\lambda_5 - \lambda_8) S_H^* + (\lambda_6 - \lambda_5) \beta_{DH} S_H^* I_D^*}{B3}\right), \nu_{H\max}\right\} S_H - (m_H + \alpha_H) R_H + \min\left\{\max\left(0, \frac{(\lambda_6 - \lambda_8) E_H^* + (\lambda_7 - \lambda_6) \delta_H \gamma_H E_H^*}{B4}\right), \rho_{H\max}\right\} E_H, \\
&\frac{d\lambda_1}{dt}, \frac{d\lambda_2}{dt}, \frac{d\lambda_3}{dt}, \frac{d\lambda_4}{dt}, \frac{d\lambda_5}{dt}, \frac{d\lambda_6}{dt}, \frac{d\lambda_7}{dt}, \frac{d\lambda_8}{dt}, \\
&\text{with, } \lambda_i(t_f) = 0, \quad i = 1, 2, 3, 4, 5, 6, 7, 8.
\end{aligned} \tag{60}$$

5.1. Numerical Simulations of the Optimality System. To determine the control strategies ν_D, ρ_D, ν_H, and ρ_H, as given in the objective functional, we began an iteration of the model until convergence is achieved. The results of the simulation of the control strategies are displayed below. We consider equal weights of $(A_1 = 1,\ A_2 = 1,\ A_3 = 1,\ A_4 = 1)$ for both exposed and infected classes. We varied the cost associated with the objective functional, which indicate that, with low cost of vaccination, the rate at which individuals will seek for vaccination of their susceptible dogs will increase, and this could result in low transmission of rabies in a heterogeneous population. We consider the various cost of preexposure prophylaxis and postexposure prophylaxis to be ($B1 = 1$,

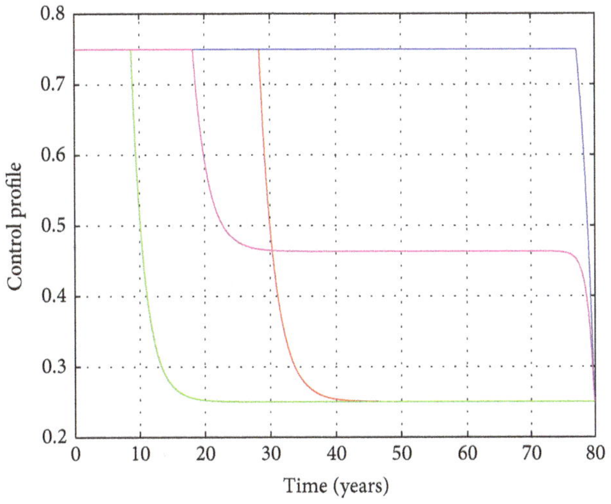

FIGURE 4: The simulation effect of the controls.

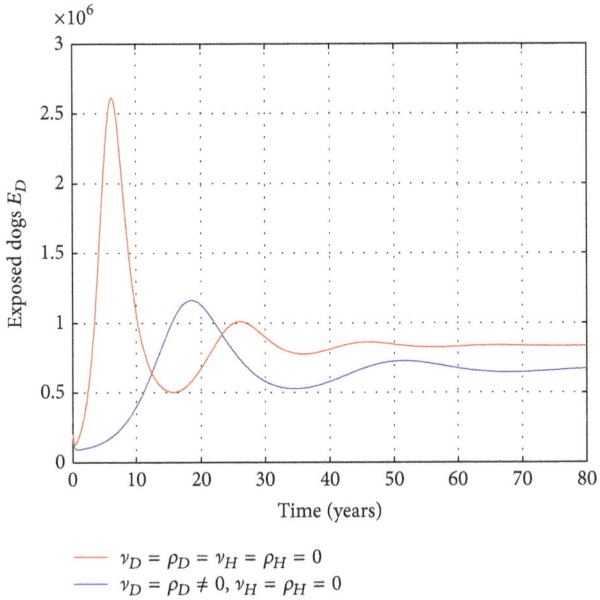

(a) A plot of $\nu_D = \rho_D \neq 0$ and $\nu_H = \rho_H = 0$

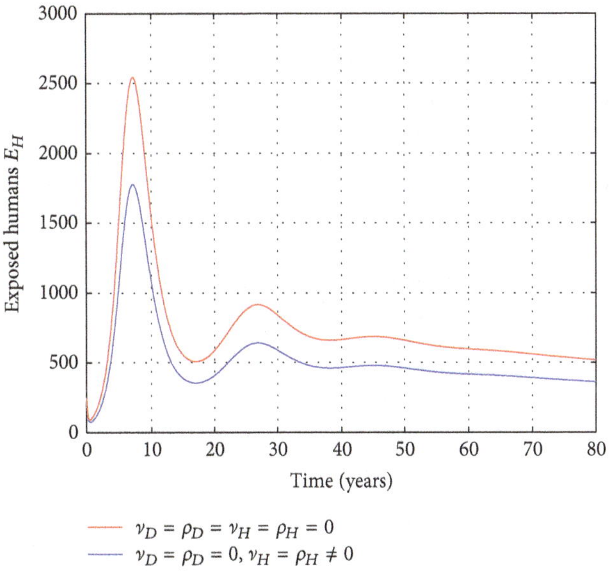

(b) A plot of $\nu_D = \rho_D = 0$ and $\nu_H = \rho_H \neq 0$

FIGURE 5: The trajectories of the model with and without pre- and postexposure prophylaxis on exposed humans and that of the exposed dogs.

$B2 = 4$, $B3 = 1$, $B4 = 4$). We found that the optimal time in controlling the infection using preexposure prophylaxis in dogs is much better than using postexposure prophylaxis in dogs, as shown by the trajectories of the red line and blue line in Figure 4, respectively. The blue line in Figure 4 indicates that applying postexposure prophylaxis will considerably take a longer time in controlling of rabies in dogs. The green line in Figure 4 signifies that preexposure prophylaxis in humans increases the immunity levels of humans and hence reduces the rate at which individuals move to the infected stage. Figures 5 and 6 show the effect of using only one control strategy on the model. Therefore, Figure 5(a) shows that applying only postexposure prophylaxis (treatment or quarantine) of dogs has a low positive impact on the model. Figure 5(b) shows that sticking to the use of pre- and postexposure prophylaxis in human without administering pre- and postexposure

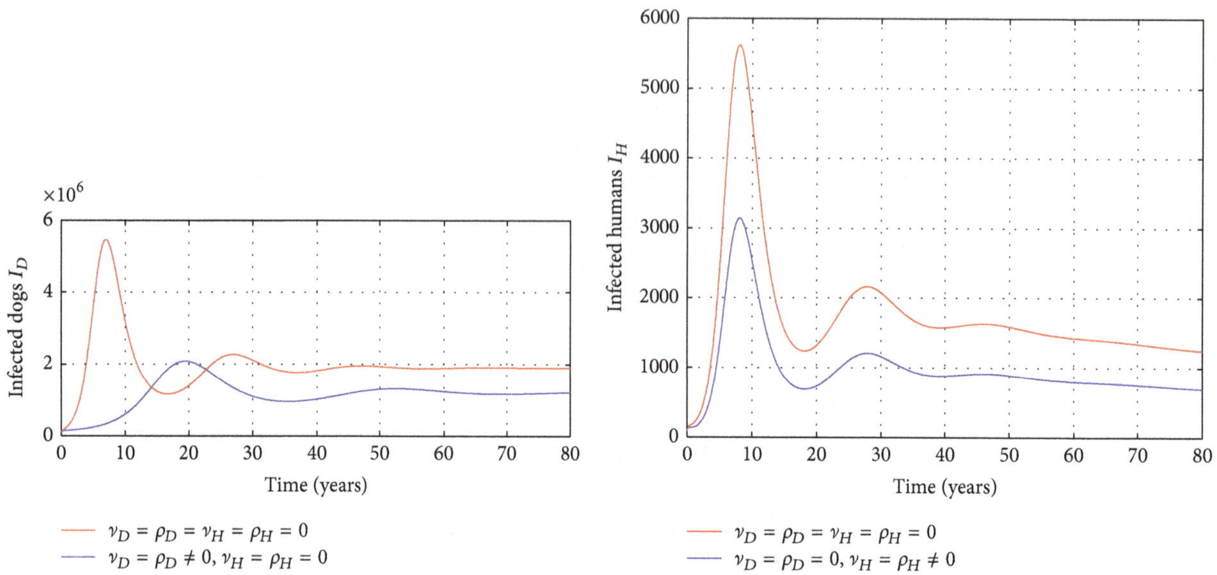

(a) A plot of $\nu_D = \rho_D \neq 0$ and $\nu_H = \rho_H = 0$

(b) A plot of $\nu = \rho = 0$ and $\nu_H = \rho_H \neq 0$

FIGURE 6: The trajectories of the model with and without pre- and postexposure prophylaxis on infected humans and that of the infected dogs.

prophylaxis in the dog population will result in a high of the rabies infection in the human population. Figure 6(a) also shows that combining pre- and postexposure prophylaxis (vaccination and treatment/quarantine) in the dog compartment will reduce the spread of the rabies virus, thereby reducing the using of pre- and postexposure prophylaxis (vaccination and treatment) in humans. Figure 6(b) indicates that a rapid use of pre- and postexposure prophylaxis in the human population will reduce the number of rabies deaths in the human population. Figure 7 shows the simulation effects of applying both controls on the model. Figure 7(a) shows that, with the use of the optimal control strategies, the rate of the infection in the susceptible dogs will reduce significantly. Figures 7(b) and 7(c) show that there is a proportional decrease in the number of exposed and infected dogs when the control measures are applied. Similarly, Figures 7(e) and 7(f) show a significant decrease in the number of infected and exposed humans when the control measures are applied. Figure 7(d) shows that there is a proportional increase in the number of recovered dogs when the control measures are applied. Finally, Figures 8(a)–8(h) show the simulation effect of corresponding adjoint functions.

6. Discussion

The numerical simulations of the resulting optimality system show that, during the case where it is more expensive to vaccinate than treatment, more resources should be invested in treating affected individuals until the disease prevalence begins to fall. This option, however, does not reduce the number of individuals expose to the disease quickly enough, thus resulting in an overall increase in the infected human population. On the other hand, if it is more expensive to treat than to vaccinate, then more susceptible dogs should be vaccinated, so as to lower the rate at which newborn dogs get infected. Nevertheless, in the case where both measures are equally expensive, the simulation shows that the optimal way to drive the epidemic towards eradication within any specified period is to use more preexposure prophylaxis in both compartments.

7. Conclusion

We studied an optimal control model of rabies transmission dynamics in dogs and the best way of reducing death rate of rabies in humans. The stability analysis shows that the disease-free equilibrium is locally and globally asymptotically stable. We also obtained an optimal control solution for the model which predicts that the optimal way of eliminating deaths from canine rabies as projected by the global alliance for rabies control [30] is using more of preexposure prophylaxis in both dogs and humans and public education; however, the results show that the effective and optimal consideration of preexposure prophylaxis and postexposure prophylaxis in humans without an optimal use of vaccination in the dog population is not beneficial if total elimination of the disease is desirable in Africa and Asia. Any combination strategy which involves vaccination in the dogs' population gives a better result and hence it may be beneficial in eliminating the disease in Asia, Africa, and Latin America.

Disclosure

The authors fully acknowledge that this paper was developed as a result of the first author's thesis work submitted to the Department of Mathematics, Kwame Nkrumah University of

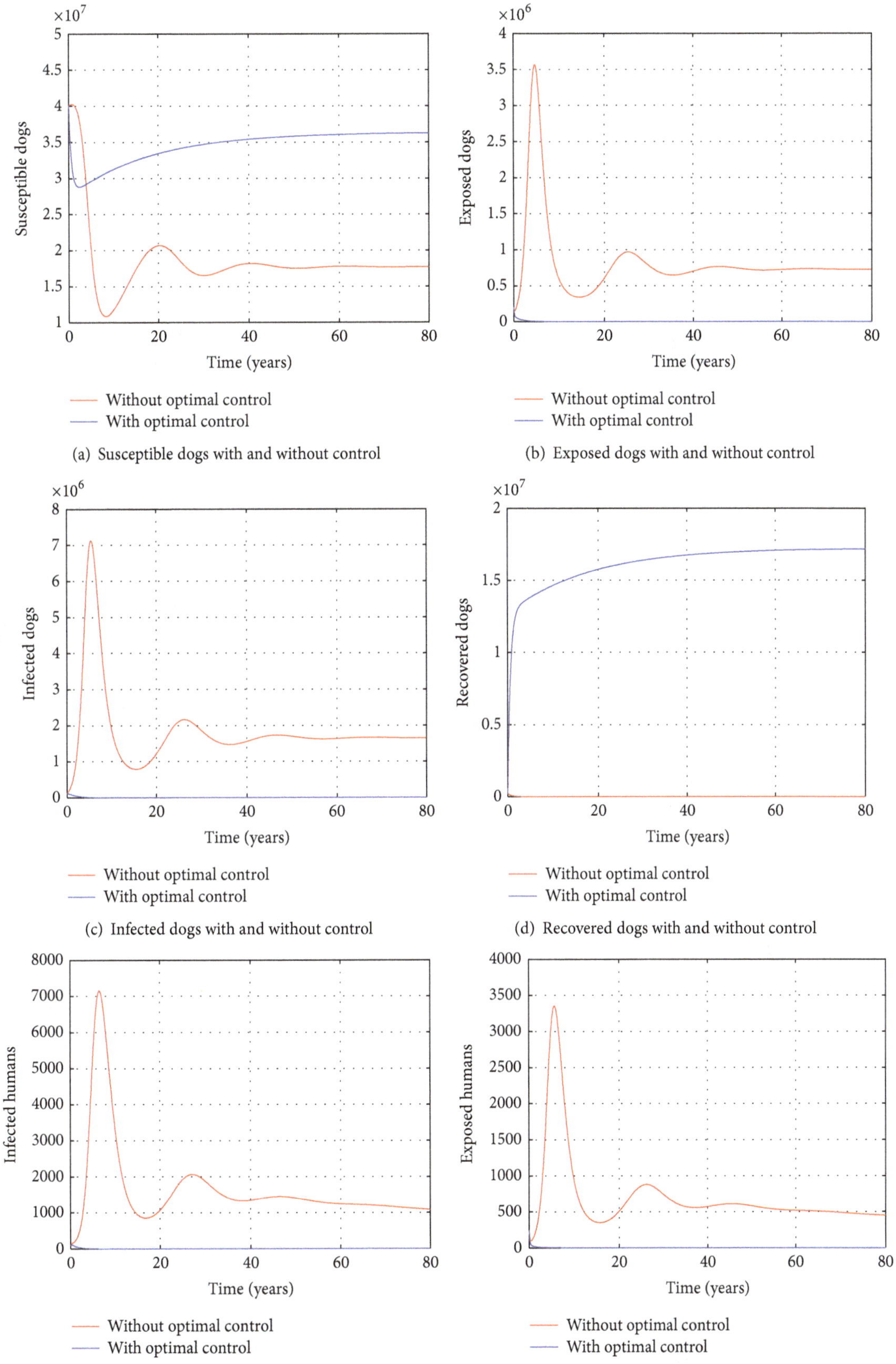

(a) Susceptible dogs with and without control

(b) Exposed dogs with and without control

(c) Infected dogs with and without control

(d) Recovered dogs with and without control

(e) Infected humans with and without control

(f) Exposed humans with and without control

FIGURE 7: The trajectories of the model with and without optimal control on individual compartments.

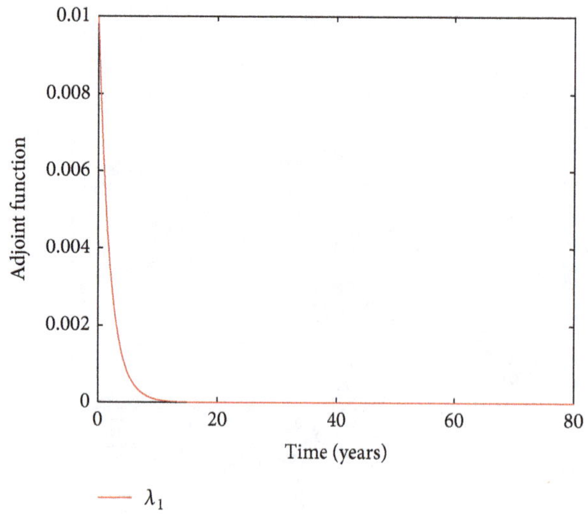

(a) The cost function λ_1 for $A_1 = A_2 = A_3 = A_4 = 1$ and $B1 = 1$, $B2 = 4$, $B3 = 1$, $B4 = 4$

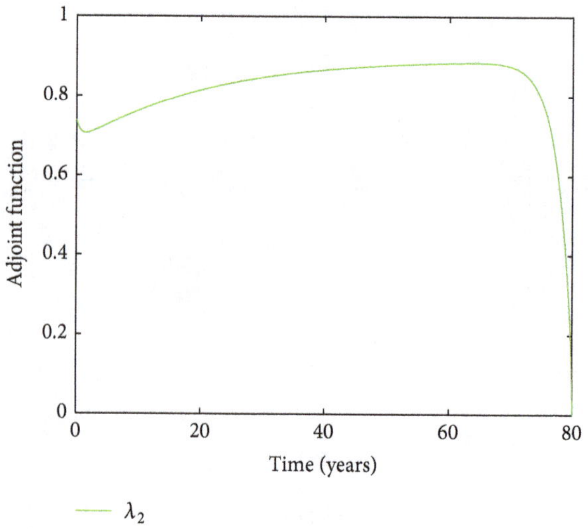

(b) The cost function λ_2 for $A_1 = A_2 = A_3 = A_4 = 1$ and $B1 = 1$, $B2 = 4$, $B3 = 1$, $B4 = 4$

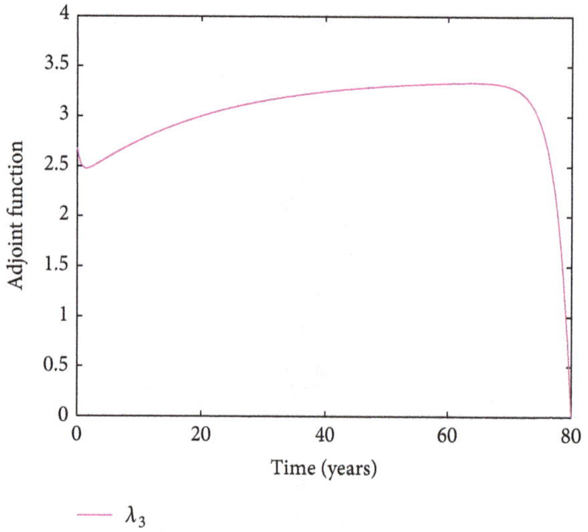

(c) The cost function λ_3 for $A_1 = A_2 = A_3 = A_4 = 1$ and $B1 = 1$, $B2 = 4$, $B3 = 1$, $B4 = 4$

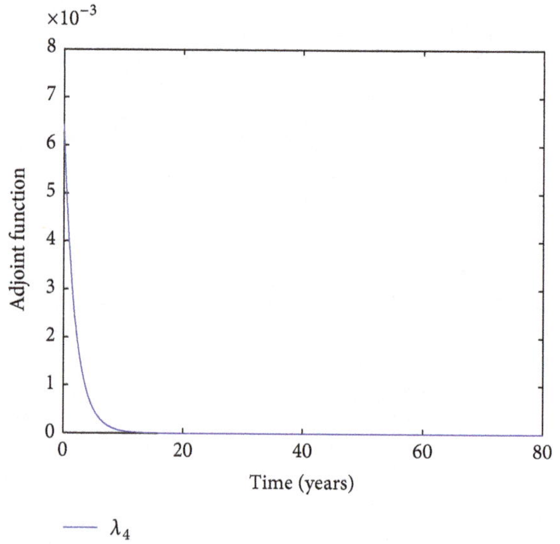

(d) The cost function λ_4 for $A_1 = A_2 = A_3 = A_4 = 1$ and $B1 = 1$, $B2 = 4$, $B3 = 1$, $B4 = 4$

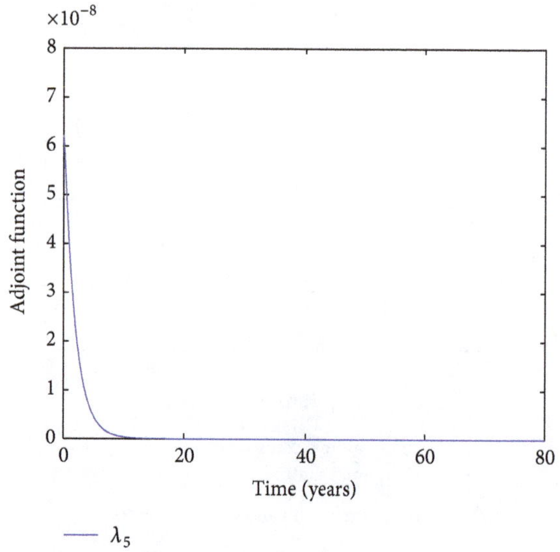

(e) The cost function λ_5 for $A_1 = A_2 = A_3 = A_4 = 1$ and $B1 = 1$, $B2 = 4$, $B3 = 1$, $B4 = 4$

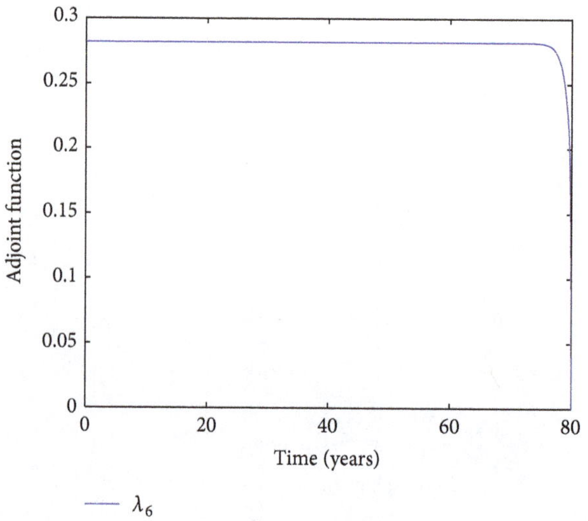

(f) The cost function λ_6 for $A_1 = A_2 = A_3 = A_4 = 1$ and $B1 = 1$, $B2 = 4$, $B3 = 1$, $B4 = 4$

FIGURE 8: Continued.

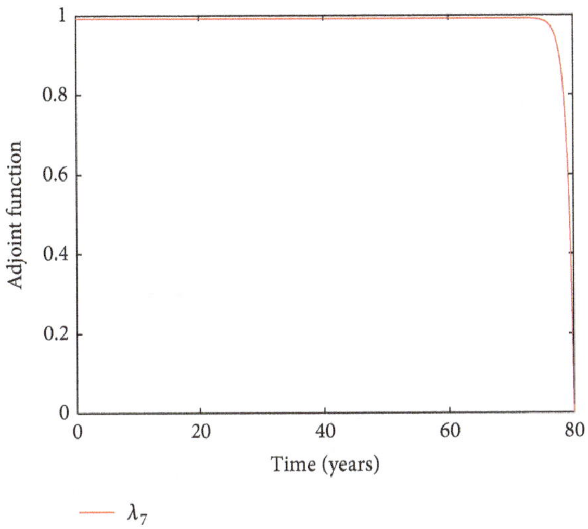

(g) The cost function λ_7 for $A_1 = A_2 = A_3 = A_4 = 1$ and $B1 = 1$, $B2 = 4$, $B3 = 1$, $B4 = 4$

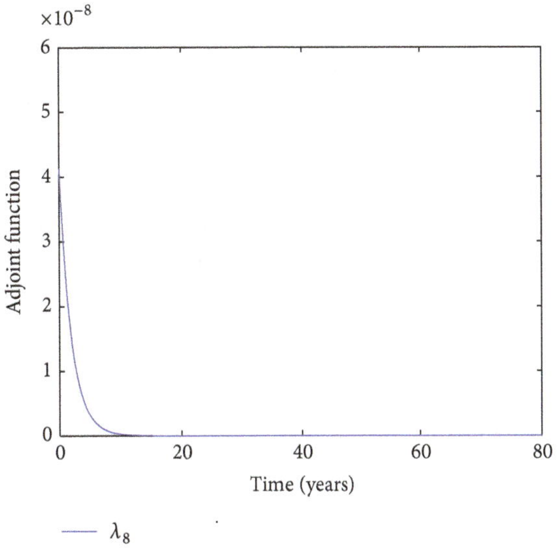

(h) The cost function λ_8 for $A_1 = A_2 = A_3 = A_4 = 1$ and $B1 = 1$, $B2 = 4$, $B3 = 1$, $B4 = 4$

FIGURE 8: The trajectories of the model with and without optimal control on individual compartments and corresponding adjoint function.

Science and Technology (see http://ir.knust.edu.gh/xmlui/handle/123456789/10053).

Conflicts of Interest

The authors declare that they have no conflicts of interest.

Acknowledgments

The first author's work (thesis) was partly supported by the Government of Ghana Annual Research Grant for Postgraduate Studies.

References

[1] D. T. S. Hayman, N. Johnson, D. L. Horton et al., "Evolutionary history of rabies in Ghana," *PLoS Neglected Tropical Diseases*, vol. 5, no. 4, Article ID e1001, 2011.

[2] "Rabies fact sheet," 2016, http://www.who.int/mediacentre/factsheets/fs099/en/.

[3] C. E. Rupprecht, D. Briggs, C. M. Brown, R. Franka, S. L. Katz, H. D. Kerr et al., *Use of A Reduced (4-Dose) Vaccine Schedule for Postexposure Prophylaxis to Prevent Human Rabies: Recommendations of The Advisory Committee on Immunization Practices*, Department of Health and Human Services, Centers for Disease Control and Prevention, 2010.

[4] R. L. M. Neilan, *Optimal control applied to population and disease models [P.hD. thesis]*, University of Tennessee, Knoxville, Tenn, USA, 2009.

[5] A. O. Isere, J. E. Osemwenkhae, and D. Okuonghae, "Optimal control model for the outbreak of cholera in Nigeria," *African Journal of Mathematics and Computer Science Research*, vol. 7, no. 2, pp. 24–30, 2014.

[6] H. S. Rodrigues, "Optimal control and numerical optimization applied to epidemiological models," 2014, https://arxiv.org/abs/1401.7390?context=math.

[7] A. A. Lashari, *Mathematical Modeling and Optimal Control of A Vector*, National University of Modern Languages, Islamabad, Pakistan, 2012.

[8] A. Abdelrazec, *Modeling, Dynamics and Optimal Control of West Nile Virus with Seasonality*, York University, Toronto, Canada, 2014.

[9] D. Greenhalgh, "Age-structured models and optimal control in mathematical equidemiology: a survey," 2010.

[10] A. Stashko, *The Effects of Prevention and Treatment Interventions in A Microeconomic Model of HIV Transmission*, Duke University, Durham, NC, USA, 2012.

[11] B. Seidu and O. D. Makinde, "Optimal control of HIV/AIDS in the workplace in the presence of careless individuals," *Computational and Mathematical Methods in Medicine*, vol. 2014, Article ID 831506, 19 pages, 2014.

[12] S. D. D. Njankou and F. Nyabadza, "An optimal control model for Ebola virus disease," *Journal of Biological Systems*, vol. 24, no. 1, pp. 29–49, 2016.

[13] M. Aubert, "Costs and benefits of rabies control in wildlife in France," *Revue Scientifique et Technique de l'OIE*, vol. 18, no. 2, pp. 533–543, 1999.

[14] R. M. Anderson and R. M. May, "Population biology of infectious diseases," in *Proceedings of the Dahlem Workshop*, Springer, Berlin, Germany, March 1982.

[15] G. Bohrer, S. Shem-Tov, E. Summer, K. Or, and D. Saltz, "The effectiveness of various rabies spatial vaccination patterns in a simulated host population with clumped distribution," *Ecological Modelling*, vol. 152, no. 2-3, pp. 205–211, 2002.

[16] S. A. Levin, T. G. Hallam, and L. J. Gross, Eds., *Applied Mathematical Ecology*, vol. 18 of *Biomathematics*, Springer, Berlin, Germany, 2012.

[17] M. J. Coyne, G. Smith, and F. E. McAllister, "Mathematic model for the population biology of rabies in raccoons in the mid-Atlantic states," *American journal of veterinary research*, vol. 50, no. 12, pp. 2148–2154, 1989.

[18] J. E. Childs, A. T. Curns, M. E. Dey et al., "Predicting the local dynamics of epizootic rabies among raccoons in the United

States," *Proceedings of the National Academy of Sciences of the United States of America*, vol. 97, no. 25, pp. 13666–13671, 2000.

[19] K. Hampson, J. Dushoff, J. Bingham, G. Brückner, Y. H. Ali, and A. Dobson, "Synchronous cycles of domestic dog rabies in sub-Saharan Africa and the impact of control efforts," *Proceedings of the National Academy of Sciences of the United States of America*, vol. 104, no. 18, pp. 7717–7722, 2007.

[20] M. J. Carroll, A. Singer, G. C. Smith, D. P. Cowan, and G. Massei, "The use of immunocontraception to improve rabies eradication in urban dog populations," *Wildlife Research*, vol. 37, no. 8, pp. 676–687, 2010.

[21] X. Wang and J. Lou, "Two dynamic models about rabies between dogs and human," *Journal of Biological Systems*, vol. 16, no. 4, pp. 519–529, 2008.

[22] W. Yang and J. Lou, "The dynamics of an interactional model of rabies transmitted between human and dogs," *Bollettino della Unione Matematica Italiana. Serie 9*, vol. 2, no. 3, pp. 591–605, 2009.

[23] J. Zinsstag, S. Dürr, M. A. Penny et al., "Transmission dynamics and economics of rabies control in dogs and humans in an African city," *Proceedings of the National Academy of Sciences of the United States of America*, vol. 106, no. 35, pp. 14996–15001, 2009.

[24] W. Ding, L. J. Gross, K. Langston, S. Lenhart, and L. A. Real, "Rabies in raccoons: optimal control for a discrete time model on a spatial grid," *Journal of Biological Dynamics*, vol. 1, no. 4, pp. 379–393, 2007.

[25] G. C. Smith and C. L. Cheeseman, "A mathematical model for the control of diseases in wildlife populations: culling, vaccination and fertility control," *Ecological Modelling*, vol. 150, no. 1-2, pp. 45–53, 2002.

[26] J. M. Tchuenche and C. T. Bauch, "Can culling to prevent monkeypox infection be counter-productive? Scenarios from a theoretical model," *Journal of Biological Systems*, vol. 20, no. 3, pp. 259–283, 2012.

[27] J. Zhang, Z. Jin, G.-Q. Sun, T. Zhou, and S. Ruan, "Analysis of rabies in China: transmission dynamics and control," *PLoS ONE*, vol. 6, no. 7, p. e20891, 2011.

[28] O. A. Grace, "Modelling of the spread of rabies with pre-exposure vaccination of humans," *Mathematical Theory and Modeling*, vol. 4, no. 8, 2014.

[29] E. D. Wiraningsih, F. Agusto, L. Aryati et al., "Stability analysis of rabies model with vaccination effect and culling in dogs," *Applied Mathematical Sciences*, vol. 9, no. 77-80, pp. 3805–3817, 2015.

[30] "Global alliance for rabies control," 2016, https://rabiesalliance.org/.

[31] CIA world factbook and other sources, 2016, http://www.theodora.com/wfbcurrent/ghana/ghana_people.html.

[32] G. Birkhoff and G. Rota, *Ordinary differential equations*, Ginn & Co, New York, NY, USA, 1962.

[33] V. Lakshmikantham, S. Leela, and S. Kaul, "Comparison principle for impulsive differential equations with variable times and stability theory," *Nonlinear Analysis. Theory, Methods & Applications*, vol. 22, no. 4, pp. 499–503, 1994.

[34] O. Diekmann, J. A. Heesterbeek, and J. A. Metz, "On the definition and the computation of the basic reproduction ratio R_0 in models for infectious diseases in heterogeneous populations," *Journal of Mathematical Biology*, vol. 28, no. 4, pp. 365–382, 1990.

[35] E. C. Pielou, *An introduction to Mathematical Ecology*, Wiley-InterScience, Hoboken, NJ, USA, 1969.

[36] R. C. May, "Stable limit cycles in pre-predator population. A comment," *Science*, p. 1074, 1973.

[37] M. Martcheva, *An Introduction to Mathematical Epidemiology*, vol. 61 of *Texts in Applied Mathematics*, Springer, New York, NY, USA, 2015.

[38] T. T. Yusuf and F. Benyah, "Optimal control of vaccination and treatment for an SIR epidemiological model," *World Journal of Modelling and Simulation*, vol. 8, no. 3, pp. 194–204, 2012.

[39] S. D. Hove-Musekwa, F. Nyabadza, and H. Mambili-Mamboundou, "Modelling hospitalization, home-based care, and individual withdrawal for people living with HIV/AIDS in high prevalence settings," *Bulletin of Mathematical Biology*, vol. 73, no. 12, pp. 2888–2915, 2011.

[40] C. V. De León, "Constructions of Lyapunov functions for classics SIS, SIR and SIRS epidemic model with variable population size," *Foro-Red-Mat: Revista Electrónica De Contenido Matemático*, vol. 26, no. 5, article 1, 2009.

[41] J. LaSalle, "The stability of dynamical systems," in *Proceedings of the CBMS-NSF Regional Conference Series in Applied Mathematics 25*, SIAM, Philadelphia, Pa, USA, 1976.

[42] T. Zhang, K. Wang, X. Zhang, and Z. Jin, "Modeling and analyzing the transmission dynamics of HBV epidemic in Xinjiang, China," *PLoS ONE*, vol. 10, no. 9, Article ID e138765, 2015.

[43] O. Sharomi and T. Malik, "Optimal control in epidemiology," *Annals of Operations Research*, vol. 251, no. 1-2, pp. 55–71, 2017.

5

Viscosity Solution of Mean-Variance Portfolio Selection of a Jump Markov Process with No-Shorting Constraints

Moussa Kounta

The College of the Bahamas, School of Mathematics, Physics and Technology, P.O. Box 4912, Nassau, Bahamas

Correspondence should be addressed to Moussa Kounta; moussakounta@gmail.com

Academic Editor: Jinde Cao

We consider the so-called mean-variance portfolio selection problem in continuous time under the constraint that the short-selling of stocks is prohibited where all the market coefficients are random processes. In this situation the Hamilton-Jacobi-Bellman (HJB) equation of the value function of the auxiliary problem becomes a coupled system of backward stochastic partial differential equation. In fact, the value function V often does not have the smoothness properties needed to interpret it as a solution to the dynamic programming partial differential equation in the usual (classical) sense; however, in such cases V can be interpreted as a viscosity solution. Here we show the unicity of the viscosity solution and we see that the optimal and the value functions are piecewise linear functions based on some Riccati differential equations. In particular we solve the open problem posed by Li and Zhou and Zhou and Yin.

1. Introduction

The mean-variance approach proposed in 1952 by the Nobel prize winning economist Markowitz [1] has become the foundation of modern finance by discovering the static mean-variance portfolio selection formulation in a market in which shorting is not allowed. This theory has inspired numerous extensions and applications. For instance, Li and Ng [2] and Zhou and Li [3] successfully extended the unconstrained mean-variance portfolio selection formulation to the multi-period setting. Zhou and Yin [4] consider the mean-variance portfolio selection problem in continuous time where the market parameters including the bank interest rate and the appreciation and volatility rates of the stocks depend on the market mode that switches among a finite number of states where random regime switching is assumed to be independent of the underlying Brownian motion. This essentially renders the underlying market incomplete. A Markov chain modulated diffusion formulation is employed to model the problem and Zhou and Yin [4] use the techniques of stochastic linear quadratic (LQ) control to derive mean-variance efficient portfolios and efficient frontier based on solutions of two systems of linear ordinary differential equations.

After Li and Ng published [2], Markowitz suggested that one of them extends the results to the dynamic mean-variance formulation with no-shorting constraint and proposed a conjecture of a piecewise quadratic value function for such a situation. Influenced by Markowitz's comments, Li et al. [5] formulated the LQ control problem by constraining the control portfolio to take nonnegative values due to the no-shorting restriction on the market mode (not random processes). They derived the optimal portfolio policy for the continuous-time mean-variance model with no-shorting constraint using the duality method [6].

However, there are several interesting problems that deserve further investigation; for instance, Li et al. [5] open a problem by stating in their conclusion that "an immediate open problem is to extend the results in this paper to the case where all the market coefficients are random processes." In this paper we solve this problem.

By making use of the techniques of LQ control, we see that, in an attempt to pursue the method of dynamic programming in the auxiliary problem, the value function which is a generalized solution to the Hamilton-Jacobi equation coupled is not smooth enough to satisfy the dynamic programming equations in the classical or usual sense.

A difficulty with the concept of generalized solution is that the dynamic programming together with the boundary data typically has many generalized solutions. Among them, there is one provided by Crandall and Lions [7], called the viscosity solution, which is the natural generalized solution. This unique viscosity solution turns out to coincide with the value function V [8]. The central component of our solution to the problem of Li et al. [5] is the proof of the unicity of the viscosity solution of the value function of the auxiliary problem, which we establish by adapting the techniques of [9]. By making use of the duality method, we also derive a solution for efficient portfolio. The value function of the auxiliary problem depends on a set of Riccati differential equations and we use the Magnus approach to provide the solution. A work is in progress to develop numerical implementation. This will be subject of a future publication.

2. Viscosity Solutions for Weakly Coupled Systems of Second-Order Hamilton-Jacobi-Bellman Equation

2.1. Notation. We make use of the following notations:

(i) $(\Omega, \mathcal{F}, P)$: a fixed probability space on which we defined standard n-dimensional Brownian motion $W(t) \equiv (W_1(t), \ldots, W_n(t))'$ and continuous-time stationary Markov chain $\alpha(t)$ taking value in a finite state space $\mathcal{M} = \{1, 2, \ldots, m\}$ such that $W(t)$ and $\alpha(t)$ are independent of each other. The Markov chain has a generator $Q = (q_{ij})_{m\times m}$ and stationary transition probabilities:

$$p_{ij}(t) = p(\alpha(t) = j \mid \alpha(0) = i), \qquad t \ge 0,\ i, j = 1, 2, \ldots, m. \tag{1}$$

(ii) Define $\mathcal{F}_t = \sigma\{(W(s), \alpha(s)) : 0 \le s \le t\}$.

(iii) $\mathcal{B}(\Sigma) = \sigma$-algebra of Borel sets of Σ.

(iv) Consider the following:

$$E_{tx}\Phi(x(s)) = \int_\Sigma \Phi(y)\widehat{P}(t, x, s, dy), \quad t < s.$$
$$\widehat{P}(t, x, s, B) = P(x(s) \in B \mid x(t) = x), \qquad \forall B \in \mathcal{B}(\Sigma). \tag{2}$$

(v) Hilbert space $\mathcal{H}$ with the norm $\|\cdot\|_{\mathcal{H}}$: define the Banach space

$$L^2_{\mathcal{F}}(0, T; \mathcal{H}) = \Big\{\varphi(\cdot) \mid \varphi(\cdot) \text{ is an } \mathcal{F}_t\text{-adapted, } \mathcal{H}\text{-valued measurable process on } [a, b], \ E\int_a^b \|\varphi(t,\omega)\|_{\mathcal{H}}\, dt < \infty\Big\}, \tag{3}$$

with norm

$$\|\varphi(\cdot)\|_{\mathcal{F},2} = E\int_a^b \|\varphi(t,\omega)\|_{\mathcal{H}}\, dt < \infty. \tag{4}$$

(vi) M': the transpose of any vector or matrix.

(vii) M_j: the jth component of any vector M; we will use indifferently this notation $M(t, x, u, i) \equiv M_i(t, x, u)$.

(viii) $C([0, T]; X)$: the Banach space of X-valued continuous functions on $[0, T]$.

(ix) $C^2([0, T] \times \mathbb{R}^n)$: the space of all twice continuously differentiable functions on $[0, T] \times \mathbb{R}^n$.

(x) Consider $D_x = \partial(\cdot)/\partial x$, $D_x^2 = \partial^2(\cdot)/\partial x^2$, $D^\alpha(\cdot) = \partial^{|\alpha|}(\cdot)/\partial x_1^{\alpha_1} \cdots \partial x_n^{\alpha_n}$, and $\dot{f}(t) = (d/dt)f(t)$, $\partial_t = \partial/\partial t$.

(xi) Consider

$$\mathcal{Q} = [0, T] \times \mathcal{D}, \quad \mathcal{D} \subset \mathbb{R}. \tag{5}$$

(xii) Kronecker delta symbol:

$$\delta_{ij}(t) = \begin{cases} 0 & \text{if } i \ne j \\ 1 & \text{if } i = j. \end{cases} \tag{6}$$

(xiii) $[A, B] \equiv AB - BA$ (Lie bracket), A, B matrices with appropriate dimension.

(xiv) Consider $W^{k,p}(\Omega) = \{u \in L^p(\Omega) : D^\alpha u \in L^p(\Omega),\ 1 \le p \le \infty,\ \forall |\alpha| \le k\}$.

(xv) $C_b(\Sigma)$ is bounded function in Σ.

(xvi) If F is a real-valued function on a set U which has a minimum on U, then

$$\operatorname*{argmin}_{v\in U} F(v) = \{v^* \in U : F(v^*) \le F(v),\ \forall v \in U\}. \tag{7}$$

3. Notion of Viscosity Solution

We consider the following coupled system of backward PDEs:

$$\partial_t V_i(t, x) + \inf_{u\ge 0}\Big\{\frac{1}{2}\overline{g}_i(t, x(t), u)\overline{g}_i(t, x(t), u)' D_x^2 V_i(t, x) + \overline{f}_i(t, x(t), u) D_x V_i(t, x)\Big\} - \sum_{j\ne i} q_{ij}\big[V_i(t, x) - V_j(t, x)\big] = 0,$$
$$V_i(T, x) = g(x) \qquad (t, x) \in \mathcal{Q} \text{ where } u(t) = (u_1(t), \ldots, u_m(t))', \tag{8}$$

and the conditions on matrix $(q_{ij})_{1\le i,j\le n}$ are

$$q_{kl} > 0, \quad \text{for } k \ne l, \ q_{kk} < 0,$$
$$\sum_{l=1}^m q_{kl} = 0 \quad \text{for } k = 1, \ldots, m. \tag{9}$$

We suppose

$$\overline{f}_i(t, x(t); u),\ \overline{g}_i(t, x(t), u) \in W^{2,\infty}(\mathcal{Q}). \tag{10}$$

Under appropriate regularity assumptions on $\partial\mathcal{Q}$ and the coefficients, we define and prove existence and uniqueness results of the viscosity solutions to (8).

3.1. Viscosity Solution Definition. It is well known that (8) does not in general have classical smooth solutions. We define a generalized concept of solution called a viscosity solution [7].

Definition 1 ($w \in (C_b^2(\mathcal{Q}))^m$). w is a viscosity subsolution (supersolution) of system (8), if, for all $\phi_i \in C^2(\mathcal{Q})$,

$$\begin{aligned}&\partial_t\phi_i(t_0, x_0)\\&+\inf_{u\geq 0}\left\{\frac{1}{2}\overline{g}_i(t_0, x_0, u)\overline{g}_i(t_0, x_0, u)' D_x^2\phi_i(t_0, x_0)\right.\\&\left.+\overline{f}_i(t_0, x_0, u) D_x\phi_i(t_0, x_0)\right\} - \sum_{j\neq i} q_{ij}\left[w_i(t_0, x_0)\right.\\&\left.- w_j(t_0, x_0)\right] \geq 0,\end{aligned} \tag{11}$$

$$\begin{aligned}&\partial_t\phi_i(t_0, x_0)\\&+\inf_{u\geq 0}\left\{\frac{1}{2}\overline{g}_i(t_0, x_0, u)\overline{g}_i(t_0, x_0, u)' D_x^2\phi_i(t_0, x_0)\right.\\&\left.+\overline{f}_i(t_0, x_0, u) D_x\phi_i(t_0, x_0)\right\} - \sum_{j\neq i} q_{ij}\left[w_i(t_0, x_0)\right.\\&\left.- w_j(t_0, x_0)\right] \leq 0,\end{aligned} \tag{12}$$

respectively, whenever $w_i - \phi_i$ has a local maximum (minimum) at $(t_0, x_0) \in \mathcal{Q}$; w is a viscosity solution if it is both a viscosity subsolution and supersolution.

3.2. Uniqueness Result. Next, we can let

(i)

$$\begin{aligned}&\mathcal{H}_i(t, x, p, A)\\&= \inf_{u\geq 0}\left\{\frac{1}{2}\overline{g}_i(t, x, u)\overline{g}_i(t, x, u)' A + \overline{f}_i(t, x, u) p\right\}\\&= -\sup_{u\geq 0}\left\{-\frac{1}{2}\overline{g}_i(t, x, u)\overline{g}_i(t, x, u)' A - \overline{f}_i(t, x, u) p\right\},\end{aligned} \tag{13}$$

and we assume

(ii)

(a) $|(\overline{f}_i)_t| + |(\overline{f}_i)_x| \leq C$, $|(\overline{g}_i)_t| + |(\overline{g}_i)_x| \leq C$;
(b) $|\overline{f}_i(t, x, u)| \leq C(1 + |x| + |u|)$;
(c) $|\overline{g}_i(t, x, u)| \leq C(1 + |x| + |u|)$.

Lemma 2 (see [8], let $\mathcal{H}_i$ be as in (13)). *Assume ((ii)(a)–(c)).*

Then, there exists a continuous function $\omega : [0,\infty) \to [0,\infty)$ that satisfies $\omega(0) = 0$ such that

$$\begin{aligned}&\mathcal{H}_i(t, y, \beta(x - y), B) - \mathcal{H}_i(t, x, \beta(x - y), A)\\&\leq \omega\left(\beta|x - y|^2 + |x - y|\right),\end{aligned} \tag{14}$$

for every $(t, x), (t, y) \in \mathcal{Q}$, $\beta > 0$, and symmetric matrices A, B satisfying

$$-3\beta\begin{pmatrix} I & 0\\ 0 & I\end{pmatrix} \leq \begin{pmatrix} A & 0\\ 0 & -B\end{pmatrix} \leq 3\beta\begin{pmatrix} I & -I\\ -I & I\end{pmatrix}, \tag{15}$$

where I is the identity matrix with appropriate dimension.

Proposition 3. *Suppose assumptions (9) and (10) hold and w_k is a viscosity subsolution of (8) and v_k is a supersolution of (8).*

If $w_k(t, x) \leq v_k(t, x)$ on $\partial\mathcal{Q}$, $k = 1, \ldots, m$, then $w_k(t, x) \leq v_k(t, x)$ on $\mathcal{Q}$, $k = 1, \ldots, m$.

Proof. Suppose that there does not exist an index, s and $(l, z) \in \mathcal{Q}$, such that

$$(w_s - v_s)(l, z) = \max_{x,t,k}\{(w_k - v_k)(t, x)\} > 0. \tag{16}$$

(i) If $(l, z) \in \partial\mathcal{Q}$, we are done.

(ii) Assume $(l, z) \in \mathcal{Q}$; let

$$\begin{aligned}\phi_r^\varepsilon(t, x, y) &= w_k(t, x) - v_k(t, y) - \frac{1}{\varepsilon^2}|x - y|^2\\&\quad - \varepsilon|t - l|^2.\end{aligned} \tag{17}$$

There exists an index r and $(t_0, x_0, y_0) = (t_0^\varepsilon, x_0^\varepsilon, y_0^\varepsilon)$ such that

$$\phi_r^\varepsilon(t_0, x_0, y_0) = \max_{x,t,y,k}\{\phi_k^\varepsilon(t, x, y)\}. \tag{18}$$

We now show

$$w_r(t_0, x_0) - v_r(t_0, y_0) \geq 0. \tag{19}$$

But

$$\phi_s^\varepsilon(l, z, z) = w_s(l, z) - v_s(l, z) \leq \phi_r^\varepsilon(t_0, x_0, y_0) \tag{20}$$

implies

$$\begin{aligned}&\frac{1}{\varepsilon^2}|x_0 - y_0|^2 + \varepsilon|t_0 - l|^2\\&\leq w_r(t_0, x_0) - v_r(t_0, y_0) - (w_s(l, z) - v_s(l, z)).\end{aligned} \tag{21}$$

Since $w_s(l, z) - v_s(l, z) > 0$,

$$\begin{aligned}0 &\leq \frac{1}{\varepsilon^2}|x_0 - y_0|^2 + \varepsilon|t_0 - l|^2\\&\leq w_r(t_0, x_0) - v_r(t_0, y_0) - (w_s(l, z) - v_s(l, z))\\&\leq w_r(t_0, x_0) - v_r(t_0, y_0).\end{aligned} \tag{22}$$

(i) Since w is a viscosity subsolution of (8) and the function

$$\Psi_{r,1}^{\varepsilon} : (x,t) \longmapsto \phi_r^{\varepsilon}(t,x,y_0) \tag{23}$$

has a maximum at (t_0, x_0), set

$$\varphi_{r,1}^{\varepsilon}(t,x) = v_r(t,y_0) + \frac{1}{\varepsilon^2}|x-y_0|^2 + \varepsilon|t-l|^2, \tag{24}$$

then $w_r(t,x) - \varphi_{r,1}^{\varepsilon}(t,x)$, has a maximum at (t_0, x_0), and hence

$$\begin{gathered}
\left.\frac{\partial\left(w_r - \varphi_{r,1}^{\varepsilon}\right)}{\partial t}\right|_{(t_0,x_0)} = \left.\frac{\partial\left(w_r - \varphi_{r,1}^{\varepsilon}\right)}{\partial x}\right|_{(t_0,x_0)} = 0 \Longrightarrow \\
\left.\frac{\partial \varphi_{r,1}^{\varepsilon}}{\partial t}\right|_{(t_0,x_0)} = \left.\frac{\partial w_r}{\partial t}\right|_{(t_0,x_0)}, \\
\left.\frac{\partial \varphi_{r,1}^{\varepsilon}}{\partial x}\right|_{(t_0,x_0)} = \left.\frac{\partial w_r}{\partial x}\right|_{(t_0,x_0)}, \\
\left.\frac{\partial \Psi_{r,1}^{\varepsilon}}{\partial x}\right|_{(t_0,x_0)} = 0 \Longleftrightarrow \\
\left.\frac{\partial w_r}{\partial x}\right|_{(t_0,x_0)} - \frac{2}{\varepsilon^2}(x_0 - y_0) = 0, \\
\left.\frac{\partial \Psi_{r,1}^{\varepsilon}}{\partial t}\right|_{(t_0,x_0)} = 0 \Longleftrightarrow \\
\left.\frac{\partial w_r}{\partial t}\right|_{(t_0,x_0)} - \left.\frac{\partial v_r(t_0,x_0)}{\partial t}\right|_{(t_0,x_0)} - 2\varepsilon(t_0 - l) = 0,
\end{gathered} \tag{25}$$

$$\begin{aligned}
0 &\le \frac{\partial \varphi_{r,1}^{\varepsilon}(t_0,x_0)}{\partial t} \\
&\quad - \mathscr{H}_r\left(t_0, x_0, \frac{2}{\varepsilon^2}(x_0 - y_0), D_x^2\varphi_{r,1}^{\varepsilon}(t_0,x_0)\right) \\
&\quad - \sum_{k\neq r} q_{rk}\left[w_r(t_0,x_0) - w_k(t_0,x_0)\right].
\end{aligned} \tag{26}$$

(ii) Since v is a viscosity supersolution of (8) and the function

$$\Psi_{r,2}^{\varepsilon} : (x,t) \longmapsto -\phi_r^{\varepsilon}(t,x_0,y) \tag{27}$$

has a minimum at (t_0, y_0), set

$$\varphi_{r,2}^{\varepsilon}(t,y) = w_r(t,x_0) - \frac{1}{\varepsilon^2}|x_0-y|^2 - \varepsilon|t-l|^2 \tag{28}$$

then $\varphi_{r,2}^{\varepsilon}(t,x) - v_r(t,x)$, has a maximum at (t_0, x_0), and hence

$$\begin{gathered}
\left.\frac{\partial\left(\varphi_{r,2}^{\varepsilon} - v_r\right)}{\partial t}\right|_{(t_0,y_0)} = \left.\frac{\partial\left(\varphi_{r,2}^{\varepsilon} - v_r\right)}{\partial x}\right|_{(t_0,y_0)} = 0 \Longrightarrow \\
\left.\frac{\partial \varphi_{r,2}\varepsilon}{\partial t}\right|_{(t_0,y_0)} = \left.\frac{\partial v_r}{\partial t}\right|_{(t_0,y_0)}, \\
\left.\frac{\partial \varphi_{r,2}^{\varepsilon}}{\partial x}\right|_{(t_0,y_0)} = \left.\frac{\partial v_r}{\partial x}\right|_{(t_0,y_0)}, \\
\left.\frac{\partial \Psi_{r,2}^{\varepsilon}}{\partial y}\right|_{(t_0,y_0)} = 0 \Longleftrightarrow \\
+\left.\frac{\partial v_r}{\partial y}\right|_{(t_0,y_0)} - \frac{2}{\varepsilon^2}(x_0 - y_0) = 0, \\
\left.\frac{\partial \Psi_{r,2}^{\varepsilon}}{\partial t}\right|_{(t_0,x_0)} = 0 \Longleftrightarrow \\
-\left.\frac{\partial w_r}{\partial t}\right|_{(t_0,x_0)} + \left.\frac{\partial v_r(t_0,x_0)}{\partial t}\right|_{(t_0,x_0)} + 2\varepsilon(t_0 - l) = 0,
\end{gathered} \tag{29}$$

$$\begin{aligned}
&\frac{\partial \varphi_{r,2}^{\varepsilon}(t_0,y_0)}{\partial t} \\
&\quad - \mathscr{H}_r\left(t_0, x_0, \frac{2}{\varepsilon^2}(x_0 - y_0), D_y^2\varphi_{r,2}^{\varepsilon}(t_0,y_0)\right) \\
&\quad - \sum_{k\neq r} q_{rk}\left[v_r(t_0,y_0) - v_k(t_0,y_0)\right] \le 0.
\end{aligned} \tag{30}$$

By combining (26) and (30),

$$\begin{aligned}
0 \le{}& \frac{\partial w_r(t_0,x_0)}{\partial t} - \frac{\partial v_r(t_0,y_0)}{\partial t} \\
&+ \mathscr{H}_r\left(t_0, y_0, \frac{2}{\varepsilon^2}(x_0 - y_0), D_y^2\varphi_{r,2}^{\varepsilon}(t_0,y_0)\right) \\
&- \mathscr{H}_r\left(t_0, x_0, \frac{2}{\varepsilon^2}(x_0 - y_0), D_x^2\varphi_{r,1}^{\varepsilon}(t_0,x_0)\right) \\
&- \sum_{k\neq r} q_{rk}\left[w_r(t_0,x_0) - w_k(t_0,x_0)\right] \\
&+ \sum_{k\neq r} q_{rk}\left[v_r(t_0,y_0) - v_k(t_0,y_0)\right],
\end{aligned} \tag{31}$$

and by Lemma 2

$$\begin{aligned}
0 \le{}& 2\varepsilon|t_0 - l| + \omega\left(\frac{2}{\varepsilon^2}|x_0 - y_0|^2 + |x_0 - y_0|\right) \\
&- \sum_{k\neq r} q_{rk}\left[w_r(t_0,x_0) - w_k(t_0,x_0)\right] \\
&+ \sum_{k\neq r} q_{rk}\left[v_r(t_0,y_0) - v_k(t_0,y_0)\right],
\end{aligned} \tag{32}$$

we obtain

$$0 \le 2\varepsilon (t_0 - l) + \omega\left(\frac{2}{\varepsilon^2}|x_0 - y_0|^2 + |x_0 - y_0|\right) - \sum_{k\neq r} q_{rk}\left[w_r(t_0, x_0) - v_r(t_0, y_0)\right] + \sum_{k\neq r} q_{rk}\left[w_k(t_0, x_0) - v_k(t_0, y_0)\right]. \quad (33)$$

We have $-q_{rk} < 0$ and $w_r(t_0, x_0) - v_r(t_0, y_0) > 0$ (19) and hence

$$0 \le 2\varepsilon (t_0 - l) + \omega\left(\frac{2}{\varepsilon^2}|x_0 - y_0|^2 + |x_0 - y_0|\right) + \sum_{k\neq r} q_{rk}\left[w_k(t_0, x_0) - v_k(t_0, y_0)\right]. \quad (34)$$

Since $w_k(t_0, x_0) - v_k(t_0, y_0) \le w_r(t_0, x_0) - v_r(t_0, y_0)$ and $\sum_{k\neq r} q_{rk} = -q_{kk}$, we obtain

$$0 \le 2\varepsilon (t_0 - l) + \omega\left(\frac{2}{\varepsilon^2}|x_0 - y_0|^2 + |x_0 - y_0|\right) - q_{kk}\left[w_r(t_0, x_0) - v_r(t_0, y_0)\right]. \quad (35)$$

Thus,

$$\begin{aligned} w_s(l, z) - v_s(l, z) &\le \phi_r^{\varepsilon}(t_0, x_0, y_0) \\ &\le w_r(t_0, x_0) - v_r(t_0, y_0) \\ &\le \frac{1}{q_{kk}} 2\varepsilon |t_0 - l| \\ &\quad + \frac{1}{q_{kk}}\omega\left(\frac{2}{\varepsilon^2}|x_0 - y_0|^2 + |x_0 - y_0|\right). \end{aligned} \quad (36)$$

To finish the proof, we need to show

$$\omega\left(\frac{2}{\varepsilon^2}|x_0 - y_0|^2 + |x_0 - y_0|\right) \longrightarrow 0, \quad \text{as } \varepsilon \longrightarrow 0. \quad (37)$$

Let

$$h(q) = \operatorname{Sup}\left\{|v_r(t, x) - v_r(t, y)| : t \in [0, T],\ (x, y) \in \overline{\mathcal{Q}}\ |x - y|^2 \le q\right\}, \quad (38)$$

so that for any (t, x) and $(t, y) \in \overline{\mathcal{Q}}$

$$|v_r(t, x) - v_r(t, y)| \le h\left(|x - y|^2\right). \quad (39)$$

Since (x_0, t_0, y_0, r) maximizes ϕ_r^{ε} over $\mathcal{Q}$,

$$\begin{aligned} &w_r(t_0, x_0) - v_r(t_0, y_0) - \frac{1}{\varepsilon^2}|x_0 - y_0|^2 - \varepsilon |t_0 - l|^2 \\ &\ge w_r(t_0, x_0) - v_r(t_0, x_0) - \frac{1}{\varepsilon^2}|x_0 - x_0|^2 \\ &\quad - \varepsilon |t_0 - l|^2. \end{aligned} \quad (40)$$

We obtain

$$\frac{1}{\varepsilon^2}|x_0 - y_0|^2 \le v_r(t_0, x_0) - v_r(t_0, y_0) \le h\left(|x_0 - y_0|^2\right). \quad (41)$$

Since h is bounded by some constant K, this implies that

$$\frac{2}{\varepsilon^2}|x_0 - y_0|^2 \le K. \quad (42)$$

The definition of h yields

$$\frac{2}{\varepsilon^2}|x_0 - y_0|^2 + |x_0 - y_0| \le h\left(\frac{K\varepsilon^2}{2}\right) + \varepsilon\sqrt{\frac{K}{2}}, \quad (43)$$

and we obtain

$$\omega\left(\frac{2}{\varepsilon^2}|x_0 - y_0|^2 + |x_0 - y_0|\right) \le \omega\left(h\left(\frac{K\varepsilon^2}{2}\right) + \varepsilon\sqrt{\frac{K}{2}}\right), \quad (44)$$

and we obtain

$$w_s(l, z) - v_s(l, z) \le 0 \quad \text{as } \varepsilon \longrightarrow 0, \quad (45)$$

which is a contradiction to (16).

Corollary 4. *The viscosity solution satisfying the boundary condition is unique.*

Proof. If v_k^1 and v_k^2 are 2 viscosity solutions such that $v_k^1 = v_k^2$ on $\partial\mathcal{Q}$, then

(i) $v_k^1 \le v_k^2$ on $\partial\mathcal{Q} \Rightarrow v_k^1 \le v_k^2$ on $\mathcal{Q}$ accordingly (Proposition 3);

(ii) $v_k^1 \ge v_k^2$ on $\partial\mathcal{Q} \Rightarrow v_k^1 \ge v_k^2$ on $\mathcal{Q}$ accordingly (Proposition 3).

Hence $v_k^1 = v_k^2$ on $\mathcal{Q}$.

4. Application in Finance: Continuous-Time Mean-Variance Model without Shorting where the Market Parameters Are Random

We now briefly recall the results of the continuous-time mean-variance model without shorting [5] and the mean-variance portfolio selection problem in continuous time where the market parameters are random processes [4].

We study the intersection of the both cases [4, 5], that is, continuous-time mean-variance model without shorting where the market parameters are random.

Consider a market in which $n + 1$ assets are traded continuously on a finite time horizon $[0, T]$. One of the assets is a bank account whose price $P_0(t)$ is subject to the stochastic ODE (ordinary differential equation)

$$\begin{aligned} dP_0(t) &= r(t, \alpha(t)) P_0(t)\, dt, \quad t \in [0, T] \\ P_0(0) &= p_0 > 0, \quad t \in [0, T],\ \alpha(t) = i \in \mathbb{N}, \end{aligned} \quad (46)$$

where $r(t,i) \geq 0$, $i = 1,2,\ldots,m$, are given as interest rate processes corresponding to different market modes. The other n assets are stocks whose price processes $P_m(t)$ $m = 1,2,\ldots,n$ satisfy the system of SDE (system of differential equation)

$$dP_m(t) = P_m(t) \cdot \left\{ b_m(t,\alpha(t))\,dt + \sum_{p=1}^{n} \sigma_{mp}(t,\alpha(t))\,dW_p(t) \right\}, \quad t \in [0,T] \tag{47}$$

$$P_m(0) = p_m > 0, \quad t \in [0,T], \quad \alpha(t) = i \in \mathbb{N},$$

where for each $i = 1,2,\ldots,n$ $b_m(t,i)$ is the appreciation rate process and $\sigma_m(t,i) = (\sigma_{m1}(t,i),\ldots,\sigma_{mn}(t,i))$ is the volatility or the dispersion rate process of the mth stock, corresponding to $\alpha(t) = i$.

Define the volatility matrix

$$\sigma(t,i) \equiv \left(\sigma_{mp}(t,i)\right)_{n\times n} \quad \text{for each } i = 1,\ldots,n. \tag{48}$$

We assume

$$\sigma(t,i)\,\sigma(t,i)' \geq \delta I \quad \forall t \in [0,T],\ \delta > 0 \tag{49}$$

and $r(t,i)$, $b_m(t,i)$, $\sigma_{mn}(t,i)$ are measurable and uniformly bounded in t.

Denote by $y(t)$ the total wealth of the agent with $y(0) = y_0 > 0$ being his initial wealth; $y(t)$ satisfies

$$dy(t) = \left[r(t,i)\,y(t) + B(t,i)\,u(t) + \sum_{m=1}^{n} \left[b_m(t,i) - r(t,i)\right] \right] u_m(t)\,dt + \sum_{p=1}^{n}\sum_{m=1}^{n} \sigma_{mp}(t,i)\,u_m(t)\,dW_p(t), \quad t \in [s,T] \tag{50}$$

$$y(0) = y_0 > 0, \quad \alpha(0) = i_0, \text{ the initial market mode,}$$

where $u_m(t)$ is the total market value of the agent's wealth in the mth asset and $m = 0,1,\ldots,n$ at time t.

$u(\cdot) = (u_1(\cdot),\ldots,u_n(\cdot))'$ is called a portfolio of the agent.

$u_0(\cdot)$, the asset in the bank account, is completely specified since $u_0(t) = y(t) - \sum_{i=1}^{n} u_i(t)$. Thus, in our analysis to follow, only $u(\cdot)$ is considered.

Setting

$$B(t,i) = \left(b_1(t,i) - r(t,i),\ldots,b_n(t,i) - r(t,i)\right), \quad i \in \mathcal{M}, \tag{51}$$

wealth equation (50) satisfies

$$dy(t) = \left[r(t,i)\,y(t) + B(t,i)\,u(t)\right]dt + u(t)'\,\sigma(t,i)\,dW(t), \quad t \in [s,T] \tag{52}$$

$$y(0) = y_0 > 0, \quad \alpha(0) = i_0.$$

The objective of the agent is to find an admissible portfolio $u(\cdot) \geq 0$, whose expected terminal wealth is $E_{ty}y(T) = d$ for a given $d \in \mathbb{R}$, so that the risk is measured by the variance of the terminal wealth. Namely, the goal of the agent is to solve the following constrained stochastic optimization problem, parameterized by $d \in \mathbb{R}$:

$$\begin{aligned} &\text{minimize} \quad J_{\mathrm{MV}}(y_0, i_0, u(\cdot)) = E_{ty}\left[y(T) - d\right]^2, \\ &\text{subject to} \quad E_{ty}y(T) = d, \\ &\qquad (y(\cdot), u(\cdot)) \text{ admissible,} \end{aligned} \tag{53}$$

called mean-variance portfolio.

Formula (53) is a convex optimization problem; by using a Lagrange multiplier $\mu \in \mathbb{R}$, we can attach the equality constraint $E_{ty}y(T) = d$ to the first equation of (53). In this way, the portfolio problem can be solved via the following optimal stochastic control problem:

$P(d)$:

$$\begin{aligned} &\text{minimize} \quad E_{ty}\left\{\left[y(T) - d\right]^2 + 2\mu\left[E_{ty}y(T) - d\right]\right\}, \\ &\text{subject to} \quad E_{ty}y(T) = d, \\ &\qquad (y(\cdot), u(\cdot)) \text{ admissible,} \end{aligned} \tag{54}$$

where factor 2 in front of the multiplier μ is introduced in the objective function just for convenience.

This problem is equivalent to the following:

$(A(\mu))$:

$$\begin{aligned} &\text{minimize} \quad E_{ty}\left[\frac{1}{2}\left[y(T) - (d-\mu)\right]^2\right], \\ &\text{subject to} \quad u(\cdot) \in L^2_{\mathcal{F}}(0,T;\mathbb{R}^m_+) \\ &\qquad (y(\cdot), u(\cdot)) \text{ admissible,} \end{aligned} \tag{55}$$

in the sense that two problems have exactly the same optimal control [5].

Next, we let $x(t) = y(t) - (d - \mu)$.

Consider $(A(\mu))$:

$$\begin{aligned} &\text{minimize} \quad E_{tx}\left[\frac{1}{2}\left[x(T)\right]^2\right], \\ &\text{subject to} \quad u(\cdot) \in L^2_{\mathcal{F}}(0,T;\mathbb{R}^m_+) \\ &\qquad (x(\cdot), u(\cdot)) \text{ admissible,} \end{aligned} \tag{56}$$

and (52) is equivalent to

$$\begin{aligned} dx(t) &= \left[A(t,i)\,x(t) + B(t,i)\,u(t) + A(t,i)(d-\mu)\right]dt + \sum_{p=1}^{n} G_p(t,i)\,u(t)\,dW_p(t), \quad t \in [s,T] \\ x(s) &= y(s) - (d-\mu) \in \mathbb{R} \end{aligned} \tag{57}$$

$$\text{where } G_p(t,i) = \left(\sigma_{1p}(t,i),\ldots,\sigma_{np}(t,i)\right)$$

$$A(t,i) \in \mathbb{R}.$$

Problem $A(\mu)$ is a stochastic optimal linear quadratic coupled (LQC) control problem, and we can get the solution of $(A(\mu))$ by guessing the solution as a quadratic function. By making use of the duality relationship between $(P(d))$ and $(A(\mu))$, see Appendix A.2; we obtain the solution of the original problem $(P(d))$.

4.1. A General Constrained Stochastic Linear Quadratic Problem. Consider controlled linear stochastic differential equation (57).

We assume that the matrix $\sum_{p=1}^{n} G_p(t,i)' G_p(t,i)$ is nonsingular. Our objective is to find an optimal control $u(\cdot)$ that minimizes the quadratic terminal cost function. Set

$$\mathcal{U}[s,T] = L^2_{\mathcal{F}}(s,T;\mathbb{R}^m_+). \tag{58}$$

Given $u(\cdot) \in \mathcal{U}[s,T]$, the pair $(x(\cdot),u(\cdot))$ is admissible if $x(\cdot) \in L^2_{\mathcal{F}}(s,T;\mathbb{R})$ is a solution of (57). Let

$$J_i(s,x;u(\cdot)) = E_{tx}\left\{\frac{1}{2}x(T)^2\right\}. \tag{59}$$

The value function associated with LQC problem (57) and (59) is defined by

$$V_i(s,x) = \inf_{u(\cdot)\in\mathcal{U}[s,T]} J_i(s,x;u(\cdot)). \tag{60}$$

In Appendix A.3, and also [8], value function (60) satisfies (8). Next, we will provide an explicit viscosity solution of (8).

Definition 5.

(i) A portfolio $u(\cdot)$ is said to be admissible if $u(\cdot) \in L^2_{\mathcal{F}}(0,T;\mathbb{R}^n_+)$ and the SDE (57) has a unique solution $x(\cdot)$ corresponding to $u(\cdot)$. In this case, we refer to $(x(\cdot),u(\cdot))$ as an admissible (wealth, portfolio) pair.

(ii) The problem is called *feasible* if there is at least one portfolio satisfying all the constraints.

(iii) The problem is called finite if it is *feasible* and the infimum of $J_{MV}(x_0,i_0,u(\cdot))$ is finite.

(iv) An optimal portfolio to the above problem, if it ever exists, is called an *efficient portfolio corresponding to* d, and the corresponding $(\operatorname{Var} x(T),d) \in \mathbb{R}^2$

and $(\sigma_{x(T)},d) \in \mathbb{R}^2)$ are interchangeably called an efficient point, and the set of all the efficient points is called the efficient frontier.

Next, we let

$$\overline{f}_i(t,x;u) = A(t,i)\,x(t) + B(t,i)\,u + f(t,i),$$
$$\overline{g}_i(t,x;u) = G(t,i)\,u, \tag{61}$$
$$\text{where } f(t,i) = A(t,i)(d-\mu).$$

4.2. Viscosity Solution of the Coupled System. By guessing the value function of (8) as

$$V_i(t,x) = \frac{1}{2}P(t,i)\,x^2 + M(t,i)\,x + R(t,i) \tag{62}$$

we will see that the coefficients of (8) satisfy the following Riccati equation.

Definition 6. We define the system of Riccati equations as follows

$$\frac{d\widehat{P}(t,i)}{dt} = \left[-2A(t,i) - \left\|\overline{\xi}(t,i)\right\|^2 - 2B_i(t)\,G^{-1}(t,i)\,\overline{\xi}(t,i)\right] \cdot \widehat{P}(t,i) + \sum_{j\neq i}^{n} q_{i,j}\left[\widehat{P}(t,i) - \widehat{P}(t,j)\right], \tag{63}$$

$$\widehat{P}(T,k) = 1,$$

$$\frac{d\widehat{M}(t,i)}{dt} = \left[-A(t,i) - \left\|\overline{\xi}(t,i)\right\|^2 - 2B(t,i)\,G^{-1}(t,i)\,\overline{\xi}(t,i)\right] \cdot \widehat{M}(t,i) - \widehat{P}(t,i)\,f(t,i) + \sum_{j\neq i}^{n} q_{i,j}\left[\widehat{M}(t,i) - \widehat{M}(t,j)\right], \tag{64}$$

$$\widehat{M}(T,k) = 0,$$

$$\frac{d\widehat{R}(t,i)}{dt} = -\widehat{M}(t,i)\,f(t,i) - B(t,i)\,G^{-1}(t,i)\,\overline{\xi}(t,i) \cdot \widehat{M}^2(t,i)\,\widehat{P}(t,i)^{-1} - \frac{1}{2}\left\|\overline{\xi}(t,i)\right\|^2 \widehat{M}^2(t,i)\,\widehat{P}(t,i)^{-1} + \sum_{j\neq i}^{n} q_{i,j}\left[\widehat{R}(t,i) - \widehat{R}(t,j)\right], \tag{65}$$

$$\widehat{R}(T,k) = 0,$$

$$\frac{d\widetilde{P}(t,i)}{dt} = -2A(t,i)\,\widetilde{P}(t,i) + \sum_{j\neq i}^{n} q_{i,j}\left[\widehat{P}(t,i) - \widehat{P}(t,j)\right], \tag{66}$$

$$\widehat{P}(T,k) = 1,$$

$$\frac{d\widetilde{M}(t,i)}{dt} = -A(t,i)\,\widetilde{M}(t,i) + \widetilde{P}(t,i)\,f(t,i) + \sum_{j\neq i}^{n} q_{i,j}\left[\widetilde{M}(t,i) - \widetilde{M}(t,j)\right], \tag{67}$$

$$\widehat{M}(T,k) = 0,$$

$$\frac{d\widetilde{R}(t,i)}{dt} = -\widetilde{M}(t,i)\, f(t,i) + \sum_{j\neq i}^{n} q_{i,j}\left[\widetilde{R}(t,i) - \widetilde{R}(t,j)\right], \qquad (68)$$

$$\widetilde{R}(T,k) = 0,$$

where $\overline{\xi}(t,i)$ is as in Lemma A.1

Remark 7. By letting

$$\begin{aligned}\widetilde{\alpha}(t,i) &= -2A(t,i) - \left\|\overline{\xi}(t,i)\right\|^2 - 2B(t,i)\, G^{-1}(t,i)\,\overline{\xi}(t,i),\\ \widetilde{\beta}(t,i) &= -2A(t,i) - \left\|\overline{\xi}(t,i)\right\|^2 - 2B(t,i)\, G^{-1}(t,i)\,\overline{\xi}(t,i),\\ \widetilde{\gamma}(t,i) &= -\widehat{M}(t,i)\, f(t,i) - B(t,i)\, G^{-1}(t,i)\,\overline{\xi}(t,i)\,\widehat{M}^2(t,i)\,\widehat{P}(t,i) - \frac{1}{2}\left\|\overline{\xi}(t,i)\right\|^2 \widehat{M}^2(t,i)\,\widehat{P}(t,i).\end{aligned} \qquad (69)$$

We see that (63) is equivalent to

$$\begin{aligned}\frac{d\widehat{P}(t)}{dt} &= M(t)\,\widehat{P}(t),\\ M(t) &= \left[\widetilde{\alpha}(t,i)\,\delta_{ij} + q_{ij}\right]_{1\le i,j\le n},\\ \widehat{P}(T) &= 1,\end{aligned} \qquad (70)$$

where $\widehat{P}(t) = \left(\widehat{P}(t,i)\right)_{1\le i\le m}$;

(64) is equivalent to

$$\begin{aligned}\frac{d\widehat{M}(t)}{dt} &= N(t)\,\widehat{M}(t) + G(t),\\ N(t) &= \left[\widetilde{\beta}(t,i)\,\delta_{ij} + q_{ij}\right]_{1\le i,j\le n},\\ G(t) &= \left(\widehat{P}_1 f_1(t)\cdots \widehat{P}_n f_n(t)\right),\\ \widehat{M}(T) &= 0,\end{aligned} \qquad (71)$$

where $\widehat{M}(t) = \left(\widehat{M}(t,i)\right)_{1\le i\le m}$;

(65) is equivalent to

$$\begin{aligned}\frac{d\widehat{R}(t)}{dt} &= Q(t)\,\widehat{R}(t) + \widetilde{\gamma}(t),\\ Q(t) &= \left[q_{ij}\right]_{1\le i,j\le n},\\ \widehat{R}(T) &= 0,\end{aligned} \qquad (72)$$

where $\widehat{R}(t) = \left(\widehat{R}(t,i)\right)_{1\le i\le m}$;

(66) is equivalent to

$$\begin{aligned}\frac{d\widetilde{P}(t)}{dt} &= H(t)\,\widetilde{P}(t),\\ H(t) &= \left[-2A_i(t)\,\delta_{ij} + q_{ij}\right]_{1\le i,j\le n},\\ \widetilde{P}(T) &= 0,\end{aligned} \qquad (73)$$

where $\widetilde{P}(t) = \left(\widetilde{P}(t,i)\right)_{1\le i\le m}$;

(67) is equivalent to

$$\begin{aligned}\frac{d\widetilde{M}(t)}{dt} &= L(t)\,\widetilde{M}(t) + K(t),\\ L(t) &= \left[-A_i(t)\,\delta_{ij} + q_{ij}\right]_{1\le i,j\le n},\\ K(t) &= \left(\widetilde{P}_1 f_1(t)\cdots \widetilde{P}_n f_n(t)\right),\\ \widetilde{P}(T) &= 0,\end{aligned} \qquad (74)$$

where $\widetilde{M}(t) = \left(\widetilde{M}(t,i)\right)_{1\le i\le m}$;

(68) is equivalent to

$$\begin{aligned}\frac{d\widetilde{R}(t)}{dt} &= Q(t)\,\widetilde{R}(t) + O(t),\\ Q(t) &= \left[q_{ij}\right]_{1\le i,j\le n},\\ O(t) &= \left(\widetilde{M}_1 f_1(t)\cdots \widetilde{M}_n f_n(t)\right),\\ \widetilde{R}(t) &= 0,\end{aligned} \qquad (75)$$

where $\widetilde{R}(t) = \left(\widetilde{R}(t,i)\right)_{1\le i\le m}$.

4.3. Riccati Equation Magnus Approach. We will show how to provide the solutions of (70)–(75) by making use of Magnus method.

Proposition 8 (see [10]). *Given the $n \times n$ coefficient matrix $A(t)$,*

$$\begin{aligned}\frac{dY(t)}{dt} &= A(t)\,Y(t),\\ Y(t_0) &= Y_0\end{aligned} \qquad (76)$$

where $Y(t) = (Y(t,i))_{1\le i\le m}$

and then $Y(t) = \exp((\Omega(t,t_0))Y_0)$ which is subsequently constructed as a series expansion

$$\begin{aligned}\Omega(t,t_0) &= \sum_{k=1}^{\infty}\Omega_k(t,t_0) \quad \text{where } \Omega_1 = \int_{t_0}^{t} A(\tau)\,d\tau\\ \Omega_n(t,t_0) &= \sum_{j=1}^{n-1}\frac{B_j}{j!}\int_{t_0}^{t} S_n^{(j)}(\tau)\,d\tau, \quad n\ge 2,\end{aligned} \qquad (77)$$

where S_n^j is defined recursively by

$$\begin{aligned} S_n^{(j)} &= \sum_{m=1}^{n-j}\left[\Omega_m, S_{n-m}^{(j-1)}\right], \quad 2 \le j \le n-1, \\ S_n^{(1)} &= \left[\Omega_{n-1}, A\right], \\ S_n^{(n-1)} &= ad_{\Omega_1}^{n-1}(A), \end{aligned} \tag{78}$$

ad_Ω^k iterated commutator

$$\begin{aligned} ad_\Omega A &= [\Omega, A], \\ ad_\Omega^{k+1} A &= \left[\Omega, ad_\Omega^k A\right], \\ ad_\Omega^0 A &= A, \\ & k \in \mathbb{N}, \end{aligned} \tag{79}$$

and B_j is the Bernoulli numbers.

Proposition 9 (see [11]).

$$\begin{aligned} \frac{dY(t)}{dt} &= M(t)Y(t) + Y(t)N(t) + F(t), \\ Y(t_0) &= Y_0 \\ & t \in [t_0, T], \end{aligned} \tag{80}$$

where $Y(t), F(t) \in \mathbb{C}^{p\times q}$, $M(t) \in \mathbb{C}^{p\times p}$, and $N(t) \in \mathbb{C}^{q\times q}$. The solution of (80) is given by

$$Y(t) = \Phi_M(t,t_0)\, Y_0 \Phi_N^*(t,t_0) + \Psi(t,t_0) \tag{81}$$

with

$$\Psi(t,t_0) = \int_{t_0}^{t} \Psi_M(t,s)\, F(s)\, \Phi_N^*(t,s)\, ds, \tag{82}$$

where $\Phi_m(t,t_0)$ and $\Phi_N^*(t,t_0)$ are the fundamental solution matrices of the associated homogeneous equations

$$\begin{aligned} \Phi_M'(t,t_0) &= M(t)\Phi_M(t,t_0), \\ \Phi_M(t_0,t_0) &= I_p, \\ \Phi^{*\prime}(t,t_0) &= \Phi^*(t,t_0)\, N(t), \\ \Phi_N(t_0,t_0) &= I_q. \end{aligned} \tag{83}$$

Remark 10. By making use of Proposition 8 we get (70) and (71)–(75) are special case of Proposition 9 when $N = [0_{ij}]_{n\times n}$.

Theorem 11. *The value function of (60) is given by*

$$V_i(t,x) = \begin{cases} \overline{V}_i(t,x) = \dfrac{1}{2}\widehat{P}(t,i)x^2 + \widehat{M}(t,i)x + \widehat{R}(t,i), & \text{if } x + \overline{\eta}(t,i) \le 0, \\ \widetilde{V}_i(t,x) = \dfrac{1}{2}\widetilde{P}(t,i)x^2 + \widetilde{M}(t,i)x + \widetilde{R}(t,i), & \text{if } x + \overline{\eta}(t,i) > 0 \end{cases} \tag{84}$$

and the optimal control is given by

$$u^* = \begin{cases} -\left(G(t,i)'\right)^{-1}\overline{\xi}(t,i)\left[x + \overline{\eta}(t,i)\right], & \text{if } x + \overline{\eta}(t,i) \le 0, \\ 0, & \text{if } x + \overline{\eta}(t,i) > 0, \end{cases} \tag{85}$$

where

$$\overline{\eta}(t,i) = \frac{\widehat{M}(t,i)}{\widehat{R}(t,i)}. \tag{86}$$

Proof. Let

$$\begin{aligned} \Gamma_1^i &= \left\{(t,x,i) \in [0,T]\times\mathbb{R}\times\mathbb{N} \mid x + \overline{\eta}(t,i) \le 0\right\}, \\ \Gamma_2^i &= \left\{(t,x,i) \in [0,T]\times\mathbb{R} \mid x + \overline{\eta}(t,i) > 0\right\}. \end{aligned} \tag{87}$$

(i) In Γ_1^i, V as given by (62) is well defined, with

$$\begin{aligned} \frac{\partial V_i(t,x)}{\partial t} &= \frac{1}{2}\dot{\widehat{P}}(t,i)x^2 + \dot{\widehat{M}}(t,i)x + \dot{\widehat{R}}(t,i), \\ \frac{\partial V_i(t,x)}{\partial x} &= \widehat{P}(t,i)x + \widehat{M}(t,i), \\ \frac{\partial^2 V_i(t,x)}{\partial x^2} &= \widehat{P}(t,i). \end{aligned} \tag{88}$$

Substituting them into the left-hand side (LHS) of (8), we obtain

$$\begin{aligned} \text{LHS} = &\left(\frac{1}{2}\dot{\widehat{P}}(t,i) + \widehat{P}(t,i)A(t) + \frac{1}{2}\right. \\ &\left.\cdot \sum_{j\ne i} q_{ij}\left(\widehat{P}(t,i) - \widehat{P}(t,j)\right)\right)x^2 \left(\dot{\widehat{M}}(t,i) + \widehat{P}(t,i)\right. \\ &\cdot f(t,i) + \widehat{M}(t,i)A_i(t) \end{aligned}$$

$$+\sum_{j\neq i} q_{ij}\left(\widehat{R}(t,i)-\widehat{R}(t,i)\right)\Bigg)x+\Bigg(\dot{\widehat{R}}(t,i)+\widehat{M}(t,i)$$

$$\cdot f(t,i)+\sum_{j\neq i} q_{ij}\left(\widehat{R}(t,i)-\widehat{R}(t,i)\right)\Bigg)$$

$$\cdot \inf_{u\geq 0}\left[\frac{1}{2}u'G(t,i)'G(t,i)u\right.$$

$$\left.+B(t,i)\left(x+\frac{\widehat{M}(t,i)}{\widehat{P}(t,i)}\right)u\right]. \tag{89}$$

Let $\overline{\eta}(t,i) = \widehat{M}(t,i)/\widehat{P}(t,i)$ and, by using Lemma A.1 $\overline{\alpha} = -[x+\overline{\eta}(t,i)] > 0$, it follows that the minimizer of (89) is achieved by

$$u^{*} = -\left(G(t,i)'\right)^{-1}\overline{\xi}(t,i)\left[x+\overline{\eta}(t,i)\right]. \tag{90}$$

Substituting $u^{*}(t,x)$ back into (8) and noting (63)–(65), it immediately follows that LHS = 0.

Now, we will show that $\overline{V}$ is a viscosity subsolution.

Let $\varphi_i \in C^2(\mathcal{Q})$ and choose $(\overline{t},\overline{x}) \in \operatorname{argmax}\{(\overline{V}_i - \varphi_i)(t,x) \mid (t,x) \in \overline{\mathcal{Q}}\} \cap \mathcal{Q}$; then,

$$\frac{\partial \varphi_i(\overline{t},\overline{x})}{\partial t} = \frac{\partial \overline{V}_i(\overline{t},\overline{x})}{\partial t},$$

$$\frac{\partial \varphi_i(\overline{t},\overline{x})}{\partial x} = \frac{\partial \overline{V}_i(\overline{t},\overline{x})}{\partial x}, \tag{91}$$

$$\frac{\partial^2\left(\overline{V}_i-\varphi_i\right)}{\partial x^2}(\overline{t},\overline{x}) \leq 0 \Longrightarrow$$

$$\overline{P}_i = \frac{\partial^2 \overline{V}_i(\overline{t},\overline{x})}{\partial x^2} \leq \frac{\partial^2 \varphi_i(\overline{t},\overline{x})}{\partial x^2}$$

and we obtain

$$0 = \partial_t \overline{V}_i(\overline{t},\overline{x})$$

$$+\inf_{u\geq 0}\left\{\frac{1}{2}g_i(\overline{t},\overline{x},u)\,g_i(\overline{t},\overline{x},u)'D_x^2\overline{V}_i(\overline{t},\overline{x})\right.$$

$$\left.+\overline{f}_i(\overline{t},\overline{x},u)D_x\overline{V}_i(\overline{t},\overline{x})\right\}-\sum_{j\neq i} q_{ij}\left[\overline{V}_i(\overline{t},\overline{x})\right.$$

$$\left.-\overline{V}_j(\overline{t},\overline{x})\right] \leq \partial_t\varphi_i(\overline{t},\overline{x}) \tag{92}$$

$$+\inf_{u\geq 0}\left\{\frac{1}{2}g_i(\overline{t},\overline{x},u)\,g_i(\overline{t},\overline{x},u)'D_x^2\varphi_i(\overline{t},\overline{x})\right.$$

$$\left.+\overline{f}_i(\overline{t},\overline{x},u)\varphi_i(\overline{t},\overline{x})\right\}-\sum_{j\neq i} q_{ij}\left[\overline{V}_i(\overline{t},\overline{x})\right.$$

$$\left.-\overline{V}_j(\overline{t},\overline{x})\right].$$

Hence, $\overline{V}_i$ is a viscosity subsolution.

(ii) In Γ_2^i, we proceed similarly with

$$\frac{\partial V_i(t,x)}{\partial t} = \frac{1}{2}\dot{\widetilde{P}}(t,i)x^2+\dot{\widetilde{M}}(t,i)x+\dot{\widetilde{R}}(t,i),$$

$$\frac{\partial V_i(t,x)}{\partial x} = \widetilde{P}(t,i)x+\widetilde{M}(t,i), \tag{93}$$

$$\frac{\partial^2 V_i(t,x)}{\partial x^2} = \widetilde{P}(t,i).$$

Substituting them into the left-hand side (LHS) of (8), we obtain

$$\mathrm{LHS} = \Bigg(\frac{1}{2}\dot{\widetilde{P}}(t,i)+\widetilde{P}(t,i)A(t,i)+\frac{1}{2}$$

$$\cdot\sum_{j\neq i} q_{ij}\left(\dot{\widetilde{P}}(t,i)-\dot{\widetilde{P}}(t,j)\right)\Bigg)x^2\Bigg(\dot{\widetilde{M}}(t,i)+\widetilde{P}(t,i)$$

$$\cdot f_i(t)+\widetilde{M}(t,i)A(t,i)$$

$$+\sum_{j\neq i} q_{ij}\left(\widetilde{R}(t,i)-\widetilde{R}(t,j)\right)\Bigg)x+\Bigg(\dot{\widetilde{R}}(t,i)+\dot{\widetilde{M}}(t,i) \tag{94}$$

$$\cdot f(t,i)+\sum_{j\neq i} q_{ij}\left(\widetilde{R}(t,i)-\widetilde{R}(t,j)\right)\Bigg)$$

$$\cdot\inf_{u\geq 0}\left[\frac{1}{2}u'G(t,i)'G_i(t)u\right.$$

$$\left.+B(t,i)\left(x+\frac{\widetilde{M}(t,i)}{\widetilde{P}(t,i)}\right)u\right].$$

Since $\alpha = -[x+\overline{\eta}(t,i)] > 0$, the minimizer of (94) is

$$u_i^{*} = 0. \tag{95}$$

Substituting u^{*} into (8), it is easy to show that $\widetilde{V}$ satisfies HJBC equation (8) in Γ_2^i.

Now, we will show that $\widetilde{V}$ is a viscosity subsolution.

Let $\phi_i \in C^2(\mathcal{Q})$ and choose $(\widetilde{t},\widetilde{x}) \in \operatorname{argmin}\{(\widetilde{V}_i - \varphi_i)(t,x) \mid (t,x) \in \overline{\mathcal{Q}}\} \cap \mathcal{Q}$; then,

$$\frac{\partial \phi_i(\widetilde{t},\widetilde{x})}{\partial t} = \frac{\partial \widetilde{V}_i(\widetilde{t},\widetilde{x})}{\partial t},$$

$$\frac{\partial \phi_i(\widetilde{t},\widetilde{x})}{\partial x} = \frac{\partial \widetilde{V}_i(\widetilde{t},\widetilde{x})}{\partial x},$$

$$\frac{\partial^2 (\widetilde{V}_i - \phi_i)}{\partial x^2} (\bar{t}, \bar{x}) \geq 0 \Longrightarrow$$

$$\widetilde{P}_i = \frac{\partial^2 \widetilde{V}_i (\bar{t}, \bar{x})}{\partial x^2} \geq \frac{\partial^2 \phi_i (\bar{t}, \bar{x})}{\partial x^2} \tag{96}$$

and we obtain

$$\begin{aligned} 0 &= \partial_t \widetilde{V}_i (\bar{t}, \bar{x}) \\ &+ \inf_{u \geq 0} \left\{ \frac{1}{2} \bar{g}_i (\bar{t}, \bar{x}, u) \bar{g}_i (\bar{t}, \bar{x}, u)' D_x^2 \widetilde{V}_i (\bar{t}, \bar{x}) \right. \\ &\left. + \bar{f}_i (\bar{t}, \bar{x}, u) D_x \widetilde{V}_i (\bar{t}, \bar{x}) \right\} - \sum_{j \neq i} q_{ij} \left[\widetilde{V}_i (\bar{t}, \bar{x}) \right. \\ &\left. - \widetilde{V}_j (\bar{t}, \bar{x}) \right] \geq \partial_t \phi_i (\bar{t}, \bar{x}) \\ &+ \inf_{u \geq 0} \left\{ \frac{1}{2} \bar{g}_i (\bar{t}, \bar{x}, u) \bar{g}_i (\bar{t}, \bar{x}, u)' D_x^2 \phi_i (\bar{t}, \bar{x}) \right. \\ &\left. + \bar{f}_i (\bar{t}, \bar{x}, u) \phi_i (\bar{t}, \bar{x}) \right\} - \sum_{j \neq i} q_{ij} \left[\widetilde{V}_i (\bar{t}, \bar{x}) \right. \\ &\left. - \widetilde{V}_j (\bar{t}, \bar{x}) \right]. \end{aligned} \tag{97}$$

Hence, $\widetilde{V}_i$ is a viscosity supersolution.

We see that the value function V is a viscosity solution.

Remark 12. we see clearly that $\partial^2 V(t,x,i)/\partial x^2$ does not exist in $\mathcal{Q}$, since $\widetilde{P}(t,i) \neq \widehat{P}(t,i)$. For this reason, we are required to work within the framework of viscosity solutions.

5. Efficient Strategies

Consider $x(t) = y(t) - (d - \mu)$. The problem $A(\mu)$ is equivalent to the following problem:

$$\begin{aligned} &\min \quad E_{tx} \left[\frac{1}{2} x(T) \right] \\ &dx(t) \\ &\quad = \left[A(t,i) x(t) + B(t,i) u + f(t,i) \right] dt \\ &\qquad + G(t,i) u dW(t), \quad t \in [s,T] \\ &x(0) = y_0 - (d - \mu), \end{aligned} \tag{98}$$

where $u(\cdot) \in L^2_{\mathcal{F}}(0,T;\mathbb{R}^m_+)$ and

$$\begin{aligned} A(t,i) &= r(t,i), \\ B(t,i) &= (b_1(t,i) - r(t,i), \ldots, b_n(t,i) - r(t,i)), \\ f(t,i) &= (d - \mu) r(t,i), \\ G(t,i) &= (\sigma_{i1}(t,i), \ldots, \sigma_{in}(t,i)). \end{aligned} \tag{99}$$

Now, corresponding to (A.3), set

$$\begin{aligned} \bar{\pi}_i(t) &= \operatorname*{argmin}_{\pi(t,i) \in [0,\infty)^m} \frac{1}{2} \left\| \sigma(t,i)^{-1} \pi(t,i) \right. \\ &\quad \left. + \sigma(t,i)^{-1} (b(t,i) - r(t,i) \mathbf{1}) \right\|^2, \\ \bar{\theta}_i(t) &= \sigma_i(t)^{-1} \pi(t,i) + \sigma_i(t)^{-1} (b(t,i) - r(t,i) \mathbf{1}). \end{aligned} \tag{100}$$

5.1. An Optimal Strategy. We present the optimal investment strategy for the problem $A(\mu)$. The optimal control obtained in (85) translates into the following strategy:

$$\begin{aligned} u^* &\equiv (u_1^*, \ldots, u_m^*)', \\ u^* &= \begin{cases} -\left(\sigma(t,i)'\right)^{-1} \bar{\theta}(t,i) \left[y + (d - \mu) + \bar{\eta}(t,i) \right], & \text{if } y + (d - \mu) + \bar{\eta}(t,i) \leq 0, \\ 0, & \text{if } y + (d - \mu) + \bar{\eta}(t,i) > 0. \end{cases} \end{aligned} \tag{101}$$

Theorem 13. *The optimal investment strategy to the problem $A(\mu)$ is given by (101).*

6. Efficient Frontier

Since $x(t) = y(t) - (d - \mu)$, we obtain the solution of the original problem $P(D)$. Hence, for every fixed μ, we have

$$\begin{aligned} &\min_{u(\cdot) \in \mathcal{U}[0,T]} E_{ty} \left\{ \frac{1}{2} [y(T) - d]^2 \right\} + \mu \left[E_{ty} y(T) - d \right] \\ &\quad = \min_{u(\cdot) \in \mathcal{U}[0,T]} E_{tx} \left\{ \frac{1}{2} x(T)^2 \right\} - \frac{1}{2} \mu^2 \\ &\quad = V_{i_0}(0,x) - \frac{1}{2} \mu^2. \end{aligned} \tag{102}$$

Hence, the value function of $P(D)$ is given:

$$\begin{aligned} V_{i_0}(0,x) - \frac{1}{2}\mu^2 &= \left(\frac{1}{2} \widehat{P}(0,i_0) \left[y_0 - (d - \mu) \right]^2 \right. \\ &\quad \left. + \widehat{M}(0,i_0) \left[y_0 - (d - \mu) \right] + \widehat{R}(0,i_0) - \frac{1}{2}\mu^2 \right) \\ &\quad \cdot 1_{y + (d-\mu) + \bar{\eta}(t,i) \leq 0} + \left(\frac{1}{2} \widetilde{P}(0,i_0) \left[y_0 - (d - \mu) \right]^2 \right. \end{aligned}$$

$$+ \widetilde{M}(0,i_0)[y_0-(d-\mu)] + \widetilde{R}(0,i_0) - \frac{1}{2}\mu^2\Big) \cdot 1_{y+(d-\mu)+\widetilde{\eta}(t,i)>0}. \quad (103)$$

Note that the above value still depends on the Lagrange multiplier μ. To obtain the optimal value function, one needs to maximize the value of μ in (103).

Proposition 14. *The efficient strategy of portfolio selection problem (50) corresponding to the expected terminal wealth $E_{ty}y(T) = d$, as a function of time t and wealth y, is*

$$u^* = \begin{cases} -\left(\sigma(t,i)'\right)^{-1}\overline{\theta}(t,i)\left[y+(d-\mu^*)+\eta(t,i)\right], & \text{if } y+(d-\mu^*)+\eta(t,i) \le 0, \\ 0, & \text{if } y+(d-\mu^*)+\eta(t,i) > 0. \end{cases} \quad (104)$$

Moreover if

$$\mu^* = \underset{\mu}{\operatorname{argmax}}\Big(\Big(\frac{1}{2}\widehat{P}(0,i_0)[y_0-(d-\mu)]^2 + \widehat{M}(0,i_0)[y_0-(d-\mu)] + \widehat{R}(0,i_0) - \frac{1}{2}\mu^2\Big) \cdot 1_{y+(d-\mu)+\widetilde{\eta}(t,i)\le 0} + \Big(\frac{1}{2}\widetilde{P}(0,i_0)[y_0-(d-\mu)]^2 + \widetilde{M}(0,i_0)[y_0-(d-\mu)] + \widetilde{R}(0,i_0) - \frac{1}{2}\mu^2\Big) \cdot 1_{y+(d-\mu)+\widetilde{\eta}(t,i)>0}\Big) \quad (105)$$

exists, the efficient frontier is given by

$$\operatorname{Var} y(T) = V(0,x,i_0) - \frac{1}{2}\mu^{*2}. \quad (106)$$

7. Concluding Remarks

We analyzed mean-variance optimal portfolio selection for a market with regime switching. The formulation allows the market to have random switching with no-shorting constraint. Using techniques of stochastic linear quadratic control and the notion of viscosity solution, mean-variance efficient portfolio and efficient frontiers are derived explicitly in closed forms in terms of some systems of Riccati equation for which the solutions are provided by making use of the Magnus approach. The numerical application is in progress and it will be the subject of a new research paper.

Appendix

A. Useful Formulas

A.1. Convex Analysis

Lemma A.1 (see [5]). *Let h be a continuous, strictly convex quadratic function*

$$h(z(t,i)) = \frac{1}{2}z(t,i)'\mathscr{D}(t,i)'\mathscr{D}(t,i)z(t,i) - \alpha\mathscr{B}z(t,i) \quad (A.1)$$

over $z(t,i) \in [0,\infty)^m$, where $\mathscr{B}' \in \mathbb{R}_+^m$, $\mathscr{D}_i \in \mathbb{R}^{m\times m}$ and $\mathscr{D}'(t,i)\mathscr{D}(t,i) > 0$.

For every $\alpha \ge 0$, h has the unique minimizer $\alpha\mathscr{D}(t,i)^{-1}\overline{\xi}(t,i) \in [0,\infty)^m$, where

$$\overline{\xi}(t,i) = \left(\mathscr{D}(t,i)'\right)^{-1}\overline{z}(t,i) + \left(\mathscr{D}(t,i)'\right)^{-1}\mathscr{B}', \quad (A.2)$$

where

$$\overline{z}(t,i) = \underset{z(t,i)\in[0,\infty)^m}{\operatorname{argmin}} \frac{1}{2}\left\|\left((\mathscr{D}(t,i))'\right)^{-1}z(t,i) + \left((\mathscr{D}(t,i))'\right)^{-1}B(t,i)'\right\|. \quad (A.3)$$

A.2. Duality Method

Lemma A.2 (see [12]). *The strong duality relationship holds between $(P(d))$ and $(\mathscr{A}(\mu))$ in the following sense,*

$$\mathscr{V}(P(d)) = \max_{\mu\in\mathbb{R}}\left\{2\mathscr{V}(\mathscr{A}(\mu)) - \mu^2\right\}, \quad (A.4)$$

where $\mathscr{V}(\cdot)$ denotes the optimal value of problem $(\cdot)$.

A.3. Dynamic Programming and Random Evolution with Markov Chain Parameters. Here we sketch a proof of equation (8); for more details please see [8].

Let $\alpha(t)$ be a finite state Markov chain, with state space a finite set $\mathscr{M}$. we regard $\alpha(t)$ as a parameter process. On any interval where $\alpha(t) = \alpha$ is constant, $x(t)$ satisfies the ordinary differential equation

$$dx = \mu(t,x(t),u(t),\alpha(t))\,dt + \sigma(t,x(t),u(t),\alpha(t))\,dw(t) \quad (A.5)$$

and we assume that $\mu(t,x(t),u(t),\alpha(t))$ and $\sigma(t,x(t),u(t),\alpha(t))$ satisfy the conditions

(i) $|\mu_t(t,x(t),u(t),\alpha(t))| + |\mu_x(t,x(t),\alpha(t))| \le C$,
$|\sigma_t(t,x(t),u(t),\alpha(t))| + |\sigma_x(t,x(t),u(t),\alpha(t))| \le C$;

(ii) $|\mu(t,x(t),u(t),\alpha(t))| \le C(1+|x|+|u|)$;

(iii) $|\sigma(t,x,u(t),\alpha(t))| \le C(1+|x|+|u|)$,

for each $\alpha \in \mathcal{M}$. Let $s \leq t \leq T$, and let $\tau_1 < \tau_2 < \cdots < \tau_m$ denote the successive jump times of the parameter process $\alpha(t)$ during $[s,T]$. We let $\tau_0 = t$, $\tau_{m+1} = T$, and define $x(t)$ by

$$dx = \mu(t, x(t), u(t), \alpha(\tau_i^+))\, dt + \sigma(t, x(t), u(t), \alpha(\tau_{i+1}))\, dw(t) \qquad \tau_i \leq t < \tau_{i+1},\ i = 0, \ldots, m, \ldots x(s) = x, \tag{A.6}$$

with the requirement that $x(\cdot)$ is continuous at each jump time τ_i. The process $x(s)$ is not Markov. However, $(x(t), \alpha(t))$ is a Markov process, with state space $\Sigma = \mathbb{R} \times \mathcal{M}$. For each $\Phi(t, x(t), \alpha(t))$ such that $\Phi(\cdot,\cdot,\alpha) \in C^2(\overline{\mathcal{Q}})$, we have

$$\begin{aligned} A^u\Phi(t,x,i) &= \lim_{x\to 0} h^{-1}\left[E_{tx}\Phi(t+h, x(t+h), i) - \Phi(t,x,i)\right] \\ &= \Phi_t(t,x,i) + \frac{1}{2}\sigma(t,x,u,i) D_x^2\Phi(t,x,i) \\ &\quad + \mu(t,x,u,i) D_x\Phi(t,x,i) \\ &\quad + \sum_{j\neq i}\rho(t,i,j)\left[\Phi(t,x,j) - \Phi(t,x,i)\right]. \end{aligned} \tag{A.7}$$

Dynkin formula is

$$E_{tx}\Phi(t,x,i) - \Phi(t,x,i) = E_{tx}\int_t^{t_1} A^{u(s)}\Phi(s, x(s), \alpha(s))\, ds, \tag{A.8}$$

where $\rho(t,x,y)$ represents an infinitesimal rate at which $x(t)$ jumps from x to y:

$$\rho(t,i,j) = \lim_{h\to 0} h^{-1} P\left[x(t+h) = j \mid x(t) = i\right] = q_{ij}(t). \tag{A.9}$$

Criterion to Be Optimized. The control problem of a finite time interval $t \leq s \leq T$ is to minimize

$$J = E_{tx}\left\{\int_t^T L(s, x(s), u(s))\, ds + \psi x(T)\right\} \tag{A.10}$$

in our case the Lagrangian $L(t,x,u) \equiv 0$, that is, the Mayer form.

The value function

$$V_i(t,x) = \inf_C J(t,x,i;\text{control}). \tag{A.11}$$

Bellman's Principe of Dynamic Programming. This states that for $t \leq t+h \leq T$

$$V_i(t,x) = \inf_C E_{tx} V(t+h, x(t+h), i). \tag{A.12}$$

If we take constant control $u(s) = v$ for $t \leq s \leq t+h$,

$$V_i(t,x) \leq E_{tx} V(t+h, x(t+h), i), \tag{A.13}$$

we substract $V(t,x,i)$ from both sides, divided by h, and let $h \to 0$:

$$\begin{aligned} &\lim_{x\to 0^+} h^{-1}\left[\mathbb{E}_{tx}V(t+h, x(t+h), i) - V(t,x,i)\right] \\ &= \lim_{x\to 0^+} h^{-1}\mathbb{E}_{tx}\int_t^{t+h} A^v V(s, x(s), i)\, ds \\ &= A^v V(t,x,i). \end{aligned} \tag{A.14}$$

Hence, for all $v \in U$,

$$0 \leq A^v V(t,x,i). \tag{A.15}$$

On the other hand, if $\underline{u}^*$ is an optimal Markov control policy, we should have

$$V_i(t,x) = E_{tx} V(t+h, x^*(t+h), i), \tag{A.16}$$

where $x^*(s)$ is the Markov process generated by $A^{\underline{u}^*}$. A similar argument gives, under sufficiently strong assumption (including continuity of $\underline{u}^*$ at (t,x)),

$$0 = A^{\underline{u}^*} V(t,x,i). \tag{A.17}$$

Inequatlities (A.15) and (A.17) are equivalent to the dynamic programming equation

$$0 = \min_{v\in U} A^v V(t,x,i). \tag{A.18}$$

Competing Interests

The author declares that no competing interests exist.

References

[1] H. M. Markowitz, "Portfolio selection," *The Journal of Finance*, vol. 7, no. 1, pp. 77–91, 1952.

[2] D. Li and W.-L. Ng, "Optimal dynamic portfolio selection: multi-period mean variance formulation," *Mathematical Finance*, vol. 10, no. 3, pp. 387–406, 2000.

[3] X. Y. Zhou and D. Li, "Continuous-time mean-variance portfolio selection: a stochastic LQ framework," *Applied Mathematics and Optimization*, vol. 42, no. 1, pp. 19–33, 2000.

[4] X. Y. Zhou and G. Yin, "Markowitz's mean-variance portfolio selection with regime switching: a continuous-time model," *SIAM Journal on Control and Optimization*, vol. 42, no. 4, pp. 1466–1482, 2003.

[5] X. Li, X. Y. Zhou, and A. E. Lim, "Dynamic mean-variance portfolio selection with no-shorting constraints," *SIAM Journal on Control and Optimization*, vol. 40, no. 5, pp. 1540–1555, 2002.

[6] G.-L. Xu and S. E. Shreve, "A duality method for optimal consumption and investment under short-selling prohibition. I. General market coefficients," *The Annals of Applied Probability*, vol. 2, no. 1, pp. 87–112, 1992.

[7] M. G. Crandall and P.-L. Lions, "Viscosity solutions of Hamilton-Jacobi equations," *Transactions of the American Mathematical Society*, vol. 277, no. 1, pp. 1–42, 1983.

[8] W. H. Fleming and H. M. Soner, *Controlled Markov Processes and Viscosity*, Springer, 2nd edition, 2006.

[9] S. M. Lenhart, "Viscosity solutions for weakly coupled systems of first-order partial differential equations," *Journal of Mathematical Analysis and Applications*, vol. 131, no. 1, pp. 180–193, 1988.

[10] S. Blanes, F. Casas, J. A. Oteo, and J. Ros, "The Magnus expansion and some of its applications," *Physics Reports*, vol. 470, no. 5-6, pp. 151–238, 2009.

[11] S. Blanes and E. Ponsoda, "Time-averaging and exponential integrators for non-homogeneous linear IVPs and BVPs," *Applied Numerical Mathematics*, vol. 62, no. 8, pp. 875–894, 2012.

[12] R. C. Merton, "An analytical derivation of the efficient portfolio frontier," *The Journal of Financial and Quantitative Analysis*, vol. 7, no. 4, pp. 1851–1872, 1972.

6

Contagious Criminal Career Models Showing Backward Bifurcations: Implications for Crime Control Policies

Silvia Martorano Raimundo,[1] Hyun Mo Yang,[2] and Eduardo Massad[1,3]

[1] *Faculdade de Medicina da Universidade de São Paulo, MLS and LIM01-HCFMUSP, São Paulo, SP, Brazil*
[2] *Universidade de Campinas, DMA-IMECC, Campinas, SP, Brazil*
[3] *School of Applied Mathematics, Fundação Getúlio Vargas, Rio de Janeiro, RJ, Brazil*

Correspondence should be addressed to Eduardo Massad; edmassad@dim.fm.usp.br

Academic Editor: Zhidong Teng

We provide a theoretical framework to study how criminal behaviors can be treated as an infectious phenomenon. There are two infectious diseases like models that mimic the role of convicted criminals in contaminating individuals not yet engaged in the criminal career. Equilibrium analyses of each model are studied in detail. The models proposed in this work include the social, economic, personal, and pressure from peers aspects that can, theoretically, determine the probability with which a susceptible individual with criminal propensity engages in a criminal career. These crime-inducing parameters are treated mathematically and their inclusion in the model aims to help policy-makers design crime control strategies. We propose, to the best of our knowledge by the first time in quantitative criminology, the existence of thresholds for the stability of crime-endemic equilibrium which are the equivalent to the "basic reproduction number" widely used in the mathematical epidemiology literature. Both models presented the phenomena of backward bifurcation and breaking-point when the contact rates are chosen as bifurcation parameters. The finding of backward bifurcation in both models implies that there is an endemic equilibrium of criminality even when the threshold parameter for contagion is below unit, which, in turn, implies that control strategies are more difficult to achieve considerable impact on crime control.

1. Introduction

There is now substantial support in the specialized literature on economics, sociology, criminology, and social psychology to the attempts to explain how and why an individual's propensity to engage in criminal behavior is influenced by his/her social context [1]. One interesting metaphor is the one that states that criminal behavior is contagious or that individuals can be susceptible to what economists call endogenous effects [2]. According to this effect, the social milieu in which individuals live may change the individual's propensity to engage in that same criminal behavior as their peers [3, 4]. In addition, the individual's criminal behavior can be affected by other attributes of his/her neighbors, like in the "role model" theory [5] or the peer pressure to maintain local order [6]. Moreover, the institutional or other characteristics of neighborhoods, including crime prevalence, may induce criminal behavior in susceptible individuals [7].

However, in spite of the large theoretical literature on whether or not crime is contagious, the empirical support for this hypothesis is still limited. Some authors (see [8, 9]) reported an excess in the variation in crime rates across areas which cannot be explained only by the variation in standard sociodemographic determinants of criminal behavior. This suggests that social interactions are more important for less serious than more serious crimes. In a famous study, Crane (1991) (see [10]) showed that in the presence of endogenous residential sorting such reports may be biased by the causal effects of environmental and some individual or family characteristics on the selection of the neighborhood. As pointed out by Kling and Ludwig (see [1]), however, even in the absence of the biased selection problem, it would be very difficult to determine which of the theoretical perspectives above are responsible for any observed neighborhood effects on criminal behavior.

The influence of others' behavior on criminal offences of susceptible individuals is called "behavioral contagion" [11, 12]. Behavioral contagion has been defined as the spread of any attitude or behavior from one individual or group of individuals to another individual or group of individuals throughout a social network of varying structures [11].

Social contagion arises among people interacting in social structures of diverse nature [13–15]. In such interactions information, behavioral innovation, belief, or meme is transmitted in a similar way to infectious diseases spread in groups of susceptible individuals [16]. Contagion occurs when susceptible individuals interact with contagious people in such a way that this interaction results in a new case [17]. In social networks, contact is defined by the communication and/or imitation of influential processes that make transmission potentially effective (see [13] pp. 1288-1289).

The social contagion theory of violence describes the spread of criminal behavior as similar to the spread of infectious diseases [18]. In such a context, the contagious nature of criminal behavior can be understood, described, and analyzed with the tools developed for studying infectious diseases spreading, in particular the use of mathematical models [19]. However, in contrast with directly transmitted diseases, the infectivity of violence does not require direct contact between susceptible and infected individuals.

In this paper, we provide a different approach to the contagious effect of criminal behavior in susceptible individuals. Rather than considering direct contagious effect by the social environment, we consider a contagious effect by criminals already convicted and who control criminal activities from inside prison, a common effect of some countries like Brazil.

The advent of organized crime in Brazilian prisons, especially in the state of São Paulo, and its role in the contagion of criminality to susceptible individuals outside the prisons constitute the object of this article. The gang leaders the Capital's First Command (PCC, Primeiro Comando da Capital) (see [20]) unleashed a series of attacks in May 2006, resulting in deaths, brought cities to a halt, and cornered authorities in charge preventing them from applying law and order and these are the starting as well as reference points taken. In addition, the gang leaders were sustained by an organization maintained by a hierarchical structure of disciplined and obedient employees capable of executing orders without questioning them. Operating from a base of support networks disseminated in distinct mobile points throughout the state, they revealed that they had an able and agile communication system among leaders, followers, and those who took orders, through protected channels barely permeated by external interference by means of cell phones, telephone exchanges, and carrier pigeons [21].

Criminal contagion from inside the prison system of Brazil to outside susceptible individuals with criminal propensity will be considered as an infectious event in a dynamical system context. The models proposed in this work include the social, economic, personal, and pressure from peers aspects that can, theoretically, determine the probability with which a susceptible individual with criminal propensity engage in a criminal career. These crime-inducing parameters are treated mathematically and their inclusion in the model aims to help policy-makers design crime control strategies.

2. The Models

The models presented here are based on the criminal activity within a population in an effort to understand how the criminal careers change and evolve over time. We assume a population-based approach, similar to those models of the spreading of infections that confer temporary/permanent immunity. By introducing the key epidemiological concept of a threshold from mass action law [22, 23], we illustrate the fundamental relationship between incarceration and recidivism within a population and use it to show how the criminal activity could be controlled to reduce the likelihood of an individual to engage in a criminal career. We compute and analyze this threshold considering the spread of the crime and the dynamics of incarceration and recidivism. Keeping in mind both the threshold and the perspective of criminal dynamics, it is possible to evaluate what programs of rehabilitation or prevention contribute to the reduction of the recidivism and the number of contacts among individuals susceptible to crime, offenders, and ex-offenders.

In epidemiology theory, the core groups are conceptualized as being the individuals in a population who will infect more than one person over the duration of infection. Core groups are recognized as playing a central role in sustaining the infection in a population and interventions targeting these groups are central to an effective prevention response. Building on this, we extend the concept of core groups to the inmate population.

Especially in inmate population, most criminality is found among core groups and the criminality will only become more generalized if the contact spreads throughout other networks. Hence, it is very likely that incarcerated individuals will contact those susceptible individuals who have never been incarcerated but have an intrinsic criminal propensity. The basic problem is to find out when this contact occurs, who regulates contact, what types of contact are feasible and desirable, and what are the effects of contact (or lack thereof) on susceptible individuals.

In this way, the total population size, denoted by $N(t)$, is then characterized by three classes: susceptible (S), incarcerated (C), and desisting (D) individuals. We specify first-time (i = 1) and multiple-time (i = 2, 3,. . .,n) incarceration by adding subscripts to model variables and parameters.

2.1. The Partially Contagious Criminality Model (PCCM) Formulation. Let the susceptible in core group population be divided into two categories: S_0, those individuals who have a criminal propensity but have never been incarcerated, not criminally active but susceptible to crime, and S_i, those individuals susceptible to criminal activities who were once incarcerated (S_1) and those susceptible who were multiple-time incarcerated (S_i, $i = 2, 3, \ldots, n$) and became criminally active again. Similarly, let the incarcerated population be divided into two categories: C_1, those who are first-time incarcerated, and C_i (i = 2,3,. . .,n), those who were multiple-time incarcerated at a given time. Finally, D_0 represents those individuals who have a criminal propensity but desisted from criminal life either by their own or as a result of early interventions when discharged out of prison (desisting offenders/criminal desisters), and D_i (i = 1,2,3,. . .,n) represents those individuals who were either first-time or multiple-time incarcerated but desisted from criminal behavior (desisting offenders/criminal desisters) due to variety of reasons.

It is assumed that individuals susceptible to the crime S_0 move to either incarcerated (C_1) or desisting (D_0) class. The rate of initial participation in crime (δ_0) at which individuals move from state S_0 (not criminally active) to C_1 (criminally active and incarcerated) is proportional to intrinsic criminal propensity of individuals. It is also assumed that special intervention programmes (γ_1) may change their basic propensity traits and affect the decisions to engage in crime [24]. As a consequence, we should expect that certain interventions discourage participation in crime, resulting in desistance from criminal activity and reintegration back into society. Primary prevention is an attempt to reduce the risk of behaviors that potentially lead to incarceration. This point is one of the great interests here.

The average length of the primary incarceration term is given by $1/\tau_1$; that is, the rate at which inmates move from state C_1 to S_1 (formerly incarcerated but not criminally active, i.e., ex-offenders) is τ_1. In the same way, the average length of the multiple incarceration term is given by $1/\tau_2$, with $1/\tau_2 > 1/\tau_1$.

It is worth mentioning that the representation of the cycles of criminal dynamics, with a focus on modeling the recidivism process (criminal careers), could be extended including S_i, C_i, and D_i (i = 2,3,. . .,n). However, we do not intend to develop complex models that account for these cycles and we will explore two cycles only (i = 1,2).

In addition, it is assumed that the recidivism prevention takes place during incarceration and after release back into society. Its purpose is to reduce the risk of an individual reoffending and eventually returning to the prison system. However, some individuals may resume criminal activity very soon after being released from prisons depending on having contact with those individuals still incarcerated and criminally active. It should be mentioned that Walsh and Graig (see [25]) indicated that recidivism is also called falling back into a previous criminal behavior.

We also define β_i (i=1,2) as the rate of imprisonment, which captures the return to criminal activity of the individuals released from prison, such that β_1 and β_2 describe the rates with which individuals engage into criminal activity, depending on having had any contagious contact with those incarcerated criminals. They are analogous to the effective contact rate in infectious diseases models.

Finally, μ is the natural mortality rate and a_i (i=1,2) is the incarceration-related additional deaths rate (inmate mortality rate can be caused by illness, such as AIDS-related, suicide, accidental self-injury, execution, or any other unspecified cause). Moreover, since the model monitors human populations, all parameters are assumed as nonnegative. We also assumed homogeneous population without any differences in

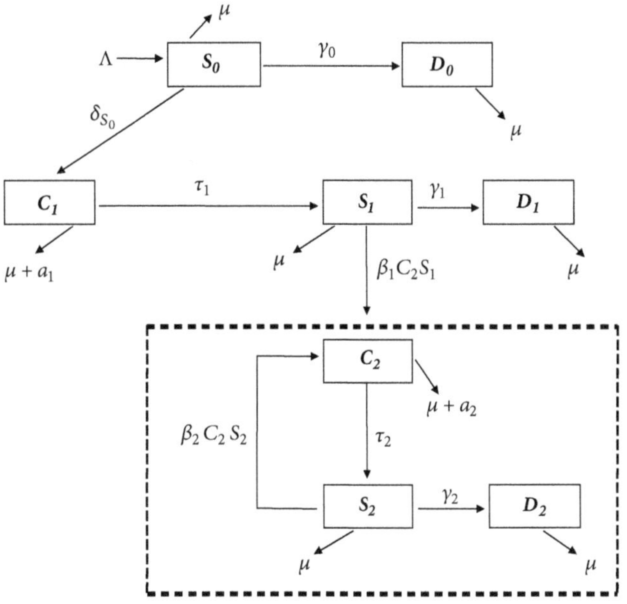

FIGURE 1: The flow diagram for the low-high criminality model (1).

age or in crime type occurrences. In addition, we used the number of incarcerations as a proxy for reoffending.

The flow diagram of the PCCM for two stages is depicted in Figure 1. The variables and parameters are described in Tables 1 and 2, respectively.

Combining the above derivations and assumptions, it follows that the model for transmission dynamics of criminality for two stages is given by the following nonlinear system of differential equations:

$$
\begin{aligned}
\frac{dS_0}{dt} &= \Lambda - (\delta_0 + \gamma_0 + \mu) S_0 \\
\frac{dD_0}{dt} &= \gamma_0 S_0 - \mu D_0 \\
\frac{dC_1}{dt} &= \delta_0 S_0 - (\tau_1 + \mu + a_1) C_1 \\
\frac{dS_1}{dt} &= \tau_1 C_1 - (\gamma_1 + \mu) S_1 - \beta_1 C_2 S_1 \\
\frac{dD_1}{dt} &= \gamma_1 S_1 - \mu D_1 \\
\frac{dC_2}{dt} &= (\beta_1 S_1 + \beta_2 S_2) C_2 - (\tau_2 + \mu + a_2) C_2 \\
\frac{dS_2}{dt} &= \tau_2 C_2 - \beta_2 C_2 S_2 - (\gamma_2 + \mu) S_2 \\
\frac{dD_2}{dt} &= \gamma_2 S_2 - \mu D_2
\end{aligned}
\tag{1}
$$

with generic initial conditions $S_0(0) \geq 0$, $D_0(0) \geq 0$, $C_1(0) \geq 0$, $S_1(0) \geq 0$, $D_1(0) \geq 0$, $C_2(0) \geq 0$, $S_2(0) \geq 0$, and $D_2(0) \geq 0$.

Table 1: Models variables and their biological meaning.

Variables	Description
S_0	Individuals not criminally active but susceptible to crime (core group)
C_1	First-time incarcerated individuals
S_1	First-time ex-offenders individuals who are again susceptible to crime
D_i	Individuals who desist from criminal behavior (i = 0,1,2) (desisting offenders/criminal desisters)
C_2	Individuals who were two or multiple times incarcerated (recidivists, reoffenders)
S_2	Second-time ex-offenders individuals and susceptible to crime (at least two incarcerations)

Table 2: Models parameters and their biological meaning.

Parameters	Description
Λ	Rate of recruitment of individuals into the core group
δ_0	Basic flow to the criminality ($time^{-1}$) (criminal propensity)
β_1	Rate of cooptation of first-time ex-offenders (contact rate between C_2 and S_1)
β_2	Rate of cooptation of second-time ex-offenders (contact rate between C_2 and S_2)
γ_0	Rate of early desistance from crime
γ_1	Rate of desistance from crime when in the first cycle
γ_2	Rate of desistance from crime when in at least the second cycle
μ	Natural inmate mortality rate
a_1	Additional inmate mortality rate when in C_1
a_2	Additional inmate mortality rate when in C_2
τ_1	Release rate from incarceration of first-time offenders
τ_2	Release rate from incarceration of at least second-time offenders

By summing up the above equations, the total population size $N(t)$ is variable with

$$\frac{dN}{dt} = \Lambda - \mu N - a_1 C_1 - a_2 C_2 \tag{2}$$

Thus, in the absence of additional inmate mortalities, that is, $a_1 = a_2 = 0$, the population size evolves as an immigration model with natural mortality, that is, according to $dN/dt = \Lambda - \mu N$. This equation has a single equilibrium $N = N(0) = \Lambda/\mu$ for any initial value of $N(0)$. Thus, in the long run, the population size settles to this constant value. It follows from (2) that $\lim_{t\longrightarrow\infty} N(t) \le \Lambda/\mu = N(0)$.

The differential equation for N implies that solutions of (1), starting in the positive orthant $\mathbb{R}^8_+$, either approach, enter, or remain in the subset $\mathbb{R}^8_+$ defined by

$$\Omega = \{(S_0, D_0, C_1, S_1, D_1, C_2, S_2, D_2) \in \mathbb{R}^8_+ : S_0 + D_0 + C_1 + S_1 + D_1 + C_2 + S_2 + D_2 \le N(0)\}. \tag{3}$$

Thus it suffices to consider solutions in the region Ω. Solutions of the initial value problem starting in Ω and defined by (1) exist and are unique on a maximal interval [26]. Since solutions remain bounded in the positively invariant region Ω, the initial value problem is then both mathematically and epidemiologically well posed [27]. Hence, it is sufficient to consider the dynamics of the flow generated by model (1) in Ω.

2.2. Analysis of the PCCM Model. In this section, system (1) is qualitatively analyzed to investigate the existence of its equilibria [28] and the control strategies of its dynamical behavior.

From system (1), with the right-hand size equal to zero, it can be seen from the first five equations that the coordinates of the equilibrium point are given, respectively, by

$$\begin{aligned} S_0 &= \frac{\Lambda}{(\delta_0 + \gamma_0 + \mu)}, \\ D_0 &= \frac{\gamma_0}{\mu} S_0, \\ C_1 &= \frac{\delta_0}{(\tau_1 + \mu + a_1)} S_0, \\ S_1 &= \frac{\tau_1 \delta_0}{(\tau_1 + \mu + a_1)(\mu + \gamma_1 + \beta_1 C_2)} S_0, \\ D_1 &= \frac{\gamma_1}{\mu} S_1. \end{aligned} \tag{4}$$

Moreover, from seventh and eighth equations of system (1), we obtain

$$\begin{aligned} S_2 &= \frac{\tau_2 C_2}{\mu + \gamma_2 + \beta_2 C_2}, \\ D_2 &= \frac{\gamma_2}{\mu} S_2. \end{aligned} \tag{5}$$

From the sixth equation of system (1), one gets

(i) $C_2 = 0$,

(ii) $C_2 \neq 0$, which implies $\beta_1 S_1 + \beta_2 S_2 - (\tau_2 + \mu + a_2) = 0$.

If $C_2 = 0$, model (1) has a low-criminality equilibrium $P_l = (S_0, D_0, C_1, S_1, D_1, 0, 0, 0)$ which indicates the existence of offenders who are incarcerated only once in life (C_1), given by

$$\begin{aligned} S_0 &= \frac{\Lambda}{(\delta_0 + \gamma_0 + \mu)}, \\ D_0 &= \frac{\gamma_0}{\mu} S_0, \\ C_1 &= \frac{\delta_0}{(\tau_1 + \mu + a_1)} S_0, \\ S_1 &= \frac{\tau_1 \delta_0}{(\tau_1 + \mu + a_1)(\mu + \gamma_1)} S_0, \\ D_1 &= \frac{\gamma_1}{\mu} S_1. \end{aligned} \tag{6}$$

To determine the stability of this equilibrium, the Jacobian of system (1) is computed and evaluated at P_l. Hence, the low-criminality equilibrium $P_l = (S_0, D_0, C_1, S_1, D_1, 0, 0, 0)$ is locally asymptotically stable if $R_1^* < 1$, where R_1^*, defined as Criminality Reproduction Number (CRN), is given by

$$R_1^* = \frac{\beta_1}{\beta_1^*} \tag{7}$$

with

$$\beta_1^* = \frac{(\tau_1 + \mu + a_1)(\mu + \gamma_1)(\tau_2 + \mu + a_2)}{\tau_1 \delta_0 S_0}. \tag{8}$$

Note that, in analogy to the spread of infectious diseases models [29], the CRN, R_1^*, represents the "average expected number of new offenders originated by a single persisting offender C_2^*, whilst in a criminal career." In other words, one person C_2^* who was incarcerated two times gets into contact with S_1 susceptible individuals, just released from first imprisonment, successfully and induces R_1^* persons to commit crime. In other words, R_1^* is the average number of individuals who commit crime influenced by one inmate C_2^*.

For $C_2 \neq 0$, that is, for $\beta_1 S_1 + \beta_2 S_2 - (\tau_2 + \mu + a_2) = 0$, replacing both expressions for S_1 given by (4) and S_2 given by (5), an expression for $C_2 = C_2^* > 0$ is obtained as

$$b_2 (C_2^*)^2 + b_1 C_2^* + b_0 = 0, \tag{9}$$

where

$$\begin{aligned} b_2 &= \beta_1 \beta_2 (\mu + a_2). \\ b_1 &= (\mu + \gamma_1)(\tau_2 + \mu + a_2) \\ &\quad \cdot \beta_2 \left\{ \left[\frac{\beta_1 (\mu + \gamma_2)}{\beta_2 (\mu + \gamma_1)} + \frac{(\mu + a_2)}{(\tau_2 + \mu + a_2)} \right] - R_1^* \right\}, \\ b_0 &= (\mu + \gamma_1)(\mu + \gamma_2)(\tau_2 + \mu + a_2)(1 - R_1^*), \end{aligned} \tag{10}$$

Let us now determine the conditions under which the quadratic equation (9) has positive real roots; that is, we search for the existence of multiple equilibria of system (1). However, the expression for the discriminant of the quadratic equation (9) is very complex, so we will analyze the signs of its coefficients to ensure the existence of real solutions. Thus, the conditions under which this equation has either one or two positive real roots can be determined, and these results translated into nontrivial equilibrium of system (1) which is biologically feasible (i.e., positive).

Hence, substituting the positive real solutions of the quadratic equation (9) (i.e., positive values of C_2^*) into the expressions in (5), model (1) has a high-criminality equilibrium, $P_h = (S_0, D_0, C_1, S_1^*, D_1^*, C_2^*, S_2^*, D_2^*)$, where there is coexistence of both offenders C_1 and C_2^*, given by

$$\begin{aligned} S_1^* &= \frac{\tau_1 \delta_0}{(\tau_1 + \mu + a_1) + (\mu + \gamma_1) + \beta_1 C_2^*} S_0 \\ D_1^* &= \frac{\gamma_1}{\mu} S_1^* \\ S_2^* &= \frac{\tau_2 C_2^*}{(\mu + \gamma_2) + \beta_2 C_2^*} \\ D_2^* &= \frac{\gamma_2}{\mu} S_2^*, \end{aligned} \tag{11}$$

with S_0, D_0, and C_1 given by (6). Thus, the following result is then established.

Theorem 1. *Model (1) has*

(i) *a unique positive equilibrium P_h if $b_0 < 0 \iff R_1^* > 1$;*

(ii) *a unique positive equilibrium P_h if $b_0 = 0$ and $b_1 < 0$;*

(iii) *two positive equilibria, P_h, if $b_0 > 0$ and $b_1 < 0$ and $b_1^2 - 4b_2 b_0 > 0$;*

(iv) *no positive equilibrium, otherwise.*

Since all model parameters are assumed as nonnegative, it follows from (10) that the coefficient b_2 is always positive, $b_0 < 0$ for $R_1^* > 1$ and $b_0 > 0$ for $R_1^* < 1$. Thus, it is clear from Theorem 1 that model (1) has a unique positive equilibrium, P_h, when $b_0 < 0$, that is, when $R_1^* > 1$ (case (i)).

Now, for $b_0 > 0$ and $R_1^* < 1$, the quadratic equation (9) has two positive solutions if $b_1 < 0$ and $b_1^2 - 4b_2 b_0 > 0$ (case (iii)). Hence, assuming that (9) has two positive real solutions, let C_2^- and C_2^+ be the smaller and higher value of C_2^*, respectively. Translating it into equilibrium of system (1), the question is to address what means the positive high-criminality equilibria P_h for $R_1^* < 1$. It is important to note that, in this case, system (1) can have two equilibria, which are biologically feasible, even though $R_1^* < 1$. This idea is explored more deeply below.

It is instructive at this point to explore some qualitative features for $R_1^* = 1$ and $R_1^* > 1$. Firstly, for $R_1^* = 1$, it follows that $b_0 = 0$; the quadratic equation (9) has either a unique positive root (if $b_1 < 0$) or no positive root (if $b_1 > 0$). In other words, for $R_1^* = 1$ and $b_1 < 0$, model (1) has a unique positive high-criminality equilibrium given by P_h (case (ii)).

Moreover, note that when $\beta_2 = 0$, we have $b_2 = 0$ and $b_1 = (\mu + \gamma_2)(\tau_2 + \mu + a_2)\beta_1$. Thus, if $R_1^* > 1$, then

$$C_2^* (\beta_2 = 0) = \frac{(\mu + \gamma_1)}{\beta_1} (R_1^* - 1), \tag{12}$$

and model (1) has a unique positive high-criminality equilibrium, P_h, for $\beta_2 = 0$.

For $\beta_2 \longrightarrow \infty$, we have $b_1 = (\mu + \gamma_1)(\tau_2 + \mu + a_2)\beta_2[(\mu + a_2)/(\tau_2 + \mu + a_2) - R_1^*]$ and $b_2 = \beta_1\beta_2(\mu + a_2)$ with $b_0 \ll |b_1|$ and $b_0 \ll b_2$ such that

$$C_2^* (\beta_2 \longrightarrow \infty) = \frac{(\mu + \gamma_1)(\tau_2 + \mu + a_2)}{\beta_1 (\mu + a_2)} \left[R_1^* - \frac{(\mu + a_2)}{(\tau_2 + \mu + a_2)} \right], \tag{13}$$

and $C_2^* > 0$ if only if $R_1^* > (\mu + a_2)/(\tau_2 + \mu + a_2)$. Thus, model (1) has a unique high-criminality equilibrium P_h if $\beta_2 \longrightarrow \infty$.

Unfortunately, this high-criminality equilibrium cannot be studied from its closed form, so we carried out its local stability using numerical methods. The results are provided next.

Finally, if (i), (ii), and (iii) do not occur, then there are no endemic equilibria for system (1).

In what follows, model (1) admits two realistic scenarios: the best-case scenario, where offenders were incarcerated once and it is still possible to fight crime and recidivism is given by the low-criminality equilibrium $P_l = (S_0, D_0, C_1, S_1, D_1, 0, 0, 0)$, and the worst-case scenario, where the offenders were incarcerated at least once (recidivism), which can potentially lead to increased criminal activities, given by the high-criminality equilibrium $P_h = (S_0, D_0, C_1, S_1^*, D_1^*, C_2^*, S_2^*, D_2^*)$.

Having found the scenarios in which there exist the equilibria for system (1), it is instructive to analyze whether or not these equilibria are stable under any of these scenarios. Moreover, together with the CRN, R_1^* (see (7)), and the parameter β_2, we see that each scenario can be used as a check for the existence and the stability of the equilibria. Another crucial question is which of the two incidence rates β_1 and β_2 is more likely to affect the criminal dynamics in general and criminal contacts in particular. These points are of great interest here and we then explore how our model behaves under control strategies when $R_1^* < 1$. We fix β_2 and explore the system behavior by varying β_1. We also explore what happens when we fix β_1 while β_2 is varied. These control measures are designed to fight recidivism. In both cases, the recidivism will depend upon the initial sizes of the subpopulation C_2 and indicate the possibility of backward bifurcation.

Firstly, it is worth remembering that the quantity β_1 measures the average number of new contacts generated by a typical incarcerated individual C_2 with those susceptible individuals S_1 who are ex-offenders and can become susceptible to the crime again. β_2, in turn, measures the average number of new contacts generated by C_2 with those individuals S_2 who were reoffenders (recidivist behavior) and are susceptible to the crime once again.

In this way, there are two groups of offenders: those who are under much lower risk, most of whom will go to prison once and not come back (ex-offenders), and those who repeatedly do crimes and come back multiple times to the prison (reoffenders). The strong implication of the findings is that individuals who are incarcerated are extremely likely to reoffend once they are free.

Note that when system (1) has a small influx of reoffenders (C_2), it does not generate high criminality rates and, for $R_1^* < 1$, it is still possible to minimize the spreading of crime (best-case scenario). We will see that, in this case, the corresponding low-criminality equilibrium P_l could be locally asymptotically stable. On the other hand, the criminality will persist and increase if $R_1^* > 1$ (worst-case scenario). In this case, the corresponding high-criminality equilibrium P_h could be locally asymptotically stable. This phenomenon, where the possibility of fighting the spread of the crime is lost and the criminality is potentially active, that is, where P_l loses its stability and a unique P_h appears as R_1^* increases through one, is known as forward bifurcation in epidemiology.

For models that exhibit this type of bifurcation, the requirement $R_1^* < 1$ is necessary and sufficient for the high-criminality elimination. In contrast, other models undergo another type of bifurcation, known as "backward bifurcation" in epidemiology, where two equilibria P_h^+ and P_h^- coexist with the low-criminality equilibrium, P_l; that is, there are three steady states when R_1^* is immediately less than one. Thus, the requirement $R_1^* < 1$ is necessary but is not sufficient to fight the spread of crime.

2.2.1. Analytic Strategy: β_2 Fixed with β_1 Increasing. The physical implication of backward bifurcation is that the C_2 population can engage in a high criminality level even when R_1^* crosses unity downwards. In other words, in the presence of recidivism, decreasing R_1^* below one is not a sufficient condition to make the criminality level decrease. This has very important consequence for crime control, as will be discussed later in this paper.

To check the possibility of backward bifurcation in model (1), it is necessary to know other subthreshold, and we will refer to this limit point, as expressed on the R_1^* scale, as R_1^{thr}. Hence, for $R_1^* < R_1^{thr}$, system (1) will present only the locally asymptotically stable low-criminality equilibrium P_l. For $R_{thr}^* < R_1^* < 1$, system (1) will present the locally asymptotically stable low-criminality equilibrium P_l plus two positive high-criminality equilibria, P_h^+ and P_h^-, which will correspond to the solutions of (9): C_2^+, the higher solution, which corresponds to the stable equilibrium, and C_2^-, the smaller solution, which corresponds to the unstable equilibrium.

Although the critical value of the bifurcation could not be found analytically due to the high dimension of system (1), this task can be performed numerically. In this way, our simulations show that there exists a critical value $R_{thr}^* < R_1^* = 1$, where model (1) undergoes backward bifurcation.

A schematic diagram of the backward and the forward bifurcations for system (1) is given in Figure 2, where β_1 is chosen as a bifurcation parameter; that is, we fixed β_2, whereas β_1 increases.

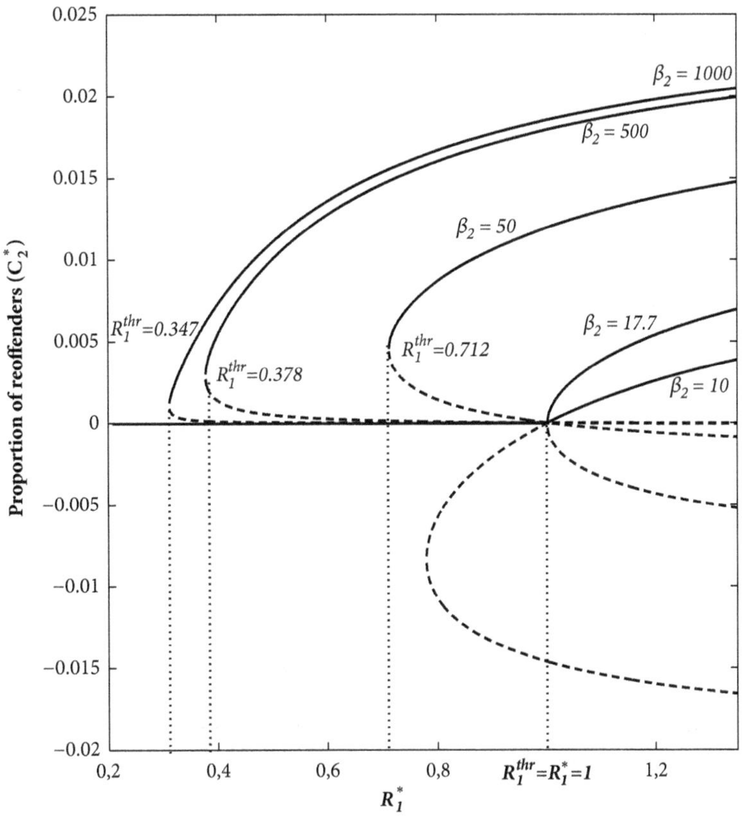

FIGURE 2: Backward and forward bifurcations diagram for proportion of reoffenders (C_2^*) for model (1), where β_1 is chosen as a bifurcation parameter (β_2 fixed and β_1 increasing). The higher solution corresponds to the stable equilibrium (solid curve); the smaller solution corresponds to the unstable equilibrium (dashed curve). Backward bifurcation for (a) $\beta_2 = 1000$, $\beta_1 = 5.371$, and $R_1^{thr} = 0.347$; (b) $\beta_2 = 500$, $\beta_1 = 5.845$, and $R_1^{thr} = 0.378$; and (c) $\beta_2 = 50$, $\beta_1 = 11$, and $R_1^{thr} = 0.712$. Forward bifurcation for (a) $\beta_2 = 17.7$, $\beta_1 = 15.45$, and $R_1^{thr} = R_1^* = 1$ and (b) $\beta_2 = 10.0$, $\beta_1 = 15.45$, and $R_1^{thr} < R_1^* = 1$. Parameters' values used are as given in Table 3.

Figure 2 shows the profile of the proportion of both reoffenders C_2^+ (solid curve) and C_2^- (dashed curve) as a function of R_1^* with decreasing values of $\beta_2 = 1000; 500; 50; 17.7; 10\ (year^{-1})$ as R_1^* increases (i.e., as β_1 increases). The solid curve stands for the stable high-criminality equilibrium, C_2^+, and the dashed curve stands for the unstable high-criminality equilibrium, C_2^-. For $\beta_2 = 1000$ and $\beta_1 = 5.371$, one has $R_1^{thr} = 0.347$. Thus, for $0.347 < R_1^* < 1$, model (1) has two positive high-criminality equilibria, while for $R_1^* > 1$, model (1) has one positive high-criminality equilibrium. Similarly, for $\beta_2 = 500$ and $\beta_1 = 5.845$, one gets $R_1^{thr} = 0.378$. Thus, for $0.378 < R_1^* < 1$, model (1) has two positive high-criminality equilibria, while for $R_1^* > 1$, model (1) has one positive high-criminality equilibrium. For $\beta_2 = 50$ and $\beta_1 = 11$, one gets $R_1^{thr} = 0.712$. Thus, for $0.712 < R_1^* < 1$, model (1) has two positive high-criminality equilibria, while for $R_1^* > 1$, model (1) has one positive high-criminality equilibrium. Finally, for $\beta_2 = 17.7$ and $\beta_1 = 15.45$ and for $\beta_2 = 10.0$ and $\beta_1 = 15.45$, we have $R_1^{thr} \leq R_1^* = 1$, and the model has one positive high-criminality equilibrium for $R_1^* > 1$ and no positive equilibrium for $R_1^* < 1$. Consequently, model (1) exhibits the forward bifurcation at $R_1^* = 1$. As it should be expected, it can be seen in Figure 2 that R_1^{thr} increases with increasing β_1 and a greater reduction in recidivism prevalence is recorded for decreasing values of β_2.

TABLE 3: Baseline values for model (1).

Variable	Description
Λ	0.015 $(years^{-1})$
δ_0	0.02 $(years^{-1})$
β_1	*variable* $(years^{-1})$
β_2	*variable* $(years^{-1})$
γ_0	0.2 $(years^{-1})$
γ_1	0.1 $(years^{-1})$
γ_2	0.08 $(years^{-1})$
μ_0	0.015 $(years^{-1})$
a_1	0.03 $(years^{-1})$
a_2	0.025 $(years^{-1})$
τ_1	0.2 $(years^{-1})$
τ_2	0.1 $(years^{-1})$

As stated earlier, the physical significance of the phenomenon of backward bifurcation is that the classical requirement of $R_1^* < 1$ is no longer sufficient for avoiding the

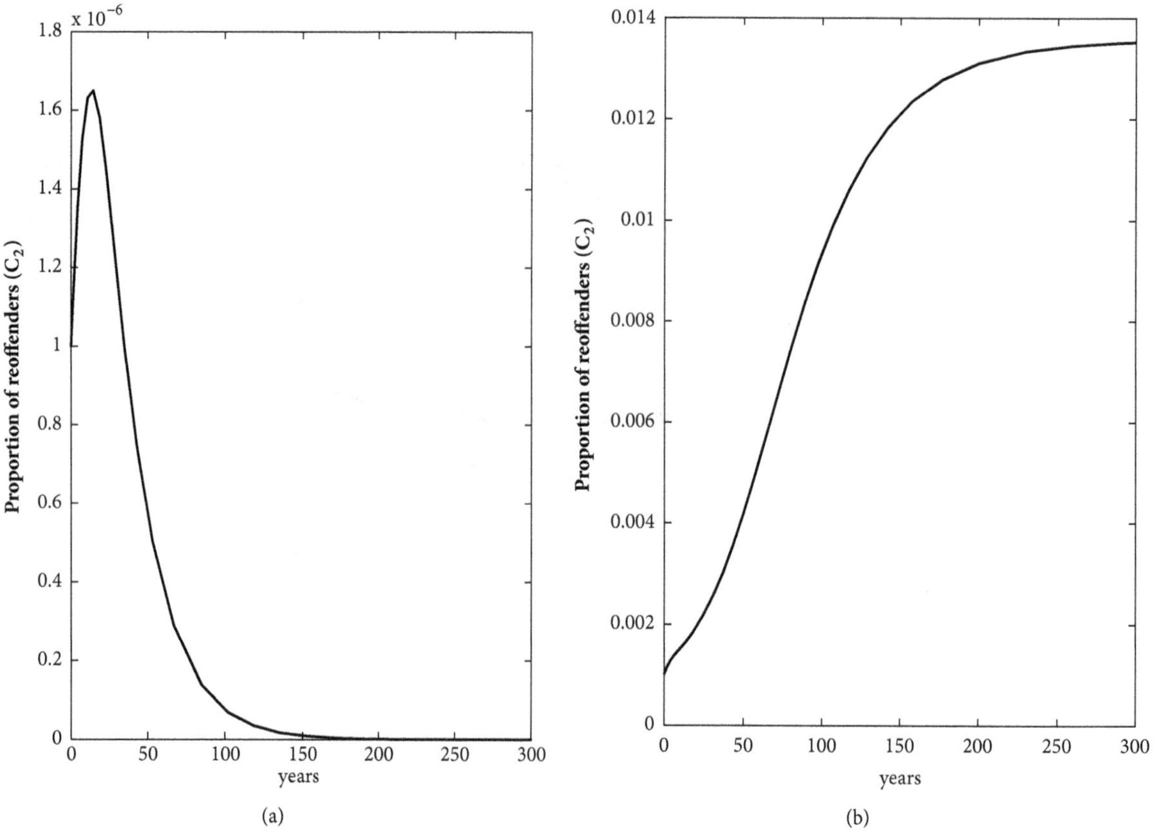

FIGURE 3: For $\beta_1 = 11$, $\beta_2 = 50$, and $R_1^{thr} < R_1^* < 1$. The other parameters' values are given in Table 3. Profile of the population of reoffenders (C_2). (a) If the recidivism is low (or $C_2(0)$ small), then the equilibrium point P_l is locally asymptotically stable; (b) if the recidivism is higher (or $C_2(0)$ large), then the equilibrium point P_h is locally asymptotically stable.

recidivism prevalence, as it is for the forward bifurcation. In such a scenario, the recidivism would depend on the initial sizes of the subpopulation C_2 of the model. That is, the presence of backward bifurcation in model (1) suggests that the possibility of avoiding the recidivism event when $R_1^* < 1$ could be dependent on the initial sizes of the subpopulation C_2. This scenario is illustrated numerically in Figure 3. Thus, if the recidivism is low, then $C_2(0)$ is small, such that the low-criminality equilibrium P_l is stable (Figure 3(a)). In contrast, if the recidivism is higher, then $C_2(0)$ is large, so the high-criminality equilibrium P_h is stable (Figure 3(b)). Therefore, the stability of these equilibrium points depends on the initial condition of system (1). This clearly indicates the coexistence of two locally asymptotically stable equilibria when $R_1^* < 1$, confirming that model (1) undergoes the phenomenon of backward bifurcation with one stable high-criminality equilibrium P_h^+ (higher, solid curve in Figure 2), one unstable high-criminality equilibrium P_h^- (lowest dashed curve in Figure 2), and one low-criminality equilibrium P_l.

2.2.2. Analytic Strategy: β_1 Fixed with β_2 Increasing. Alternatively, from now on we explore the implications of the parameter β_2 on the criminal dynamics. It is worth mentioning that, from expression (10), it is easy to verify that $b_1 < 0$ if and only if

$$\beta_2 > \frac{\beta_1(\mu + \gamma_2)}{(\mu + \gamma_1)\left[R_1^* - (\mu + a_2)/(\tau_2 + \mu + a_2)\right]} > 0, \qquad (14)$$

and thus, in such case, system (1) has two positive high-criminality equilibria (case (iii), see Theorem 1.) for $b_0 > 0$, $b_1^2 - 4b_2b_0 > 0$, and $R_1^* > (\mu + a_2)/(\tau_2 + \mu + a_2)$.

A schematic diagram of this bifurcation phenomenon for system (1) is given in Figure 4, where β_2 is chosen as a bifurcation parameter. As mentioned previously, this bifurcation phenomenon of model (1) is only illustrated numerically.

Figure 4 shows the profile of the proportion of both reoffenders C_2^+ (solid curve) and C_2^- (dashed curve) as a function of β_2 with decreasing values of $\beta_1 = 30$; 15.449; 11; 7.5; 5.845 ($year^{-1}$) as β_2 increases. The solid curve stands for the stable equilibrium and the dashed curve stands for the unstable equilibrium. Note that, for $\beta_1 = 30$ and $R_1^* > 1$, $\beta_2 = 0$ such that system (1) has a unique positive high-criminality equilibrium (see case (i) of Theorem 1 and (12)).

It should be also noticed in Figure 4 that R_1^* increases with decreasing β_2, such that greater reduction in recidivism prevalence is recorded for decreasing values of β_1. One aspect to differentiate both parameters β_1 and β_2 is the way that they account for the transition to criminal activity.

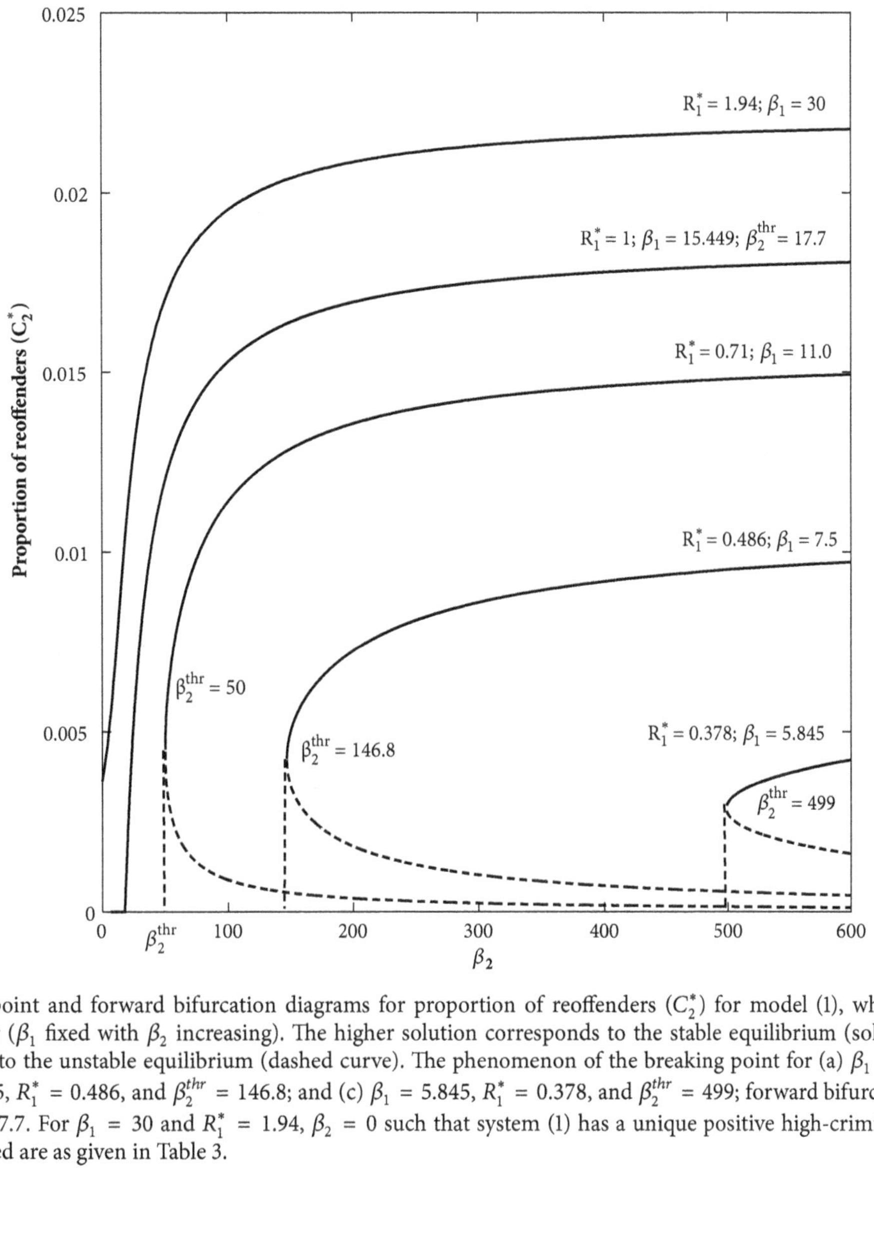

FIGURE 4: Breaking point and forward bifurcation diagrams for proportion of reoffenders (C_2^*) for model (1), where β_2 is chosen as a bifurcation parameter (β_1 fixed with β_2 increasing). The higher solution corresponds to the stable equilibrium (solid curve); the smaller solution corresponds to the unstable equilibrium (dashed curve). The phenomenon of the breaking point for (a) $\beta_1 = 11$, $R_1^* = 0.71$, and $\beta_2^{thr} = 50$; (b) $\beta_1 = 7.5$, $R_1^* = 0.486$, and $\beta_2^{thr} = 146.8$; and (c) $\beta_1 = 5.845$, $R_1^* = 0.378$, and $\beta_2^{thr} = 499$; forward bifurcation for $\beta_1 = 15.449$, $R_1^* = 1$, and $\beta_2^{thr} = 17.7$. For $\beta_1 = 30$ and $R_1^* = 1.94$, $\beta_2 = 0$ such that system (1) has a unique positive high-criminality equilibrium P_h. Parameters' values used are as given in Table 3.

In the low-criminality scenario, the only way someone becomes criminally active is on his/her own, without interacting with another person who is already criminally active. This restriction is necessary in order to illustrate the dynamics of a system which would lead to low-criminality equilibrium. If new criminal could only emerge on his/her own, then reducing the number of individuals who are not criminally active (core group) through either intervention or spontaneous desistance could lead to the lowest criminal scenario. For this, we calculated the threshold between the low-criminality equilibrium and the high-criminality equilibrium. Therefore, for $R_1^* < R_1^{thr}$, system (1) presented the locally asymptotically stable low-criminality equilibrium P_l. Although not a realistic possibility, understanding the intrinsic criminal propensity that leads to such kind of deviant behavior is a necessary step for understanding the high-criminality prevalence in the model, which will be presented in the next section.

In addition, in the high-criminality scenario, the way someone can become criminally active is to interact with another person who is already criminally active. Thus, if new reoffenders could only emerge by interaction between reoffenders (C_2^*) and those second-time offenders susceptible to the crime (S_2^*) at a rate β_2, then reducing the number of contacts between C_2^* and S_1, that is, reducing β_1, could lead to the low-criminality scenario if the proportion of reoffenders C_2^* is small (see Figure 3). Hence, to reach the low-criminality equilibrium, a community needs to reduce β_1. However, if β_1 is large, then the proportion of reoffenders C_2^*, as well as S_2^*, increases, which leads to higher crime prevalence, such that the high-criminality scenario emerges.

Figure 5(a) shows the long-term behavior of $\ln \beta_2^{thr}$ plotted versus β_1. The threshold β_2^{thr} is expressed as a minimum value where system (1) has two positive equilibrium points. Thus, for each fixed β_1, there is a corresponding unique

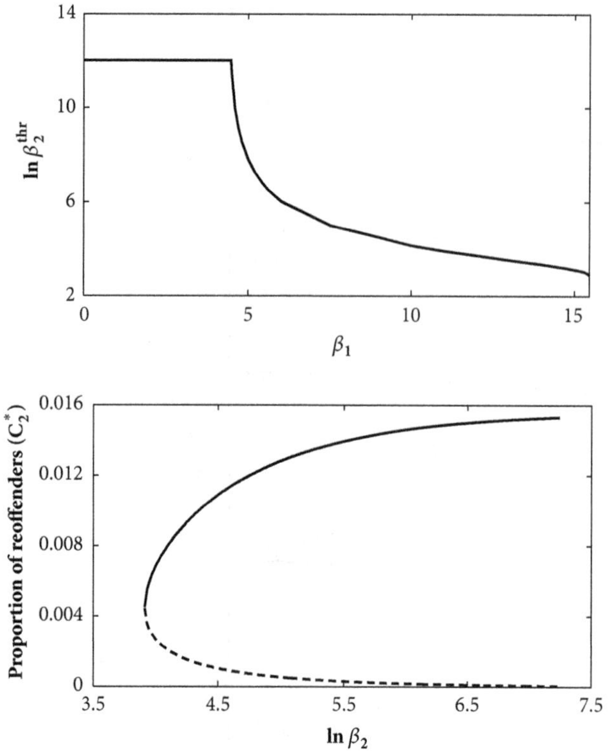

FIGURE 5: (a) The long-term behavior of $\ln \beta_2^{thr}$ plotted versus β_1. (b) For $\beta_1 = 11$, $R_1^* = 0.71$, and $\ln \beta_2 = 3.91$, system (1) has two positive equilibrium points. The solid curve stands for the stable equilibrium and the dashed curve stands for the unstable equilibrium. The other parameters' values are given in Table 3.

threshold β_2^{thr}. For one such example (see Figure 5(b)), $\beta_1 = 11$, $R_1^* = 0.71$, and $\beta_2^{thr} = 50$ (or $\ln 50 = 3.91$) are the minimum values where system (1) has two positive value. For $\beta_2 > \beta_2^{thr}$, we have $C_2^+ > 0$ and $C_2^- > 0$. In this case, system (1) has a stable equilibrium point, C_2^+ (the solid curve in Figure 5(b)), and an unstable equilibrium point, C_2^- (the dashed curve in Figure 5(b)). On the other hand, for $\beta_2 < \beta_2^{thr}$, we have $C_2^+ = C_2^- = 0$, which means that there are no positive equilibrium points for system (1). Moreover, for $\beta_1^{thr} = 15.4492917$, we have $R_1^* = 1$ and $\beta_2^{thr} = 17.867$ (or $\ln 17.867 = 2.883$) such that, for $\beta_2 < \beta_2^{thr}$, one has $C_2^+ = 0$ and $C_2^- < 0$. Otherwise, for $\beta_2 > \beta_2^{thr}$, one has $C_2^+ > 0$ and $C_2^- = 0$ (see Figure 6).

2.3. The Full Contagion Criminal Model (FCCM) Formulation. In this section, we relax the assumption made in the PCCM model (1) that the susceptible individuals (S_0) can enter to crime only on their own, so we make $\delta_0 = 0$. Hence, in the modified model, the flow of the susceptible individuals S_0 into the criminal activity depends only on the contact with those individuals who are incarcerated, C_1. We also define β_i (i = 0,1,2) as the rate of imprisonment; it captures the return to criminal activity of those released from prison, such that β_0, β_1, and β_2 are the flows into criminal activity which depend on having had contact with those first-time and second-time incarcerated individuals (C_1 and C_2), respectively. Following the idea of the previous model (1), here β_i (i = 0,1,2) is analogous to the effective contact rate in infectious diseases model. This derivation adopts a standard incidence formulation, where the contact rate is assumed to be constant, unlike the case of the mass action formulation, where the contact rate depends on the size of the total population (see, e.g., [27, 30], for detailed derivation of these incidences' functions).

The flow diagram of the FCCM is depicted in Figure 7. The variables and parameters' values are given in Table 3, except for β_0 that is the contact rate between C_1 and S_0 (cooptation rate).

The model is represented by the following nonlinear system of differential equations:

$$\begin{aligned}
\frac{dS_0}{dt} &= \Lambda - \beta_0 C_1 S_0 - (\gamma_0 + \mu) S_0 \\
\frac{dD_0}{dt} &= \gamma_0 S_0 - \mu D_0 \\
\frac{dC_1}{dt} &= \beta_0 C_1 S_0 - (\tau_1 + \mu + a_1) C_1 \\
\frac{dS_1}{dt} &= \tau_1 C_1 - (\gamma_1 + \mu) S_1 - \beta_1 C_2 S_1 \\
\frac{dD_1}{dt} &= \gamma_1 S_1 - \mu D_1 \\
\frac{dC_2}{dt} &= (\beta_1 S_1 + \beta_2 S_2) C_2 - (\tau_2 + \mu + a_2) C_2 \\
\frac{dS_2}{dt} &= \tau_2 C_2 - \beta_2 C_2 S_2 - (\gamma_2 + \mu) S_2 \\
\frac{dD_2}{dt} &= \gamma_2 S_2 - \mu D_2
\end{aligned} \qquad (15)$$

2.4. The Existence and Local Stability of Equilibria

2.4.1. Crime-Free Equilibrium. In the absence of crime, that is, for $C_1 = C_2 = 0$, model (15) has a crime-free equilibrium $P_0 = (S_0^0, D_0^0, 0, 0, 0, 0, 0, 0)$ which is obtained by setting the right-hand sides of system (15) to zero, where

$$\begin{aligned}
S_0^0 &= \frac{\Lambda}{(\mu + \gamma_0)}, \\
D_0 &= \frac{\gamma_0}{\mu} S_0^0.
\end{aligned} \qquad (16)$$

To analyze the local stability of this equilibrium, the Jacobian of system (15) is computed and evaluated at P_0, which is locally asymptotically stable if the real parts of the eigenvalues of the Jacobian matrix are all negative.

Thus, the local stability of the crime-free equilibrium P_0 is governed by the Jacobian matrix

$$M_{P_0} = \begin{bmatrix} A_{P_0} & 0 \\ 0 & B_{P_0} \end{bmatrix}, \qquad (17)$$

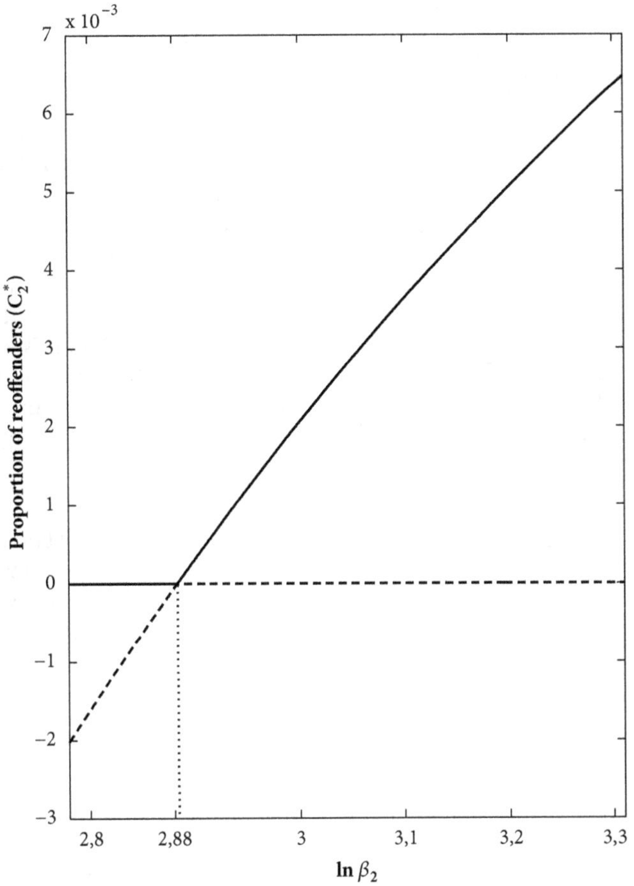

FIGURE 6: Long-term physical outcome C_2^* plotted versus $\ln \beta_2$. For $\beta_1^{thr} = 15.449$, $R_1^* = 1$, and $\beta_2 > \beta_2^{thr}$, system (1) has one positive equilibrium point. The solid curve stands for the stable equilibrium and the dashed curve stands for the unstable equilibrium. The other parameters' values are given in Table 3.

where

$$A_{P_0} = \begin{bmatrix} -(\mu + \gamma_0) & -\beta_0 S_0^0 \\ 0 & \beta_0 S_0^0 - (\tau_1 + \mu + a_1) \end{bmatrix}, \quad (18)$$

$$B_{P_0} = \begin{bmatrix} -(\mu + \gamma_1) & 0 & 0 \\ 0 & -(\tau_2 + \mu + a_2) & 0 \\ 0 & \tau_2 & -(\mu + \gamma_2) \end{bmatrix}. \quad (19)$$

It is easy to verify that the three eigenvalues of matrix (19) are always negative. In the same way, it is also straightforward to verify that one of the eigenvalues of matrix (18) is always negative, while the other is negative whenever

$$R_0 = \frac{\beta_0}{(\tau_1 + \mu + a_1)} S_0^0 < 1. \quad (20)$$

Hence, all the eigenvalues of matrix (17) are negative or have negative real parts if and only if $R_0 < 1$. In summary, the crime-free equilibrium P_0 of system (15) is locally asymptotically stable if the Basic Criminality Reproduction Number $R_0 < 1$. Therefore, we have established the following result.

Lemma 2. *The crime-free equilibrium P_0 of system (15) is locally asymptotically stable if $R_0 < 1$ and is unstable if $R_0 > 1$.*

2.4.2. Low-Criminality Equilibrium. In the absence of the reoffenders, that is, for $C_2 = 0$, model (15) has a low-criminality equilibrium $P_l^* = (S_0, D_0, C_1, S_1, D_1, 0, 0, 0)$, where

$$\begin{aligned} S_0 &= \frac{(\tau_1 + \mu + a_1)}{\beta_0} \\ D_0 &= \frac{\gamma_0}{\mu} S_0 \\ C_1 &= \frac{\Lambda\beta_0 - (\mu + \gamma_0)(\tau_1 + \mu + a_1)}{(\tau_1 + \mu + a_1)\beta_0} \\ &= \frac{(\mu + \gamma_0)}{\beta_0}(R_0 - 1) \\ S_1 &= \frac{\tau_1 C_1}{(\mu + \gamma_1)} \\ D_1 &= \frac{\gamma_1}{\mu} S_1. \end{aligned} \quad (21)$$

Note that P_l^* exists if $C_1 > 0$, that is, if $R_0 > 1$. Now consider the resulting model (15). The local stability

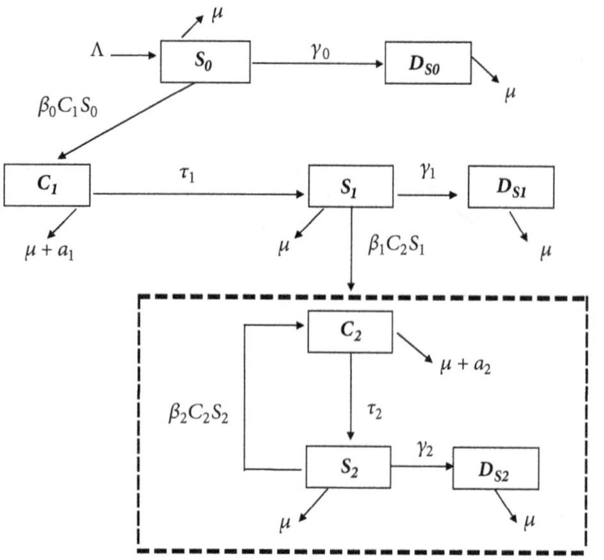

FIGURE 7: The flow diagram for the free-low-high criminality model (15).

of the low-criminality equilibrium P_l, which is examined by linearizing system (15) around P_l, is governed by the Jacobian matrix

$$M_{P_l} = \begin{bmatrix} M_{11} & -\beta_0 S_0 & 0 & 0 & 0 \\ \beta_0 C_1 & M_{22} & 0 & 0 & 0 \\ 0 & \tau_1 & M_{33} & -\beta_1 S_1 & 0 \\ 0 & 0 & 0 & M_{44} & 0 \\ 0 & 0 & 0 & \tau_2 & M_{55} \end{bmatrix}, \quad (22)$$

with $M_{11} = -\beta_0 C_1 - (\mu + \gamma_0)$; $M_{22} = \beta_0 S_0 - (\tau_1 + \mu + a_1)$; $M_{33} = -(\mu+\gamma_1)$; $M_{44} = \beta_1 S_1 - (\tau_2+\mu+a_2)$; and $M_{55} = -(\mu+\gamma_2)$.

The eigenvalues of matrix (22) are $\lambda_1 = M_{55}$, $\lambda_2 = M_{44}$, $\lambda_3 = M_{33}$, and the roots of

$$\phi^2 + z_1\phi + z_0 = 0, \quad (23)$$

with $z_1 = \Lambda\beta_0/(\tau_1 + \mu + a_1) > 0$ and $z_0 = \Lambda\beta_0 - (\mu + \gamma_0)(\tau_1 + \mu + a_1) > 0 (\Longleftrightarrow R_0 > 1)$.

The eigenvalues λ_1 and λ_3 are real negative from definition of the parameters μ, γ_1, and γ_2. The eigenvalue λ_2 is negative if

$$\beta_1 < \frac{(\tau_2 + \mu + a_2)}{S_1} = \frac{\beta_0 (\tau_2 + \mu + a_2)(\mu + \gamma_1)}{\tau_1 (\mu + \gamma_0)(R_0 - 1)} \quad (24)$$

or, equivalently, if

$$R_1 = \frac{\beta_1 \tau_1 (\mu + \gamma_0)(R_0 - 1)}{(\tau_2 + \mu + a_2)(\mu + \gamma_1)\beta_0} < 1 \quad (25)$$

Finally, the eigenvalues λ_4 and λ_5 are real negative if $z_0 > 0$ and $z_1 > 0$ by applying the Routh-Hurwitz criteria [31, 32] on polynomial (23). In this sense, $z_0 > 0$ and $z_1 > 0$ if and only if $R_0 > 1$.

Thus, we have established the following result.

Lemma 3. *The low-criminality equilibrium P_l^* of system (15) exists and it is locally asymptotically stable if $R_0 > 1$ and $R_1 < 1$. Otherwise, P_l^* is unstable.*

It should be mentioned that the consequence of the above result is that when P_l^* becomes unstable, two scenarios emerge: one for the case where the criminality is eliminated from population and P_0 is stable and the other where new criminals could emerge and there is high-criminality prevalence in the population such that model (15) has a high-criminality equilibrium with coexistence of both offenders: $C_1 \neq 0$ and $C_2 \neq 0$. In this way, it is instructive to determine the possible interventions on β_0 and β_1 in order to reduce R_0 and R_1 below one, that is, to guarantee the conditions under which the criminality is eliminated, or at least having its incidence reduced.

2.4.3. High-Criminality Equilibrium. In what follows, if $R_1 > 1$, such that the low-criminality equilibrium P_l^* is unstable, system (15) has the positive high-criminality equilibrium, $P_h^* = (S_0, D_0, C_1, S_1^*, D_1^*, C_2^*, S_2^*, D_2^*)$, where

$$\begin{aligned} S_1^* &= \frac{\tau_1 C_1}{\mu + \gamma_1 + \beta_1 C_2^*}, \\ S_2^* &= \frac{\tau_2 C_2^*}{\mu + \gamma_2 + \beta_2 C_2^*}, \\ D_1^* &= \frac{\gamma_1}{\mu} S_1^*, \\ D_2^* &= \frac{\gamma_2}{\mu} S_2^*. \end{aligned} \quad (26)$$

Remembering from (21) that $C_1 > 0 \Longleftrightarrow R_0 > 1$, so P_h^* exists if and only if $R_0 > 1$. Now replacing the expressions for S_1^* and S_2^* in the sixth equation of system (15), the positive high-criminality equilibrium, which cannot be expressed cleanly in closed form, can then be obtained by solving for C_2^* the following expression:

$$\eta_2 (C_2^*)^2 + \eta_1 C_2^* + \eta_0 = 0, \quad (27)$$

where

$$\begin{aligned} \eta_2 &= \beta_1 \beta_2 (\mu + a_2), \\ \eta_1 &= (\mu + \gamma_1)(\tau_2 + \mu + a_2) \\ &\quad \cdot \beta_2 \left\{ \left[\frac{\beta_1 (\mu + \gamma_2)}{\beta_2 (\mu + \gamma_1)} + \frac{(\mu + a_2)}{(\tau_2 + \mu + a_2)} \right] - R_1 \right\} \\ \eta_0 &= (\mu + \gamma_1)(\mu + \gamma_2)(\mu + \gamma_2 + a_2)(1 - R_1). \end{aligned} \quad (28)$$

Since all the model's parameters are nonnegative, it follows from (28) that the coefficient η_2 is always positive and $\eta_0 < 0$ for $R_1 > 1$. Thus, it is clear that model (15) has a unique positive equilibrium P_h^+ when $R_1 > 1$ and $R_0 > 1$. For $\eta_0 = 0$ and $\eta_1 < 0$, model (15) also has a unique positive equilibrium

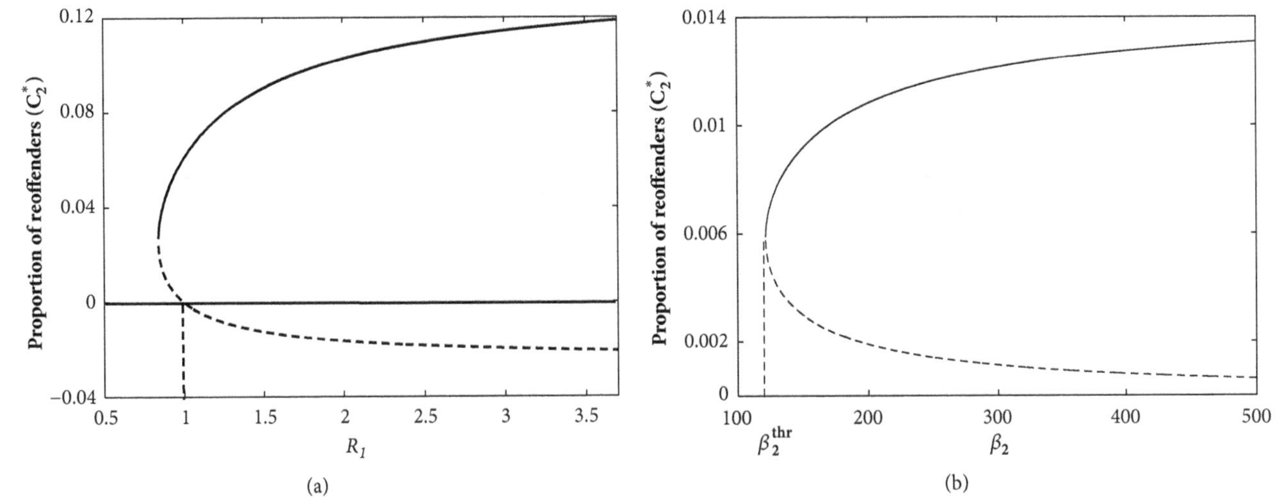

FIGURE 8: The phenomena of backward bifurcation and breaking point, when β_1 and β_2 are chosen as a bifurcation parameter. (a) Backward bifurcation for $\beta_2 = 5$ with β_0 and β_1 increasing. (b) Breaking point for $\beta_0 = \beta_1 = 5$ with β_2 increasing, where $\beta_2^{thr} = 121.2$ and $R_0 = 1.139$ and $R_1 = 0.464$. The higher solution corresponds to the stable equilibrium (solid curve); the smaller solution corresponds to the unstable equilibrium (dashed curve). Parameters' values used are as given in Table 3.

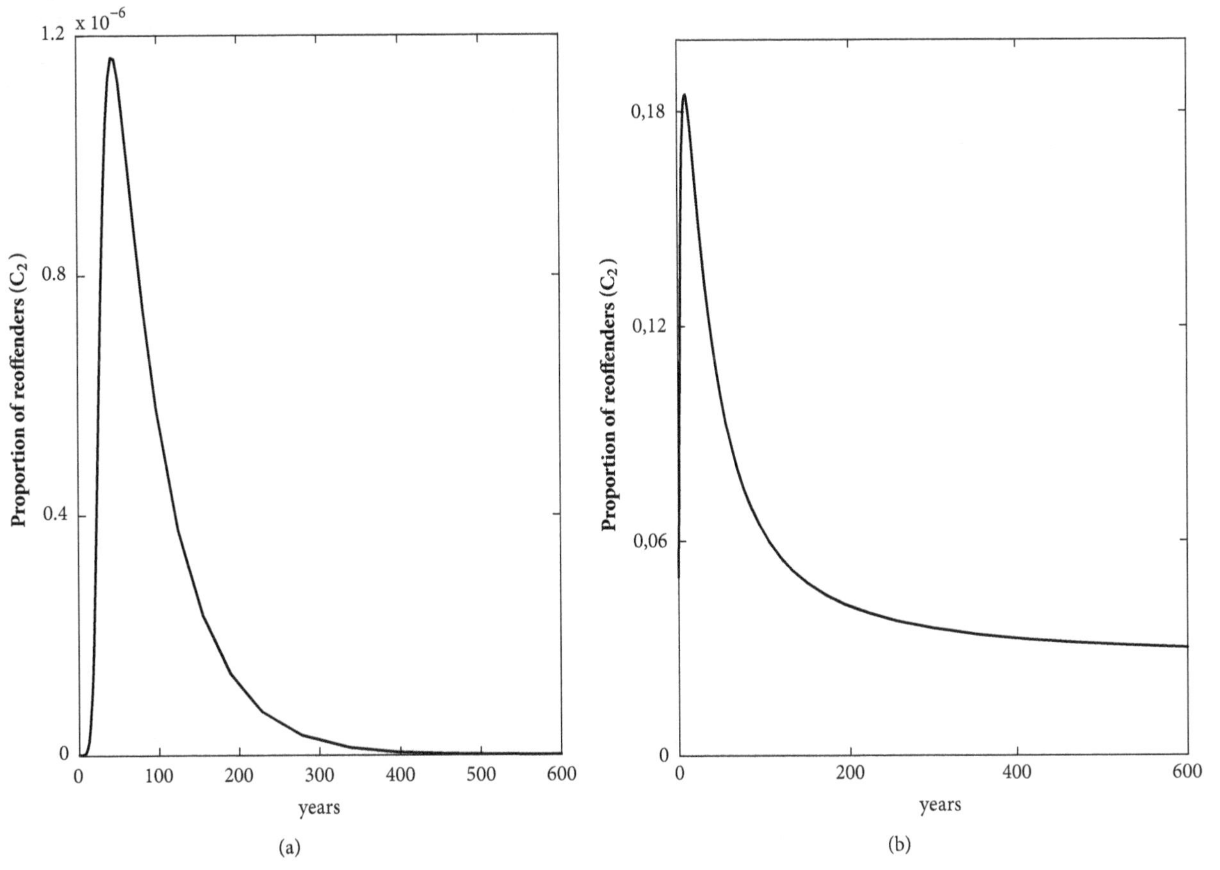

FIGURE 9: For $\beta_1 = 2.0$, $\beta_2 = 4.4$ ($\beta_2 > \beta_2^{thr}$), $R_0 = 2.39$, and $R_1 = 0.89$. The other parameters' values are given in Table 3. Profile of population of reoffenders (C_2). (a) If the recidivism is low (or $C_2(0)$ small), then the equilibrium point P_l^* is locally asymptotically stable. (b) If the recidivism is higher (or $C_2(0)$ large), then the equilibrium point P_h^+ is locally asymptotically stable.

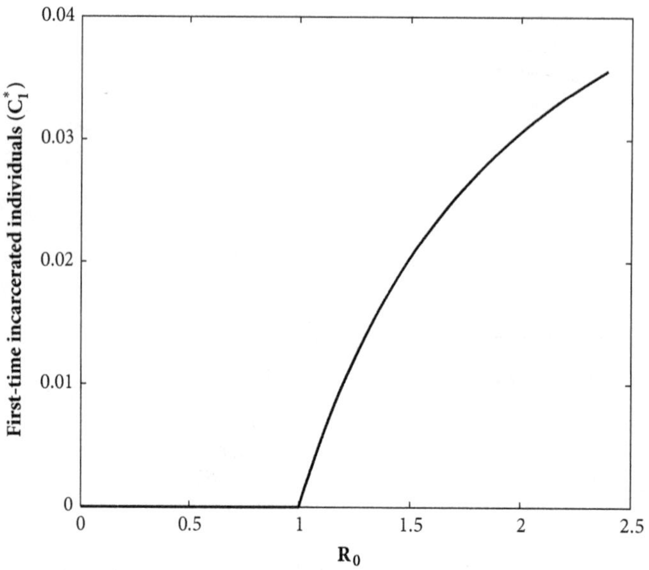

FIGURE 10: The phenomenon of forward bifurcation when β_0 is chosen as a bifurcation parameter. The crime-free equilibrium P_0 of system (15) is locally asymptotically stable if $R_0 < 1$ and is unstable if $R_0 > 1$. Parameters' values used are as given in Table 3.

P_h^+. Moreover, whenever $R_0 > 1$, if $R_1 = 1$, then $\eta_0 = 0$ and $\eta_1 < 0$ if only if

$$\beta_2 > \frac{(\tau_2 + \mu + a_2)(\mu + \gamma_2)\beta_1}{\tau_2(\mu + \gamma_1)}, \quad (29)$$

such that model (15) also has a unique positive equilibrium P_h^*. Finally, for $\eta_0 > 0$, $\eta_1 < 0$, and $\eta_1^2 - 4\eta_2\eta_0 > 0$, that is, for $R_0 > 1$ and $R_1 < 1$, model (15) has two positive equilibria P_h^+ and P_h^-. Note that these equilibrium points cannot be studied in its closed form, so we carried out its local stability using numerical methods.

Thus, the following result is then established.

Theorem 4. *Model (15) has*

(i) *a unique positive equilibrium P_h^+ if $\eta_0 < 0 \iff R_1 > 1$;*

(ii) *a unique positive equilibrium P_h^+ if $\eta_0 = 0$ and $\eta_1 < 0$;*

(iii) *two positive equilibria, P_h^+ and P_h^-, if $\eta_0 > 0$ and $\eta_1 < 0$ and $\eta_1^2 - 4\eta_2\eta_0 > 0$;*

(iv) *no positive equilibrium, otherwise.*

As explored in model (1), model (15) also exhibits the phenomenon of backward bifurcation (Figure 8(a)) and breaking point (Figure 8(b)) when β_1 and β_2 are chosen as bifurcation parameters.

Figure 8(a) shows the backward bifurcation for $\beta_2 = 5$ and $\beta_0 = 8.4$ ($R_0 = 2.39$) with β_1 increasing. Figure 8(b) shows the phenomenon of breaking point for $\beta_1 = \beta_0 = 5$, with β_2 increasing, where $\beta_2^* = 121.2$ and $R_0 = 1.139$ and $R_1 = 0.464$. As it should be expected, Figures 8(a) and 8(b) present the same results shown in Figures 2 and 4. In summary, if $R_0 > 1$ and $R_1 > 1$, model (15) has the locally asymptotically stable high-criminality equilibrium P_h^+. Finally, if $R_0 > 1$ and $R_1 < 1$ and $\beta_2 > \beta_2^{thr}$, then there is coexistence of two locally asymptotically stable equilibria, P_l^* and P_h^+, such that the stability of these equilibrium points depends on the initial condition of system (15) (see Figure 9).

In contrast to model (1), when β_0 is chosen as a bifurcation parameter, model (15) has the crime-free equilibrium P_0 given by (16), which indicates the possibility of the forward bifurcation (Figure 10). If $R_0 < 1$, then the crime-free equilibrium P_0 of system (15) is locally asymptotically stable; if $R_0 > 1$, P_0 becomes unstable and P_l^* is locally asymptotically stable if $R_1 < 1$. Thus, the substitution of the assumption that the susceptible individuals (S_0) can enter to crime only on their own (i.e., δ_0) by the standard incidence (i.e., $\beta_1 C_1 S_0$) in model (1) includes the forward bifurcation phenomenon to model (15).

3. Results

To illustrate the theoretical results contained in this paper, models (1) and (15) are simulated using baseline parameters values/ranges summarized in Table 3 (unless otherwise stated). The parameters are chosen for simulations purposes only, so we could illustrate our qualitative results. Moreover, it is worth mentioning that if the assumption made in the PCCM model (1), that is, susceptible individuals can enter to crime only on their own (i.e., $\delta_0 \neq 0$), and the assumptions made in FCCM model (15), that is, susceptible individuals get into criminal behavior depending only on the contact with those individuals who are incarcerated, C_1 (i.e., $\beta_0 C_1 S_0$), were considered in a single model, this new model would be more realistic. However, the analytical results and the phenomena of backward bifurcation and breaking point of this new model would be similar to model (1), except for the fact that R_0 would no longer exist. For this reason, we did not study this more realistic mixed model.

4. Discussion

In this paper, we attempt to present theoretical models of criminal careers using the dynamical system approach traditionally used in the study of infectious diseases spreading in a homogeneous population. The models consider crime dynamics as contagious phenomena in which a susceptible cohort of individuals with criminal propensity is "infected" by criminal individuals who have been convicted for their crimes and are arrested in a prison. The contagion occurs by several ways, in particular through the sending of messages by mobile phones (a very widespread habit in Brazilian prisons) and through "carrier pigeons" represented by relatives and lawyers.

We propose, to the best of our knowledge by the first time in quantitative criminology, the existence of thresholds for the stability of crime-endemic equilibrium which are the equivalent to the "basic reproduction number" widely used in the mathematical epidemiology literature [22], as shown in Lemmas 2 and 3. Both model (1) and model (15), however, exhibit the phenomena of backward bifurcation and breaking point when the contact rates β_1 and β_2 are chosen as bifurcation parameters. Since data strongly suggests that standard incidence formulation is more suited for modeling human diseases [22, 23, 27] and we adopted this formulation in the current study, the above results show that the phenomena of the backward bifurcation and the breaking point could be important properties of the criminality model.

As in other criminal career models, our models seek to provide a theoretical framework to analyse the longitudinal behavior of individuals who commit criminal offenses [33]. In addition, we centrered our analysis, although in an implicit way, in the parameter considered to be the most important for the analysis of criminal career, namely, the rate at which offenders commit crime, denoted in the specialized literature by the Greek letter λ, coincidently, the same symbol used to denote the force-of-infection in epidemiology of transmissible diseases.

Other dynamical system models proposed the analysis of criminal dynamics including differential equations, like the works by Farrington synthesized in Refs. [16, 34]. However, these models are either linear (the former) or related to other kinds of infectious contagion (the latter).

Finally, it is noteworthy that the models proposed in the paper are intended only to provide a theoretical framework upon which other works can provide empirical support for the assumptions and values for the parameters determinant of the dynamical behavior of the systems here studied. The present work, therefore, is intended to provide the first step in the study of criminal careers as determined by contagious events like the ones related to the phenomenon of incarcerated criminals influencing the behavior of susceptible juveniles outside prisons with a criminal propensity. The finding of backward bifurcation in both models, however, implies that there is an endemic equilibrium of criminality even when the threshold parameter for contagion is below unit which, in turn, implies that control strategies are more difficult to achieve any considerable impact on crime control in situations similar to the ones here analyzed.

Conflicts of Interest

The authors declare that they have no conflicts of interest.

Acknowledgments

This work has been funded by LIM01-HCFMUSP, FAPESP, CNPq and by the Office of Naval Research, USA, Grant Number 2059558. An earlier version of this work has been presented as at EMAp Seminars according to the following link: https://emap.fgv.br/seminarios/contagious-criminal-career-models-showing-backward-bifurcations-implications-crime.

References

[1] J. R. Kling and J. Ludwig, "Is Crime Contagious?" *The Journal of Law and Economics*, vol. 50, no. 3, pp. 491–518, 2007.

[2] C. F. Manski, "Identification of endogenous social effects: the reflection problem," *Review of Economic Studies*, vol. 60, no. 3, pp. 531–542, 1993.

[3] K. A. Goss and P. Cook A, *A Selective Review of the Social-Contagion Literature*, Terry Sanford Instiute Working Paper, Duke University, Durham, NC, USA, 1996.

[4] C. F. Manski, "Economic analysis of social interactions," *Journal of Economic Perspectives (JEP)*, vol. 14, no. 3, pp. 115–136, 2000.

[5] W. J. Wilson, *The Truly Disadvantaged: The Inner City, the Underclass, and Public Policy*, The University of Chicago Press, Chicago, Illinois, 1987.

[6] R. J. Sampson, S. W. Raudenbush, and F. Earls, "Neighborhoods and violent crime: a multilevel study of collective efficacy," *Science*, vol. 277, no. 5328, pp. 918–924, 1997.

[7] C. Jencks and S. E. Mayer, "The Social Consequences of Growing Up," in *Inner-City Poverty in the United States*, L. Lynn and M. McGeary, Eds., National Academy of Sciences, Washington, DC, USA, 1990.

[8] E. L. Glaeser, B. Sacerdote, and J. A. Scheinkman, "Crime and social interactions," *The Quarterly Journal of Economics*, vol. 111, no. 2, pp. 507–548, 1996.

[9] E. L. Glaeser, B. I. Sacerdote, and J. A. Scheinkman, "The social multiplier," *Journal of the European Economic Association*, vol. 1, no. 2-3, pp. 345–353, 2003.

[10] J. Crane, "The epidemic theory of ghettos and neighborhood effects on dropping out and teenage childbearing," *American Journal of Sociology*, vol. 96, no. 5, pp. 1226–1259, 1991.

[11] D. A. Levy and P. R. Nail, "Contagion: a theoretical and empirical review and reconceptualization," *Genetic Social and General Psychology Monographs*, vol. 119, no. 2, pp. 233–284, 1993.

[12] J. Fagan, D. L. Wilkinson, and G. Davies, "Social contagion of violence," in *The Cambridge Handbook of Violent Behavior and Aggression*, D. Flannery, A. T. Vazsonyi, and I. D. Waldman, Eds., Cambridge University Press, Cambridge, UK, 2007.

[13] R. S. Burt, "Social Contagion and Innovation: Cohesion versus Structural Equivalence," *American Journal of Sociology*, vol. 92, no. 6, pp. 1287–1335, 1987.

[14] L. O. Gostin, "The interconnected epidemics of drug dependency and AIDS," *Harvard Civil Rights Civil Liberties Law Review*, vol. 26, no. 1, pp. 113–184, 1991.

[15] J. L. Rodgers and D. C. Rowe, "Social contagion and adolescent sexual behavior: A developmental EMOSA model," *Psychological Review*, vol. 100, no. 3, pp. 479–510, 1993.

[16] S. B. Patten, "Epidemics of violence," *Medical Hypotheses*, vol. 53, no. 3, pp. 217–220, 1999.

[17] E. Massad, A. F. Rocha, F. A. B. Coutinho, and L. F. Lopez, "Modelling the spread of memes: How inovations are transmitted from brain to brain," *Applied Mathematical Sciences*, vol. 7, no. 45-48, pp. 2295–2306, 2013.

[18] K. Kirkpatrick, "The Social Contagion of Violence: A Theoretical Exploration of the Nature of Violence in Society," in *Social Science*, vol. 461, 462, California Polytechnic State University, San Luis Obispo, CA, USA, 2018, http://digitalcommons.calpoly.edu/socssp/78.

[19] S. B. Patten and J. A. Arboleda-Flórez, "Epidemic theory and group violence," *Social Psychiatry and Psychiatric Epidemiology*, vol. 39, no. 11, pp. 853–856, 2004.

[20] C. C. N. Dias, *PCC: Hegemonia nas prisões e monopólio da violência (Portuguese)*, Editora Saraiva, São Paulo, Brazil, 2013.

[21] S. Adorno and F. Salla, "Criminalidade organizada nas prisões e os ataques do PCC," *Estudos Avançados*, vol. 21, no. 61, pp. 7–29, 2007.

[22] R. M. Anderson and R. M. May, *Population Biology of Infectious Diseases*, Springer-Verlag, Berlin, Heidelberg, New York, 1982.

[23] R. M. Anderson and R. M. May, *Infectious Diseases of Humans: Dynamics and Control*, Oxford University Press, Oxford, England, 2nd edition, 1991.

[24] S. Machin, M. Olivier, and V. Suncica, "The crime reducing effect of education," *Economic Journal*, vol. 121, no. 552, pp. 463–484, 2011.

[25] M. Maguire, "Merry Morash: Understanding Gender, Crime and Justice," *Critical Criminology*, vol. 16, no. 3, pp. 225–227, 2008.

[26] J. K. Hale, *Ordinary Differential Equations*, Krieger, Basel, Switzerland, 1980.

[27] H. W. Hethcote, "The mathematics of infectious diseases," *SIAM Review*, vol. 42, no. 4, pp. 599–653, 2000.

[28] V. Lakshmikantham, S. Leela, and A. A. Martynyuk, *Stability Analysis of Nonlinear Systems*, Marcel Dekker, New York, NY, USA, 1989.

[29] E. Massad and F. A. B. Coutinho, "Vectorial capacity, basic reproduction number, force of infection and all that: Formal notation to complete and adjust their classical concepts and equations," *Memórias do Instituto Oswaldo Cruz*, vol. 107, no. 4, pp. 564–567, 2012.

[30] O. Sharomi, C. N. Podder, A. B. Gumel, E. H. Elbasha, and J. Watmough, "Role of incidence function in vaccine-induced backward bifurcation in some HIV models," *Mathematical Biosciences*, vol. 210, no. 2, pp. 436–463, 2007.

[31] L. Fernandez Lopez, F. A. Bezerra Coutinho, M. Nascimento Burattini, and E. Massad, "Threshold conditions for infection persistence in complex host-vectors interactions," *Comptes Rendus Biologies*, vol. 325, no. 11, pp. 1073–1084, 2002.

[32] S. M. Raimundo, H. M. Yang, E. Venturino, and E. Massad, "Modeling the emergence of HIV-1 drug resistance resulting from antiretroviral therapy: Insights from theoretical and numerical studies," *BioSystems*, vol. 108, no. 1-3, pp. 1–13, 2012.

[33] A. Blumstein, J. Cohen, J. A. Roth, and C. A. Visher, *Criminal Careers and Career Criminals*, vol. I, The National Academies Press, Washington, DC, USA, 1986.

[34] J. F. MacLeod, P. Grove, and D. Farrington, *Explaining Criminal Careers: Implications for Justice Policy*, Oxford University Press, Oxford, UK, 2012.

7

An Efficient Numerical Method for a Class of Nonlinear Volterra Integro-Differential Equations

M. H. Daliri Birjandi, J. Saberi-Nadjafi, and A. Ghorbani

Department of Applied Mathematics, School of Mathematical Sciences, Ferdowsi University of Mashhad, Mashhad, Iran

Correspondence should be addressed to J. Saberi-Nadjafi; najafi141@gmail.com

Academic Editor: Mehmet Sezer

We investigate an efficient numerical method for solving a class of nonlinear Volterra integro-differential equations, which is a combination of the parametric iteration method and the spectral collocation method. The implementation of the modified method is demonstrated by solving several nonlinear Volterra integro-differential equations. The results reveal that the developed method is easy to implement and avoids the additional computational work. Furthermore, the method is a promising approximate tool to solve this class of nonlinear equations and provides us with a convenient way to control and modify the convergence rate of the solution.

1. Introduction

Many physical phenomena in different fields of sciences and engineering have been formulated using integro-differential equations. The nonlinear integro-differential equations play a crucial role to describe many process like fluid dynamics, biological models and chemical kinetics, population, potential theory, polymer theology, and drop wise condensation (see [1–4] and the references cited therein). In fact analytical solutions of integro-differential equations either do not exist or they are hard to compute. Eventually an exact solution is computable, the required calculations may be tedious, or the resulting solution may be difficult to interpret. Due to this, it is required to obtain an efficient numerical solution. In literature there exist several numerical methods for solving integro-differential equations such as successive approximation method, meshless method [5], Taylor polynomial [6], Tau method [4], wavelet-Galerkin method [7], Adomain decomposition method [8], Homotopy perturbation method [9], Homotopy analysis method [10], Sinc collocation [11], Legendre polynomials [12], and Taylor collocation method [13]. The monograph by Bruner [14] includes a wealth of material on the theory and numerical methods for Volterra integro-differential equations.

The parametric iteration method (PIM) is an analytic approximate method that provides the solution of linear and nonlinear problem as a sequence of iterations. In fact, the PIM as a fixed- point iteration method is a reconstruction of variational iteration method [15]. The PIM, however, suffers from a number of restrictive measures, such as the resulting integrals in its iterative relation which may not be performed analytically. Also, the implementation of the PIM generally leads to calculation of unneeded terms, in which more time is consumed in repeated calculations for series solutions.

In order to overcome these shortcomings, a useful improvement of the PIM was proposed in [16]. Therefore, the strategy that will be pursued in this work rests mainly on establishing a simple algorithm, requiring no tedious computational work, based on the improved PIM and the spectral collocation technique for obtaining an accurate solution for the following nonlinear Volterra integro-differential equation (VIDE):

$$u'(t) = f(t) + \int_0^t k(t,s) G(u(s))\, ds, \quad t \in [0,T]$$
$$u(0) = u_0, \tag{1}$$

where the kernels $k(t,s)$, $f(t)$ and $G(u(s))$ are smooth functions. The existence and uniqueness of the solution for (1) are presented in [17].

To demonstrate the utility of the proposed method, some examples of the nonlinear VIDEs are given, which are solved using the established method. The obtained results are compared with the numerical solutions. In all cases, the present algorithm performed excellently.

2. The Basic Idea of the PIM

The PIM gives a rapidly convergent approach by using successive approximations of the exact solution if such a solution exists; otherwise the approximations can be used for numerical purposes. The idea of the PIM is very simple and straightforward. To explain the PIM, consider (1) as below:

$$L[u(t)] + N[u(t)] = f(t), \quad (2)$$

where L with the property $Lv \equiv 0$ when $v \equiv 0$ and it denotes the auxiliary linear operator with respect to u. In (2) N is a nonlinear continuous operator with respect to u and $f(t)$ is the source term.

According to [15, 16], we construct the following family of the explicit PIM for (2) as

$$L[u_{k+1}(t) - u_k(t)] = hH(t)A[u_k(t)], \quad (3)$$

where

$$\begin{aligned} A[u_k(t)] &= L[u_k(t)] + N[u_k(t)] - f(t) \\ &= u_k'(t) - \int_0^t k(t,s)G(u_k(s))\,ds - f(t), \end{aligned} \quad (4)$$

with the initial condition

$$u_{k+1}(0) = u_0. \quad (5)$$

Also we can construct a family of the implicit PIM for (2) as follows:

$$\begin{aligned} &L[u_{k+1}(t) - u_k(t)] \\ &= hH(t)\{L[u_k(t)] + N[u_{k+1}(t)] - f(t)\}, \end{aligned} \quad (6)$$

with the above initial condition.

$u_0(t)$ is the initial guess which can be freely found from solving its corresponding linear equation ($L[u_0(t)] = 0$ or $L[u_0(t)] = f(t)$) and the subscript k denotes the kth iteration. Accordingly the approximations of $u_k(t), k \geq 0$ for the PIM iterative relation will be obtained readily in the auxiliary parameter h. Consequently, the exact solution can be obtained by using

$$u(t) = \lim_{k\to\infty} u_k(t). \quad (7)$$

The parametric iteration formula (3) makes a recurrence sequence $u_k(t)$. Obviously, the limit of the sequence will be the solution of (1) if the sequence is convergent. In the following, we give a proof of convergence of the PIM. Here we assume that for every k, $u_k \in C^1[0,T]$ and $\{u_k'\}$ is uniformly convergent.

Theorem 1. *If the sequence $u_k(t)$ converges, where $u_k(t)$ is produced by the parametric iteration formulation of (3), then it must be the exact solution of (1).*

Proof. If the sequence $\{u_k(t)\}$ converges, we define

$$U(t) = \lim_{k\to\infty} u_k(t), \quad (8)$$

and it holds

$$U(t) = \lim_{k\to\infty} u_{k+1}(t). \quad (9)$$

From (16) and (9) and the definition of L, we can easily acquire

$$\begin{aligned} &\lim_{k\to\infty} L[u_{k+1}(t) - u_k(t)] = L\lim_{k\to\infty}[u_{k+1}(t) - u_k(t)] \\ &= 0. \end{aligned} \quad (10)$$

From (10) and according to (3), we obtain

$$hH(t)\lim_{k\to\infty} A[u_k(t)] = L\lim_{k\to\infty}[u_{k+1}(t) - u_k(t)] = 0. \quad (11)$$

Since $h \neq 0$ and also $H(t) \neq 0$ for all t, the relation (11) gives us

$$\lim_{k\to\infty} A[u_k(t)] = 0. \quad (12)$$

From (12) and the continuity property of the operator G, it follows that

$$\begin{aligned} &\lim_{k\to\infty} A[u_k(t)] \\ &= \lim_{k\to\infty}\left(u_k'(t) - \int_0^t k(t,s)G(u_k(s))\,ds - f(t)\right) \\ &= \left(\lim_{k\to\infty} u_k(t)\right)' - \int_0^t k(t,s)G\left(\lim_{k\to\infty} u_k(s)\right)ds \\ &- f(t) = U'(t) - \int_0^t k(t,s)G(U(s))\,ds - f(t). \end{aligned} \quad (13)$$

From (12) and (13), we get

$$U'(t) - \int_0^t k(t,s)G(U(s))\,ds - f(t) = 0, \quad 0 \leq s,t \leq T. \quad (14)$$

On the other hand, in view of the initial condition of the $(k+1)$th order PIM and (9), it holds that

$$U(0) = \lim_{k\to\infty} u_{k+1}(0) = u(0) = u_0. \quad (15)$$

Hence, according to the expressions (14) and (15), $U(t)$ must be the exact solution of (1) and this ends the proof.

It is obvious that the convergence of the sequence (16) depends upon the initial guess $u_0(t)$, the auxiliary linear operator L, the auxiliary parameter h, and the auxiliary function $H(t)$. Fortunately, the PIM provides us with the great freedom of choosing these items. Thus, as long as $u_0(t)$, L, h, and $H(t)$ are property chosen so that the sequence (16) converges in a region $0 \leq t \leq T$, it should converge to the exact solution in this region. Therefore, the combination of the convergence theorem and the freedom of the choice of the above factors establishes the cornerstone of the validity and flexibility of the PIM.

Remark 2. In the case of failure of convergence of the PIM, the presence of the parameter h in (3) or (6) could play a very important role in the frame of the PIM. Although we can find a valid region of h for every physical problem by plotting the solution or its derivatives versus the parameter h in some points, an approximate optimal value of the convergence accelerating parameter h can be determined at the order of approximation by the residual error [15]

$$Res(h) = \int_0^T \{L[u_k(t;h)] + N[u_k(t;h)] - f(t)\}^2 dx. \tag{16}$$

One can minimize (16) by imposing the requirement $dRes(h)/dh = 0$.

3. A Spectral Collocation PIM

In general, the application of the PIM to solve the nonlinear VIDEs leads to the calculation of unneeded and repeated terms. The unneeded and repeated calculations may or may not lead to faster convergence. Also, since the PIM provides the solution as a sequence of iterates, its successive iterations may be very complex so that the resulting integrals in its iterative relation may not be performed analytically. In this section, we will overcome this shortcoming of the original PIM for solving (1) by suggesting a spectral collocation PIM. As will be shown in this paper later, the proposed method will be very simple to implement and save time and calculations.

Consider the basis functions ϕ_j which are polynomials of degree $N-1$ satisfying $\phi_j(t_k) = \delta_{j,k}$ for the shifted Chebyshev nodes (note that $t_1 = T$ and $t_N = 0$)

$$t_k = \frac{T}{2}\left[\cos\left(\frac{(k-1)\pi}{N-1}\right) + 1\right], \quad k = 1, \ldots, N. \tag{17}$$

The unknown function $u(t)$ is approximated as a truncated series of polynomials. The polynomial

$$p(t) \cong u(t) = \sum_{j=1}^{N} u_j \phi_j(t), \tag{18}$$

interpolates the points (t_j, u_j), $j = 1, \ldots, N$; that is, $p(\mathbf{t}) = \mathbf{u}$, where $\mathbf{t} = (t_1, \ldots, t_N)$ and $\mathbf{u} = (u_1, \ldots, u_N)$. The values of the interpolating polynomial's first derivative at the nodes are $p'(\mathbf{t}) = D^{(1)}\mathbf{u}$, and the value of integral at the nodes is defined by $\int_0^t k(t,s)u(s)ds = V \cdot \mathbf{u}$, where V is the Volterra integration matrix [18, 19].

Generally, in order to solve problem (1) using a spectral collocation scheme, the interpolating polynomial $p(t)$ is required to satisfy the equation at the interior nodes. The values of the interpolating polynomial at the interior nodes $t_2, \ldots, t_N$ are $p(\mathbf{t}_m) = (\mathbf{u})_m = I_{m,:}\mathbf{u}$ $(m = 1 : N-1)$ and the derivative value is $p'(\mathbf{t}_m) = D^{(1)}_{m,:}\mathbf{u}$. The initial condition that involves the interpolating polynomial can be handled by using the formula $p(\mathbf{t}_N) = (\mathbf{u})_N = I_{N,:}\mathbf{u}$, where $I_{N,:}$ denotes the last row of the $(N \times N)$ identity matrix.

For the interpolating polynomial to satisfy the nonlinear VIDE of (1) at each interior node, the collocation equation

$$p'(\mathbf{t}_m) = f(\mathbf{t}_m) + \int_0^{\mathbf{t}} k(\mathbf{t}, s) G(p(s)) ds, \qquad p(\mathbf{t}_N) = u_0, \tag{19}$$

should be satisfied. Substituting the differentiation and integration matrix relations into equation (19), we get

$$\begin{bmatrix} D^{(1)}_{m,:} \\ I_{N,:} \end{bmatrix} \mathbf{u} = \begin{bmatrix} \mathbf{f}_m \\ u_0 \end{bmatrix} + \begin{bmatrix} I_{m,:}(V \cdot G(\mathbf{u})) \\ 0 \end{bmatrix}, \tag{20}$$

where $\mathbf{f}_m = \{f(t_1), \ldots, f(t_{N-1})\}$. Now, in view of (3) and the definitions of L and A, by substituting the differentiation and integration matrix relations, we will have the following explicit PIM for solving (1) which is called the spectral PIM (SPIM):

$$\mathbf{u}_{k+1} = \mathbf{u}_k + h \begin{bmatrix} D^{(1)}_{m,:} \\ I_{N,:} \end{bmatrix}^{-1} \cdot \left(\begin{bmatrix} D^{(1)}_{m,:} \\ I_{N,:} \end{bmatrix} \mathbf{u}_k - \begin{bmatrix} \mathbf{f}_m \\ u_0 \end{bmatrix} - \begin{bmatrix} I_{m,:}(V \cdot G(\mathbf{u}_k)) \\ 0 \end{bmatrix} \right), \tag{21}$$

where for simplicity we chose $H(t) \equiv 1$. If we define $\mathbf{L} = [D^1_{m,:}, I_{N,:}]^T$, $\mathbf{f} = [\mathbf{f}_m, u_0]^T$, and $\mathbf{N}\mathbf{u}_k = [I_{m,:}(V \cdot G(\mathbf{u}_k)), 0]^T$, then we will have the following explicit iterative relation for finding the solution vector $\mathbf{u}_{k+1}$:

$$\mathbf{u}_{k+1} = \mathbf{u}_k + h\mathbf{L}^{-1}(\mathbf{L}\mathbf{u}_k - \mathbf{f} - \mathbf{N}\mathbf{u}_k). \tag{22}$$

Here the vector $\mathbf{u}_{k+1}$ is defined as

$$\mathbf{u}_{k+1} = \{u_{k+1}(t_1), \ldots, u_{k+1}(t_{N-1})\}. \tag{23}$$

In using the SPIM algorithm above, we begin by choosing the best possible initial approximation that satisfies the initial condition. To this end, we may determine the initial approximation by solving $\mathbf{L}\mathbf{u}_0 = 0$ or $\mathbf{L}\mathbf{u}_0 = \mathbf{f}$. Thus, starting from the initial approximation $\mathbf{u}_0(t)$, we can use the recurrence formula (22) to successively obtain directly $\mathbf{u}_{k+1}(t)$ for $k \geq 0$.

4. Test Problems

In this section, we demonstrate the effectiveness of the SPIM by applying the method to three nonlinear NVIDs. All of the numerical computations have been performed in MATLAB R2014a and terminated when the current iterate satisfies $\|\mathbf{u}_k - \mathbf{u}_{k-1}\| \leq 10^{-16}$, where $\mathbf{u}_k$ is the solution vector of the kth SPIM iteration.

Example 1. Consider the following nonlinear VIDE [20]:

$$u'(t) = \frac{1}{\varepsilon}\left(u - u^2 - u\int_0^t u(s)ds\right), \quad t \in [0,1] \tag{24}$$

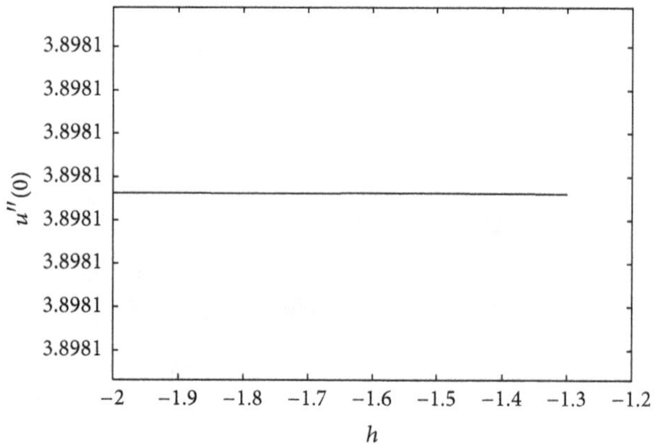

FIGURE 1: The valid region of h for the explicit spectral PIM when $N = 15$ for Example 1.

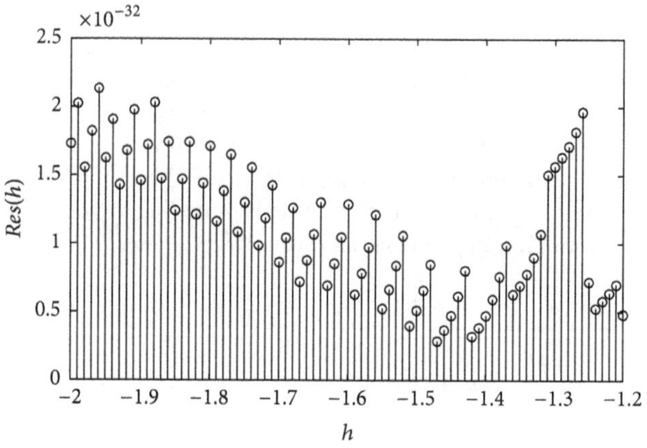

FIGURE 2: The approximate optimal h ($h = -1.47$) for $N = 15$ for Example 1.

with the initial condition $u(0) = 0.1$. Here we aim to solve the above Volterra population equation for the value $\varepsilon = 1/10$. To use the proposed method in this paper, i.e., (22), we could choose

$$L[u(t)] = u'(t),$$

$$N[u(t)] = \frac{1}{\varepsilon}\left(u - u^2 - u\int_0^t u(s)\,ds\right), \qquad (25)$$

$$g(t) \equiv 0.$$

To investigate the valid region h of the solution obtained via the explicit spectral PIM algorithm (22) for $N = 15$ of (24) with $\varepsilon = 1/10$, we try to plot the curve of $u''(0)$ with respect to h, as shown in Figure 1. According to this curve, it is easy to discover the valid region of h. It is usually convenient to investigate the valid region of h for the PIM by means of such kinds of the curves.

According to Figure 1, it could be seen that the explicit spectral PIM for $h = -1$ and $N = 15$ (even for large N) is not a convergent approach for solving (24). The presence of the auxiliary parameter h in the framework of the explicit spectral PIM could play a very important role. As mentioned above, we can find an approximate optimal value for h from (16) by estimating the residual error $Res(h)$ in a sequence of values h, as the value of h with the lowest residual will be the approximate optimal h. Figure 2 shows the approximate optimal value of h for the explicit spectral PIM for $N = 15$, i.e., $h = -1.47$ with two decimal digits.

Figure 3(b) shows the absolute error of the explicit spectral PIM for $N = 15$ and $h = -1.47$. Also the behavior of the numerical and explicit spectral PIM solutions of this example for $N = 15$ and $h = -1.47$ is presented in Figure 3(a).

Example 2. Consider the following nonlinear VIDE:

$$u'(t) = 1 - \frac{1}{2}t + \frac{1}{2}e^{-t^2} + \int_0^t tse^{-u^2(s)}ds, \qquad (26)$$

with initial condition $u(0) = 0$ and the exact solution $u(t) = t$, [21].

To investigate the valid region h for the solution obtained via the explicit spectral PIM algorithm (22) for $N = 10$ of (26), here we plot the curve of $u''(0)$ with respect to h, as shown in Figure 4.

Figure 5 shows the approximate optimal value of h of the explicit spectral PIM when $N = 10$, i.e., $h = -1.1$ with one decimal digit.

Figure 6 shows the absolute error of the explicit spectral PIM for $N = 10$ and $h = -1.1$.

Example 3. Consider the following nonlinear VIDE [22]:

$$u'(t) = f(t) + \int_0^t (t-s)\ln(1 + u(s))\,ds, \qquad (27)$$

where

$$f(t) = \frac{1}{24}\left(8 + 9t^2 + \frac{12}{\sqrt{1+t}} - 8\sqrt{1+t} - 4t\left(-6 + 5\sqrt{1+t}\right) - 12t^2\ln\left(1 + \sqrt{1+t}\right)\right) \qquad (28)$$

with the initial condition $u(0) = 1$ and the corresponding exact solution is given by $u(t) = \sqrt{1+t}$.

To investigate the valid region h of the solution obtained via the explicit spectral PIM algorithm (22) for $N = 10$ of (27), here we plot the curve of $u''(0)$ with respect to h, as shown in Figure 7.

Figure 8 shows the approximate optimal value of h for the explicit spectral PIM when $N = 10$, i.e., $h = -0.8$ with one decimal digit.

Figure 9 shows the absolute error of the explicit spectral PIM for $N = 10$ and $h = -0.8$.

5. Conclusion

In this paper, we presented a new application of the spectral parametric iteration method (PIM) for solving a class of nonlinear Volterra integro-differential equations (VIDEs).This new method is easy to implement and is accurate when

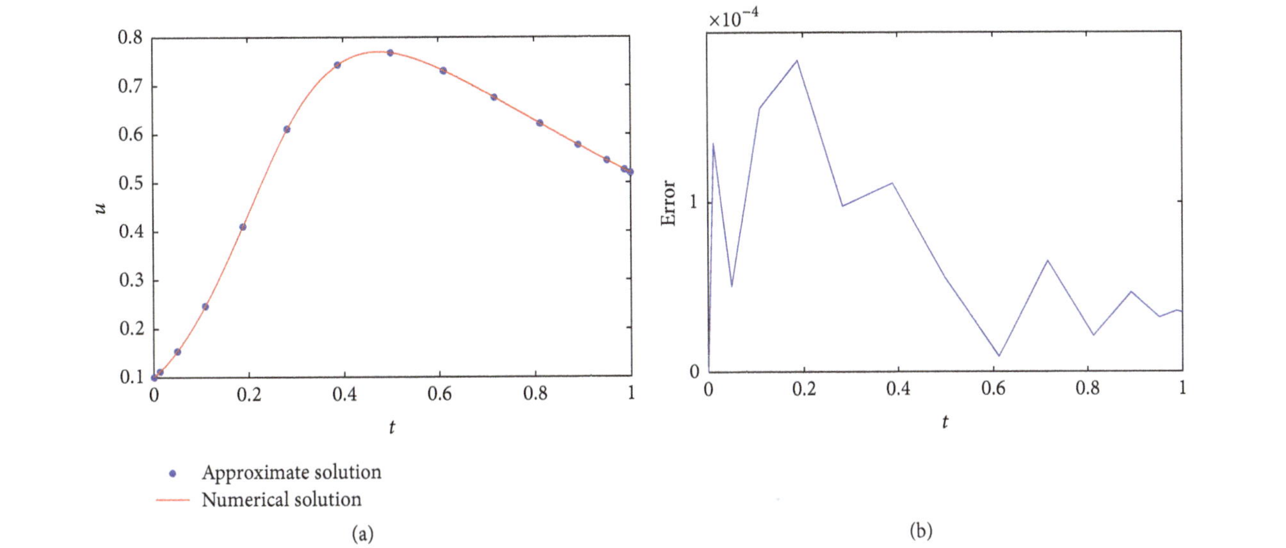

FIGURE 3: (a) Approximate solution of the explicit spectral PIM for $N = 15$. (b) Absolute error of the explicit spectral PIM for $N = 15$ and $h = -1.47$ for Example 1.

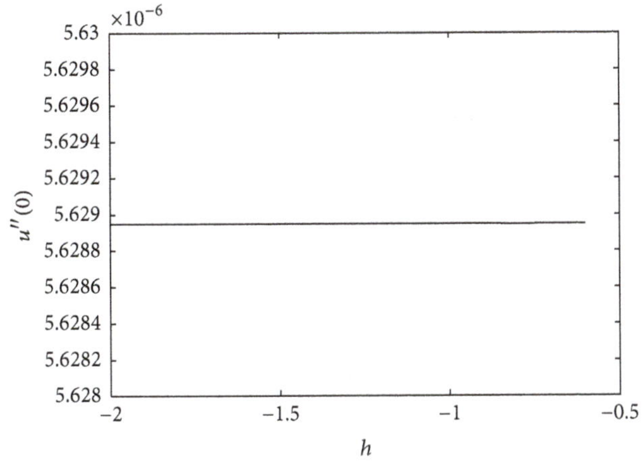

FIGURE 4: The valid region h for the explicit spectral PIM when $N = 10$ for Example 2.

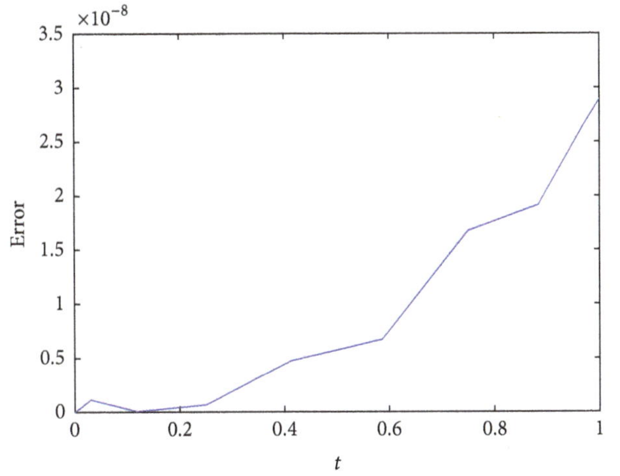

FIGURE 6: Absolute error of the explicit spectral PIM for $N = 10$ and $h = -1.1$ for Example 2.

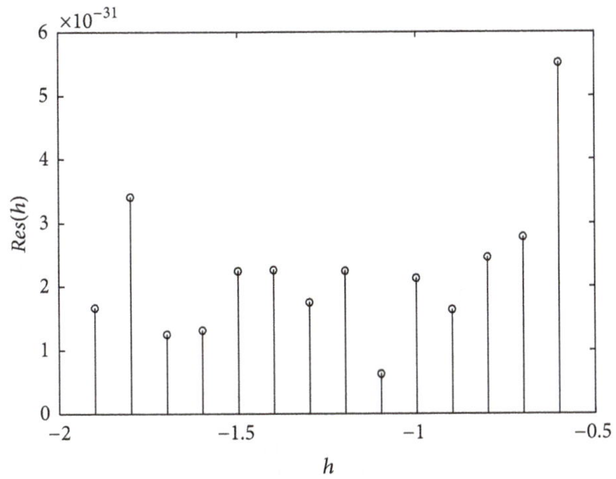

FIGURE 5: The approximate optimal h ($h = -1.1$) for $N = 10$ for Example 2.

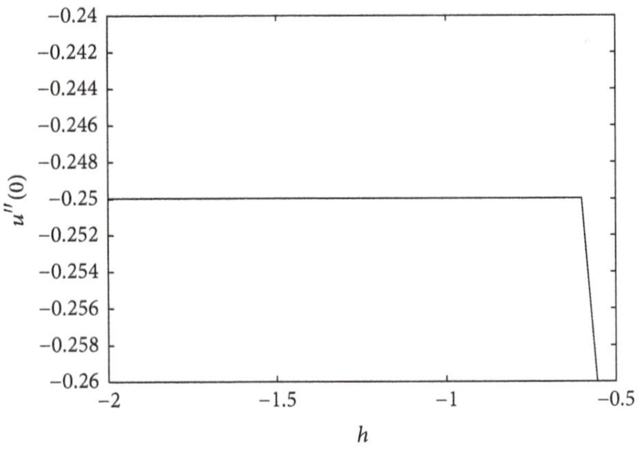

FIGURE 7: The valid region h of the explicit spectral PIM when $N = 10$ for Example 3.

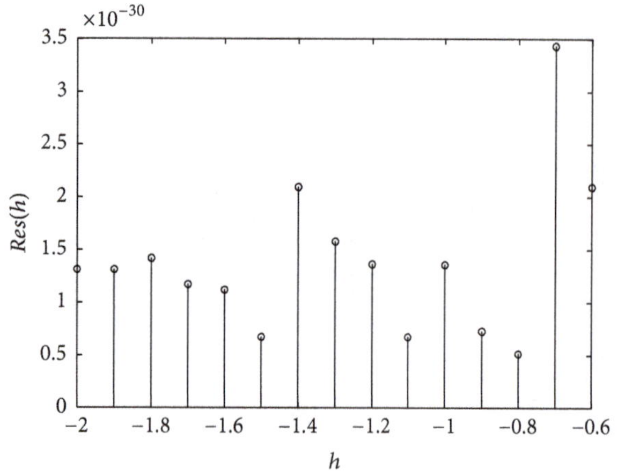

FIGURE 8: The approximate optimal h ($h = -0.8$) for $N = 10$ for Example 3.

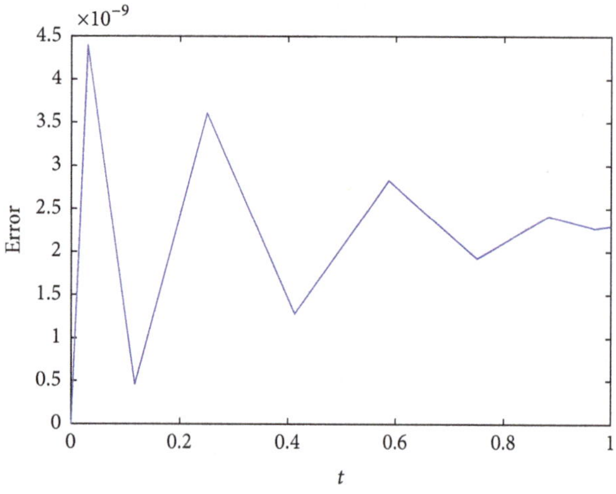

FIGURE 9: Absolute error of the explicit spectral PIM for $N = 10$ and $h = -0.8$ for Example 3.

applied to the nonlinear VIDEs. The numerical results of the spectral PIM were compared with the exact solutions and excellent agreement was obtained. This could confirm the validity of the proposed spectral PIM as a suitable method for solving this class of the nonlinear VIDEs.

Conflicts of Interest

M. H. Daliri Birjandi, J. Saberi-Nadjafi, and A. Ghorbani declare that there are no conflicts of interest regarding the publication of this paper.

References

[1] I. Abdul, *Introduction to Integral Equations with Application*, Wiley, NewYork, NY, USA, 1999.

[2] S. Shaw and J. R. Whiteman, "Adaptive space-time finite element solution for Volterra equations arising in viscoelasticity problems," *Journal of Computational and Applied Mathematics*, vol. 125, no. 1-2, pp. 337–345, 2000.

[3] V. Volterra, *Theory of Functionals And of Integral And Integro-Differential Equations*, Dover Publications, Inc., New York, NY, USA, 1959.

[4] A. J. Jerri, *Introduction to Integral Equations with Applications*, Marcel Dekker, New Yourk, NY, USA, 1999.

[5] M. Dehghan and R. Salehi, "The numerical solution of the non-linear integro-differential equations based on the meshless method," *Journal of Computational and Applied Mathematics*, vol. 236, no. 9, pp. 2367–2377, 2012.

[6] K. Maleknejad and Y. Mahmoudi, "Taylor polynomial solution of high-order nonlinear Volterra-Fredholm integro-differential equations," *Applied Mathematics and Computation*, vol. 145, no. 2-3, pp. 641–653, 2003.

[7] K. Maleknejad and F. Mirzaee, "Using rationalized Haar wavelet for solving linear integral equations," *Applied Mathematics and Computation*, vol. 160, no. 2, pp. 579–587, 2005.

[8] S. M. El-Sayed and M. R. Abdel-Aziz, "A comparison of Adomian's decomposition method and wavelet-Galerkin method for solving integro-differential equations," *Applied Mathematics and Computation*, vol. 136, no. 1, pp. 151–159, 2003.

[9] J.-H. He, "The homotopy perturbation method nonlinear oscillators with discontinuities," *Applied Mathematics and Computation*, vol. 151, no. 1, pp. 287–292, 2004.

[10] S. Behzadi, S. Abbasbandy, S. T. Allahviranloo, and A. Yildirim, "Application of homotopy analysis method for solving a class of nonlinear Volterra-Fredholm integro-differential equations," *Journal of Applied Analysis and Computation*, pp. 1–14, 2012.

[11] M. Zarebnia, "Sinc numerical solution for the Volterra integro-differential equation," *Communications in Nonlinear Science and Numerical Simulation*, vol. 15, no. 3, pp. 700–706, 2010.

[12] E. Tohidi and O. R. N. Samadi, "Optimal control of nonlinear Volterra integral equations via Legendre polynomials," *IMA Journal of Mathematical Control and Information*, vol. 30, no. 1, pp. 67–83, 2013.

[13] A. Karamete and M. Sezer, "A Taylor collocation method for the solution of linear integro-differential equations," *International Journal of Computer Mathematics*, vol. 79, no. 9, pp. 987–1000, 2002.

[14] H. Brunner, *Collocation Method for Volterra Integral and Related Functional Equations*, Cambridge University Press, Cambridge, London, UK, 2004.

[15] M. H. Daliri Birjandi, J. Saberi-Nadjafi, and A. Ghorbani, "A novel method for solving nonlinear volterra integro-differential equation systems," *Abstract and Applied Analysis*, vol. 2018, Article ID 3569139, 6 pages, 2018.

[16] A. Ghorbani and J. Saberi-Nadjafi, "A piecewise-spectral parametric iteration method for solving the nonlinear chaotic Genesio system," *Mathematical and Computer Modelling*, vol. 54, no. 1-2, pp. 131–139, 2011.

[17] P. Linz, *Analytical and Numerical Methods for Volterra Equations*, vol. 7 of *SIAM Studies in Applied Mathematics*, Society for Industrial and Applied Mathematics (SIAM), Philadelphia, Pa, USA, 1985.

[18] J. A. C. Weideman and S. C. Reddy, "A MATLAB diferentiation matrix suite," *ACM Transactions on Mathematical Software*, vol. 26, no. 4, pp. 465–519, 2000.

[19] T. Damian, *Matrix Based Operatorial Approach to Differential and Integral Problems*, Babes-Bolyai University of Cluj-Napoca Romania, Romania, 2011.

[20] R. D. Small, "Population Growth in a Closed System," *SIAM Review*, vol. 25, no. 1, pp. 93–95, 1983.

[21] K. Maleknejad and M. Tamamgar, "A New Reconstruction of Variational Iteration Method and Its Application to Nonlinear Volterra Integrodifferential Equations," *Abstract and Applied Analysis*, vol. 2014, pp. 1–6, 2014.

[22] K. Kim and B. Jang, "A novel semi-analytical approach for solving nonlinear Volterra integro-differential equations," *Applied Mathematics and Computation*, vol. 263, pp. 25–35, 2015.

8

Exponentially Fitted and Trigonometrically Fitted Explicit Modified Runge-Kutta Type Methods for Solving $y'''(x) = f(x, y, y')$

N. Ghawadri,[1] N. Senu,[1,2] F. Ismail,[1,2] and Z. B. Ibrahim[1,2]

[1]*Institute for Mathematical Research, Universiti Putra Malaysia (UPM), 43400 Serdang, Selangor, Malaysia*
[2]*Department of Mathematics, Universiti Putra Malaysia (UPM), 43400 Serdang, Selangor, Malaysia*

Correspondence should be addressed to N. Senu; norazak@upm.edu.my

Academic Editor: Igor Andrianov

Exponentially fitted and trigonometrically fitted explicit modified Runge-Kutta type (MRKT) methods for solving $y'''(x) = f(x, y, y')$ are derived in this paper. These methods are constructed which exactly integrate initial value problems whose solutions are linear combinations of the set functions $e^{\omega x}$ and $e^{-\omega x}$ for exponentially fitted and $\sin(\omega x)$ and $\cos(\omega x)$ for trigonometrically fitted with $\omega \in R$ being the principal frequency of the problem and the frequency will be used to raise the accuracy of the methods. The new four-stage fifth-order exponentially fitted and trigonometrically fitted explicit MRKT methods are called EFMRKT5 and TFMRKT5, respectively, for solving initial value problems whose solutions involve exponential or trigonometric functions. The numerical results indicate that the new exponentially fitted and trigonometrically fitted explicit modified Runge-Kutta type methods are more efficient than existing methods in the literature.

1. Introduction

This work deals with exponentially fitted and trigonometrically fitted modified Runge-Kutta type methods for solving third-order ordinary differential equations (ODEs)

$$\begin{aligned} y'''(x) &= f(x, y(x), y'(x)), \\ y(x_0) &= y_0, \\ y'(x_0) &= y'_0, \\ y''(x_0) &= y''_0, \end{aligned} \qquad x \geq x_0. \tag{1}$$

This sort of problems is often found in numerous physical problems like thin film flow, gravity-driven flows, electromagnetic waves, and so on. In the past and recent years many researchers constructed exponentially fitted and trigonometrically fitted explicit Runge-Kutta methods for solving first-order and second-order ordinary differential equations. Paternoster [1] developed Runge-Kutta-Nyström methods for ODEs with periodic solutions based on trigonometric polynomials. Vanden Berghe et al. [2] developed exponentially fitted Runge-Kutta methods. Simos [3] extended exponentially fitted Runge-Kutta methods for the numerical solution of the Schrodinger equation and related problems. Kalogiratou et al. [[4, 5]] constructed trigonometrically and exponentially fitted Runge-Kutta-Nyström methods for the numerical solution of the Schrodinger equation and related problems which is eighth algebraic order. Next Simos et al. [6] constructed exponentially fitted Runge-Kutta-Nyström method for the numerical solution of initial value problems with oscillating solutions. Sakas et al. [7] developed a fifth algebraic order trigonometrically fitted modified Runge-Kutta Zonneveld method for the numerical solution of orbital problems. Van de Vyver [8] in 2005 constructed Runge-Kutta-Nyström pair for the numerical integration of perturbed oscillators. Then Yang et al. [9] constructed trigonometrically fitted adapted Runge-Kutta-Nyström methods for perturbed oscillators. Recently, Demba et al. [10] constructed an explicit trigonometrically fitted Runge-Kutta-Nyström method using Simos technique.

In this paper we construct explicit exponentially fitted and trigonometrically fitted modified Runge-Kutta type methods with four-stage fifth-order, called EFMRKT5 and TFMRKT5, respectively. Section 2 discussed the oscillatory and nonoscillatory properties of the third-order linear differential equation. In Section 3, the necessary conditions and the derivation for exponentially fitted and trigonometrically fitted modified Runge-Kutta type methods for solving third-order ODEs are given. The error analysis of the new EFMRKT5 and TFMRKT5 methods was discussed in Section 4, respectively. The effectiveness of the new methods when compared with existing methods is given in Section 5. The thin film flow problem is discussed in Section 6.

2. Third-Order Linear Differential Equation with Oscillating and Nonoscillating Solutions

This section discusses the oscillatory and nonoscillatory properties of the third-order linear differential equation

$$y'''(x) + p(x)y' + q(x)y = 0. \tag{2}$$

A solution of (2) will be said to be oscillatory if it changes signs for arbitrarily large values of x. The other solutions will be said to be nonoscillatory.

If $p(x) < 0$ and $q(x) < 0$ are constants, then it is easy to show that if (2) has an oscillatory solution, then there are two linearly independent oscillatory solutions of (2) whose zeroes separate and such that any oscillatory solution of (2) is a linear combination of them. Assuming that $p(x)$, $p'(x)$, and $q(x)$ are continuous on $[0, +\infty)$ the following will be established (see [11–14]).

Definition 1. A solution of (2) will be called oscillatory iff it has an infinity of zeroes in$(0, +\infty)$ and nonoscillatory iff it has but a finite number of zeroes in this interval. Equation (2) is said to be oscillatory iff it has at least one (nontrivial) oscillatory solution and nonoscillatory iff all of its (nontrivial) solutions are nonoscillatory.

Particularly, this paper deals with two cases based on (2) when $q(x) = 0$, as follows:

(i) $y'''(x) = py'$, $(p > 0)$; it is clear that the characteristic roots equations are real and one of them is zero; then solutions will consist of exponential functions.

(ii) $y'''(x) = -py'$, $(p > 0)$; one of the characteristic roots equations is zero and another two are conjugate roots and the solutions are in oscillatory form,

where p is constant.

3. Exponentially Fitted and Trigonometrically Fitted MRKT Methods

In this section, we will determine the conditions and develop exponentially fitted and trigonometrically fitted MRKT methods. In order to construct the exponentially fitted and trigonometrically fitted MRKT methods, the extra γ_i and $\hat{\gamma}_i$ are absolutely necessary to insert at each stage and the MRKT methods is given as follows:

$$y_{n+1} = y_n + hy'_n + \frac{h^2}{2}y''_n + h^3\sum_{i=1}^{s} b_i k_i, \tag{3}$$

$$y'_{n+1} = y'_n + hy''_n + h^2\sum_{i=1}^{s} b'_i k_i, \tag{4}$$

$$y''_{n+1} = y''_n + h\sum_{i=1}^{s} b''_i k_i, \tag{5}$$

where

$$k_1 = f\left(x_n, y_n, y'_n\right), \tag{6}$$

$$k_i = f\left(x_n + c_i h, \gamma_i y_n + hc_i y'_n + \frac{h^2}{2}c_i^2 y''_n + h^3\sum_{j=1}^{s} a_{ij}k_j, y'_n + \hat{\gamma}_i hc_i y''_n + h^2\sum_{j=1}^{s}\hat{a}_{ij}k_j\right) \tag{7}$$

for $i = 2, 3, \ldots, s$.

The parameters of the MRKT methods are $c_i, a_{ij}, \hat{a}_{ij}, b_i, b'_i, b''_i, \gamma_i$ and $\hat{\gamma}_i$ for $i = 1, 2, \ldots, s$ and $j = 1, 2, \ldots, s$ are assumed to be real. If $a_{ij} = 0$ and $\hat{a}_{ij} = 0$ for $i \leqslant j$, it is an explicit method and otherwise implicit method.

The MRKT method can be expressed in Butcher notation using the table of coefficients as follows (see Table 1).

Table 1: The Butcher tableau MRKT method.

c	γ	$\hat{\gamma}$	A	$\hat{A}$	
			b	b'	b''

3.1. Exponentially Fitted MRKT Method. To construct the exponentially fitted Runge-Kutta type four-stage fifth-order method the functions $e^{\omega x}$ and $e^{-\omega x}$ need to integrate exactly at each stage; therefore the following four equations are obtained:

$$e^{\pm c_i v} = \gamma_i \pm c_i v + \frac{1}{2}c_i^2 v^2 \pm v^3\sum_{j=1}^{s} a_{ij}e^{\pm c_j v}, \tag{8}$$

$$e^{\pm c_i v} = 1 \pm \hat{\gamma}_i c_i v \pm v^2\sum_{j=1}^{s}\hat{a}_{ij}e^{\pm c_j v}, \tag{9}$$

and six more equations corresponding to y, y', and y'':

$$e^{\pm v} = 1 \pm v + \frac{1}{2}v^2 \pm v^3\sum_{i=1}^{s} b_i e^{\pm c_i v}, \tag{10}$$

$$e^{\pm v} = 1 \pm v + v^2\sum_{i=1}^{s} b'_i e^{\pm c_i v}, \tag{11}$$

$$e^{\pm v} = 1 \pm v\sum_{i=1}^{s} b_i'' e^{\pm c_i v}, \tag{12}$$

where $v = \omega h$, $\omega \in$ R. The relations $\cosh(v) = (e^{v} + e^{-v})/2$ and $\sinh(v) = (e^{v} - e^{-v})/2$ will be used in the derivation process. The following order conditions are obtained:

$$\cosh(vc_i) = \gamma_i + \frac{1}{2}v^2c_i^2 + v^3\sum_{j=1}^{i-1} a_{ij}\sinh(vc_j), \tag{13}$$

$$\sinh(vc_i) = vc_i + v^3\sum_{j=1}^{i-1} a_{ij}\cosh(vc_j), \tag{14}$$

$$\cosh(vc_i) = 1 + v^2\sum_{j=1}^{i-1} \hat{a}_{ij}\cosh(vc_j), \tag{15}$$

$$\sinh(vc_i) = \hat{\gamma}_i c_i v + v^2\sum_{j=1}^{i-1} \hat{a}_{ij}\sinh(vc_j), \tag{16}$$

and six equations corresponding to y, y', and y'':

$$\cosh(v) = 1 + \frac{1}{2}v^2 + v^3\sum_{i=1}^{s} b_i\sinh(vc_i), \tag{17}$$

$$\sinh(v) = v + v^3\sum_{i=1}^{s} b_i\cosh(vc_i), \tag{18}$$

$$\cosh(v) = 1 + v^2\sum_{i=1}^{s} b_i'\cosh(vc_i), \tag{19}$$

$$\sinh(v) = v + v^2\sum_{i=1}^{s} b_i'\sinh(vc_i), \tag{20}$$

$$\cosh(v) = 1 + v\sum_{i=1}^{s} b_i''\sinh(vc_i), \tag{21}$$

$$\sinh(v) = v\sum_{i=1}^{s} b_i''\cosh(vc_i). \tag{22}$$

Solving (13) to (16), we find $a_{i,i-1}$, $\hat{a}_{i,i-1}$, γ_i, and $\hat{\gamma}_i$.

$$\gamma_i = \cosh(vc_i) - \frac{1}{2}v^2c_i^2 - v^3\sum_{j=1}^{i-1} a_{i,j}\sinh(vc_j), \tag{23}$$

$$a_{i,i-1} = \frac{\sinh(vc_i) - vc_i - v^3\sum_{j=1}^{i-2} a_{i,j}\cosh(vc_j)}{v^3\cosh(vc_{i-1})}, \tag{24}$$

$$\hat{a}_{i,i-1} = \frac{\cosh(vc_i) - 1 - v^2\sum_{j=1}^{i-2} \hat{a}_{i,j}\cosh(vc_j)}{v^2\cosh(vc_{i-1})}, \tag{25}$$

$$\hat{\gamma}_i = \frac{\sinh(vc_i) - v^2\sum_{j=1}^{i-2} \hat{a}_{i,j}\sinh(vc_j)}{vc_i}, \tag{26}$$

$$i = 2, \ldots, s.$$

Referring to the following fifth-order four-stage method developed by Fawzi et al. [15]:

$$\begin{aligned}
c_1 &= 0, \\
c_2 &= \frac{1}{5}, \\
c_3 &= \frac{2}{3}, \\
c_4 &= 1, \\
a_{21} &= 0, \\
a_{31} &= -\frac{49}{4860}, \\
a_{41} &= \frac{7}{50}, \\
a_{42} &= -\frac{1}{50}, \\
\hat{a}_{31} &= -\frac{1}{27}, \\
\hat{a}_{41} &= \frac{3}{10}, \\
\hat{a}_{42} &= -\frac{2}{35}, \\
b_3 &= \frac{3}{112}, \\
b_4 &= 0, \\
b_3' &= \frac{9}{56}, \\
b_4' &= 0, \\
b_3'' &= \frac{7}{56}, \\
b_4' &= \frac{5}{48},
\end{aligned} \tag{27}$$

we solve (23) to (26) and let $\hat{a}_{21}$, $\hat{a}_{32}$, $\hat{a}_{43}$, a_{32}, a_{43}, γ_2, γ_3, γ_4, $\hat{\gamma}_2$, $\hat{\gamma}_3$, and $\hat{\gamma}_4$ be free parameters and yields.

$$\hat{a}_{21} = \frac{\cosh(v/5) - 1}{v^2},$$

$$\hat{a}_{32} = \frac{\cosh(2v/3) - 1 + (1/27)v^2}{v^2\cosh(v/5)},$$

$$
\begin{aligned}
\widehat{a}_{43} &= \frac{\cosh(v) - 1 - v^2 \left(3/10 - (2/35)\cosh(v/5)\right)}{v^2 \cosh(2v/3)}, \\
a_{32} &= \frac{\sinh(2v/3) - 2v/3 + (49/4860)\, v^3}{v^3 \cosh(c_2 v)}, \\
a_{43} &= \frac{\sinh(v) - v - (7/50)\, v^3 + (1/50)\, v^3 \cosh(v/5)}{v^3 \cosh(2v/3)}, \\
\gamma_2 &= \cosh\left(\frac{v}{5}\right) - \frac{v^2}{50}, \\
\gamma_3 &= \cosh\left(\frac{2v}{3}\right) - \frac{2v^2}{9} - \frac{301v^3}{4860} \sinh\left(\frac{v}{5}\right), \\
\gamma_4 &= \cosh(v) - \frac{v^2}{2} - v^3 \left(-\frac{1}{50} \sinh\left(\frac{v}{5}\right) + \frac{1}{25} \sinh\left(\frac{2v}{3}\right)\right), \\
\widehat{\gamma}_2 &= \frac{5}{v} \sinh\left(\frac{v}{5}\right), \\
\widehat{\gamma}_3 &= \frac{2}{3v} \sinh\left(\frac{2v}{3}\right) - \frac{2\left(\cosh(2v/3) - 1 + v^2/27\right) \sinh(v/5)}{3\cosh(v/5)\, v^2}, \\
\widehat{\gamma}_4 &= \frac{1}{v} \sinh(v) + \frac{2v}{35} \sinh\left(\frac{v}{5}\right) - \frac{\left(\cosh(v) - 1 - \left(3v^2/10 - \left(2v^2/35\right)\cosh(v/5)\right)\right) \sinh(2v/3)}{v \cosh(2v/3)}.
\end{aligned}
\tag{28}
$$

Next, we solve (17) to (22) and use the above coefficients to find $b_1, b_2, b_1', b_2', b_1''$, and b_2''.

$$
\begin{aligned}
b_1 &= \frac{3}{112} \frac{\cosh(v/5)\sinh(2v/3) - \cosh(2v/3)\sinh(v/5)}{\sinh(v/5)} + \frac{-2\cosh(v/5)\cosh(v) + 2\cosh(v/5) + \cosh(v/5)\, v^2 + 2\sinh(v)\sinh(v/5) - 2v\sinh(v/5)}{2v^3 \sinh(v/5)}, \\
b_2 &= -\frac{3}{112} \frac{\sinh(2v/3)}{\sinh(v/5)} - \frac{-2\cosh(v) + 2 + v^2}{2v^3 \sinh(v/5)}, \\
b_1' &= \frac{9}{56} \frac{\cosh(v/5)\sinh(2v/3) - \cosh(2v/3)\sinh(v/5)}{\sinh(v/5)} - \frac{\cosh(v/5)\sinh(v) - \cosh(v/5)\, v - \cosh(v)\sinh(v/5) + \sinh(v/5)}{v^2 \sinh(v/5)}, \\
b_2' &= -\frac{9}{56} \frac{\sinh(2v/3)}{\sinh(v/5)} + \frac{\sinh(v) - v}{v^2 \sinh(v/5)}, \\
b_2'' &= -\frac{1 + (27/56)\, v\sinh(2v/3) + (5/48)\, v\sinh(v) - \cosh(v)}{v\sinh(v/5)}, \\
b_1'' &= \frac{\left((27/56)\cosh(v/5)\, v\sinh(2v/3) + (5/48)\cosh(v/5)\, v\sinh(v) - \cosh(v/5)\cosh(v) + \cosh(v/5) - (27/56)\, v\cosh(2v/3)\sinh(v/5) - (5/48)\, v\cosh(v)\sinh(v/5) + \sinh(v)\sinh(v/5)\right)}{\left(v\sinh(v/5)\right)}.
\end{aligned}
\tag{29}
$$

These lead to our new exponentially fitted Runge-Kutta type four-stage fifth-order explicit MRKT method denoted as EFMRKT5. The corresponding Taylor series expansion of the solution is given by

$$
\begin{aligned}
b_1 &= \frac{1}{48} + \frac{1}{2160} v^2 + \frac{101}{136080000} v^4 \\
&\quad - \frac{5713}{183708000000} v^6 \\
&\quad - \frac{11330339}{81841914000000000} v^8 \\
&\quad - \frac{57722879}{134057055132000000000} v^{10} + \cdots, \\
b_2 &= \frac{5}{42} - \frac{1}{2160} v^2 - \frac{2921}{136080000} v^4 \\
&\quad - \frac{1361}{45927000000} v^6 \\
&\quad + \frac{54293587}{40920957000000000} v^8 \\
&\quad + \frac{6964030429}{670285275660000000000} v^{10} + \cdots,
\end{aligned}
$$

$$b_1' = \frac{1}{24} - \frac{17}{283500}\nu^4 - \frac{149}{218700000}\nu^6 - \frac{1055069}{341007975000000}\nu^8 - \frac{23025689}{2538959377500000000}\nu^{10} + \cdots,$$

$$b_2' = \frac{25}{84} + \frac{241}{2268000}\nu^4 + \frac{22871}{6123600000}\nu^6 + \frac{112778137}{2728063800000000}\nu^8 + \frac{13599351683}{5585710630500000000}\nu^{10} + \cdots,$$

$$b_1'' = \frac{1}{24} + \frac{1}{21600}\nu^4 + \frac{167}{58320000}\nu^6 + \frac{528389}{16533720000000}\nu^8 + \frac{967343}{5845851000000000}\nu^{10} + \cdots$$

$$b_2'' = \frac{125}{336} - \frac{1}{21600}\nu^4 - \frac{2867}{408240000}\nu^6 - \frac{3022109}{16533720000000}\nu^8 - \frac{1457821}{730731375000000}\nu^{10} + \cdots,$$

$$\widehat{a}_{21} = \frac{1}{50} + \frac{1}{15000}\nu^2 + \frac{1}{11250000}\nu^4 + \frac{1}{15750000000}\nu^6 + \frac{1}{35437500000000}\nu^8 + \frac{1}{116943750000000000}\nu^{10} + \cdots,$$

$$\widehat{a}_{32} = \frac{7}{27} + \frac{37}{12150}\nu^2 + \frac{287}{6561000}\nu^4 - \frac{68827}{516678750000}\nu^6 + \frac{7097417}{1674039150000000}\nu^8 - \frac{39720321233}{6214870344375000000000}\nu^{10} + \cdots,$$

$$\widehat{a}_{43} = \frac{9}{35} - \frac{43}{3000}\nu^2 + \frac{3323}{1350000}\nu^4 - \frac{111136579}{255150000000}\nu^6 + \frac{26973882539}{344452500000000}\nu^8 - \frac{721361598388001}{51151196250000000000}\nu^{10} + \cdots,$$

$$a_{32} = \frac{289}{4860} - \frac{67}{729000}\nu^2 + \frac{26141}{2755620000}\nu^4 - \frac{4671541}{39858075000000}\nu^6 + \frac{2216008103}{1104865839000000000}\nu^8 - \frac{15664491766661}{4847598868612500000000000}\nu^{10} + \cdots,$$

$$a_{43} = \frac{7}{150} - \frac{221}{135000}\nu^2 + \frac{114463}{637875000}\nu^4 - \frac{1010472889}{34445250000000}\nu^6 + \frac{222576819697}{42625996875000000}\nu^8 - \frac{84318883418716333}{89770349418750000000000}\nu^{10} + \cdots,$$

$$\gamma_2 = 1 + \frac{1}{15000}\nu^4 + \frac{1}{11250000}\nu^6 + \frac{1}{15750000000}\nu^8 + \frac{1}{35437500000000}\nu^{10} + \cdots,$$

$$\gamma_3 = 1 - \frac{59}{225}\nu^2 - \frac{9569}{1215000}\nu^4 - \frac{74831}{911250000}\nu^6 + \frac{26104429}{11481750000000}\nu^8 - \frac{495937229}{25833937500000000}\nu^{10} + \cdots$$

$$\gamma_4 = 1 - \nu^2 + \frac{19}{1000}\nu^4 + \frac{2267}{4050000}\nu^6 - \frac{4858267}{255150000000}\nu^8 + \frac{65060629}{344452500000000}\nu^{10} + \cdots,$$

$$\widehat{\gamma}_2 = 1 + \frac{1}{150}\nu^2 + \frac{1}{75000}\nu^4 + \frac{1}{78750000}\nu^6 + \frac{1}{141750000000}\nu^8 + \frac{1}{389812500000000}\nu^{10} + \cdots,$$

$$\widehat{\gamma}_3 = 1 - \frac{1}{270}\nu^2 + \frac{13}{60750}\nu^4 - \frac{2032}{717609375}\nu^6 + \frac{136109}{2906317968750}\nu^8 - \frac{2726807}{3596568486328125}\nu^{10} + \cdots,$$

$$\widehat{\gamma}_4 = 1 + \frac{1}{150}\nu^2 + \frac{79}{15000}\nu^4 - \frac{2161951}{2126250000}\nu^6 + \frac{78384961}{425250000000}\nu^8 - \frac{5664933289891}{170503987500000000}\nu^{10} + \cdots \tag{30}$$

where $\gamma_1 = 1, \widehat{\gamma}_1 = 1$.

This results in the new method called EFMRKT5. As $v \longrightarrow 0$, the coefficients $b_1, b_2, b_1', b_2', b_1'', b_2'', a_{32}, a_{42}, \hat{a}_{21}, \hat{a}_{32}, \hat{a}_{43}, \gamma_2, \gamma_3, \gamma_4, \hat{\gamma}_2, \hat{\gamma}_3$, and $\hat{\gamma}_4$ of the new method EFMRKT5 reduce to the coefficients of the original method RKT5. That is to say, $b_1(0), b_2(0), b_1'(0), b_2'(0), b_1''(0), b_2''(0), a_{32}(0), a_{42}(0), \hat{a}_{21}(0), \hat{a}_{32}(0), \hat{a}_{43}(0), \gamma_2(0), \gamma_3(0), \gamma_4(0), \hat{\gamma}_2(0), \hat{\gamma}_3(0)$, and $\hat{\gamma}_4(0)$ are identical to $b_1, b_2, b_1', b_2', b_1'', b_2'', a_{32}, a_{42}, \hat{a}_{21}, \hat{a}_{32}, \hat{a}_{43}, \gamma_2, \gamma_3, \gamma_4, \hat{\gamma}_2, \hat{\gamma}_3$, and $\hat{\gamma}_4$ of RKT5 method. Other than that, $v \longrightarrow 0$, as EFMRKT5 method will have the same error constant as RKT5 method.

3.2. Trigonometrically Fitted MRKT Method. Exponentially fitted method leads to trigonometrically fitted method when replacing $v = wh$ with iv and solving (8) to (9) to find $a_{i,i-1}$, $\hat{a}_{i,i-1}$, γ_i, and $\hat{\gamma}_i$.

$$\gamma_i = \cos(vc_i) - \frac{1}{2}v^2 c_i^2 - v^3 \sum_{j=1}^{i-1} a_{i,j} \sin(vc_j), \quad (31)$$

$$\hat{a}_{i,i-1} = \frac{1 - \cos(vc_i) - v^2 \sum_{j=1}^{i-2} \hat{a}_{i,j} \cos(vc_j)}{v^2 \cos(vc_{i-1})}, \quad (32)$$

$$a_{i,i-1} = \frac{-\sin(vc_i) + v \cdot c_i - v^3 \sum_{j=1}^{i-2} a_{i,j} \cos(vc_j)}{v^3 \cos(vc_{i-1})}, \quad (33)$$

$$\hat{\gamma}_i = \frac{\sin(vc_i) + v^2 \sum_{j=1}^{i-2} \hat{a}_{i,j} \sin(vc_j)}{vc_i}, \quad i = 2, \ldots, s. \quad (34)$$

Consider the same coefficients of fifth-order four-stage method developed by Fawzi et al.[15] as in Section 3.1. Solving the (31) to (34) and letting $\hat{a}_{21}, \hat{a}_{32}, \hat{a}_{43}, a_{32}, a_{43}, \gamma_2, \gamma_3, \gamma_4, \hat{\gamma}_2, \hat{\gamma}_3$, and $\hat{\gamma}_4$ be free parameters will give

$$\begin{aligned}
\hat{a}_{21} &= \frac{1 - \cos(v/5)}{v^2}, \\
\hat{a}_{32} &= \frac{1 - \cos(2v/3) + (1/27)v^2}{v^2 \cos(v/5)}, \\
\hat{a}_{43} &= \frac{1 - \cos(v) - v^2(3/10 - (2/35)\cos(v/5))}{v^2 \cos(2v/3)}, \\
a_{32} &= \frac{-\sin(2v/3) + 2v/3 + (49/4860)v^3}{v^3 \cos(v/5)}, \\
a_{43} &= \frac{v - \sin(v) - v^3(7/50 - (1/50)\cos(v/5))}{v^3 \cos(2v/3)}, \\
\gamma_2 &= \cos\left(\frac{v}{5}\right) - \frac{v^2}{50}, \\
\gamma_3 &= \cos\left(\frac{2v}{3}\right) - \frac{2v^2}{9} - \frac{301v^3}{4860}\sin\left(\frac{v}{5}\right), \\
\gamma_4 &= \cos(v) - \frac{v^2}{2} - v^3\left(-\frac{1}{50}\sin\left(\frac{v}{5}\right) + \frac{1}{25}\sin\left(\frac{2v}{3}\right)\right), \\
\hat{\gamma}_2 &= \frac{5}{v}\sin\left(\frac{v}{5}\right), \\
\hat{\gamma}_3 &= \frac{2}{3v}\sin\left(\frac{2v}{3}\right) + \frac{2\left(1 - \cos(2v/3) + v^2/27\right)\sin(v/5)}{3v^2\cos(v/35)}, \\
\hat{\gamma}_4 &= \frac{\sin(v)}{v} - \frac{2v}{35}\sin\left(\frac{v}{5}\right) \\
&\quad + \frac{\left(1 - \cos(v) - \left(3v^2/10 - \left(2v^2/5\right)\cos(v/5)\right)\right)\sin(2v/3)}{v\cos(2v/3)}.
\end{aligned} \quad (35)$$

Next, solving (10) to (12), and using the above Fawzi coefficients to find b_1, b_2, b_1', b_2', b_1'', and b_2'',

$$\begin{aligned}
b_1 &= \frac{3\cos(v/5)\sin(2v/3) - 3\cos(2v/3)\sin(v/5)}{112\sin(v/5)} - \frac{2\cos(v/5)\cos(v) - 2\cos(v/5) + \cos(v/5)v^2 + 2\sin(v)\sin(v/5) - 2v\sin(v/5)}{2v^3\sin(v/5)}, \\
b_2 &= -\frac{3}{112}\frac{\sin(2v/3)}{\sin(v/5)} + \frac{2\cos(v) - 2 + v^2}{2v^3\sin(v/5)}, \\
b_1' &= \frac{9\cos(v/5)\sin(2v/3) - 9\cos(2v/3)\sin(v/5)}{56\sin(v/5)} + \frac{\cos(v/5)\sin(v) - v\cos(v/5) - \cos(v)\sin(v/5) + \sin(v/5)}{v^2\sin(v/5)}, \\
b_1'' &= -\frac{(25v/243)\sin(v) + (80/81)\cos(v) - 80/81 + (23v/48)\sin(2v/3)}{(80v/81)\sin(v/5) - (v/125)\sin(2v/3)}, \\
b_2'' &= \frac{-1 + (27/56)v\sin(2v/3) + (5/48)v\sin(v) + \cos(v)}{v\sin(v/5)}, \\
b_2' & \\
&= -\frac{((27/56)\cos(v/5)v\sin(2v/3) + (5/48)\cos(v/5)v\sin(v) + \cos(v/5)\cos(v) - \cos(v/5) - (27/56)v\cos(2v/3)\sin(v/5) - (5/48)v\cos(v)\sin(v/5) + \sin(v)\sin(v/5))}{v\sin(v/5)}
\end{aligned} \quad (36)$$

These lead to our new explicit trigonometrically fitted MRKT which is called TFMRKT5 method. The corresponding Taylor series expansion of the solution is given by

$$\begin{aligned}
b_1 &= \frac{1}{48} - \frac{1}{2160}v^2 + \frac{101}{136080000}v^4 \\
&\quad + \frac{5713}{183708000000}v^6 \\
&\quad - \frac{11330339}{81841914000000000}v^8 \\
&\quad + \frac{57722879}{134057055132000000000}v^{10} + \cdots, \\
b_2 &= \frac{5}{42} + \frac{1}{2160}v^2 - \frac{2921}{136080000}v^4 \\
&\quad + \frac{1361}{45927000000}v^6
\end{aligned}$$

$$+ \frac{54293587}{40920957000000000}\nu^8$$

$$- \frac{6964030429}{670285275660000000000}\nu^{10} + \cdots,$$

$$b_1' = \frac{1}{24} - \frac{17}{283500}\nu^4 + \frac{149}{218700000}\nu^6$$

$$- \frac{1055069}{341007975000000}\nu^8$$

$$+ \frac{23025689}{253895937750000000}\nu^{10} + \cdots,$$

$$b_2' = \frac{25}{84} + \frac{241}{2268000}\nu^4 - \frac{22871}{6123600000}\nu^6$$

$$+ \frac{112778137}{2728063800000000}\nu^8$$

$$- \frac{13599351683}{558571063050000000000}\nu^{10} + \cdots,$$

$$b_1'' = \frac{1}{24} + \frac{1}{21600}\nu^4 - \frac{167}{58320000}\nu^6$$

$$+ \frac{528389}{16533720000000}\nu^8$$

$$- \frac{967343}{5845851000000000}\nu^{10} + \cdots,$$

$$b_2'' = \frac{125}{336} - \frac{1}{21600}\nu^4 + \frac{2867}{408240000}\nu^6$$

$$- \frac{3022109}{16533720000000}\nu^8$$

$$+ \frac{1457821}{730731375000000}\nu^{10} + \cdots,$$

$$\widehat{a}_{21} = \frac{1}{50} - \frac{1}{15000}\nu^2 + \frac{1}{11250000}\nu^4$$

$$- \frac{1}{15750000000}\nu^6 + \frac{1}{35437500000000}\nu^8$$

$$- \frac{1}{116943750000000000}\nu^{10} + \cdots,$$

$$\widehat{a}_{32} = \frac{7}{27} - \frac{37}{12150}\nu^2 + \frac{287}{6561000}\nu^4$$

$$+ \frac{68827}{516678750000}\nu^6$$

$$+ \frac{7097417}{1674039150000000}\nu^8$$

$$+ \frac{39720321233}{621487034437500000000}\nu^{10} + \cdots,$$

$$\widehat{a}_{43} = \frac{9}{35} + \frac{43}{3000}\nu^2 + \frac{3323}{1350000}\nu^4 + \frac{111136579}{255150000000}\nu^6$$

$$+ \frac{26973882539}{344452500000000}\nu^8$$

$$+ \frac{721361598388001}{51151196250000000000}\nu^{10} + \cdots,$$

$$a_{32} = \frac{289}{4860} + \frac{67}{729000}\nu^2 + \frac{26141}{2755620000}\nu^4$$

$$+ \frac{4671541}{39858075000000}\nu^6$$

$$+ \frac{2216008103}{1104865839000000000}\nu^8$$

$$+ \frac{15664491766661}{484759886861250000000000}\nu^{10} + \cdots,$$

$$a_{43} = \frac{7}{150} + \frac{221}{135000}\nu^2 + \frac{114463}{637875000}\nu^4$$

$$+ \frac{1010472889}{34445250000000}\nu^6$$

$$+ \frac{222576819697}{42625996875000000}\nu^8$$

$$+ \frac{84318883418716333}{89770349418750000000000}\nu^{10} + \cdots,$$

$$\gamma_2 = 1 - \frac{1}{25}\nu^2 + \frac{1}{15000}\nu^4 - \frac{1}{11250000}\nu^6$$

$$+ \frac{1}{15750000000}\nu^8 - \frac{1}{35437500000000}\nu^{10}$$

$$+ \cdots,$$

$$\gamma_3 = 1 - \frac{59}{225}\nu^2 - \frac{9569}{1215000}\nu^4 - \frac{74831}{911250000}\nu^6$$

$$+ \frac{26104429}{11481750000000}\nu^8$$

$$- \frac{495937229}{25833937500000000}\nu^{10} + \cdots,$$

$$\gamma_4 = 1 - \nu^2 + \frac{19}{1000}\nu^4 + \frac{2267}{4050000}\nu^6$$

$$- \frac{4858267}{255150000000}\nu^8$$

$$+ \frac{65060629}{344452500000000}\nu^{10} + \cdots,$$

$$\widehat{\gamma}_2 = 1 - \frac{1}{150}\nu^2 + \frac{1}{75000}\nu^4 - \frac{1}{78750000}\nu^6$$

$$+ \frac{1}{141750000000}\nu^8$$

$$- \frac{1}{389812500000000}\nu^{10} + \cdots,$$

$$\widehat{\gamma}_3 = 1 + \frac{1}{270}\nu^2 + \frac{13}{60750}\nu^4 + \frac{2032}{717609375}\nu^6$$

$$+ \frac{136109}{2906317968750}\nu^8$$

$$+\frac{2726807}{3596568486328125}v^{10}+\cdots,$$

$$\widehat{\gamma}_4 = 1-\frac{1}{150}v^2+\frac{79}{15000}v^4+\frac{2161951}{2126250000}v^6 + \frac{78384961}{425250000000}v^8 + \frac{5664933289891}{170503987500000000}v^{10}+\cdots, \quad (37)$$

where $\gamma_1 = 1, \widehat{\gamma}_1 = 1$.

This results in the new method called TFMRKT5. As $v \longrightarrow 0$, the coefficients $b_1, b_2, b_1', b_2', b_1'', b_2'', a_{32}, a_{42}, \widehat{a}_{21}, \widehat{a}_{32}, \widehat{a}_{43}, \gamma_2, \gamma_3, \gamma_4, \widehat{\gamma}_2, \widehat{\gamma}_3$, and $\widehat{\gamma}_4$ of the new method TFMRKT5 reduce to the coefficients of the original method RKT5. That is to say, $b_1(0), b_2(0), b_1'(0), b_2'(0), b_1''(0), b_2''(0), a_{32}(0), a_{42}(0), \widehat{a}_{21}(0), \widehat{a}_{32}(0), \widehat{a}_{43}(0), \gamma_2(0), \gamma_3(0), \gamma_4(0), \widehat{\gamma}_2(0), \widehat{\gamma}_3(0)$, and $\widehat{\gamma}_4(0)$ are identical to $b_1, b_2, b_1', b_2', b_1'', b_2'', a_{32}, a_{42}, \widehat{a}_{21}, \widehat{a}_{32}, \widehat{a}_{43}, \gamma_2, \gamma_3, \gamma_4, \widehat{\gamma}_2, \widehat{\gamma}_3$, and $\widehat{\gamma}_4$ of RKT5 method. Other than that, $v \longrightarrow 0$, as TFMRKT5 method will have the same error constant as RKT5 method.

4. Error Analysis

In this section, we will find the principal local truncation errors for y, y', and y'' (i.e., $\tau_{n+1}, \tau_{n+1}', \tau_{n+1}''$) of the new exponentially fitted and trigonometrically fitted explicit modified Runge-Kutta type methods, respectively. We first find the Taylor series expansion of the actual solution $y(x_n + h)$, the first derivative of the actual solution $y'(x_n+h)$, and the second derivative of the actual solution $y''(x_n + h)$, the approximate solution y_{n+1}, the first derivative of the approximate solution y_{n+1}', and the second derivative of the approximate solution y_{n+1}''. The local truncation errors of y, y', and y'' are given as

$$\begin{aligned}\tau_{n+1} &= y_{n+1} - y(x_n + h),\\ \tau_{n+1}' &= y_{n+1}' - y'(x_n + h),\\ \tau_{n+1}'' &= y_{n+1}'' - y''(x_n + h)\end{aligned} \quad (38)$$

The τ_{n+1}, τ_{n+1}', and τ_{n+1}'' of the methods are given in the Appendix.

Notes: from τ_{n+1}, τ_{n+1}', and τ_{n+1}'', we can see that the order of TFMRKT5 is order 5 because all of the coefficients up to h^5 vanished.

5. Problems Tested and Numerical Results

In this section, we will apply the new explicit exponentially fitted modified Runge-Kutta type method to some $y''' = f(x, y, y')$ ODEs for problems (1)-(4) which consist of exponential solutions and the new trigonometrically fitted modified Runge-Kutta type method to some ODEs problems (5)-(8) with trigonometric functions solutions. The numerical results are compared with the results obtained when the same set of problems are reduced to a system of first-order equations and is solved using the existing Runge-Kutta of the same order.

(i) h: step sizes.

(ii) **TFMRKT5:** the four-stage fifth-order trigonometrically fitted RK type method derived in this paper.

(iii) **EFMRKT5:** the four-stage fifth-order exponentially fitted RK type method derived in this paper.

(iv) **RKT5:** the four-stage fifth-order RK type method given by Fawzi et al. [15].

(v) **RK5B:** the six-stage fifth-order RK method given in Butcher [16].

(vi) **RKF5:** the six-stage fifth-order RK method given in Lambert [17].

(vii) **TFRK:** the six-stage fifth-order trigonometrically fitted RK method given in Anastassi et al. [18].

Problem 2 (homogeneous linear problem).

$$y'''(x) = 2y'(x), \qquad y(0) = 0,\ y'(0) = 1,\ y''(0) = 0, \quad (39)$$

exact solution is

$$y(x) = \frac{\sqrt{2}e^{\sqrt{2}x}}{4} - \frac{\sqrt{2}e^{-\sqrt{2}x}}{4}. \quad (40)$$

Estimated frequency $\omega = \sqrt{2}$.

Problem 3 (homogeneous linear system).

$$\begin{aligned} y_1'''(x) &= 8y_3'(x), & y_1(0) = 2,\ y_1'(0) = 4,\ y_1''(0) = 8,\\ y_2'''(x) &= 8y_1'(x), & y_2(0) = 4,\ y_2'(0) = 8,\ y_2''(0) = 16,\\ y_3'''(x) &= y_2'(x), & y_3(0) = 1,\ y_3'(0) = 2,\ y_3''(0) = 4,\end{aligned} \quad (41)$$

exact solutions are

$$\begin{aligned} y_1(x) &= 2e^{2x},\\ y_2(x) &= 4e^{2x},\\ y_3(x) &= e^{2x}.\end{aligned} \quad (42)$$

Estimated frequency $\omega = 2$.

Problem 4 (inhomogeneous linear system).

$$\begin{aligned}
&y_1'''(x) = y_3'(x) + 1, \\
&\qquad y_1(0) = 2,\ y_1'(0) = 3,\ y_1''(0) = 5, \\
&y_2'''(x) = y_1'(x) + 2, \\
&\qquad y_2(0) = 1,\ y_2'(0) = 2,\ y_2''(0) = 5, \\
&y_3'''(x) = y_2'(x) + 3, \\
&\qquad y_3(0) = 0,\ y_3'(0) = 4,\ y_3''(0) = 5,
\end{aligned} \tag{43}$$

exact solutions are

$$\begin{aligned}
y_1(x) &= 5e^x - 2x - 3, \\
y_2(x) &= 5e^x - 3x - 4, \\
y_3(x) &= 5e^x - x - 5.
\end{aligned} \tag{44}$$

Estimated frequency $\omega = 1$.

Problem 5 (inhomogeneous linear problem).

$$\begin{aligned}
&y'''(x) = 5y'(x) + \sinh(x), \\
&\qquad y(0) = -\frac{1}{4},\ y'(0) = 0,\ y''(0) = -\frac{1}{4},
\end{aligned} \tag{45}$$

exact solution is

$$y(x) = -\frac{e^x}{8} - \frac{e^{-x}}{8}. \tag{46}$$

Estimated frequency $\omega = 1$.

Problem 6 (homogeneous linear problem).

$$\begin{aligned}
&y'''(x) = -25y'(x), \\
&\qquad y(0) = 0,\ y'(0) = 0,\ y''(0) = 1,
\end{aligned} \tag{47}$$

exact solution is

$$y(x) = \frac{1}{25} - \frac{1}{25}\cos(5x). \tag{48}$$

Estimated frequency $\omega = 5$.

Problem 7 (inhomogeneous linear problem).

$$\begin{aligned}
&y'''(x) = -27y'(x) + \sin(x), \\
&\qquad y(0) = 1,\ y'(0) = -1,\ y''(0) = 0,
\end{aligned} \tag{49}$$

exact solution is

$$\begin{aligned}
y(x) = {} & \frac{\sqrt{3}}{702}\cos\left(3\sqrt{3}x\right) - \frac{\sqrt{3}}{9}\sin\left(3\sqrt{3}x\right) \\
& - \frac{1}{26}\cos(x) + \frac{28}{27}.
\end{aligned} \tag{50}$$

Estimated frequency $\omega = 3\sqrt{3}$.

Problem 8 (inhomogeneous linear system).

$$\begin{aligned}
&y_1'''(x) = -27y_1'(x), \\
&\qquad y_1(0) = 0,\ y_1'(0) = -1,\ y_1'(0) = 0, \\
&y_2'''(x) = -27y_2'(x) + \cos(x), \\
&\qquad y_2(0) = 1,\ y_2'(0) = -1,\ y_2''(0) = 0, \\
&y_3'''(x) = -27y_3'(x), \\
&\qquad y_3(0) = 1,\ y_3'(0) = 0,\ y_3''(0) = -1,
\end{aligned} \tag{51}$$

exact solutions are

$$\begin{aligned}
y_1(x) &= -\frac{\sqrt{3}}{9}\sin\left(3\sqrt{3}x\right), \\
y_2(x) &= 1 - \frac{3\sqrt{3}}{26}\sin\left(3\sqrt{3}x\right) + \frac{1}{26}\sin(x), \\
y_3(x) &= \frac{26}{27} + \frac{1}{27}\cos\left(3\sqrt{3}x\right).
\end{aligned} \tag{52}$$

Estimated frequency $\omega = 3\sqrt{3}$.

Problem 9 (inhomogeneous linear system).

$$\begin{aligned}
&y_1'''(x) = -7y_2'(x) - \cos(x), \\
&\qquad y_1(0) = \frac{1}{7},\ y_1'(0) = 0, y_1''(0) = -1, \\
&y_2'''(x) = -7y_1'(x) - \cos(x), \\
&\qquad y_2(0) = 1,\ y_2'(0) = 0,\ y_2''(0) = -1, \\
&y_3'''(x) = -7y_3'(x) - \cos(x), \\
&\qquad y_3(0) = 0,\ y_3'(0) = 1,\ y_3''(0) = 0,
\end{aligned} \tag{53}$$

exact solutions are

$$\begin{aligned}
y_1(x) &= -\frac{1}{6}\sin(x) + \frac{\sqrt{7}}{42}\sin\left(\sqrt{7}x\right) + \frac{1}{7}\cos\left(\sqrt{7}x\right), \\
y_2(x) &= -\frac{1}{6}\sin(x) + \frac{\sqrt{7}}{42}\sin(x)\sqrt{7} + \frac{6}{7} \\
&\quad + \frac{1}{7}\cos\left(\sqrt{7}x\right), \\
y_3(x) &= \frac{\sqrt{7}}{6}\sin\left(\sqrt{7}x\right) - \frac{1}{6}\sin(x).
\end{aligned} \tag{54}$$

Estimated frequency $\omega = 1$.

6. An Application to a Problem in Thin Film Flow

Here, we will use the suggested method to a famous problem in engineering and physics based on the thin film flow of a liquid. Many researchers in the literature explain this problem more. Momoniat and Mahomed[19] constructed symmetry reduction and numerical solution of a third-order ODE from thin film flow. Tuck and Schwartz [20] discussed the movement of a thin film of viscous fluid over a solid surface and taken into account tension and gravity, as well as viscosity. The problem was evaluated and solved using third-order ODE as follows:

$$\frac{d^3 y}{dx^3} = f(y). \tag{55}$$

Many forms of the function were studied by [20] for the drainage dry surface; it has the form of $f(y)$ which can be stated as

$$\frac{d^3 y}{dx^3} = -1 + \frac{1}{y^2}. \tag{56}$$

When the surface is prewetted by a thin film with thickness $\delta > 0$ (where $\delta > 0$ is very small), the function f is given by

$$f(y) = -1 + \frac{1 + \delta + \delta^2}{y^2} - \frac{\delta + \delta^2}{y^3} \tag{57}$$

Problems concerning the flow of thin films of viscous fluid with a free surface in which surface tension effects play a role typically lead to third-order ODEs governing the shape of the free surface of the fluid,$y = y(x)$. As indicated by [20], one such equation is

$$y'''(x) = y^{-k}, \quad x \ge x_0 \tag{58}$$

with initial conditions

$$\begin{aligned} y(x_0) &= y_0, \\ y'(x_0) &= y'_0, \\ y''(x_0) &= y''_0, \end{aligned} \tag{59}$$

where y_0, y'_0, and y''_0 are constants, which is of specific significance since it portrays the dynamic balance amongst surface and gooey strengths in a thin fluid layer in disregard of gravity. For compare and contrast, we utilized Runge-Kutta methods which are fifth-order (RKT5, RK5B, RKF5, and TFRKT) strategies, individually. To utilize Runge-Kutta techniques we write (1) as a system of three first-order equations. Biazar et al. [21] we can write (58) as the following system:

$$\begin{aligned} \frac{dy_1}{dx} &= y_2(x), \\ \frac{dy_2}{dx} &= y_3(x), \\ \frac{dy_2}{dx} &= y_1^{-k}, \end{aligned} \tag{60}$$

where

$$\begin{aligned} y_1(0) &= 1, \\ y_2(0) &= 1, \\ y_3(0) &= 1. \end{aligned} \tag{61}$$

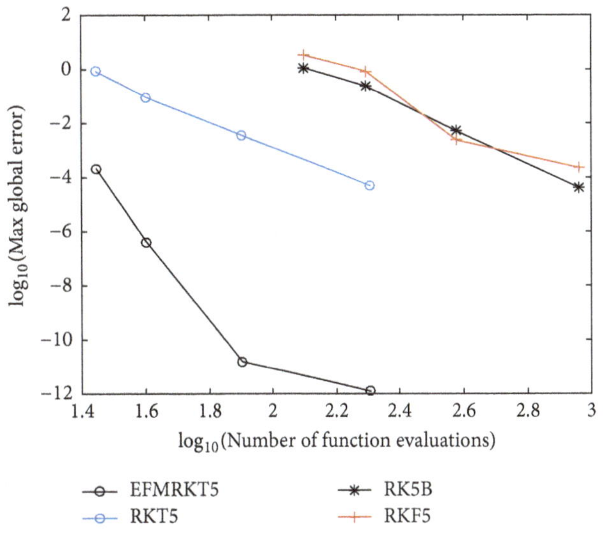

FIGURE 1: The efficiency curve for EFMRKT5, RKT5, RK5B, and RKF5 for Problem 2 with $x_{end} = 5$ and $h = 0.1, 0.25, 0.5, 0.75$.

We have taken $x_0 = 0$ and $y_0 = y'_0 = y''_0 = 1$. Unfortunately, for general k, (58) cannot be solved analytically. However, we can use these reductions to determine an efficient way to solve (1) numerically. Here, we are focusing on the cases $k = 2$ and $k = 3$ (see Mechee et al.[22]).

7. Discussion and Conclusion

In this research, we have derived exponentially fitted and trigonometrically fitted explicit modified Runge-Kutta type methods for solving $y'''(x) = f(x, y, y')$ with application to thin film flow problem. Consequently, the new four-stage fifth-order exponentially-fitted and trigonometrically-fitted methods which are denoted as EFMRKT5 and TFMRKT5, respectively, were constructed and we used in numerical comparison the criteria based on computing the maximum error in the solution $(\max(|y(t_n) - y_n|))$ which is equal to the maximum between absolute errors of the actual solutions and computed solutions. The numerical outcomes are plotted in Figures 1–8. Figures 1–8 demonstrate that the new TFMRKT5 and EFMRKT5 methods require less capacity assessments than the RKT5, RK5B, RKF5, and TFRK methods. The figures showed the efficiency of the new methods where the common logarithm of the maximum global error throughout the integration versus computational cost was measured by the number of function evaluations. The numerical results obtained showed clearly that the global error for a short period of integration for the new exponentially fitted method and for a large period of integration for the new trigonometrically fitted explicit modified Runge-Kutta type method is smaller than that of the other existing methods. The new EFMRKT5 and TFMRKT5 methods are much more efficient than the other existing methods when solving third-order ODEs of the form $y''' = f(x, y, y')$ straightforwardly. For Tables 2 and 3 we observed that the numerical results using TFMRKT5 and EFMRKT5 methods are correct to five decimal places. Applying RK5B, RKF5, TFRK, and RKT5 to

Table 2: Numerical results for problem in Thin Film Flow (58) taking $h = 0.1$ and $k = 2$.

x	Exact Solution	RK5B	RKF5	RKT5	EMFRKT5	TFRK	TFMRKT5
0.0	1.000000000	1.000000000	1.000000000	1.000000000	1.000000000	1.000000000	1.000000000
0.2	1.221211030	1.2212100068	1.2212100097	1.2212100039	1.2212100052	1.2212100218	1.2212100052
0.4	1.488834893	1.4888347851	1.4888347895	1.4888347797	1.4888347885	1.4888348090	1.4888347885
0.6	1.807361404	1.8073614063	1.8073614114	1.8073613988	1.8073614237	1.8073614357	1.8073614237
0.8	2.179819234	2.1798192463	2.1798192513	2.1798192371	2.1798192873	2.1798192788	2.1798192873
1.0	2.608275822	2.6082748841	2.6082748883	2.6082748735	2.6082749587	2.6082749176	2.6082749587

Table 3: Numerical results for problem in thin film flow (58) taking $h = 0.01$ and $k = 2$.

x	Exact Solution	RK5B	RKF5	RKT5	EFMRKT5	TFRK	TFMRKT5
0.0	1.000000000	1.000000000	1.000000000	1.000000000	1.000000000	1.000000000	1.000000000
0.2	1.221211030	1.2212100045	1.2212100045	1.2212100045	1.2212100045	1.2212100045	1.2212100045
0.4	1.488834893	1.4888347799	1.4888347799	1.4888347799	1.4888347799	1.4888347799	1.4888347799
0.6	1.807361404	1.8073613977	1.8073613977	1.8073613977	1.8073613977	1.8073613977	1.8073613977
0.8	2.179819234	2.1798192339	2.1798192339	2.1798192339	2.1798192340	2.1798192339	2.1798192340
1.0	2.608275822	2.6082748676	2.6082748676	2.6082748676	2.6082748677	2.6082748676	2.6082748677

(58) for $k = 2$ also yields five-decimal place accuracy. Tables 4 and 5 show the numerical results for the case $k = 3$ with $h = 0.1$ and $h = 0.01$ since for $k = 3$, Problem (58) cannot be solved analytically. Table 4 shows that TFMRKT5 and EFMRKT5 manage to achieve the numerical results which agree to seven decimal places when compared to RK5B, RKF5, TFRK, and RKT5 for $h = 0.1$. In Table 5 the numerical results for TFMRKT5 and EFMRKT5 agree to nine decimal places when compared to RK5B, RKF5, TFRK, and RKT5 for $h = 0.01$. For Table 7 we observe that RK5B, RKF5, RKT5, TFRK, TFMRKT5, and EFMRKT5 have similar order of accuracy. In Table 6 values of the error are different. Therefore it is consistent with results displayed in Tables 2 and 3. Figures 9 and 10 show that the new EFMRKT5 and TFMRKT5 methods require less function evaluations than the RK5B, RKF5, TFRK, and RKT5 methods. This is because when problem (58) is solved using RK5B, RKF5, TFRK, and RKT5 methods, it needs to be reduced to a system of first-order equations which is three times the dimension.

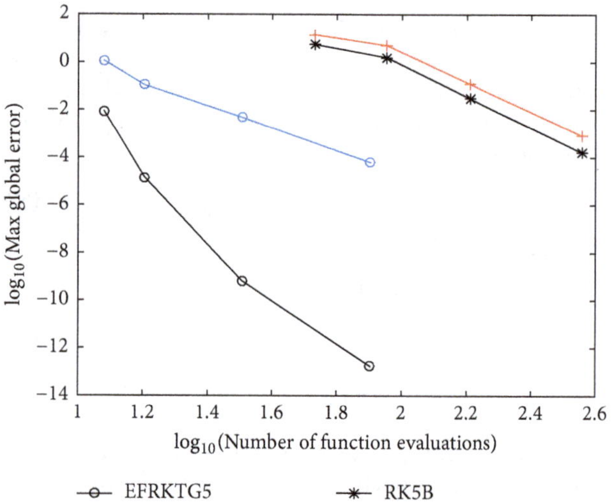

Figure 2: The efficiency curve for EFMRKT5, RKT5, RK5B, and RKF5 for Problem 3 with $x_{end} = 2$ and $h = 0.1, 0.25, 0.5, 0.75$.

Appendix

The principal local truncation errors for y, y', and y'' (i.e., $\tau_{n+1}, \tau'_{n+1}, \tau''_{n+1}$) for EFMRKT5 are as follows:

$$\begin{aligned}\tau_{n+1} = \Big(&-\frac{1}{3600}y_{xx}{}^2Fyz - \frac{1}{10800}y_x{}^3Fyyy \\ &-\frac{1}{3600}y_xFyyy_{xx} + \frac{527}{162000}y_xFyzF \\ &+\frac{527}{162000}Fzzy_{xx}F - \frac{1}{1800}y_xFxyzy_{xx} \\ &-\frac{1}{3600}y_x{}^2Fzyyy_{xx} - \frac{1}{3600}y_xy_{xx}{}^2Fzyz \\ &+\frac{971}{907200}FzFyy_x - \frac{1}{10800}Fxxx + \frac{1}{10800}w^2Fx \\ &-\frac{1}{3600}y_{xx}Fxy - \frac{1}{144}FyF - \frac{1}{3600}y_xFyxx \\ &-\frac{1}{3600}Fxzzy_{xx}{}^2 + \frac{527}{162000}FxzF \\ &-\frac{1}{10800}y_{xx}{}^3Fzzz + \frac{971}{907200}FzFx \\ &+\frac{971}{907200}Fz^2y_{xx} - \frac{1}{3600}y_x{}^2Fxyy \\ &+\frac{1}{10800}w^2Fzy_{xx} - \frac{1}{3600}Fzxxy_{xx} \\ &+\frac{1}{10800}w^2Fyy_x\Big)h^6 + O\left(h^7\right) \qquad \text{(A.1)}\end{aligned}$$

TABLE 4: Numerical results for problem in thin film flow (58) taking $h = 0.1$ and $k = 3$.

x	RK5B	RKF5	RKT5	EMFRKT5	TFRK	TFMRKT5
0.0	1.000000000	1.000000000	1.000000000	1.000000000	1.000000000	1.000000000
0.2	1.2211551491	1.2211551546	1.2211551394	1.2211551412	1.2211551831	1.2211551412
0.4	1.4881052974	1.4881053065	1.4881052807	1.4881052926	1.4881053519	1.4881052926
0.6	1.8042625677	1.8042625794	1.8042625459	1.8042625786	1.8042626364	1.8042625786
0.8	2.1715228242	2.1715228376	2.1715227987	2.1715228633	2.1715229031	2.1715228633
1.0	2.5909582923	2.5909583063	2.5909582638	2.5909583715	2.5909583783	2.5909583715

TABLE 5: Numerical results for problem in thin film flow (58) taking $h = 0.01$ and $k = 3$.

x	RK5B	RKF5	RKT5	EMFRKT5	TFRK	TFMRKT5
0.0	1.000000000	1.000000000	1.000000000	1.000000000	1.000000000	1.000000000
0.2	1.2211551424	1.2211551424	1.2211551424	1.2211551424	1.2211551424	1.2211551424
0.4	1.4881052842	1.4881052842	1.4881052842	1.4881052842	1.4881052842	1.4881052842
0.6	1.8042625481	1.8042625481	1.8042625481	1.8042625482	1.8042625481	1.8042625482
0.8	2.1715227981	2.1715227981	2.1715227981	2.1715227982	2.1715227981	2.1715227982
1.0	2.5909582591	2.5909582591	2.5909582591	2.5909582592	2.5909582591	2.5909582592

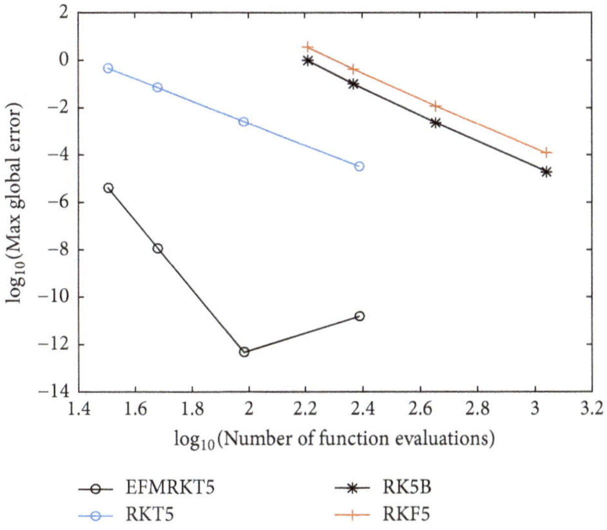

FIGURE 3: The efficiency curve for EFMRKT5, RKT5, RK5B, and RKF5 for Problem 4 with $x_{end} = 6$ and $h = 0.1, 0.25, 0.5, 0.75$.

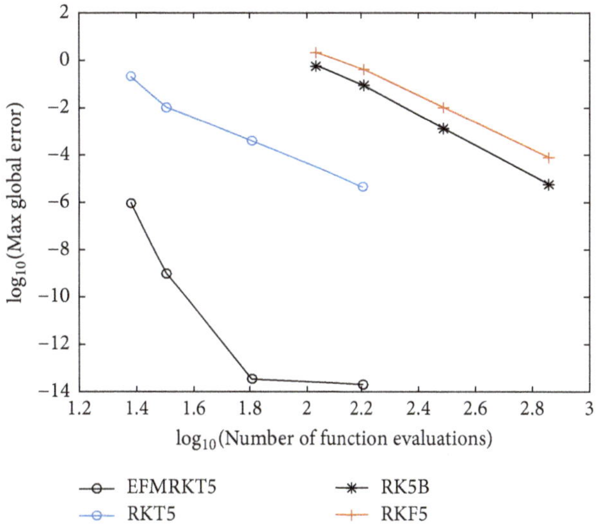

FIGURE 4: The efficiency curve for EFMRKT5, RKT5, RK5B, and RKF5 Problem 5 with $x_{end} = 4$ and $h = 0.1, 0.25, 0.5, 0.75$.

$$\begin{aligned}
\tau'_{n+1} = \Big(& \frac{1}{21600}Fxxxx + \frac{1811}{756000}Fzy_xFxy \\
& + \frac{1063}{162000}y_{xx}{}^2FzzzF + \frac{971}{226800}Fyy_xFxz \\
& + \frac{24853}{4536000}Fzy_{xx}{}^2Fzz + \frac{15143}{2268000}Fzy_{xx}Fxz \\
& + \frac{1}{3600}y_x{}^2y_{xx}{}^2Fzzyy + \frac{1811}{1512000}Fzy_x{}^2Fyy \\
& + \frac{971}{226800}Fzzy_{xx}Fx + \frac{971}{226800}y_xFyzFx \\
& - \frac{2087}{162000}y_{xx}FyzF + \frac{1}{1800}y_xy_{xx}Fxxyz \\
& + \frac{1}{1800}y_xy_{xx}{}^2Fxyzz + \frac{1}{1800}y_x{}^2y_{xx}Fxyyz \\
& + \frac{1}{5400}y_x{}^3Fzyyyy_{xx} + \frac{1}{5400}y_xy_{xx}{}^3Fzzzy \\
& + \frac{971}{226800}Fyy_x{}^2Fyz + \frac{1063}{81000}y_xFxyzF \\
& + \frac{1063}{162000}y_x{}^2FzyyF + \frac{1}{3600}y_x{}^2Fyyyy_{xx} \\
& - \frac{8689}{1512000}FzFyy_{xx} + \frac{1}{1800}y_xFxyyy_{xx} \\
& + \frac{1063}{81000}Fxzzy_{xx}F + \frac{317}{22680000}FzFw^2
\end{aligned}$$

TABLE 6: Comparison of error for problem in thin film flow (58) taking $h = 0.1$ and $k = 2$.

x	RK5B	RKF5	RKT5	EFMRKT5	TFRK	TFMRKT5
0.0	0.0000(0)	0.0000(0)	0.0000(0)	0.0000(0)	0.0000(0)	0.0000(0)
0.2	1.0230 (-6)	1.0200(-6)	1.2600(-6)	1.0250(-6)	1.0080(-6)	1.0250(-6)
0.4	1.0800(-7)	1.0300(-7)	1.1300(-7)	1.0500(-7)	8.4100(-7)	1.0500(-7)
0.6	2.0000(-9)	7.0000(-9)	5.0000(-8)	2.0000(-8)	3.2000(-8)	2.0000(-8)
0.8	1.2000(-8)	1.7000(-8)	3.0000(-9)	5.300(-8)	4.5000(-8)	5.3000(-8)
1.0	9.3800(-7)	9.3400(-7)	9.4800 (-7)	8.6300 (-7)	9.0400(-7)	8.6300(-7)

TABLE 7: Comparison of error for problem in thin film flow (58) taking $h = 0.01$ and $k = 2$.

x	RK5B	RKF5	RKT5	EFMRKT5	TFRK	TFMRKT5
0.0	0.0000(0)	0.0000(0)	0.0000(0)	0.0000(0)	0.0000(0)	0.0000(0)
0.2	1.0260(-6)	1.0260(-6)	1.0260(-6)	1.0260(-6)	1.0260(-6)	1.0260(-6)
0.4	6.0000(-7)	6.0000(-7)	6.0000(-7)	6.0000(-7)	6.0000(-7)	6.0000(-7)
0.6	9.0000(-9)	9.0000(-9)	9.0000(-9)	9.0000(-9)	9.0000(-9)	9.0000(-9)
0.8	0.0000(0)	0.0000(0)	0.0000(0)	0.0000(0)	0.0000(0)	0.0000(0)
1.0	9.5400(-7)	9.5400(-7)	9.5400(-7)	9.5400(-7)	9.5400(-7)	9.5400(-7)

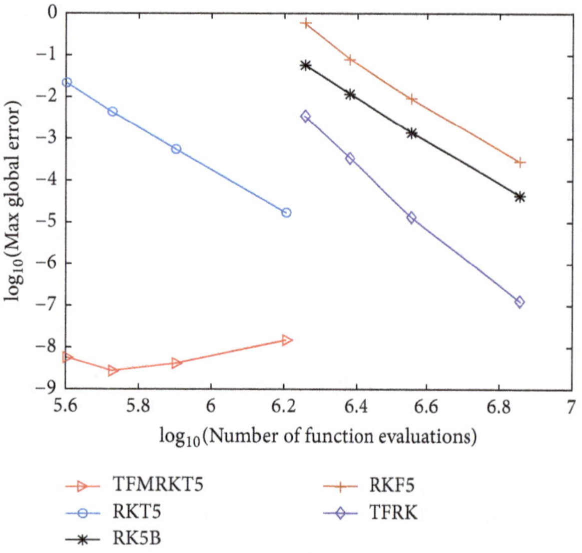

FIGURE 5: The efficiency curve for TFMRKT5, RKT5, RK5B, RKF5, and TFRK for Problem 6 with $x_{end} = 10000$ and $h = 0.025, 0.05, 0.075, 0.1$.

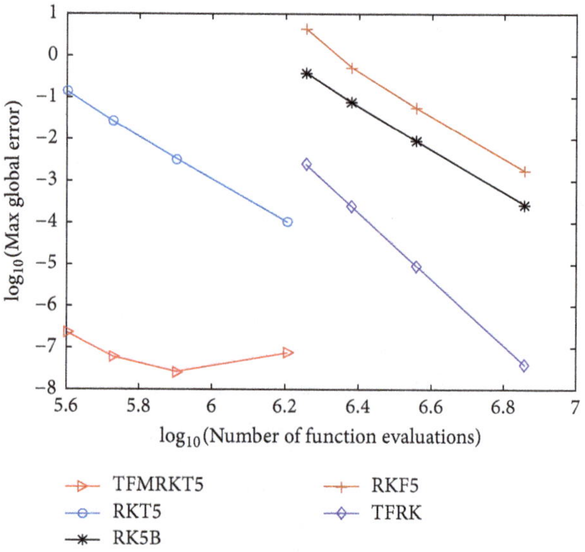

FIGURE 6: The efficiency curve for TFMRKT5, RKT5, RK5B, RKF5, and TFRK for Problem 7 with $x_{end} = 10000$ and $h = 0.025, 0.05, 0.075, 0.1$.

$$+ \frac{1}{1800} y_x Fzyyy {y_{xx}}^2 - \frac{7}{360} y_x FyyF$$
$$+ \frac{1}{5400} y_x Fxyxx + \frac{1}{21600} {y_x}^4 Fyyyy$$
$$+ \frac{1}{3600} y_{xx} Fyxx + \frac{31211}{22680000} Fz^2 F$$
$$+ \frac{1063}{162000} FzxxF - \frac{7}{360} FFxy + \frac{1}{5400} {y_x}^3 Fxyyy$$
$$- \frac{1}{144} FyFx + \frac{1}{5400} {y_{xx}}^3 Fxzzz$$
$$+ \frac{1}{21600} {y_{xx}}^4 Fzzzz + \frac{971}{226800} FxFxz$$

$$+ \frac{1811}{1512000} FzFxx + \frac{1}{3600} {y_{xx}}^2 Fxxzz$$
$$+ \frac{1}{3600} {y_x}^2 Fxyxy + \frac{1}{7200} {y_{xx}}^2 Fyy$$
$$+ \frac{57889}{14580000} FzzF^2 + \frac{1}{1800} {y_{xx}}^2 Fxyz$$
$$+ \frac{1}{3600} {y_{xx}}^3 Fzyz - \frac{1}{144} Fy^2 y_x + \frac{1}{5400} Fxxxzy_{xx}$$
$$- \frac{1}{21600} w^4 F + \frac{1063}{81000} y_x y_{xx} FzyzF$$

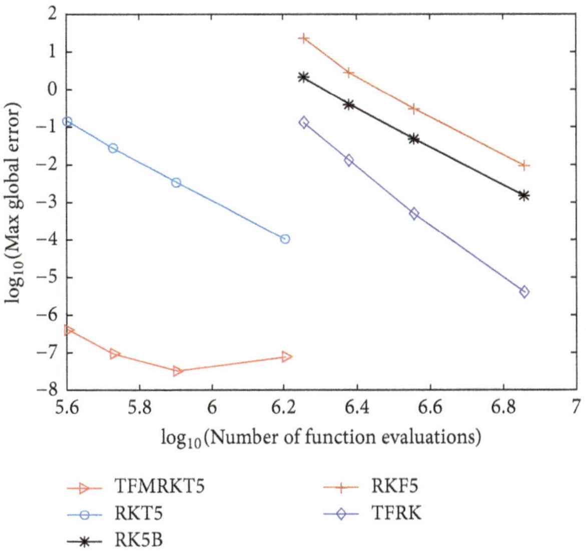

FIGURE 7: The efficiency curve for TFMRKT5, RKT5, RK5B, RKF5, and TFRK for Problem 8 with x_{end} = 10000 and h = 0.025, 0.05, 0.075, 0.1.

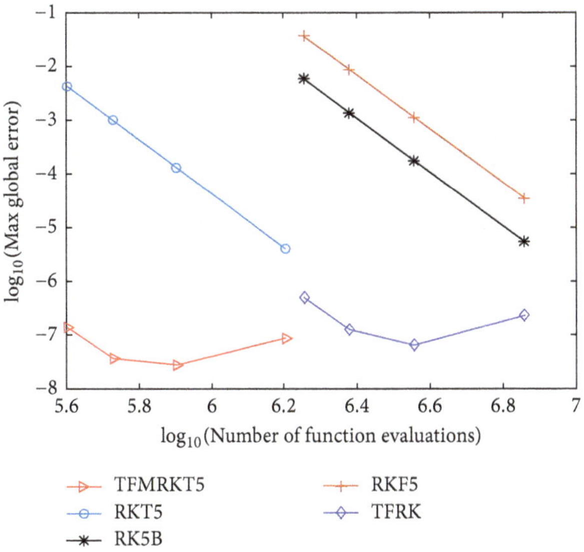

FIGURE 8: The efficiency curve for TFMRKT5, RKT5, RK5B, RKF5, and TFRK for Problem 9 with x_{end} = 10000 and h = 0.025, 0.05, 0.075, 0.1.

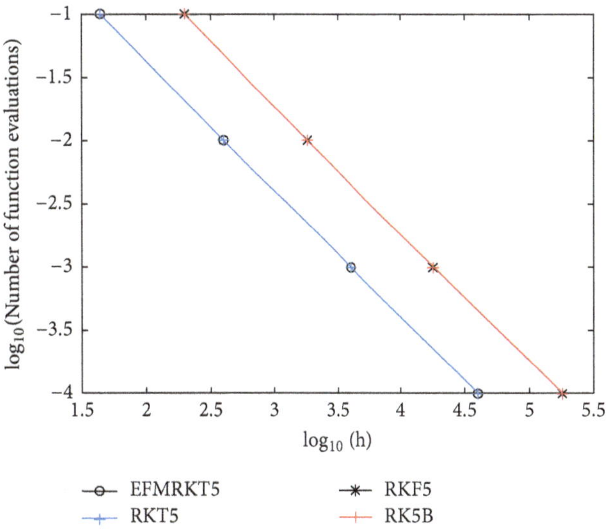

FIGURE 9: Plot of graph for function evaluations against step size h for Problem (58) taking $x_{end} = 1, h = 0.1, 0.01, 0.001, 0.0001$, and $k = 2$.

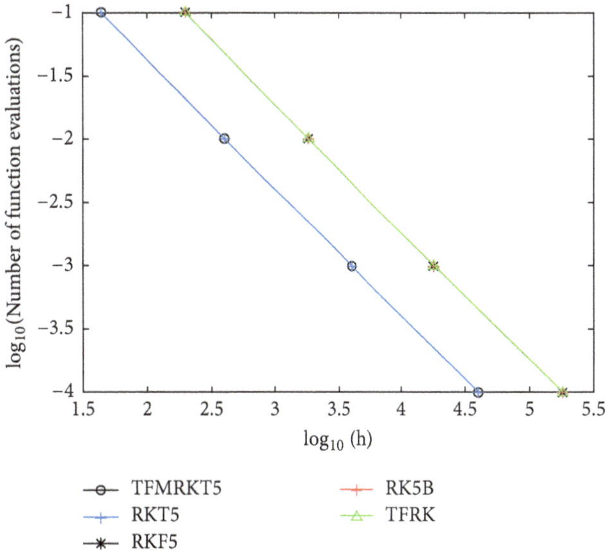

FIGURE 10: Plot of graph for function evaluations against step size h for Problem (58) taking $x_{end} = 1, h = 0.1, 0.01, 0.001, 0.0001$ and $k = 2$.

$$+ \frac{971}{226800} Fyy_x Fzzy_{xx} + \frac{15143}{2268000} Fzy_x Fyzy_{xx} \Big) \cdot h^6 + O\left(h^7\right) \tag{A.2}$$

$$\begin{aligned}
\tau''_{n+1} = \Big(&-\frac{7589}{907200} Fyy_{xx} Fxz - \frac{13777}{907200} y_{xx} FyzFx \\
&+ \frac{5011}{302400} y_x Fzzy_{xx}{}^2 Fyz + \frac{1193}{40500} y_{xx} FxxzzF \\
&+ \frac{32703767}{3061800000} FzFFxz + \frac{3551}{3189375} Fzy_x{}^3 Fyyy \\
&- \frac{1}{5400} y_x{}^3 Fxyyyzy_{xx} - \frac{1}{10800} y_{xx}{}^2 Fxxxzz \\
&+ \frac{276941}{14580000} y_{xx} FzzzF^2 + \frac{10163}{907200} y_x{}^2 FzyyFx \\
&+ \frac{10163}{453600} FxFxzzy_{xx} - \frac{1}{1800} y_x y_{xx}{}^2 Fxyyz \\
&- \frac{1}{3600} y_x y_{xx}{}^3 Fzzyy + \frac{1511}{7560000} Fw^2 Fyzy_x
\end{aligned}$$

$$-\frac{19}{720}y_xFyyFx-\frac{392059}{17010000}Fzy_xFyyy_{xx}$$
$$+\frac{5011}{907200}y_x{}^2FyyFzzy_{xx}$$
$$+\frac{989489}{68040000}Fzy_x{}^2Fzyyy_{xx}$$
$$+\frac{5011}{226800}y_xFxzFyzy_{xx}+\frac{10163}{453600}Fyy_xFxzzy_{xx}$$
$$+\frac{5011}{453600}y_xFxyFzzy_{xx}$$
$$+\frac{989489}{34020000}Fzy_xFxyzy_{xx}$$
$$+\frac{10163}{453600}y_xFxFzyzy_{xx}$$
$$+\frac{10163}{453600}Fyy_x{}^2Fzyzy_{xx}$$
$$+\frac{875857}{34020000}y_xy_{xx}{}^2FzyzFz$$
$$-\frac{1}{10800}y_x{}^2y_{xx}{}^3Fzzzyy-\frac{1}{21600}y_xy_{xx}{}^4Fzzzzy$$
$$-\frac{1}{3600}y_x{}^2Fxxyyzy_{xx}+\frac{3551}{1063125}Fzy_x{}^2Fxyy$$
$$+\frac{2513939}{204120000}y_{xx}{}^3FzzzFz$$
$$-\frac{392059}{17010000}Fzy_{xx}Fxy+\frac{5011}{302400}Fxzy_{xx}{}^2Fzz$$
$$+\frac{5011}{907200}y_xFyzFxx+\frac{875857}{34020000}Fzy_{xx}{}^2Fxzz$$
$$+\frac{10163}{907200}Fyy_xy_{xx}{}^2Fzzz+\frac{1193}{121500}y_x{}^3FzyyyF$$
$$-\frac{1}{3600}y_x{}^2y_{xx}{}^2Fxyyzz+\frac{1193}{40500}y_xFFxxyz$$
$$-\frac{1}{10800}y_x{}^3Fyyyyy_{xx}+\frac{1193}{40500}y_x{}^2FFxyyz$$
$$+\frac{4169}{34020000}Fzw^2Fx-\frac{1}{10800}y_x{}^2Fxxxyy$$
$$+\frac{5011}{907200}Fzz^2y_{xx}{}^3-\frac{1}{21600}y_x{}^4Fzyyyyy_{xx}$$
$$+\frac{5011}{453600}y_x{}^2Fyz^2y_{xx}-\frac{433}{10125}y_{xx}FxyzF$$
$$-\frac{1}{10800}y_x{}^3Fxxyyy+\frac{5011}{907200}FxzFxx$$
$$-\frac{1}{108000}y_{xx}{}^5Fzzzzz+\frac{1511}{7560000}Fw^2Fxz$$
$$-\frac{13}{360}y_{xx}FyyF-\frac{1}{3600}y_xFxyxyy_{xx}$$
$$-\frac{1}{3600}y_{xx}{}^2Fxxyz-\frac{1}{3600}y_{xx}{}^3Fxyzz$$
$$-\frac{179}{567000}FzFyF-\frac{1}{21600}y_x{}^4Fxyyyy$$
$$-\frac{1}{5400}y_xFxxxyzy_{xx}+\frac{989489}{68040000}FzxxFzy_{xx}$$
$$+\frac{21449}{1632960}FzzFFx+\frac{46577}{34992000}Fz^2Fyy_x$$
$$-\frac{1}{3600}y_x{}^2Fxyyyy_{xx}+\frac{3551}{1063125}Fzy_xFyxx$$
$$+\frac{10163}{907200}Fyy_xFzxx+\frac{276941}{14580000}y_xF^2Fzyz$$
$$-\frac{1289}{907200}Fyy_{xx}{}^2Fzz-\frac{1}{10800}y_{xx}{}^3Fxxzzz$$
$$+\frac{10163}{453600}Fyy_x{}^2Fxyz+\frac{1}{108000}w^4Fzy_{xx}$$
$$+\frac{5011}{453600}y_xFxzFxy+\frac{10163}{453600}y_xFxFxyz$$
$$+\frac{1193}{40500}y_{xx}{}^2FxzzzF+\frac{1193}{121500}y_{xx}{}^3FzzzzF$$
$$-\frac{1187}{50400}Fyy_xFyzy_{xx}+\frac{10163}{907200}FxFzxx$$
$$-\frac{1}{21600}y_xFxxxxy+\frac{276941}{14580000}FxzzF^2$$
$$-\frac{1}{7200}y_{xx}{}^2Fxyy-\frac{806011}{68040000}Fzy_{xx}{}^2Fyz$$
$$-\frac{13}{360}y_x{}^2FyyyF-\frac{13}{180}y_xFxyyF$$
$$-\frac{1}{7200}y_xFyyyy_{xx}{}^2-\frac{1}{5400}y_xy_{xx}{}^3Fxyzzz$$
$$+\frac{32703767}{3061800000}Fzy_xFyzF$$
$$+\frac{4169}{34020000}Fzw^2Fyy_x+\frac{21449}{1632960}Fyy_xFzzF$$
$$+\frac{1}{108000}w^4Fx+\frac{5011}{907200}y_x{}^3FyyFyz$$
$$-\frac{1}{10800}y_x{}^3y_{xx}{}^2Fzzyyy-\frac{29}{720}Fyy_xFxy$$
$$-\frac{539}{81000}y_{xx}{}^2FzyzF-\frac{19}{720}FxFxy$$
$$-\frac{1}{21600}y_{xx}{}^4Fxzzzz-\frac{433}{10125}y_xFzyyy_{xx}F$$
$$-\frac{1}{108000}Fxxxxx+\frac{1193}{40500}y_xy_{xx}{}^2FzzzyF$$
$$-\frac{1}{144}FyFxx+\frac{1193}{40500}y_x{}^2y_{xx}FzzyyF$$

$$+\frac{1193}{20250}y_xy_{xx}FxyzzF+\frac{1511}{7560000}Fw^2Fzzy_{xx}$$
$$+\frac{36460321}{1530900000}Fzy_{xx}FzzF+\frac{5011}{907200}Fzzy_{xx}Fxx$$
$$+\frac{46577}{34992000}Fz^2Fx+\frac{46577}{34992000}Fz^3y_{xx}$$
$$-\frac{1}{10800}F^2Fyz-\frac{1}{3600}y_xy_{xx}{}^2Fxxyzz$$
$$+\frac{5011}{453600}y_x{}^2FyzFxy-\frac{1}{10800}y_{xx}Fxyxx$$
$$-\frac{13}{360}FFyxx-\frac{1}{108000}y_x{}^5Fyyyyy$$
$$-\frac{1}{144}Fy^2y_{xx}+\frac{1193}{121500}FxxxzF$$
$$-\frac{1}{21600}Fxxxxzy_{xx}-\frac{1}{7200}y_{xx}{}^3Fzyy$$
$$+\frac{1}{108000}w^4Fyy_x+\frac{10163}{907200}Fyy_x{}^3Fzyy$$
$$-\frac{1}{30}y_x{}^2FyyFy-\frac{1}{10800}y_{xx}{}^4Fzzzy$$
$$+\frac{4169}{34020000}Fz^2w^2y_{xx}+\frac{5011}{453600}Fxz^2y_{xx}$$
$$-\frac{1}{3600}y_x{}^2Fzyyyy_{xx}{}^2+\frac{5011}{907200}y_x{}^2FyyFxz$$
$$+\frac{10163}{907200}y_{xx}{}^2FzzzFx+\frac{3551}{3189375}FzFxxx\Big)h^6$$
$$+O\left(h^7\right) \tag{A.3}$$

The principal local truncation errors for y, y', and y'' (i.e., $\tau_{n+1}, \tau'_{n+1}, \tau''_{n+1}$) for TFMRKT5 are as follows:

$$\tau_{n+1}=\Big(-\frac{1}{10800}y_x{}^3Fyyy-\frac{1}{3600}Fzxxy_{xx}$$
$$-\frac{1}{3600}y_xFyyy_{xx}-\frac{1}{3600}y_{xx}{}^2Fyz$$
$$+\frac{527}{162000}Fzzy_{xx}F-\frac{1}{1800}y_xFxyzy_{xx}$$
$$-\frac{1}{3600}y_x{}^2Fzyyy_{xx}+\frac{527}{162000}y_xFyzF$$
$$+\frac{971}{907200}FzFyy_x-\frac{1}{3600}Fxzzy_{xx}{}^2$$
$$-\frac{1}{3600}y_x{}^2Fxyy-\frac{1}{10800}y_{xx}{}^3Fzzz$$
$$+\frac{971}{907200}FzFx+\frac{971}{907200}Fz^2y_{xx}$$
$$+\frac{527}{162000}FxzF-\frac{1}{144}FyF-\frac{1}{3600}y_xy_{xx}{}^2Fzyz$$
$$-\frac{1}{3600}y_{xx}Fxy-\frac{1}{10800}w^2Fzy_{xx}-\frac{1}{10800}Fxxx$$
$$-\frac{1}{3600}y_xFyxx-\frac{1}{10800}w^2Fyy_x$$
$$-\frac{1}{10800}w^2Fx\Big)h^6+O\left(h^7\right) \tag{A.4}$$

$$\tau'_{n+1}=\Big(-\frac{2087}{162000}y_{xx}FyzF+\frac{1063}{162000}y_x{}^2FzyyF$$
$$-\frac{8689}{1512000}FzFyy_{xx}+\frac{971}{226800}Fyy_xFxz$$
$$+\frac{1811}{756000}Fzy_xFxy+\frac{1}{1800}y_x{}^2y_{xx}Fxyyz$$
$$+\frac{1}{3600}y_x{}^2y_{xx}{}^2Fzzyy+\frac{1}{1800}y_xy_{xx}Fxxyz$$
$$+\frac{1}{1800}y_xy_{xx}{}^2Fxyzz+\frac{15143}{2268000}Fzy_{xx}Fxz$$
$$+\frac{24853}{4536000}Fzy_{xx}{}^2Fzz+\frac{971}{226800}y_xFyzFx$$
$$+\frac{971}{226800}Fzzy_{xx}Fx+\frac{1}{5400}y_xy_{xx}{}^3Fzzzy$$
$$+\frac{1}{5400}y_x{}^3Fzyyyy_{xx}+\frac{1}{1800}y_xFzyyy_{xx}{}^2$$
$$+\frac{971}{226800}Fyy_x{}^2Fyz+\frac{1811}{1512000}Fzy_x{}^2Fyy$$
$$-\frac{317}{22680000}FzFw^2+\frac{1}{3600}y_x{}^2Fyyyy_{xx}$$
$$+\frac{1063}{81000}Fxzzy_{xx}F+\frac{1063}{81000}y_xFxyzF$$
$$+\frac{1}{1800}y_xFxyyy_{xx}+\frac{1063}{162000}y_{xx}{}^2FzzzF$$
$$-\frac{7}{360}y_xFyyF+\frac{1063}{81000}y_xy_{xx}FzyzF$$
$$+\frac{971}{226800}Fyy_xFzzy_{xx}+\frac{15143}{2268000}Fzy_xFyzy_{xx}$$
$$+\frac{1}{21600}Fxxxx-\frac{1}{21600}w^4F+\frac{1}{5400}Fxxxzy_{xx}$$
$$+\frac{1}{3600}y_{xx}Fyxx+\frac{1}{1800}y_{xx}{}^2Fxyz$$
$$+\frac{1}{7200}y_{xx}{}^2Fyy+\frac{57889}{14580000}FzzF^2$$
$$+\frac{1}{3600}y_{xx}{}^3Fzyz+\frac{1}{21600}y_x{}^4Fyyyy$$
$$+\frac{1}{5400}y_{xx}{}^3Fxzzz+\frac{1}{21600}y_{xx}{}^4Fzzzz$$
$$+\frac{1}{5400}y_x{}^3Fxyyy-\frac{1}{144}FyFx+\frac{1}{5400}y_xFxyxx$$

$$+\frac{1}{3600}y_{xx}{}^2Fxxzz+\frac{1}{3600}y_x{}^2Fxyxy$$
$$+\frac{971}{226800}FxzFx+\frac{1811}{1512000}FzFxx-\frac{1}{144}Fy^2y_x$$
$$+\frac{31211}{22680000}Fz^2F+\frac{1063}{162000}FzxxF$$
$$\left.-\frac{7}{360}FFxy\right)h^6+O\left(h^7\right) \tag{A.5}$$

$$\tau''_{n+1}=\left(\frac{10373}{453600}Fyy_xy_{xx}Fxzz\right.$$
$$+\frac{10373}{907200}Fyy_xy_{xx}{}^2Fzzz+\frac{1717}{75600}y_xFyzy_{xx}Fxz$$
$$+\frac{10373}{453600}y_xy_{xx}FzyzFx$$
$$+\frac{298369}{11340000}y_xy_{xx}{}^2FzyzFz$$
$$+\frac{1717}{151200}y_xFzzy_{xx}Fxy$$
$$+\frac{1717}{100800}y_xFzzy_{xx}{}^2Fyz+\frac{7811}{5832000}Fz^2Fyy_x$$
$$-\frac{27487}{648000}y_{xx}FxyzF-\frac{67}{7560000}Fzw^2Fx$$
$$-\frac{67}{7560000}Fz^2w^2y_{xx}-\frac{13}{360}y_x{}^2FyyyF$$
$$-\frac{13}{180}y_xFxyyF-\frac{13}{360}y_{xx}FyyF$$
$$-\frac{4087}{648000}y_{xx}{}^2FzyzF-\frac{61}{378000}FzFyF$$
$$-\frac{2483}{302400}Fyy_{xx}Fxz-\frac{130103}{5670000}Fzy_{xx}Fxy$$
$$+\frac{19313}{648000}y_x{}^2FFxyyz+\frac{1717}{302400}y_x{}^3FyyFyz$$
$$+\frac{9761}{8505000}Fzy_x{}^3Fyyy-\frac{1}{30}Fyy_x{}^2Fyy$$
$$+\frac{10373}{907200}y_x{}^3FzyyFy-\frac{1}{3600}y_x{}^2Fxyyyy_{xx}$$
$$+\frac{19313}{648000}y_xFFxxyz+\frac{10959589}{1020600000}FzFFxz$$
$$-\frac{13567}{907200}y_{xx}FyzFx+\frac{337511}{25515000}FzzFFx$$
$$+\frac{19313}{1944000}y_x{}^3FzyyyF-\frac{1}{10800}y_x{}^3Fyyyyy_{xx}$$
$$+\frac{19313}{648000}y_{xx}{}^2FxzzzF+\frac{19313}{1944000}y_{xx}{}^3FzzzzF$$
$$+\frac{19313}{648000}y_{xx}FxxzzF-\frac{1}{3600}y_xFxyxyy_{xx}$$
$$+\frac{1717}{302400}y_x{}^2FyyFxz+\frac{1}{108000}w^4Fzy_{xx}$$
$$+\frac{1}{108000}w^4Fyy_x+\frac{9761}{2835000}Fzy_x{}^2Fxyy$$
$$-\frac{29}{720}Fyy_xFxy+\frac{10373}{453600}y_xFxyzFx$$
$$+\frac{10373}{453600}Fxzzy_{xx}Fx+\frac{298369}{11340000}Fxzzy_{xx}{}^2Fz$$
$$+\frac{10373}{907200}y_{xx}{}^2FzzzFx+\frac{856063}{68040000}y_{xx}{}^3FzzzFz$$
$$-\frac{19}{720}y_xFyyFx+\frac{10373}{907200}y_x{}^2FzyyFx$$
$$+\frac{1717}{151200}y_xFxzFxy+\frac{10373}{907200}Fyy_xFzxx$$
$$+\frac{9761}{2835000}Fzy_xFyxx-\frac{1}{3600}y_x{}^2y_{xx}Fxxyyz$$
$$-\frac{1}{3600}y_x{}^2y_{xx}{}^2Fxyyzz-\frac{1}{10800}y_x{}^2y_{xx}{}^3Fzzzyy$$
$$-\frac{1}{5400}y_xy_{xx}Fxxxyz-\frac{1}{3600}y_xy_{xx}{}^2Fxxyzz$$
$$-\frac{1}{5400}y_xy_{xx}{}^3Fxyzzz+\frac{1717}{100800}Fxzy_{xx}{}^2Fzz$$
$$+\frac{112471}{7560000}Fzy_{xx}Fzxx+\frac{1717}{302400}y_xFyzFxx$$
$$+\frac{1717}{302400}Fzzy_{xx}Fxx-\frac{1}{21600}y_xy_{xx}{}^4Fzzzzy$$
$$-\frac{1}{5400}y_x{}^3y_{xx}Fxyyyz-\frac{1}{10800}y_x{}^3y_{xx}{}^2Fzzyyy$$
$$+\frac{1717}{151200}y_x{}^2FyzFxy+\frac{1717}{151200}y_x{}^2Fyz^2y_{xx}$$
$$+\frac{10373}{453600}Fyy_x{}^2Fxyz-\frac{1}{21600}y_x{}^4Fzyyyyy_{xx}$$
$$-\frac{1}{7200}y_xFyyyy_{xx}{}^2+\frac{1114379}{58320000}y_{xx}FzzzF^2$$
$$-\frac{1}{1800}y_xy_{xx}{}^2Fxyyz-\frac{1}{3600}y_xy_{xx}{}^3Fzzyy$$
$$-\frac{1}{3600}y_x{}^2Fzyyyy_{xx}{}^2+\frac{1114379}{58320000}y_xF^2Fzyz$$
$$-\frac{383}{302400}Fyy_{xx}{}^2Fzz-\frac{87029}{7560000}Fzy_{xx}{}^2Fyz$$
$$-\frac{1}{108000}Fxxxxx+\frac{1}{108000}w^4Fx$$
$$-\frac{1}{108000}y_x{}^5Fyyyyy+\frac{1114379}{58320000}FxzzF^2$$
$$-\frac{1}{7200}y_{xx}{}^2Fxyy+\frac{11}{43200}F^2Fyz$$
$$-\frac{1}{3600}y_{xx}{}^2Fxxyz-\frac{1}{3600}y_{xx}{}^3Fxyzz$$

$$
\begin{aligned}
&- \frac{1}{10800} y_{xx}{}^{4} Fzzzy - \frac{1}{7200} y_{xx}{}^{3} Fzyy \\
&- \frac{13}{360} FFyxx - \frac{1}{21600} Fxxxxzy_{xx} \\
&- \frac{1}{10800} y_{xx} Fxyxx - \frac{1}{144} Fy^{2} y_{xx} \\
&+ \frac{1717}{151200} Fxz^{2} y_{xx} + \frac{1717}{302400} Fzz^{2} y_{xx}{}^{3} \\
&- \frac{1}{21600} y_{x}{}^{4} Fxyyyy - \frac{1}{21600} y_{xx}{}^{4} Fxzzzz \\
&- \frac{1}{108000} y_{xx}{}^{5} Fzzzzz - \frac{19}{720} FxFxy \\
&- \frac{1}{144} FyFxx + \frac{1717}{302400} FxxFxz \\
&+ \frac{9761}{8505000} FzFxxx + \frac{7811}{5832000} Fz^{2} Fx \\
&+ \frac{7811}{5832000} Fz^{3} y_{xx} - \frac{1}{10800} y_{xx}{}^{3} Fxxzzz \\
&- \frac{1}{10800} y_{x}{}^{3} Fxxyyy - \frac{1}{21600} y_{x} Fxxxxy \\
&- \frac{1}{10800} y_{xx}{}^{2} Fxxxzz - \frac{1}{10800} y_{x}{}^{2} Fxxxyy \\
&+ \frac{10373}{907200} FxFzxx + \frac{19313}{1944000} FxxxzF \\
&- \frac{1331}{45360000} FxzFw^{2} + \frac{112471}{3780000} Fzy_{x} y_{xx} Fxyz \\
&+ \frac{1717}{302400} y_{x}{}^{2} FyyFzzy_{xx} \\
&+ \frac{112471}{7560000} Fzy_{x}{}^{2} Fzyyy_{xx} \\
&+ \frac{10373}{453600} Fyy_{x}{}^{2} Fzyzy_{xx} - \frac{27487}{648000} y_{x} FzyyFy_{xx} \\
&+ \frac{19313}{648000} y_{x}{}^{2} y_{xx} FzzyyF \\
&+ \frac{19313}{324000} y_{x} y_{xx} FxyzzF + \frac{905927}{37800000} Fzy_{xx} FzzF \\
&+ \frac{19313}{648000} y_{x} y_{xx}{}^{2} FzzzyF + \frac{337511}{25515000} Fyy_{x} FzzF \\
&+ \frac{10959589}{1020600000} Fzy_{x} FyzF - \frac{2627}{113400} Fyy_{x} Fyzy_{xx} \\
&- \frac{130103}{5670000} Fzy_{x} Fyyy_{xx} - \frac{1331}{45360000} y_{xx} FzzFw^{2} \\
&- \left. \frac{1331}{45360000} y_{x} FyzFw^{2} - \frac{67}{7560000} Fzw^{2} Fyy_{x} \right) \\
&\cdot h^{6} + O\left(h^{7}\right)
\end{aligned}
\tag{A.6}
$$

Conflicts of Interest

The authors declare that there are no conflicts of interest regarding the publication of this paper.

References

[1] B. Paternoster, "Runge-Kutta(-Nyström) methods for ODEs with periodic solutions based on trigonometric polynomials," *Applied Numerical Mathematics*, vol. 28, no. 2–4, pp. 401–412, 1998.

[2] G. Vanden Berghe, H. De Meyer, M. Van Daele, and T. Van Hecke, "Exponentially fitted Runge-Kutta methods," *Journal of Computational and Applied Mathematics*, vol. 125, no. 1-2, pp. 107–115, 2000.

[3] T. E. Simos, "Exponentially fitted Runge-Kutta methods for the numerical solution of the Schrödinger equation and related problems," *Computational Materials Science*, vol. 18, no. 3-4, pp. 315–332, 2000.

[4] Z. Kalogiratou and T. E. Simos, "Construction of trigonometrically and exponentially fitted Runge-Kutta-Nyström methods for the numerical solution of the schrödinger equation and related problems—a method of 8th algebraic order," *Journal of Mathematical Chemistry*, vol. 31, no. 2, pp. 211–232, 2002.

[5] Z. Kalogiratou, T. Monovasilis, and T. E. Simos, "Computation of the eigenvalues of the Schrödinger equation by exponentially-fitted Runge-Kutta-Nyström methods," *Computer Physics Communications*, vol. 180, no. 2, pp. 167–176, 2009.

[6] T. E. Simos, "Exponentially-fitted Runge-Kutta-Nyström method for the numerical solution of initial-value problems with oscillating solutions," *Applied Mathematics Letters*, vol. 15, no. 2, pp. 217–225, 2002.

[7] D. P. Sakas and T. E. Simos, "A fifth algebraic order trigonometrically-fitted modified Runge-Kutta Zonneveld method for the numerical solution of orbital problems," *Mathematical and Computer Modelling*, vol. 42, no. 7-8, pp. 903–920, 2005.

[8] H. Van De Vyver, "A Runge-Kutta-Nyström pair for the numerical integration of perturbed oscillators," *Computer Physics Communications*, vol. 167, no. 2, pp. 129–142, 2005.

[9] H. Yang and X. Wu, "Trigonometrically-fitted ARKN methods for perturbed oscillators," *Applied Numerical Mathematics*, vol. 58, no. 9, pp. 1375–1395, 2008.

[10] M. A. Demba, N. Senu, and F. Ismail, "Trigonometrically-fitted explicit four-stage fourth-order Runge–Kutta–Nyström method for the solution of initial value problems with oscillatory behavior," *The Global Journal of Pure and Applied Mathematics (GJPAM)*, vol. 12, no. 1, pp. 67–80, 2016.

[11] M. Hanan, "Oscillation criteria for third-order linear differential equations," *Pacific Journal of Mathematics*, vol. 11, pp. 919–944, 1961.

[12] J. Rovder, "Oscillation criteria for third-order linear differential equations," *Matematický časopis*, vol. 25, no. 3, pp. 231–244, 1975.

[13] A. C. Lazer, "The behavior of solutions of the differential equation $y''' + p(x)y' + q(x)y = 0$," *Pacific Journal of Mathematics*, vol. 17, pp. 435–466, 1966.

[14] G. D. Jones, "Properties of solutions of a class of third-order differential equations," *Journal of Mathematical Analysis and Applications*, vol. 48, pp. 165–169, 1974.

[15] F. A. Fawzi, N. Senu, F. Ismail, and Z. A. Majid, "An efficient of direct integrator of Runge-Kutta Type Method for Solving y'''=f(x, y, y') with Application to Thin Film Flow Problem,"

International Journal of Pure and Applied Mathematics, vol. 117, no. 4, accepted.

[16] J. C. Butcher, *Numerical Methods for Ordinary Differential Equations*, John Wiley & Sons, New York, NY, USA, 2nd edition, 2008.

[17] J. D. Lambert, *Numerical Methods for Ordinary Differential Systems. The Initial Value Problem*, John Wiley & Sons, New York, NY, USA, 1993.

[18] Z. A. Anastassi and T. E. Simos, "Trigonometrically fitted Runge-Kutta methods for the numerical solution of the Schrödinger equation," *Journal of Mathematical Chemistry*, vol. 37, no. 3, pp. 281–293, 2005.

[19] E. Momoniat and F. M. Mahomed, "Symmetry reduction and numerical solution of a third-order ODE from thin film flow," *Mathematical & Computational Applications*, vol. 15, no. 4, pp. 709–719, 2010.

[20] E. O. Tuck and L. W. Schwartz, "A numerical and asymptotic study of some third-order ordinary differential equations relevant to draining and coating flows," *SIAM Review. A Publication of the Society for Industrial and Applied Mathematics*, vol. 32, no. 3, pp. 453–469, 1990.

[21] J. Biazar, E. Babolian, and R. Islam, "Solution of the system of ordinary differential equations by Adomian decomposition method," *Applied Mathematics and Computation*, vol. 147, no. 3, pp. 713–719, 2004.

[22] M. Mechee, N. Senu, F. Ismail, B. Nikouravan, and Z. Siri, "A three-stage fifth-order Runge-Kutta method for directly solving special third-order differential equation with application to thin film flow problem," *Mathematical Problems in Engineering*, vol. 2013, Article ID 795397, 7 pages, 2013.

9

Infinitely Many Trees with Maximum Number of Holes Zero, One, and Two

Srinivasa Rao Kola, Balakrishna Gudla, and P. K. Niranjan

Department of Mathematical and Computational Sciences, National Institute of Technology Karnataka, Surathkal, India

Correspondence should be addressed to Srinivasa Rao Kola; srinu.iitkgp@gmail.com

Academic Editor: Ali R. Ashrafi

An $L(2,1)$-coloring of a simple connected graph G is an assignment f of nonnegative integers to the vertices of G such that $|f(u) - f(v)| \geqslant 2$ if $d(u,v) = 1$ and $|f(u) - f(v)| \geqslant 1$ if $d(u,v) = 2$ for all $u, v \in V(G)$, where $d(u,v)$ denotes the distance between u and v in G. The span of f is the maximum color assigned by f. The span of a graph G, denoted by $\lambda(G)$, is the minimum of span over all $L(2,1)$-colorings on G. An $L(2,1)$-coloring of G with span $\lambda(G)$ is called a span coloring of G. An $L(2,1)$-coloring f is said to be irreducible if there exists no $L(2,1)$-coloring g such that $g(u) \leqslant f(u)$ for all $u \in V(G)$ and $g(v) < f(v)$ for some $v \in V(G)$. If f is an $L(2,1)$-coloring with span k, then $h \in \{0, 1, 2, \ldots, k\}$ is a hole if there is no $v \in V(G)$ such that $f(v) = h$. The maximum number of holes over all irreducible span colorings of G is denoted by $H_\lambda(G)$. A tree T with maximum degree Δ having span $\Delta + 1$ is referred to as Type-I tree; otherwise it is Type-II. In this paper, we give a method to construct infinitely many trees with at least one hole from a one-hole tree and infinitely many two-hole trees from a two-hole tree. Also, using the method, we construct infinitely many Type-II trees with maximum number of holes one and two. Further, we give a sufficient condition for a Type-II tree with maximum number of holes zero.

1. Introduction

The channel assignment problem is the problem of assigning frequencies to transmitters in some optimal manner. In 1992, Griggs and Yeh [1] have introduced the concept of $L(2,1)$-coloring as a variation of channel assignment problem. The distance between two vertices u and v in a graph G, denoted by $d(u,v)$, is defined as the length of a shortest path between u and v in G. An $L(2,1)$-coloring of a graph G is an assignment $f: V(G) \longrightarrow \{0, 1, 2, \ldots, k\}$ such that, for every u, v in $V(G)$, $|f(u) - f(v)| \geqslant 2$ if u and v are adjacent and $|f(u) - f(v)| \geqslant 1$ if u and v are at distance 2. The nonnegative integers assigned to the vertices are also called colors. The span of f, denoted by *span* f, is $\max\{f(v): v \in V(G)\}$. The span of G, denoted by $\lambda(G)$, is $\min\{span\ f: f \text{ is an } L(2,1)\text{-coloring of } G\}$. An $L(2,1)$-coloring with span $\lambda(G)$ is called a span coloring. A tree is a connected acyclic graph. In the introductory paper, Griggs and Yeh [1] proved that $\lambda(P_n) = 4$ for $n \geqslant 5$; $\lambda(T)$ is either $\Delta + 1$ or $\Delta + 2$ for any tree T with maximum degree Δ. We refer to a tree as Type-I if $\lambda(T) = \Delta + 1$; otherwise it is Type-II. In a graph G with maximum degree Δ, we refer to a vertex v as a major vertex if its degree is Δ; otherwise v is a minor vertex. Wang [2] has proved that a tree with no pair of major vertices at distances 1, 2, and 4 is Type-I. Zhai et al. [3] have improved the above condition as a tree with no pair of major vertices at distances 2 and 4 is Type-I. Mandal and Panigrahi [4] have proved that $\lambda(T) = \Delta + 1$ if T has at most one pair of major vertices at distance either 2 or 4 and all other pairs are at distance at least 7. Wood and Jacob [5] have given a complete characterization of the $L(2,1)$-span of trees up to twenty vertices.

Fishburn and Roberts [6] have introduced the concept of no-hole $L(2,1)$-coloring of a graph. If f is an $L(2,1)$-coloring of a graph G with span k, then an integer $h \in \{0, 1, 2, \ldots, k\}$ is called a hole in f if there is no vertex v in G such that $f(v) = h$. An $L(2,1)$-coloring with no hole is called a no-hole coloring of G. Fishburn et al. [7] have introduced the concept of irreducibility of $L(2,1)$-coloring. An $L(2,1)$-coloring of a graph G is reducible if there exists another $L(2,1)$-coloring g of G such that $g(u) \leqslant f(u)$ for all vertices u in G and there exists a vertex v in G such that $g(v) < f(v)$. If f is not reducible then it is called irreducible.

An irreducible no-hole coloring is referred to as inh-coloring. A graph is inh-colorable if there exists an inh-coloring. For an inh-colorable graph G, the lower inh-span or simply inh-span of G, denoted by $\lambda_{inh}(G)$, is defined as $\lambda_{inh}(G) = \min\{\text{span } f : f \text{ is an inh-coloring of } G\}$. Fishburn et al. [7] have proved that paths, cycles, and trees are inh-colorable except C_3, C_4, and stars. In addition to that, they showed that $\Delta + 1 \leqslant \lambda_{inh}(T) \leqslant \Delta + 2$ where T is any nonstar tree. Laskar et al. [8] have proved that any nonstar tree T is inh-colorable and $\lambda_{inh}(T) = \lambda(T)$. The maximum number of holes over all irreducible span colorings of G is denoted by $H_\lambda(G)$. Laskar and Eyabi [9] have determined the exact values for maximum number of holes for paths, cycles, stars, and complete bipartite graphs as 2, 2, 1, and 1, respectively, and conjectured that, for any tree T, $H_\lambda(T) = 2$ if and only if T is a path P_n, $n > 4$. S. R. Kola et al. [10] have disproved the conjecture by giving a two-hole irreducible span coloring for a Type-II tree other than path.

In this article, we give a method of construction of infinitely many two-hole trees from a two-hole tree and infinitely many trees with at least one hole from a one-hole tree. Also, we find maximum number of holes for some Type-II trees given by Wood and Jacob [5] and obtain infinitely many Type-II trees of holes one and two by applying the method of construction. Further, we give a sufficient condition for a zero-hole Type-II tree.

2. Construction of Trees with Maximum Number of Holes One and Two

We start this section with a lemma which gives the possible colors to the major vertices in a two-hole span coloring of a Type-II tree.

Lemma 1. *In any two-hole span coloring of a Type-II tree T with $\Delta \geqslant 3$, all major vertices receive either the same color or the colors from any one of the sets $\{0, 2\}$, $\{0, \Delta+2\}$, or $\{\Delta, \Delta+2\}$.*

Proof. Let f be a two-hole span coloring of a Type-II tree T. Suppose that v_1 and v_2 are major vertices such that $f(v_1) \neq f(v_2)$. First, we prove that $\{f(v_1), f(v_2)\} = \{0, 2\}$ or $\{0, \Delta + 2\}$ or $\{\Delta, \Delta + 2\}$. Let $f(v_1) = l$ and $f(v_2) = l'$. Without loss of generality, we assume that $0 \leqslant l < l' \leqslant \Delta + 2$. If $l = 0$, then the color 1 must be one of the two holes in f. If $l' \neq \Delta + 2$, then $l' - 1$ and $l' + 1$ are the holes. Since $l' + 1$ cannot be 1, $l' - 1$ is 1 which implies $l' = 2$. If $l \neq 0$, then $l - 1$ and $l + 1$ are the holes in f. If $l' \neq \Delta + 2$, then $l' - 1$ and $l' + 1$ are the holes which are not possible as $l \neq l'$. If $l' = \Delta + 2$, then $\Delta + 1$ must be one of the holes in f. Since $l - 1$ cannot be $\Delta + 1$, $l + 1$ is $\Delta + 1$ which implies $l = \Delta$.

If $\{f(v_1), f(v_2)\} = \{0, 2\}$, then 1 and 3 are the holes. If any major vertex v receives a color l other than 0 and 2, then the neighbors of v cannot get the colors 1 and 3 and at least one of $l - 1$ and $l + 1$ (if $l = \Delta + 2$, then $l - 1$). This is not possible as we need $\Delta + 1$ number of colors to color a major vertex and its neighbors. Similarly, other cases can be proved.

The following lemma is a direct implication of Lemma 1.

Lemma 2. *If f is a two-hole span coloring of a Type-II tree T having two major vertices at distance less than or equal to two, then the set of holes in f is $\{1, 3\}$, $\{1, \Delta + 1\}$, or $\{\Delta - 1, \Delta + 1\}$.*

When we say connecting two trees, we mean adding an edge between them. Corresponding to the possibilities of holes given in Lemma 2, we give a list of trees which can be connected to a two-hole tree having two major vertices at distance less than or equal to two, to obtain infinitely many two-hole trees. Later, we give a list of trees which can be connected to a one-hole tree to get infinitely many one-hole trees.

Theorem 3. *If T is a tree with maximum number of holes two and having at least two major vertices at distance at most two, then there are infinitely many trees with maximum number of holes two and with maximum degree Δ same as that of T.*

Proof. Let f be an irreducible span coloring of T with two holes. Then by Lemma 2, the set of holes in f is $\{1, 3\}$ or $\{1, \Delta+1\}$ or $\{\Delta-1, \Delta+1\}$. Now, we give a method to construct trees from T using the coloring f and holes in f. For all the three possibilities of holes, we give a list of trees which can be connected to T to get a bigger tree with maximum number of holes two. Suppose 1 and 3 are the holes in f. We use Table 1 for construction.

Let u be a vertex of the tree T and c be the color received by u. Now depending on the colors of the neighbors of u, to preserve $L(2, 1)$-coloring, we connect the trees (one at a time) given in Table 1 by adding an edge between u of T and the vertex colored k of tree in the table. Note that $0 \leqslant k \leqslant \Delta + 2$ and the color k is not equal to any of the colors $c - 1, c, c + 1$, 1, and 3 and not assigned to any neighbor of u. To maintain irreducibility, we use the condition given in the last column of the table. It is easy to see that, after every step, we get a tree T' with maximum degree same as that of T and a two-hole irreducible span coloring of T'. Also, it is clear that T is a subtree of T'. Since connecting a tree to any pendant vertex is always possible, we get infinitely many trees.

Suppose 1 and $\Delta + 1$ are the holes in f. Construction is similar to the previous case using trees in Table 2.

Suppose $\Delta - 1$ and $\Delta + 1$ are the holes in f. We use trees in Table 3 for construction.

Theorem 4. *If T is a tree with $H_\lambda(T) = 1$, then there exist infinitely many trees containing T and with maximum number of holes at least 1.*

Proof. Here, we start with a one-hole irreducible $L(2, 1)$-span coloring of T having hole h. The construction of infinitely many trees is similar to that in Theorem 3 and using Table 4. Since after every step we get a tree T' with one-hole irreducible span coloring, $H_\lambda(T') \geqslant 1$.

Theorem 5. *If T is a tree with $H_\lambda(T) = 1$ and T has no two-hole span coloring, then there exist infinitely many trees with maximum number of holes one and containing T.*

Proof. Since T has no two-hole span coloring, any tree containing T having same maximum degree as that of T

TABLE 1: Trees connectable to a vertex u colored c in a two-hole tree with 1 and 3 as holes.

Color of vertex	Connectable trees	Condition
		$k = 2$
	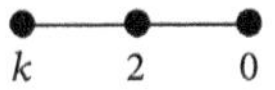	$k > 3$ and all colors greater than 3 less than k adjacent to c.
	$\mathcal{T}_1$: 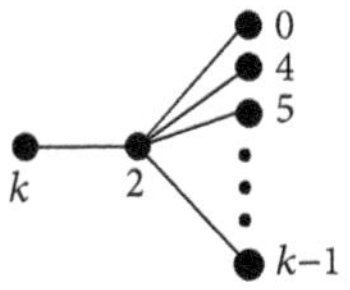	$k > 4$
	$\mathcal{T}_2$: 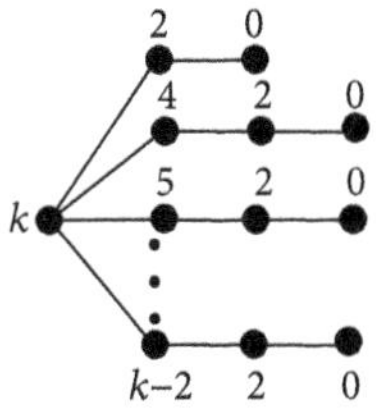	$k > 5$
$c = 0$	$\mathcal{T}_3$: 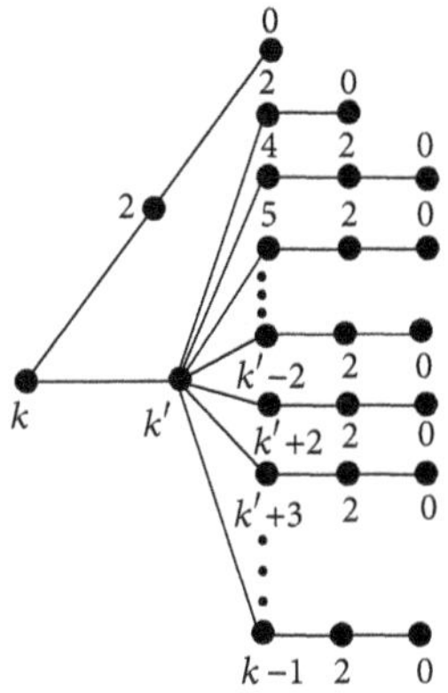	$3 < k' < k - 1$
	$\mathcal{T}_4$: 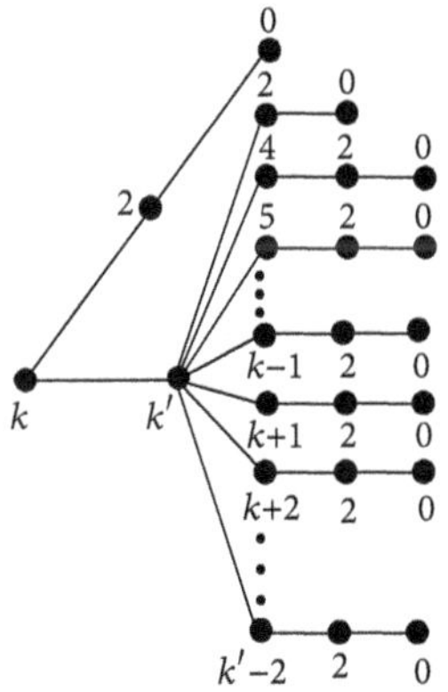	$3 < k < k' - 1$
		$k \neq 1, 2, 3$ and all colors less than k adjacent to c.
$c = 2$		$3 < k \leqslant \Delta$ and all colors less than k adjacent to c.
	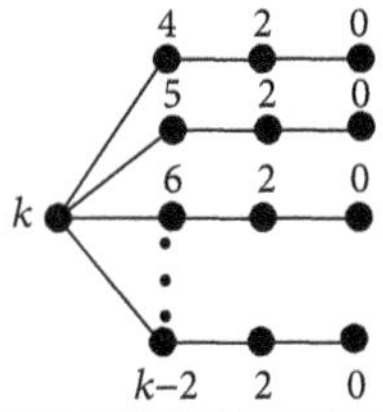	$k > 5$

Table 1: Continued.

Color of vertex	Connectable trees	Condition
	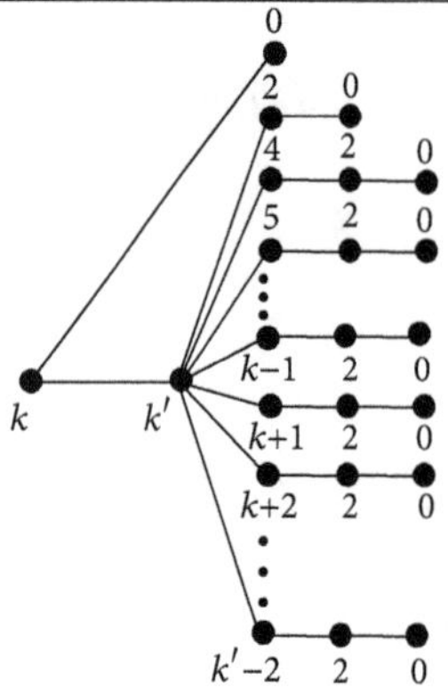	$3 < k < k' - 1$
	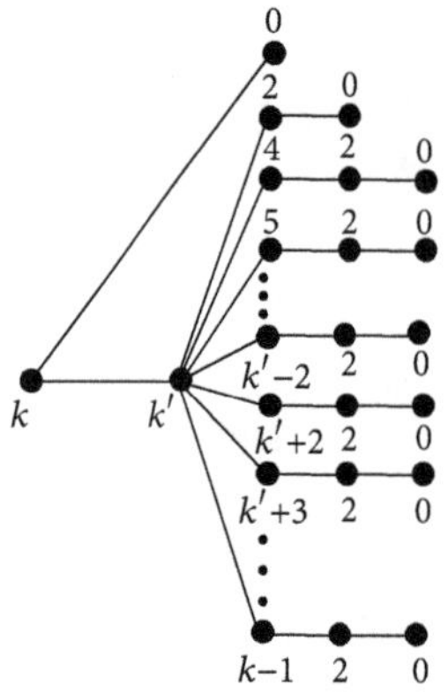	$3 < k' < k - 1$
	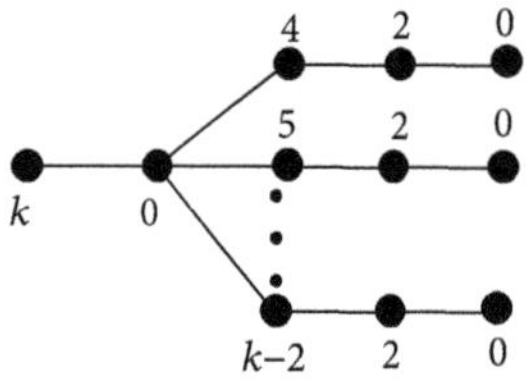	$k > 5$
	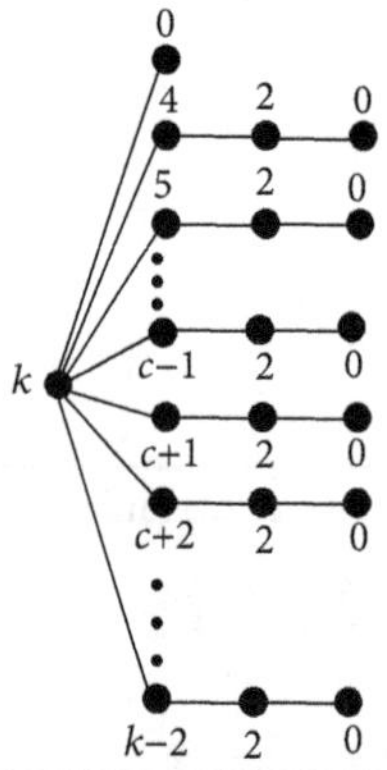	$k > c + 1$
c> 3	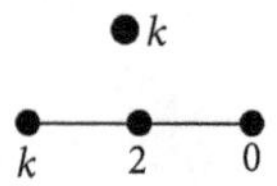	$k = 0$ $k > 3$ and all colors greater than 3 less than k adjacent to c.

TABLE 1: Continued.

Color of vertex	Connectable trees	Condition
	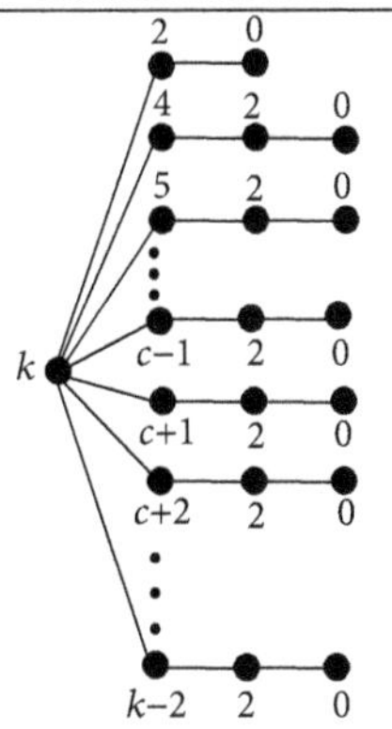	$k > c + 1$
	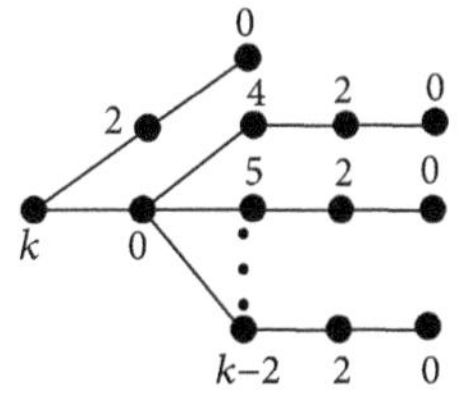	$k > 5$
	$\mathcal{T}_1$	$k > 4$
	$\mathcal{T}_2$	$5 < k < c - 1$
	$\mathcal{T}_3$	$3 < k' < k - 1$ and $c \neq k'$
	$\mathcal{T}_4$	$3 < k < k' - 1$ and $c \neq k'$

cannot have a two-hole span coloring. Therefore, every tree obtained from T using Theorem 4 has maximum number of holes one.

Corollary 6. *If T is a Type-I tree and $H_\lambda(T) = 1$, then there exist infinitely many trees with maximum number of holes one and containing T.*

3. Maximum Number of Holes in Some Type-II Trees

Recall that, in a graph G with maximum degree Δ, we refer a vertex v as a major vertex if its degree is Δ. Otherwise v is a minor vertex. Wood and Jacob [5] have given some sufficient conditions for a tree to be Type-II. We consider some of their sufficient conditions as below.

Theorem 7 (see [5]). *A tree containing any of the following subtrees is Type-II provided the maximum degrees of the subtree and the tree are the same Δ.*

(I) T_1: *a tree with an induced P_3 consisting of three major vertices.*

(II) T_2: *a tree with a minor vertex w and at least 3 major vertices adjacent to w.*

(III) T_3: *a tree with a major vertex w and at least $\Delta - 1$ major vertices at distance two from w, and T_2 is not a subtree of the tree.*

(IV) T_4: *a tree with a vertex w adjacent to $\Delta - 2$ vertices $w_1, w_2, \ldots, w_{\Delta-2}$ and two neighbors v_i, v_i' of each w_i, $1 \leqslant i \leqslant \Delta - 2$ are major.*

Since the above trees can be as small as possible, we consider the degrees of minor vertices as minimum as possible. Now, we find the maximum number of holes for the trees T_1, T_2, T_3, and T_4. For any tree T with maximum degree Δ, it is clear that $H_\lambda(T) \leqslant 2$. First, we show that $H_\lambda(T_i) \leqslant 1$, $i = 1, 2, 4$. Also, $H_\lambda(T_2) = 0$ if T_2 has a vertex adjacent to at least four major vertices. Further, we give a two-hole $L(2,1)$-irreducible span coloring of T_3 if it has exactly $\Delta - 1$ major vertices at distance two from a major vertex and we show that $H_\lambda(T_3) \leqslant 1$, if T_3 has exactly Δ major vertices at distance two from a major vertex. Later, we show that these upper bounds are the exact values by defining $L(2,1)$-irreducible span colorings with appropriate holes. Now onwards, unless we mention, tree refers to Type-II tree. In figures, we use symbol ▲ to denote a major vertex.

Theorem 8. *For the trees T_i, $i = 1, 2, 4$, $H_\lambda(T_i) \leqslant 1$.*

Proof. Let v_1, v_2, and v_3 be the major vertices of T_1. Since v_1, v_2, and v_3 receive three different colors in any $L(2,1)$-coloring, by Lemma 1, T_1 cannot have a two-hole irreducible span coloring. Similarly, we can prove that $H_\lambda(T_2) \leqslant 1$.

Now, we consider T_4 with labelling as in Figure 1.

Suppose that f is a two-hole irreducible span coloring of T_4. Then by Lemma 1, all major vertices of T_4 receive colors from $\{0, 2\}$ or $\{0, \Delta + 2\}$ or $\{\Delta, \Delta + 2\}$. Suppose the major vertices receive 0 and 2. Then 1 and 3 are holes. Without loss

TABLE 2: Trees connectable to a vertex u colored c in a two-hole tree with 1 and $\Delta + 1$ as holes.

Color of vertex	Connectable trees	Condition
$c = 0$		$1 < k \leqslant \Delta$ and all colors less than k adjacent to c.
	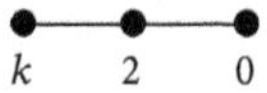	$k > 3$ and all colors greater than 3 less than k adjacent to c.
	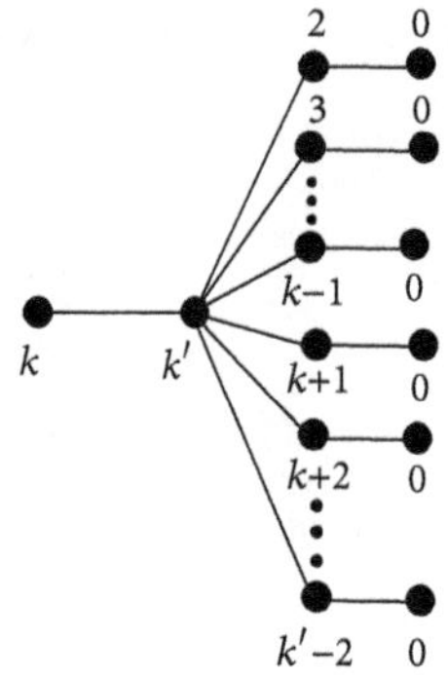	$2 < k < k' - 1 < \Delta$ or $2 < k \leqslant \Delta$ and $k' = \Delta + 2$
	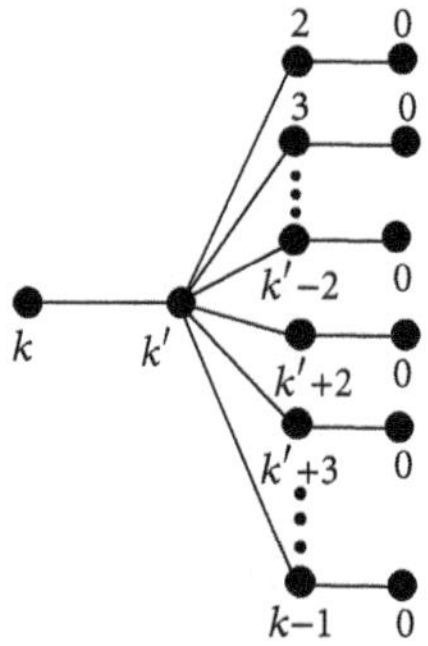	$3 < k' < k - 1 < \Delta$ or $k = \Delta + 2$ and $k' = \Delta$
	$\mathcal{T}_1$: 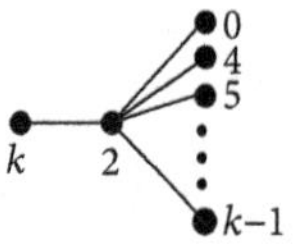	$3 < k \leqslant \Delta$
	$\mathcal{T}_2$: 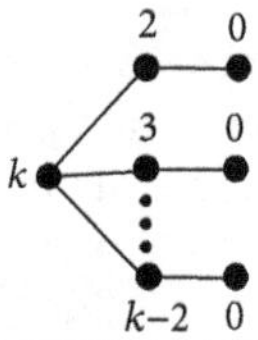	$3 < k \leqslant \Delta$ or $k = \Delta + 2$
$1 < c \leqslant \Delta$		$k = 0$
	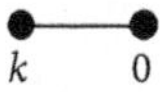	$2 \leqslant k \leqslant \Delta$ and all colors less than k adjacent to c
		$k > 3$ and all colors greater than 3 less than k adjacent to c.
	$\mathcal{T}_1$	$k > 3$ and $c \neq 2$
	$\mathcal{T}_2$	$5 \leqslant k < c - 1$
	$\mathcal{T}_3$: 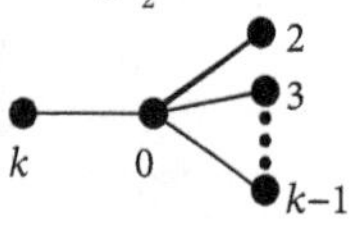	$2 < k \leqslant \Delta$ or $k = \Delta + 2$
	$\mathcal{T}_4$: 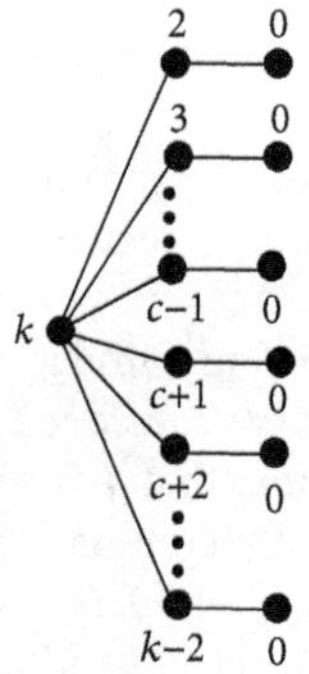	$c + 1 < k \leqslant \Delta$ or $k = \Delta + 2$

Table 2: Continued.

Color of vertex	Connectable trees	Condition
	$\mathscr{T}_5$: 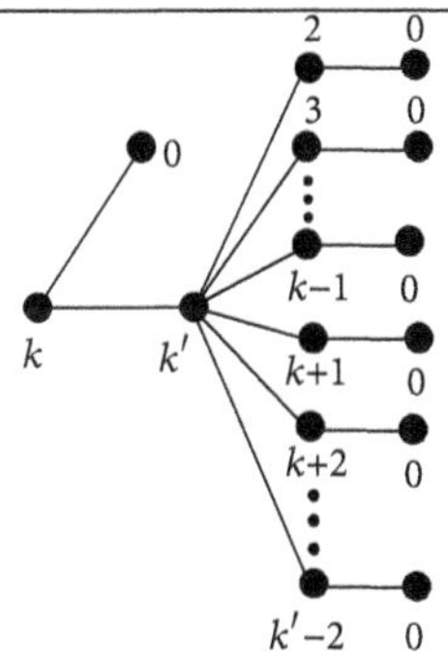	$c \neq k'$ and $1 < k < k' - 1 < \Delta$
	$\mathscr{T}_6$: 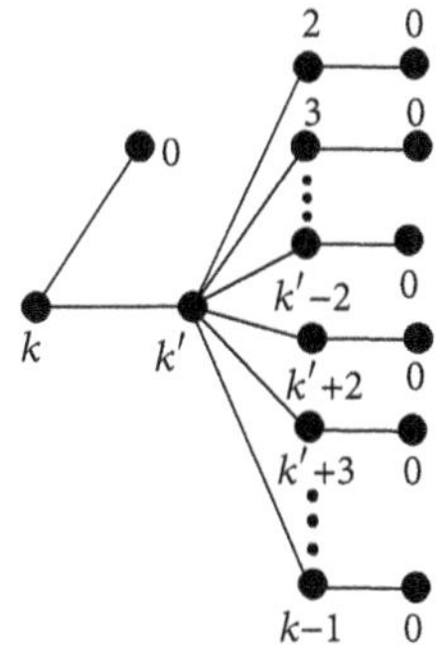	$c \neq k'$ and $1 < k' < k - 1 < \Delta$
		$k = 0$
		$2 \leqslant k \leqslant \Delta$ and all colors less than k adjacent to c
	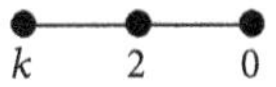	$k > 3$ and all colors greater than 3 less than k adjacent to c.
$c = \Delta + 2$	$\mathscr{T}_1$	$5 \leqslant k \leqslant \Delta$
	$\mathscr{T}_2$	$5 \leqslant k \leqslant \Delta$
	$\mathscr{T}_3$	$2 < k \leqslant \Delta$
	$\mathscr{T}_5$	$1 < k' < k - 1 < \Delta$
	$\mathscr{T}_6$	$1 < k < k' - 1 < \Delta$

T$_4$:

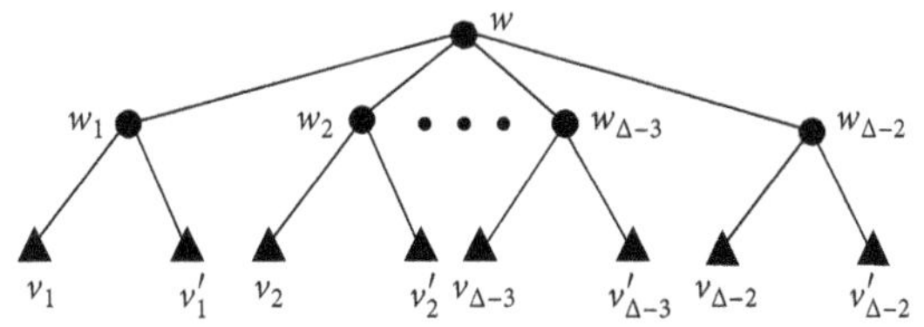

Figure 1: The tree T_4 as in Theorem 7.

of generality, we assume that $f(v_i) = 0$ and $f(v_i') = 2$. Now, one of the pendant vertices adjacent to v_i must receive a color grater than 3 which reduces to 3 giving a contradiction to the fact that f is irreducible. Similarly, we can prove the other two cases. Therefore, $H_\lambda(T_4) \leqslant 1$.

Theorem 9. *If at least four major vertices are adjacent to w in T_2, then $H_\lambda(T_2) = 0$.*

Proof. Recall that T_2 is a tree with a vertex w adjacent to at least three major vertices. Let v_1, v_2, v_3, and v_4 be four major vertices adjacent to w in T_2. Suppose that it has a one-hole irreducible $L(2,1)$-span coloring f. Let l_1, l_2, l_3, and l_4 be the colors received by v_1, v_2, v_3, and v_4, respectively. Without loss of generality, we assume that $0 \leqslant l_1 < l_2 < l_3 < l_4 \leqslant \Delta + 2$. If $l_1 \neq 0$, then except $l_1 - 1$ and $l_1 + 1$ all other colors are used to the neighbors of v_1. Also, except $l_3 - 1$ and $l_3 + 1$, all other colors are used to the neighbors of v_3. Since $l_1 - 1, l_1 + 1, l_3 - 1$, and $l_3 + 1$ are four different colors, f cannot have a hole which is a contradiction. So, $l_1 = 0$. Since f is one-hole coloring, the colors $l_2 - 1, l_2 + 1, l_3 - 1$, and $l_3 + 1$ cannot be four different colors and hence $l_2 + 1 = l_3 - 1$ is the hole. Now, a pendant neighbor of v_1 receives l_3 which reduces to the hole $l_3 - 1$ giving a contradiction to the fact that f is irreducible.

S. R. Kola et al. [10] have disproved the conjecture given by Laskar and Eyabi [9] by giving two-hole irreducible span

TABLE 3: Trees connectable to a vertex u colored c in a two-hole tree with $\Delta - 1$ and $\Delta + 1$ as holes.

Color of vertex	Connectable trees	Condition
$0 \leqslant c < \Delta - 1$	● k	$0 \leqslant k < \Delta - 1$ and all colors less than k adjacent to c.
	$\mathcal{T}_1$: 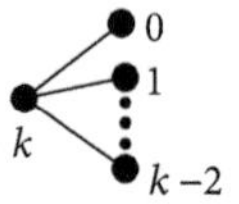	$2 \leqslant k < c - 1$
	$\mathcal{T}_2$: 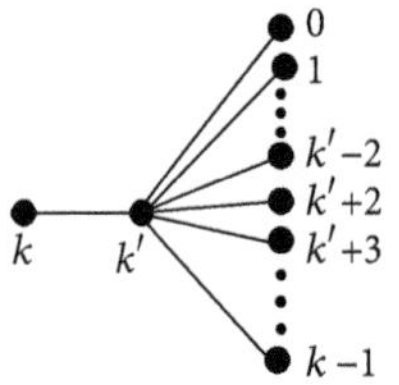	$c \neq k'$ and $0 \leqslant k' < k - 1 < \Delta - 2$ or $k = k' + 2 = \Delta + 2$ or Δ
	$\mathcal{T}_3$: 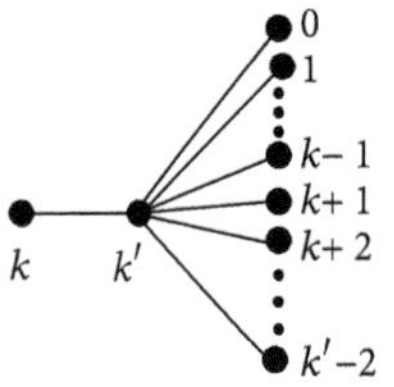	$c \neq k'$ and $0 \leqslant k < k' - 1 < \Delta - 2$ or $0 \leqslant k \leqslant \Delta - 2$ and $k' = \Delta$
	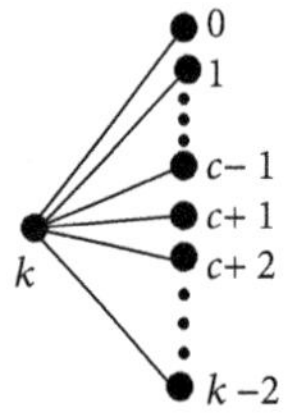	$c + 1 < k \leqslant \Delta - 2$ or $k = \Delta$
	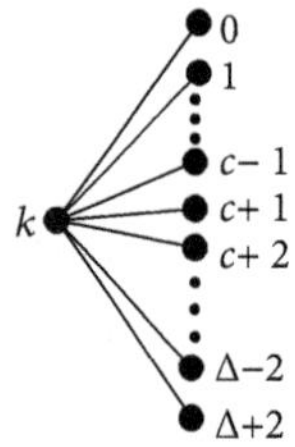	$k = \Delta$
	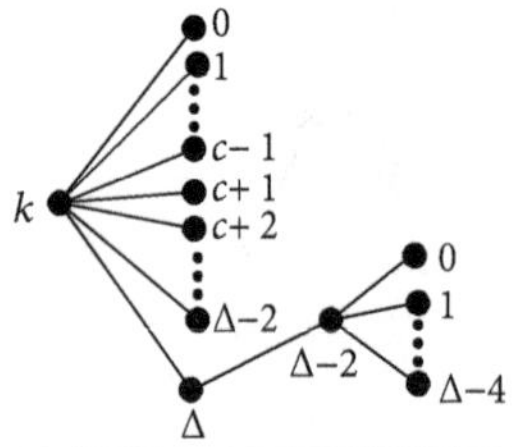	$k = \Delta + 2$
$c = \Delta$	● k	$k \neq \Delta, \Delta \pm 1$ and all colors less than k adjacent to c.
	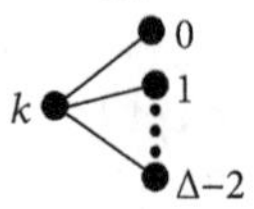	$k = \Delta + 2$
	$\mathcal{T}_1$	$2 \leqslant k < \Delta - 1$
	$\mathcal{T}_2$	$0 \leqslant k' < k - 1 < \Delta - 3$
	$\mathcal{T}_3$	$0 \leqslant k < k' - 1 < \Delta - 2$
$c = \Delta + 2$	● k	$0 \leqslant k < \Delta - 1$ and all colors less than k adjacent to c.
	$\mathcal{T}_1$	$2 \leqslant k < \Delta - 1$ or $k = \Delta$
	$\mathcal{T}_2$	$c \neq k'$ and $0 \leqslant k' < k - 1 < \Delta - 2$ or $k = k' + 2 = \Delta$
	$\mathcal{T}_3$	$c \neq k'$ and $0 \leqslant k < k' - 1 < \Delta - 2$ or $0 \leqslant k \leqslant \Delta - 2$ and $k' = \Delta$

TABLE 4: Trees connectable for a vertex colored c in a one-hole tree with hole h.

Color of vertex	Connectable trees	Condition
$0 \leqslant c < h$	• k	$0 \leqslant k < h$ and all colors less than k adjacent to c. For $c = h - 1$, $h < k \leqslant \lambda(T)$ and all colors less than k adjacent to c.
	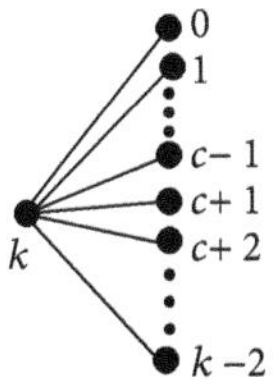	$c + 1 < k < h$ or $k = h + 1$
	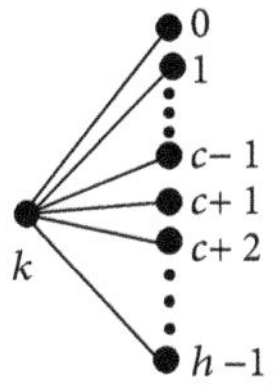	$k > h$ and all colors greater than h less than k adjacent to c.
	$\mathscr{T}_1$: 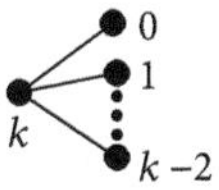	$2 \leqslant k < c - 1$
	$\mathscr{T}_2$: 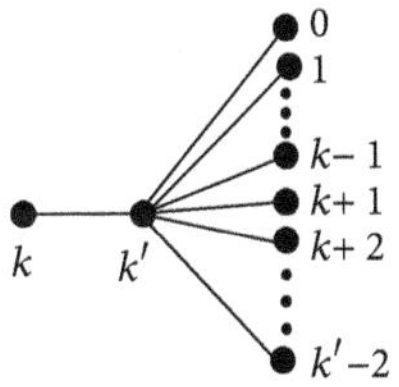	$k' \neq c$ and $0 < k < k' - 1 < h - 1$ or $0 < k < h$ and $k' = h + 1$
	$\mathscr{T}_3$: 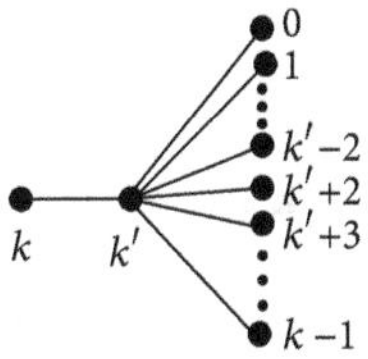	$k' \neq c$ and $0 \leqslant k' < k - 1 < h - 1$
	$\mathscr{T}_4$: 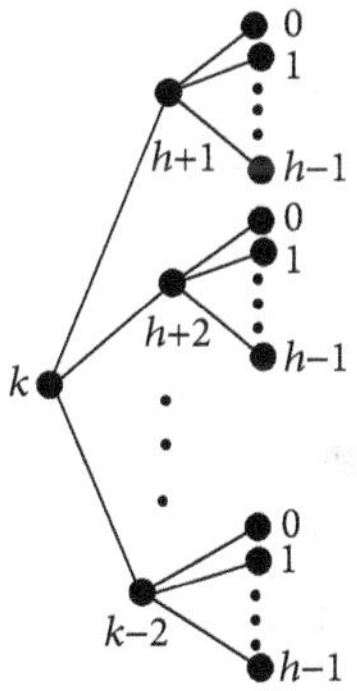	$k > h$
	$\mathscr{T}_5$: 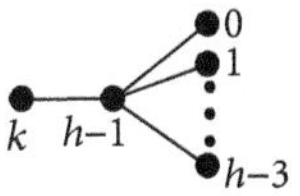	$c \neq h - 1$, $k > h$ and all colors greater than h less than k adjacent to c.
	$\mathscr{T}_6$: 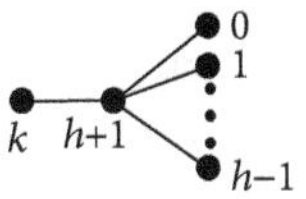	$k > h$ and all colors greater than h less than k adjacent to c.
$h \leqslant c \leqslant \lambda(T)$	• k	$0 \leqslant k < h$ and all colors less than k adjacent to c. For $c = h + 1$, $h < k \leqslant \lambda(T)$ and all colors less than k adjacent to c.
	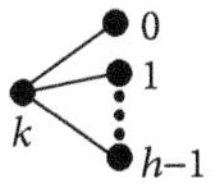	$k > h$ and all colors greater than h less than k adjacent to c and $c > h$.

TABLE 4: Continued.

Color of vertex	Connectable trees	Condition
	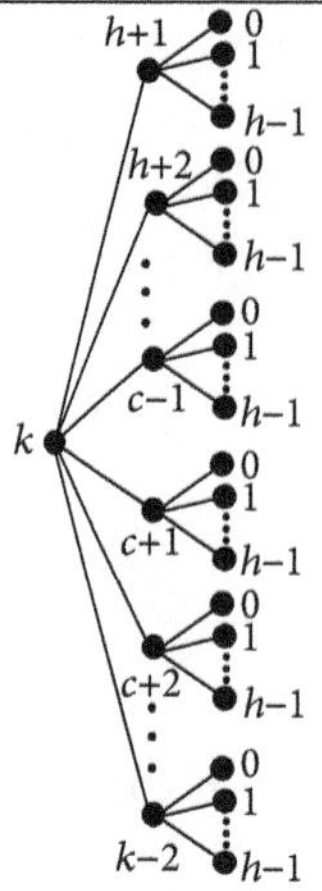	$h < c < k - 1$
	$\mathcal{T}_1$	$2 \leqslant k < h$ or $k = h + 1$
	$\mathcal{T}_2$	$0 < k < k' - 1 < h - 1$ or $0 < k < h$ and $k' = h + 1$
	$\mathcal{T}_3$	$0 \leqslant k' < k - 1 < h - 1$
	$\mathcal{T}_4$	$h + 2 < k < c - 1$
	$\mathcal{T}_5$	$k > h$ and all colors greater than h less than k adjacent to c.
	$\mathcal{T}_6$	$c \neq h + 1$, $k > h$ and all colors greater than h less than k adjacent to c.

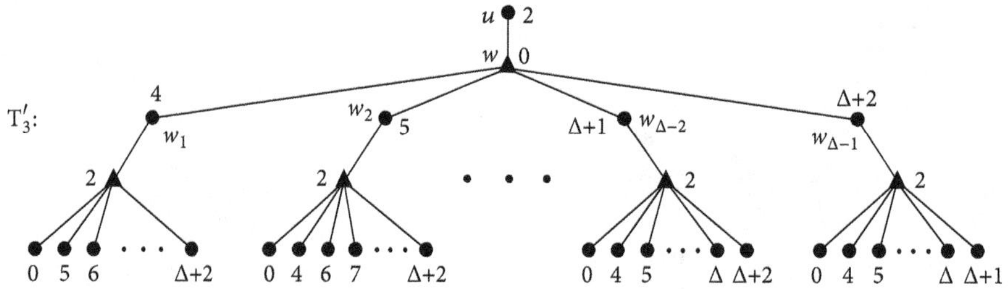

FIGURE 2: Irreducible $L(2,1)$-span coloring of T_3 with 1 and 3 as holes.

colorings for Type-II trees of maximum degrees three and four. Following theorem gives a two-hole irreducible span coloring for a tree with maximum degree Δ which is also a counterexample for the conjecture.

Theorem 10. *If exactly $\Delta - 1$ major vertices are at distance two from the major vertex w in T_3, then $H_\lambda(T_3) = 2$.*

Proof. Let T_3' be the tree T_3 with exactly $\Delta - 1$ major vertices at distance two to w. It is easy to see that the $L(2,1)$-span coloring of the tree T_3' given in Figure 2 is irreducible with 1 and 3 as holes.

Theorem 11. *If T_3'' is the tree T_3 with exactly Δ major vertices at distance two to w, then $H_\lambda(T_3'') \leqslant 1$.*

Proof. We consider T_3'' with labelling as in Figure 3.

Suppose that T_3'' has an $L(2,1)$-span coloring f with two holes. Then by Lemma 1, all major vertices of T_3'' receive colors from $\{0,2\}$ or $\{0, \Delta + 2\}$ or $\{\Delta, \Delta + 2\}$. Suppose that f assigns 0 and 2 to the major vertices. Since $\{f(w), f(v_i)\} = \{0,2\}$, $1 \leqslant i \leqslant \Delta$, it is not possible to color all w_i s as the four colors 0, 1, 2, and 3 cannot be assigned. Therefore, in this case, two-hole span coloring is not possible for T_3''. Similarly, we can prove the other two cases. Hence, $H_\lambda(T_3'') \leqslant 1$.

Let T_2' be the tree T_2 with exactly three major vertices are adjacent to a vertex.

Theorem 12. *For the trees T_1, T_2', T_3'', and T_4, the maximum number of holes is one.*

Proof. It is easy to see that the colorings of T_1, T_2', T_3'', and T_4 given in Figures 4, 5, 6, and 7, respectively, are irreducible $L(2,1)$-span colorings with hole Δ.

4. Infinitely Many Trees with Holes 0, 1, and 2

Recall that T_2' is the tree T_2 with exactly three major vertices adjacent to a vertex and T_3'' is the tree T_3 with exactly Δ major vertices at distance two from a major vertex. Let T_2'' be the tree T_2 with exactly four major vertices adjacent to the vertex

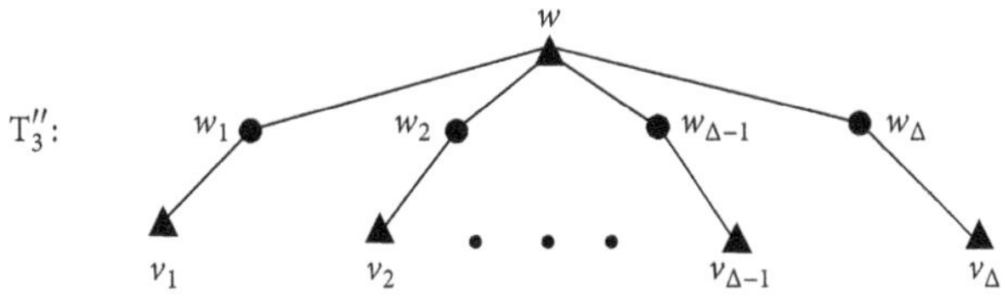

FIGURE 3: The tree T_3 with exactly Δ major vertices at distance two to w.

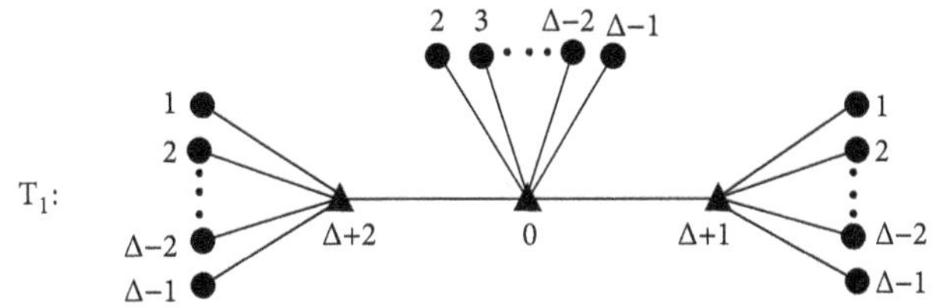

FIGURE 4: Irreducible $L(2,1)$-span coloring of T_1 with one hole.

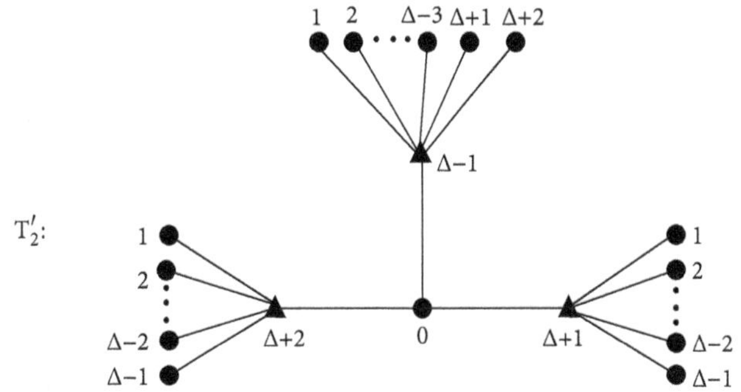

FIGURE 5: Irreducible $L(2,1)$-span coloring of T_2' with one hole.

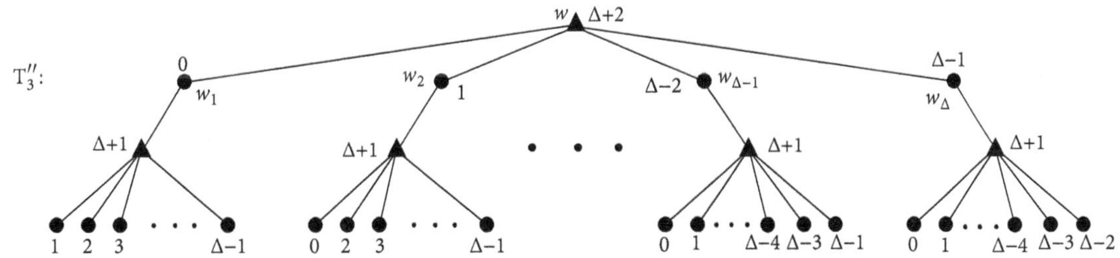

FIGURE 6: Irreducible $L(2,1)$-span coloring of T_3'' with one hole.

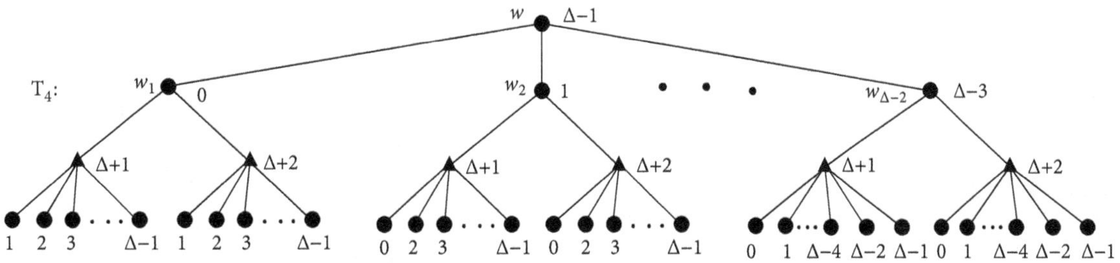

FIGURE 7: Irreducible $L(2,1)$-span coloring of T_4 with one hole.

T_2''':

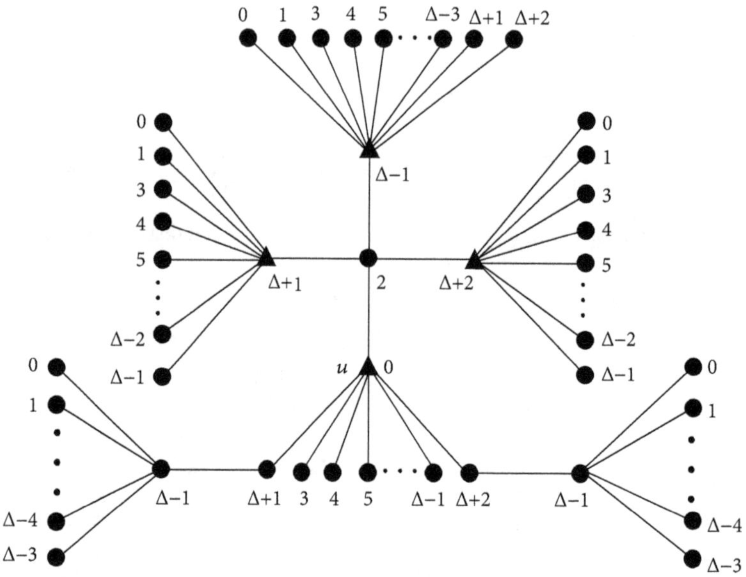

FIGURE 8: Irreducible $L(2,1)$-span coloring of T_2''' with Δ as hole.

w. In this section, we give a sufficient condition for a Type-II tree to be a zero-hole tree. Also, we construct infinitely many trees with maximum number of holes 1 from each of the trees T_1, T_2', T_2'', T_3'', and T_4 and infinitely many two-hole trees containing T_3'.

Theorem 13. *If the tree T_2 with at least five major vertices is a subtree of a tree T with maximum degree same as that of T_2, then $H_\lambda(T) = 0$.*

Proof. Let v_1, v_2, v_3, v_4, and v_5 be five major vertices adjacent to w and receive the colors l_1, l_2, l_3, l_4, and l_5, respectively, by a one-hole span coloring f of T_2. Without loss of generality, we assume that $0 \leqslant l_1 < l_2 < l_3 < l_4 < l_5 \leqslant \Delta + 2$. As in the proof of Theorem 9, we get $l_1 = 0$ and $l_2 + 1 = l_3 - 1$ is the hole. Since $l_4 < l_5 \leqslant \Delta + 2$, we have $l_4 \neq \Delta + 2$. Since $l_4 > l_3$, $l_3 - 1$ must be used to a neighbor of v_4 which is a contradiction. So, any $L(2,1)$-span coloring of T_2 with at least five major vertices is a no-hole coloring. Therefore, if a tree T contains T_2 with at least five major vertices and with maximum degree same as that of T_2, then $H_\lambda(T) = 0$.

Theorem 14. *There are infinitely many trees with maximum number of holes one and containing each of the trees T_1, T_2', T_2'', T_3'', and T_4.*

Proof. First, we prove that T_1, T_2', T_2'', T_3'', and T_4 cannot have two-hole span coloring. From Theorems 8 and 11, it is clear that T_1, T_2', T_2'', and T_3'' cannot have two-hole span colorings.

Next, we prove that T_4 cannot have a two-hole span coloring. We consider T_4 with the labelling as in Theorem 12. Suppose that T_4 has an $L(2,1)$-span coloring f with two holes. By Lemma 1, any major vertex of T_4 receives the color from $\{0, 2\}$, $\{0, \Delta + 2\}$, or $\{\Delta, \Delta + 2\}$. Suppose f assigns 0 and 2 to major vertices. Then 1 and 3 are holes, $\{f(v_i), f(v_i')\} = \{0, 2\}$, and w_i cannot receive the colors $0, 1, 2$, and 3. Therefore w_is receive $\Delta - 2$ different colors among $4, 5, 6, \ldots, \Delta + 2$ ($\Delta - 1$ in number) and so, one of these colors is not used, say c. Since either $c - 1$ or $c + 1$ (if $c = \Delta + 2$ then $c - 1$) is used to color one of the w_is, c cannot be used to w. Since 1 and 3 are holes, there is no color for w. Similarly, we can prove the other cases.

Now, to use Theorem 4, we need one-hole irreducible span coloring of T_1, T_2', T_2'', T_3'', and T_4. Since $H_\lambda(T_2'')$ is 0, first we construct a tree T_2''' from T_2'' such that $H_\lambda(T_2''') = 1$. We define a one-hole span coloring for T_2'' as in Figure 8 (T_2'' is a subtree of T_2'''). Since the colors $\Delta + 1$ and $\Delta + 2$ received by the vertices adjacent to the vertex u are reducible and there is no other color reducible, we connect star $K_{1,\Delta-2}$ to the vertices to make the colors $\Delta + 1$ and $\Delta + 2$ irreducible. The tree obtained is T_2'''.

Now, using Table 5 obtained from Table 4 corresponding to the hole $h = \Delta$ and using irreducible one-hole span colorings of T_1, T_2', T_3'', and T_4 given in Theorem 12, we construct infinitely many one-hole trees containing each of the trees T_1, T_2', T_3'', and T_4, respectively. We get infinitely many trees containing T_2'' by using irreducible one-hole coloring of T_2''' given in Figure 8 and using Table 5.

Example 15. In Figure 9, we illustrate the construction of one-hole tree as in Theorem 14 for the tree T_1 with maximum degree $\Delta = 7$. The vertex b_1 in T_1 has color 4 and its neighbor's color is 8. In Table 5, among the trees corresponding to the color $c = 4$, the pendant vertex colored 0 is connected first. Later, pendant vertices colored 1 and 2 are connected, respectively. Similarly, some trees are connected to the vertices b_2, b_3, and d_i, $1 \leqslant i \leqslant 3$.

Theorem 16. *There are infinitely many trees containing T_3' and with maximum number of holes two.*

Proof. The construction of trees is similar to the construction described in Theorem 3. For the construction, we use

TABLE 5: Trees connectable for a vertex colored c in a one-hole tree with Δ as the hole.

Color of vertex	Connectable trees	Condition
$0 \leqslant c < \Delta$	$\bullet k$	$0 \leqslant k < \Delta$ and all colors less than k adjacent to c. For $c = \Delta - 1$, $\Delta < k \leqslant \Delta + 2$ and all colors less than k adjacent to c.
	$\mathcal{T}_1$: 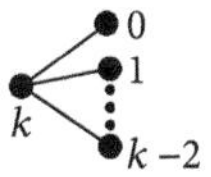	$2 \leqslant k < c - 1$
	$\mathcal{T}_2$: 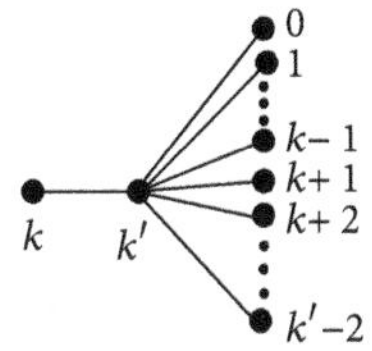	$k' \neq c$ and $0 < k < k' - 1 < \Delta - 1$ or $0 < k < \Delta$ and $k' = \Delta + 1$
	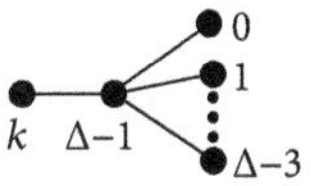	$k = \Delta + 1$ $k = \Delta + 2$ and $\Delta + 1$ is adjacent to c.
	$\mathcal{T}_3$: 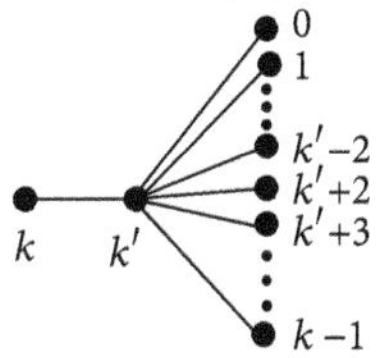	$k' \neq c$ and $0 \leqslant k' < k - 1 < \Delta - 1$ or $k = \Delta + 1$ and $k' = \Delta - 1$
	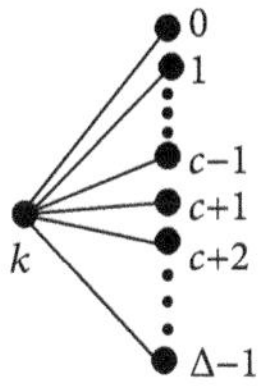	$k = \Delta + 1$ $k = \Delta + 2$ and $\Delta + 1$ is adjacent to c.
	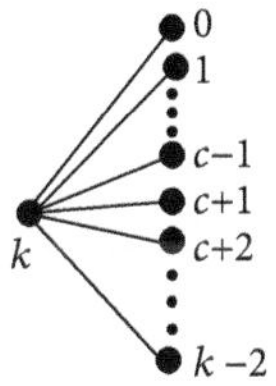	$c + 1 < k < \Delta$ or $k = \Delta + 1$
$c = \Delta + 1, \Delta + 2$	$\bullet k$	$0 \leqslant k < \Delta$ and all colors less than k adjacent to c.
	$\mathcal{T}_1$	$2 \leqslant k < \Delta$
	$\mathcal{T}_2$	$0 < k < k' - 1 < \Delta - 1$ or $0 < k < \Delta$ and $k' = \Delta + 1$
	$\mathcal{T}_3$	$0 \leqslant k' < k - 1 < \Delta - 1$

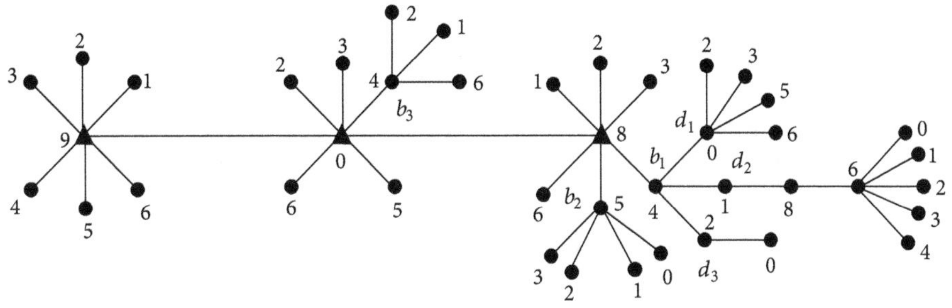

FIGURE 9: A tree with maximum number of holes one constructed from T_1.

two-hole irreducible span coloring of T_3' given in Figure 2 and Table 3.

Conflicts of Interest

The authors declare that they have no conflicts of interest.

References

[1] J. R. Griggs and R. K. Yeh, "Labeling graphs with a condition at distance 2," *SIAM Journal on Discrete Mathematics*, vol. 5, no. 4, pp. 586–595, 1992.

[2] W.-F. Wang, "The L(2,1)-labelling of trees," *Discrete Applied Mathematics*, vol. 154, no. 3, pp. 598–603, 2006.

[3] M.-q. Zhai, C.-h. Lu, and J.-l. Shu, "A note on L(2,1)-labelling of trees," *Acta Mathematicae Applicatae Sinica*, vol. 28, no. 2, pp. 395–400, 2012.

[4] N. Mandal and P. Panigrahi, "Solutions of some L(2,1)-coloring related open problems," *Discussiones Mathematicae Graph Theory*, vol. 36, no. 2, pp. 279–297, 2016.

[5] C. A. Wood and J. Jacob, "A complete L(2,1)-span characterization for small trees," *AKCE International Journal of Graphs and Combinatorics*, vol. 12, no. 1, pp. 26–31, 2015.

[6] P. C. Fishburn and F. S. Roberts, "No-hole L(2,1)-colorings," *Discrete Applied Mathematics*, vol. 130, no. 3, pp. 513–519, 2003.

[7] P. C. Fishburn, R. C. Laskar, F. S. Roberts, and J. Villalpando, "Parameters of L(2,1)-coloring," Manuscript.

[8] R. C. Laskar, G. L. Matthews, B. Novick, and J. Villalpando, "On irreducible no-hole L(2,1)-coloring of trees," *Networks. An International Journal*, vol. 53, no. 2, pp. 206–211, 2009.

[9] R. Laskar and G. Eyabi, "Holes in L(2,1)-coloring on certain classes of graphs," *AKCE International Journal of Graphs and Combinatorics*, vol. 6, no. 2, pp. 329–339, 2009.

[10] S. R. Kola, B. Gudla, and N. P.K., "Some classes of trees with maximum number of holes two," *AKCE International Journal of Graphs and Combinatorics*, 2018.

10

Accessing the Power of Tests based on Set-Indexed Partial Sums of Multivariate Regression Residuals

Wayan Somayasa

Department of Mathematics, Halu Oleo University, Indonesia

Correspondence should be addressed to Wayan Somayasa; wayan.somayasa@uho.ac.id

Academic Editor: Lucas Jodar

The intention of the present paper is to establish an approximation method to the limiting power functions of tests conducted based on Kolmogorov-Smirnov and Cramér-von Mises functionals of set-indexed partial sums of multivariate regression residuals. The limiting powers appear as vectorial boundary crossing probabilities. Their upper and lower bounds are derived by extending some existing results for shifted univariate Gaussian process documented in the literatures. The application of multivariate Cameron-Martin translation formula on the space of high dimensional set-indexed continuous functions is demonstrated. The rate of decay of the power function to a presigned value α is also studied. Our consideration is mainly for the trend plus signal model including multivariate set-indexed Brownian sheet and pillow. The simulation shows that the approach is useful for analyzing the performance of the test.

1. Introduction

Investigating the partial sums of least squares residuals has been shown to be reasonable and powerful tool for testing the adequacy of an assumed multivariate regression model; see Somayasa and et al. [1–4]. The development of the technique was motivated by the works proposed mainly for the purpose of detecting change in parameter as well as for detecting the existence of boundaries in univariate spatial regression; see [5–8] for references. The rejection region is constructed based on either Kolmogorov-Smirnov (KS) or Cramér-von Mises (CvM) functionals of the processes. It was shown in the literatures cited above that the limiting power function of the test appeared as a type of boundary crossing probability which has been involving shifted multidimensional Gaussian process.

To understand the objective considered in this paper in more detail we present below brief review how such a kind of probability appears. Let $\mathbf{Z}_p := (Z^{(i)})_{i=1}^p$ be the p–dimensional set-indexed Brownian sheet defined on a probability space $(\Omega, \mathscr{B}(\Omega), \mathbb{P})$, say with sample paths in $\mathscr{C}^p(\mathscr{B}(\mathbf{G})) := \times_{i=1}^p \mathscr{C}(\mathscr{B}(\mathbf{G}))$ and the control measure P_0, where P_0 is a probability measure on $(\mathbf{G}, \mathscr{B}(\mathbf{G}))$, $\mathbf{G} := \Pi_{k=1}^d [a_k, b_k]$, and $a_k < b_k$, for $k = 1, \ldots, d$. We refer the reader to [9, 10] for well documented notion of $\mathscr{C}(\mathscr{B}(\mathbf{G}))$. In the literature of Gaussian process $\mathbf{Z}_p$ is frequently called p–dimensional Gaussian white noise having P_0 as the control measure, cf. [11], p. 13-14. Let $\mathbf{W} := [f_1, \ldots, f_m]$ and $\mathbf{W}_{\mathscr{H}_{\mathbf{Z}_p}} := \times_{i=1}^p [S_{f_1}, \ldots, S_{f_m}]$, where for any $g \in L_2(P_0, \mathbf{G})$, S_g is defined as $S_g(A) := \int_A g dP_0$. Under mild condition, [1–3] showed after a suitable localization given to the regression function that the sequence of the partial sums of the least squares residuals obtained from the multivariate regression model

$$\mathbf{Y}(\mathbf{t}) = \mathbf{g}(\mathbf{t}) + \mathscr{E}(\mathbf{t}), \quad \mathbf{t} := (t_1, \ldots, t_d) \in \mathbf{G} \qquad (1)$$

converges, when $\mathbf{g} \notin \mathbf{W}^p := \times_{i=1}^p \mathbf{W}$, to a p–dimensional signal plus noise model defined by

$$\mathscr{Y} := \mathbf{\Sigma}^{-1/2} pr_{\mathbf{W}^{\perp}_{\mathscr{H}_{\mathbf{Z}_p}}} S_{\mathbf{g}} + pr^{*}_{\mathbf{W}^{\perp}_{\mathscr{H}_{\mathbf{Z}_p}}} \mathbf{Z}_p, \quad \mathbf{\Sigma} > 0, \qquad (2)$$

where $\mathbf{\Sigma} > 0$ means that $\mathbf{\Sigma}$ is positive definite, and for $A \in \mathscr{B}(\mathbf{G})$,

$$\left(pr_{\mathbf{W}^{\perp}_{\mathscr{H}_{\mathbf{Z}_p}}} S_{\mathbf{g}} \right)(A)$$

$$:= S_{\mathbf{g}}(A) - \sum_{j=1}^{m} \left(\langle f_j, g_i \rangle_{L_2(P_0, \mathbf{G})} \right)_{i=1}^{p} S_{f_j}(A),$$

$$\left(pr^*_{\mathbf{W}^{\perp}_{\mathscr{H}_{\mathbf{Z}_p}}}\mathbf{Z}_p\right)(A) := \mathbf{Z}_p(A) - \sum_{j=1}^{m}\left(\left(\int_{\mathbf{G}} f_j(\mathbf{t})\, dZ^{(i)}(\mathbf{t})\right)_{i=1}^{p} S_{f_j}(A)\right), \quad (3)$$

provided that $\{f_1,\ldots,f_m\}$ builds an ONB of $\mathbf{W}$ in $L_2(P_0,\mathbf{G})\cap BV_H(\mathbf{G})$. Thereby $S_{\mathbf{g}} := (S_{g_i})_{i=1}^p$ and $BV_H(\mathbf{G})$ is the space of functions on $\mathbf{G}$ with bounded variation in the sense of Hardy. It is worth mentioning that the notion of $BV_H(\mathbf{G})$ is a direct extension of the definition of $BV_H([a_1,b_1]\times[a_2,b_2])$ formulated in [12] to higher dimensional space. Here the notation $Z^{(i)}(\mathbf{t})$ stands for $Z^{(i)}(\Pi_{k=1}^d[a_k,t_k])$, cf. [8]. Throughout the paper $\mathbf{\Sigma}^{-1/2}pr_{\mathbf{W}^{\perp}_{\mathscr{H}_{\mathbf{Z}_p}}}S_{\mathbf{g}}$ will be denoted by $\varphi_{\mathbf{g}}$ and $pr^*_{\mathbf{W}^{\perp}_{\mathscr{H}_{\mathbf{Z}_p}}}\mathbf{Z}_p$ by $\mathscr{W}_{\mathbf{f},P_0}$, for the sake of brevity. It was established in [1–3] that $\mathscr{W}_{\mathbf{f},P_0}$ is a projection of $\mathbf{Z}_p$ onto the orthogonal complement of $\mathbf{W}_{\mathscr{H}_{\mathbf{Z}_p}}$ which is a finite dimensional subspace of the so-called reproducing kernel Hilbert Space (RKHS) of $\mathbf{Z}_p$, denoted by $\mathscr{H}_{\mathbf{Z}_p}$, given by

$$\mathscr{H}_{\mathbf{Z}_p} := \left\{\mathbf{h} \mid \exists \ell = (\ell_i)_{i=1}^p \in L_2^p(P_0,\mathbf{G}),\ \mathbf{h}(A) = \int_A \ell(\mathbf{t})\, P_0(d\mathbf{t})\right\}, \quad (4)$$

with $L_2^p(P_0,\mathbf{G}) := \times_{i=1}^p L_2(P_0,\mathbf{G})$. In the literatures mentioned above the process $\mathscr{W}_{\mathbf{f},P_0}$ is called the p-dimensional set-indexed residual partial sums limit process with the control measure P_0. Hence, the process $\mathbf{Z}_p$ itself and the p−dimensional set-indexed Brownian pillow $\mathbf{Z}_p^0 = (Z_i^0)_{i=1}^p$, with $\mathbf{Z}_p^0(A) := Z_i^0(A) - P_0(A)Z_i^0(\mathbf{G})$, are special cases of $\mathscr{W}_{\mathbf{f},P_0}$ that correspond to $\mathbf{W} = [f_0]$ and $\mathbf{W} = [f_1]$, respectively, with $f_0 \equiv 0$ and $f_1 \equiv 1$. The control measure P_0 appears in the process determines the design under which the experiment was constructed; see [4] for detail.

It was shown by using the well-known continuous mapping theorem that the limiting power functions of size α KS and CvM type tests for testing the hypothesis

$$H_0 : \mathbf{g} \in \mathbf{W}^p \ \textit{against}\ H_1 : \mathbf{g} \notin \mathbf{W}^p \quad (5)$$

are given, respectively, by the following complicated formulas:

$$\Psi_{\mathscr{W}_{\mathbf{f},P_0}}\left(\tilde{t}_{\mathscr{W}_{\mathbf{f},P_0}},\varphi_{\mathbf{g}}\right) := \mathbb{P}\left\{\sup_{A\in\mathscr{B}(\mathbf{G})}\left\|\varphi_{\mathbf{g}}(A) + \mathscr{W}_{\mathbf{f},P_0}(A)\right\|_{\mathbb{R}^p} \geq \tilde{t}_{\mathscr{W}_{\mathbf{f},P_0}}\right\}$$
$$\Upsilon_{\mathscr{W}_{\mathbf{f},P_0}}\left(\tilde{q}_{\mathscr{W}_{\mathbf{f},P_0}},\varphi_{\mathbf{g}}\right) := \mathbb{P}\left\{\int_{\mathbf{G}}\left\|\varphi_{\mathbf{g}}(A) + \mathscr{W}_{\mathbf{f},P_0}(A)\right\|_{\mathbb{R}^p}^2 dA \geq \tilde{q}_{\mathscr{W}_{\mathbf{f},P_0}}\right\}, \quad (6)$$

where $\|\cdot\|_{\mathbb{R}^p}^2$ stands for the Euclidean norm, whereas $\tilde{t}_{\mathscr{W}_{\mathbf{f},P_0}}$ and $\tilde{q}_{\mathscr{W}_{\mathbf{f},P_0}}$ are constants that satisfy $\Psi_{\mathscr{W}_{\mathbf{f},P_0}}(\tilde{t}_{\mathscr{W}_{\mathbf{f},P_0}},\mathbf{0}) = \Upsilon_{\mathscr{W}_{\mathbf{f},P_0}}(\tilde{q}_{\mathscr{W}_{\mathbf{f},P_0}},\mathbf{0}) = \alpha$. By the difficulty of the computation of $\tilde{t}_{\mathscr{W}_{\mathbf{f},P_0}}$ as well as $\tilde{q}_{\mathscr{W}_{\mathbf{f},P_0}}$ and the power of the test as the dimension of the experimental region and p get large, the implementation of the test in practice becomes restricted. Approximation by Monte Carlo simulation has been proposed in [1–3]. Some attempts of establishing concrete computation procedure by generalizing the principal component approach proposed, e.g., by MacNeill [5, 6] and Stute [13] for some univariate Gaussian processes on a line, have led us to incorrect result.

Since analytical computation of $\Psi_{\mathscr{W}_{\mathbf{f},P_0}}(\tilde{t}_{\mathscr{W}_{\mathbf{f},P_0}},\varphi_{\mathbf{g}})$ and $\Upsilon_{\mathscr{W}_{\mathbf{f},P_0}}(\tilde{t}_{\mathscr{W}_{\mathbf{f},P_0}},\varphi_{\mathbf{g}})$ is impossible, it is the purpose of the present paper to establish approximation procedure for that functions. As suggested in [14], p. 315, and [15], p. 423-424, studying the power function is of importance to be able to evaluate the performance of the test especially their rate of decay to α. Therefore in this paper we investigate the upper and lower bounds for (6) by considering the result for the univariate Brownian sheet and Brownian pillow presented in Janssen [17] and Hashorva [18, 19]. Upper and lower bound for the power function of goodness-of-fit test involving multiparameter Brownian process have been studied by Bass [20]. The RKHS of $\mathscr{W}_{\mathbf{f},P_0}$ is crucial for our results. By Theorem 4.1 in [11] (factorization theorem) if there exists a family $\{\mathbf{m}_A := (m_A^{(i)})_{i=1}^p \in L_2^p(P_0,\mathbf{G}) : \mathbf{A} \in \mathscr{B}(\mathbf{G})\}$, such that the covariance function of $\mathscr{W}_{\mathbf{f},P_0}$ admits the representation

$$\operatorname{Cov}\left(\mathscr{W}_{\mathbf{f},P_0}(A_1),\mathscr{W}_{\mathbf{f},P_0}(A_2)\right) = \int_{\mathbf{G}}\mathbf{m}_{A_1}^{\top}(\mathbf{t})\,\mathbf{m}_{A_2}(\mathbf{t})\,P_0(d\mathbf{t}) = \left\langle\mathbf{m}_{A_1},\mathbf{m}_{A_2}\right\rangle_{L_2^p(P_0,\mathbf{G})}, \quad (7)$$

then the corresponding RKHS is given by

$$\mathscr{H}_{\mathscr{W}_{\mathbf{f},P_0}} := \left\{\mathbf{h} \mid \exists \ell = (\ell_i)_{i=1}^p,\ \mathbf{h}(A) = \int_{\mathbf{G}}\left(m_A^{(i)}(\mathbf{t})\,\ell_i(\mathbf{t})\right)_{i=1}^p P_0(d\mathbf{t})\right\}. \quad (8)$$

Furthermore, the inner product and the corresponding norm on $\mathscr{H}_{\mathscr{W}_{\mathbf{f},P_0}}$ are denoted, respectively, by $\langle\cdot,\cdot\rangle_{\mathscr{H}_{\mathscr{W}_{\mathbf{f},P_0}}}$ and $\|\cdot\|^2_{\mathscr{H}_{\mathscr{W}_{\mathbf{f},P_0}}}$. For examples, the RKHS of $\mathbf{Z}_p^0$ is given by

$$\mathscr{H}_{\mathbf{Z}_p^0} := \left\{\mathbf{h} \mid \exists \ell \in L_2^p(P_0,\mathbf{G}),\ \mathbf{h}(A) = \int_A \ell(\mathbf{t})\,P_0(d\mathbf{t}),\ \text{s.t.}\ \mathbf{h}(\mathbf{G}) = \mathbf{h}(\emptyset) = 0\right\}, \quad (9)$$

with

$$\langle\mathbf{h}_1,\mathbf{h}_2\rangle_{\mathscr{H}_{\mathbf{Z}_p^0}} = \langle\ell_1,\ell_2\rangle_{L_2^p(P_0,\mathbf{G})},$$
$$\|\mathbf{h}\|_{\mathscr{H}_{\mathbf{Z}_p^0}} = \|\ell\|_{\mathscr{H}_{\mathbf{Z}_p^0}}, \quad (10)$$

such that $\mathbf{h}_j(A) = \int_A \ell_j(\mathbf{t})P_0(\mathbf{t}),\ j = 1,2$.

The rest of the present paper is organized as follows. In Section 2 we derive the upper and lower bounds for the power functions of KS and CvM tests by applying the Cameron-Martin translation formula of the multivariate process $\mathscr{W}_{\mathbf{f},P_0}$. The rate of decay of the power to α is also discussed. Alternative method of obtaining the bounds of the power function is proposed in Section 3. In Section 4 we propose Neyman-Pearson test which is a most powerful test. The comparison of the rate of decay of the obtained power to α with those of the KS and CvM tests is also investigated. Justification of the result is also studied in Section 5 by simulation. The paper is closed with a concluding remark in Section 6.

2. Rate of Decay of the Power of Tests

Our final goal in this section is to obtain an expression for the rate of decay of both $\Psi_{\mathscr{W}_{\mathbf{f},P_0}}(\tilde{t}_{\mathscr{W}_{\mathbf{f},P_0}}, \varphi_{\mathbf{g}})$ and $\Upsilon_{\mathscr{W}_{\mathbf{f},P_0}}(\tilde{q}_{\mathscr{W}_{\mathbf{f},P_0}}, \varphi_{\mathbf{g}})$ to the preassigned number $\alpha \in (0,1)$ representing the size of the test. First we derive their upper and lower bounds by generalizing the method proposed in [21] concerning bounds for the probability of shifted event; see also Theorem 7.3. in [11] for comparison. Second, we apply the technique studied in [17] to get the result. As reported in [17] and the references cited therein, they studied the upper and lower bounds for the power of signal detection test by applying Cameron-Martin density formula for a shifted measure. The rate of decay was obtained by means of mean value theorem.

Throughout this work let $\mathbf{P}$ be the probability distributions of $\mathscr{W}_{\mathbf{f},P_0}$ and let $\mathbf{P_h}$ be a probability measure on $\mathscr{C}^p(\mathscr{B}(\mathbf{G}))$, defined by

$$\mathbf{P_h}(\mathbf{B}) := \mathbf{P}(\mathbf{B}-\mathbf{h}), \quad \forall \mathbf{B} \in \mathscr{C}^p(\mathscr{B}(\mathbf{G})). \tag{11}$$

Then the Cameron-Martin density of $\mathbf{P_h}$ with respect to $\mathbf{P}$ for any $\mathbf{h} \in \mathscr{H}_{\mathscr{W}_{\mathbf{f},P_0}}$ is given by

$$\frac{d\mathbf{P_h}}{d\mathbf{P}}(\mathbf{x}) = \exp\left\{\mathbf{L}(\mathbf{h},\mathbf{x}) - \frac{1}{2}\|\mathbf{h}\|^2_{\mathscr{H}_{\mathscr{W}_{\mathbf{f},P_0}}}\right\}, \quad \text{for almost all } \mathbf{x} \in \mathscr{C}^p(\mathscr{B}(\mathbf{G})), \tag{12}$$

where $\mathbf{L}$ is a bilinear form, such that

$$\operatorname{Cov}\left(\mathbf{L}\left(\mathbf{h}_1, \mathscr{W}_{\mathbf{f},P_0}\right), \mathbf{L}\left(\mathbf{h}_2, \mathscr{W}_{\mathbf{f},P_0}\right)\right) = \langle \mathbf{h}_1, \mathbf{h}_2 \rangle_{\mathscr{H}_{\mathscr{W}_{\mathbf{f},P_0}}}, \quad \forall \mathbf{h}_1, \mathbf{h}_2 \in \mathscr{H}_{\mathscr{W}_{\mathbf{f},P_0}}. \tag{13}$$

This general formula can be obtained by extending the formula for the univariate model presented either in [20], Theorem 5.1 of [11], and [17] or [22] to higher dimensional set-indexed Gaussian processes.

The following theorem is already well known in the literatures mentioned above; however the proof is given only for the case of Gaussian random vector in $\mathbb{R}^n$ with zero mean and identity covariance matrix (canonical Gaussian Euclidean random vector); see [21] and [11], p. 53. In this paper we present again the theorem especially for the process $\mathscr{W}_{\mathbf{f},P_0}$ on $\mathscr{C}^p(\mathscr{B}(\mathbf{G}))$. Although the result for $\mathscr{W}_{\mathbf{f},P_0}$ is straightforward based on that of [11, 21], to give information on how the inequality for higher dimensional set-indexed Gaussian process was derived, we insist to present the proof of the theorem; see the appendix of this work.

Theorem 1 (Li and Kuelbs [21] and Lifshits [11]). *Let* $\mathbf{E}$ *be any subset of* $\mathscr{C}^p(\mathscr{B}(\mathbf{G}))$ *and* $r(\mathbf{E}) \in \mathbb{R}$ *be any constant, such that* $r(\mathbf{E}) = \Phi^{-1}(\mathbf{P}(\mathbf{E}))$. *Then for any* $\mathbf{h} \in \mathscr{H}_{\mathscr{W}_{\mathbf{f},P_0}}$, *it holds true that*

$$\Phi\left(r(\mathbf{E}) - \frac{\mathbf{L}(\mathbf{h},\mathbf{h})}{\|\mathbf{h}\|_{\mathscr{H}_{\mathscr{W}_{\mathbf{f},P_0}}}}\right) \le \mathbf{P}(\mathbf{E}-\mathbf{h}) \le \Phi\left(r(\mathbf{E}) + \frac{\mathbf{L}(\mathbf{h},\mathbf{h})}{\|\mathbf{h}\|_{\mathscr{H}_{\mathscr{W}_{\mathbf{f},P_0}}}}\right), \tag{14}$$

where Φ *is the cumulative distribution function of the standard normal distribution.*

The following corollary which gives an expression regarding the rate of decay of $\mathbf{P}(\mathbf{E}-\mathbf{h})$ to $\mathbf{P}(\mathbf{E})$, for any $\mathbf{E} \subset \mathscr{C}^p(\mathscr{B}(\mathbf{G}))$ and $\mathbf{h} \in \mathscr{H}_{\mathscr{H}_{\mathscr{W}_{\mathbf{f},P_0}}}$, is an immediate implication of Theorem 1. Rate of decay describes how fast the distance between $\mathbf{P}(\mathbf{E}-\mathbf{h})$ and $\mathbf{P}(\mathbf{E})$ vanishes, cf. [17–19].

Corollary 2. *Let* $\mathbf{E}$ *be an arbitrary subset of* $\mathscr{C}^p(\mathscr{B}(\mathbf{G}))$ *and* $r(\mathbf{E}) \in \mathbb{R}$ *be any constant, such that* $r(\mathbf{E}) = \Phi^{-1}(\mathbf{P}(\mathbf{E}))$. *Then under the assumption* $0 < \mathbf{L}(\mathbf{h},\mathbf{h})$, *we have, for any* $\mathbf{h} \in \mathscr{H}_{\mathscr{W}_{\mathbf{f},P_0}}$,

$$|\mathbf{P}(\mathbf{E}-\mathbf{h}) - \mathbf{P}(\mathbf{E})| \le \frac{1}{\sqrt{2\pi}} \frac{\mathbf{L}(\mathbf{h},\mathbf{h})}{\|\mathbf{h}\|_{\mathscr{H}_{\mathscr{W}_{\mathbf{f},P_0}}}}. \tag{15}$$

Proof. We apply the technique of proving Lemma 5 of [17]. By (14) presented in Theorem 1 and by using the symmetry of Φ, it holds that

$$\mathbf{P}(\mathbf{E}-\mathbf{h}) - \mathbf{P}(\mathbf{E}) \le \Phi\left(r(\mathbf{E}) + \frac{\mathbf{L}(\mathbf{h},\mathbf{h})}{\|\mathbf{h}\|_{\mathscr{H}_{\mathscr{W}_{\mathbf{f},P_0}}}}\right) - \Phi(r(\mathbf{E})) = \frac{\mathbf{L}(\mathbf{h},\mathbf{h})}{\|\mathbf{h}\|_{\mathscr{H}_{\mathscr{W}_{\mathbf{f},P_0}}}}\phi(\eta) \tag{16}$$

for some mean value $\eta \in (r(\mathbf{E}), r(\mathbf{E}) + \mathbf{L}(\mathbf{h},\mathbf{h})/\|\mathbf{h}\|_{\mathscr{H}_{\mathscr{W}_{\mathbf{f},P_0}}})$, where ϕ is the probability density function of $N(0,1)$. Since $\max\{\phi(t) : -\infty < t < \infty\} = 1/\sqrt{2\pi}$, then we have

$$\mathbf{P}(\mathbf{E}-\mathbf{h}) - \mathbf{P}(\mathbf{E}) \le \frac{\mathbf{L}(\mathbf{h},\mathbf{h})}{\|\mathbf{h}\|_{\mathscr{H}_{\mathscr{W}_{\mathbf{f},P_0}}}}\phi(\eta) \le \frac{1}{\sqrt{2\pi}} \frac{\mathbf{L}(\mathbf{h},\mathbf{h})}{\|\mathbf{h}\|_{\mathscr{H}_{\mathscr{W}_{\mathbf{f},P_0}}}}. \tag{17}$$

Conversely, by the inequality $\Phi(r(\mathbf{E}) - \mathbf{L}(\mathbf{h},\mathbf{h})/\|\mathbf{h}\|_{\mathscr{H}_{\mathscr{W}_{\mathrm{f},P_0}}}) \leq \mathbf{P}(\mathbf{E}-\mathbf{h})$ of (14), we can derive the following result:

$$\begin{aligned}
&\Phi\left(r(\mathbf{E}) - \frac{\mathbf{L}(\mathbf{h},\mathbf{h})}{\|\mathbf{h}\|_{\mathscr{H}_{\mathscr{W}_{\mathrm{f},P_0}}}}\right) - \mathbf{P}(\mathbf{E}) \leq \mathbf{P}(\mathbf{E}-\mathbf{h}) - \mathbf{P}(\mathbf{E}) \Longleftrightarrow \\
&\Phi\left(r(\mathbf{E}) - \frac{\mathbf{L}(\mathbf{h},\mathbf{h})}{\|\mathbf{h}\|_{\mathscr{H}_{\mathscr{W}_{\mathrm{f},P_0}}}}\right) - \Phi(r(\mathbf{E})) \\
&\leq \mathbf{P}(\mathbf{E}-\mathbf{h}) - \mathbf{P}(\mathbf{E}) \Longleftrightarrow \\
&-\frac{\mathbf{L}(\mathbf{h},\mathbf{h})}{\|\mathbf{h}\|_{\mathscr{H}_{\mathscr{W}_{\mathrm{f},P_0}}}}\phi(\kappa) \leq \mathbf{P}(\mathbf{E}-\mathbf{h}) - \mathbf{P}(\mathbf{E}),
\end{aligned} \tag{18}$$

for some mean value $\kappa \in (r(\mathbf{E}) - \mathbf{L}(\mathbf{h},\mathbf{h})/\|\mathbf{h}\|_{\mathscr{H}_{\mathscr{W}_{\mathrm{f},P_0}}}, r(\mathbf{E}))$. Since $\mathbf{L}(\mathbf{h},\mathbf{h}) > 0$, by the preceding result, we get

$$\begin{aligned}
\mathbf{P}(\mathbf{E}-\mathbf{h}) - \mathbf{P}(\mathbf{E}) &\geq -\frac{\mathbf{L}(\mathbf{h},\mathbf{h})}{\|\mathbf{h}\|_{\mathscr{H}_{\mathscr{W}_{\mathrm{f},P_0}}}}\phi(\kappa) \\
&\geq -\frac{1}{\sqrt{2\pi}}\frac{\mathbf{L}(\mathbf{h},\mathbf{h})}{\|\mathbf{h}\|_{\mathscr{H}_{\mathscr{W}_{\mathrm{f},P_0}}}},
\end{aligned} \tag{19}$$

which establises the proof.

When the model is either $\mathbf{h}_1 + \mathbf{Z}_p$, with $\mathbf{h}_1 \in \mathscr{H}_{\mathbf{Z}_p}$, or $\mathbf{h}_2 + \mathbf{Z}_p^0$, with $\mathbf{h}_2 \in \mathscr{H}_{\mathbf{Z}_p^0}$, such that $\mathbf{h}_j(A) = \int_A \ell_j(\mathbf{u})P_0(d\mathbf{u})$, for $A \in \mathscr{B}(\mathbf{G})$ and $j = 1, 2$, then

$$\begin{aligned}
\mathbf{L}(\mathbf{h}_1,\mathbf{h}_1) &= \int_{\mathbf{G}} \ell_1^{\top}(\mathbf{u})\,\ell_1(\mathbf{u})\,P_0(d\mathbf{u}) = \langle \mathbf{h}_1,\mathbf{h}_1\rangle_{\mathscr{H}_{\mathbf{Z}_p}} \\
&= \|\mathbf{h}_1\|^2_{\mathscr{H}_{\mathbf{Z}_p}} > 0 \\
\mathbf{L}(\mathbf{h}_2,\mathbf{h}_2) &= \int_{\mathbf{G}} \ell_2^{\top}(\mathbf{u})\,\ell_2(\mathbf{u})\,P_0(d\mathbf{u}) = \langle \mathbf{h}_2,\mathbf{h}_2\rangle_{\mathscr{H}_{\mathbf{Z}_p}} \\
&= \|\mathbf{h}_2\|^2_{\mathscr{H}_{\mathbf{Z}_p^0}} > 0.
\end{aligned} \tag{20}$$

Hence, when we are dealing with the p-dimensional set-indexed Brownian sheet and p-dimensional set-indexed Brownian pillow, Inequality (14), respectively, becomes

$$\begin{aligned}
\Phi\left(r(\mathbf{E}) - \|\mathbf{h}_1\|_{\mathscr{H}_{\mathbf{Z}_p}}\right) &\leq \mathbf{P}(\mathbf{E}-\mathbf{h}_1) \\
&\leq \Phi\left(r(\mathbf{E}) + \|\mathbf{h}_1\|_{\mathscr{H}_{\mathbf{Z}_p}}\right), \\
\Phi\left(r(\mathbf{E}) - \|\mathbf{h}_2\|_{\mathscr{H}_{\mathbf{Z}_p^0}}\right) &\leq \mathbf{P}(\mathbf{E}-\mathbf{h}_2) \\
&\leq \Phi\left(r(\mathbf{E}) + \|\mathbf{h}_2\|_{\mathscr{H}_{\mathbf{Z}_p^0}}\right).
\end{aligned} \tag{21}$$

The corresponding rate of decays can be obtained respectively as follows:

$$\begin{aligned}
|\mathbf{P}(\mathbf{E}-\mathbf{h}_1) - \mathbf{P}(\mathbf{E})| &\leq \frac{1}{\sqrt{2\pi}}\|\mathbf{h}_1\|_{\mathscr{H}_{\mathbf{Z}_p}}, \\
|\mathbf{P}(\mathbf{E}-\mathbf{h}_1) - \mathbf{P}(\mathbf{E})| &\leq \frac{1}{\sqrt{2\pi}}\|\mathbf{h}_1\|_{\mathscr{H}_{\mathbf{Z}_p^0}}.
\end{aligned} \tag{22}$$

In light of the preceding results, we can state the upper and lower bounds as well as the rate of decays for the power $\Psi_{\mathscr{W}_{\mathrm{f},P_0}}(\tilde{t}_{\mathscr{W}_{\mathrm{f},P_0}}, \varphi_{\mathbf{g}})$ and $\Upsilon_{\mathscr{W}_{\mathrm{f},P_0}}(\tilde{t}_{\mathscr{W}_{\mathrm{f},P_0}}, \varphi_{\mathbf{g}})$, when $\mathscr{W}_{\mathrm{f},P_0}$ is given by either $\mathbf{Z}_p$ or $\mathbf{Z}_p^0$. Let $\mathbf{E}$ be a subset of $\mathscr{C}^p(\mathscr{B}(\mathbf{G}))$, defined by

$$\mathbf{E} := \left\{\mathbf{x} \in \mathscr{C}^p(\mathscr{B}(\mathbf{G})) : \sup_{A\in\mathscr{B}(\mathbf{G})} \|\mathbf{x}(A)\|_{\mathbb{R}^p} \geq \tilde{t}_{\mathscr{W}_{\mathrm{f},P_0}}\right\}; \tag{23}$$

then for $\varphi_{\mathbf{g}} \in \mathscr{H}_{\mathscr{W}_{\mathrm{f},P_0}} \subset \mathscr{C}^p(\mathscr{B}(\mathbf{G}))$, we get

$$\begin{aligned}
\mathbf{E} - \varphi_{\mathbf{g}} &= \left\{\mathbf{x} - \varphi_{\mathbf{g}} : \mathbf{x} \in \mathbf{E}\right\} = \left\{\mathbf{x} - \varphi_{\mathbf{g}} : \sup_{A\in\mathscr{B}(\mathbf{G})} \|\mathbf{x}(A)\|_{\mathbb{R}^p} \geq \tilde{t}_{\mathscr{W}_{\mathrm{f},P_0}}\right\} \\
&= \left\{\mathbf{x} \in \mathscr{C}^p(\mathscr{B}(\mathbf{G})) : \sup_{A\in\mathscr{B}(\mathbf{G})} \left\|\varphi_{\mathbf{g}}(A) + \mathbf{x}(A)\right\|_{\mathbb{R}^p} \geq \tilde{t}_{\mathscr{W}_{\mathrm{f},P_0}}\right\}.
\end{aligned} \tag{24}$$

Since $\mathbf{P}$ is the distribution of $\mathscr{W}_{\mathrm{f},P_0}$, then $\Psi_{\mathscr{W}_{\mathrm{f},P_0}}(\tilde{t}_{\mathscr{W}_{\mathrm{f},P_0}}, \varphi_{\mathbf{g}})$ is equivalent to

$$\mathbf{P}\left(\mathbf{E} - \varphi_{\mathbf{g}}\right) = \mathbf{P}\left\{\mathbf{x} \in \mathscr{C}^p(\mathscr{B}(\mathbf{G})) : \sup_{A\in\mathscr{B}(\mathbf{G})} \left\|\varphi_{\mathbf{g}}(A) + \mathbf{x}(A)\right\|_{\mathbb{R}^p} \geq \tilde{t}_{\mathscr{W}_{\mathrm{f},P_0}}\right\}. \tag{25}$$

Analogously, let

$$\mathbf{F} := \left\{\mathbf{x} \in \mathscr{C}^p(\mathscr{B}(\mathbf{G})) : \int_{\mathbf{G}} \|\mathbf{x}(A)\|^2_{\mathbb{R}^p}\, dA \geq \tilde{q}_{\mathscr{W}_{\mathrm{f},P_0}}\right\}. \tag{26}$$

Then $\Upsilon_{\mathscr{W}_{\mathrm{f},P_0}}(\tilde{q}_{\mathscr{W}_{\mathrm{f},P_0}}, \varphi_{\mathbf{g}}) = \mathbf{P}(\mathbf{F}-\varphi_{\mathbf{g}})$. Thus by considering these two representations we have on the basis of Theorem 1 and Corollary 2 the following summary concerns the bounds for the power of the KS and CvM type tests.

Corollary 3. *Suppose that* $\varphi_{\mathbf{g}} \in \mathscr{H}_{\mathscr{W}_{\mathrm{f},P_0}}$*; then, for* $\alpha \in (0,1)$*, it holds that*

$$\Phi\left(\Phi^{-1}(\alpha) - \|\varphi_{\mathbf{g}}\|_{\mathscr{H}_{\mathscr{W}_{\mathrm{f},P_0}}}\right) \le \Psi_{\mathscr{W}_{\mathrm{f},P_0}}\left(\tilde{t}_{\mathscr{W}_{\mathrm{f},P_0}}, \varphi_{\mathbf{g}}\right) \le \Phi\left(\Phi^{-1}(\alpha) + \|\varphi_{\mathbf{g}}\|_{\mathscr{H}_{\mathscr{W}_{\mathrm{f},P_0}}}\right)$$
$$\Phi\left(\Phi^{-1}(\alpha) - \|\varphi_{\mathbf{g}}\|_{\mathscr{H}_{\mathscr{W}_{\mathrm{f},P_0}}}\right) \le \Upsilon_{\mathscr{W}_{\mathrm{f},P_0}}\left(\tilde{t}_{\mathscr{W}_{\mathrm{f},P_0}}, \varphi_{\mathbf{g}}\right) \le \Phi\left(\Phi^{-1}(\alpha) + \|\varphi_{\mathbf{g}}\|_{\mathscr{H}_{\mathscr{W}_{\mathrm{f},P_0}}}\right). \quad (27)$$

Furthermore, we have simple formulas for the rate of decay of $\Psi_{\mathscr{W}_{\mathrm{f},P_0}}(\tilde{t}_{\mathscr{W}_{\mathrm{f},P_0}}, \varphi_{\mathbf{g}})$ *and* $\Upsilon_{\mathscr{W}_{\mathrm{f},P_0}}(\tilde{t}_{\mathscr{W}_{\mathrm{f},P_0}}, \varphi_{\mathbf{g}})$ *to* α

$$\left|\Psi_{\mathscr{W}_{\mathrm{f},P_0}}\left(\tilde{t}_{\mathscr{W}_{\mathrm{f},P_0}}, \varphi_{\mathbf{g}}\right) - \alpha\right| \le \frac{1}{\sqrt{2\pi}} \|\varphi_{\mathbf{g}}\|_{\mathscr{H}_{\mathscr{W}_{\mathrm{f},P_0}}}$$
$$\left|\Upsilon_{\mathscr{W}_{\mathrm{f},P_0}}\left(\tilde{t}_{\mathscr{W}_{\mathrm{f},P_0}}, \varphi_{\mathbf{g}}\right) - \alpha\right| \le \frac{1}{\sqrt{2\pi}} \|\varphi_{\mathbf{g}}\|_{\mathscr{H}_{\mathscr{W}_{\mathrm{f},P_0}}}, \quad (28)$$

where in the context of model check, the norm of $\varphi_{\mathbf{g}}$ *related to the process* $\mathbf{Z}_p$ *and* $\mathbf{Z}_p^0$ *is given by*

$$\|\varphi_{\mathbf{g}}\|^2_{\mathscr{H}_{\mathscr{W}_{\mathrm{f},P_0}}} = \left\|\Sigma^{-1/2} pr_{\mathbf{W}^{\perp}_{\mathscr{H}_{\mathbf{Z}_p}}} S_{\mathbf{g}}\right\|^2_{\mathscr{H}_{\mathbf{Z}_p}} = \left\|pr_{\mathbf{W}^{p\perp}} \Sigma^{-1/2} \mathbf{g}\right\|^2_{L_2^p(P_0,\mathbf{G})} = \left\|\Sigma^{-1/2}\mathbf{g} - \Sigma^{-1/2} \sum_{j=1}^{m} \left(\langle f_j, g_i\rangle_{L_2(P_0,\mathbf{G})}\right)_{i=1}^{p} f_j\right\|^2_{L_2^p(P_0,\mathbf{G})}. \quad (29)$$

Corollary 3 says that the rate of decay or convergence of the power function to α in the case of $\mathbf{Z}_p$ as well as $\mathbf{Z}_p^0$ depends on the norm of the trend. A Model with small norm trend leads to faster decay. Conversely, a model with large norm trend results in slower decay. For both models, the norm can be concretely calculated. It is clear that both tests achieve their sizes as the trends vanish. Indeed the work of Samorodnitsky [23] can be incorporated in the estimation of $\Psi_{\mathscr{W}_{\mathrm{f},P_0}}(\lambda, \varphi_{\mathbf{g}})$, for any large real number λ. In Section 5 we demonstrate by simulation the behavior of the power functions of the KS and CvM tests as summarized in Corollary Corollary 3 to give empirical study regarding the rate of decay of the power functions.

3. Alternative Approaches

In this section other formulas for the upper and lower bounds of the power of KS and CvM tests involving the p-dimensional set-indexed Brownian sheet and pillow models are derived. Our results are obtained by generalizing the approach proposed in that studied in [18, 19] who confined the investigation to one-dimensional Kolmogorov type boundary noncrossing probability involving the so-called univariate ordinary Brownian sheet and pillow.

To simplify the notation we restrict the attention to the case of two-dimensional experimental region $\mathbf{G} = [a_1, b_1] \times [a_2, b_2] \subset \mathbb{R}^2$.

Theorem 4. *Let the ONB* $\{f_1, \ldots, f_m\}$ *of* $\mathbf{W}$ *be in* $L_2(P_0, \mathbf{G}) \cap BV_H(\mathbf{G})$ *and let* $\varphi_{\mathbf{g}} = \Sigma^{-1/2} pr_{\mathbf{W}^{\perp}_{\mathscr{H}_{\mathbf{Z}_p}}} S_{\mathbf{g}}$*, such that* $pr_{\mathbf{W}^{\perp}}\Sigma^{-1/2}\mathbf{g}$ *are constant on the boundary of* $\mathbf{G}$*. Then for the* $\mathbf{Z}_p$ *model it holds that*

$$1 - (1-\alpha)\mathscr{L}_{\mathbf{Z}_p} \le \Psi_{\mathbf{Z}_p}\left(\tilde{t}_{\mathbf{Z}_p}, \varphi_{\mathbf{g}}\right) \le 1 - \left(1 - \Psi_{\mathbf{Z}_p}\left(\frac{1}{2}\tilde{t}_{\mathbf{Z}_p}, \mathbf{0}\right)\right)\mathscr{U}_{\mathbf{Z}_p} \quad (30)$$

where

$$\mathscr{L}_{\mathbf{Z}_p} := \exp\left\{2\tilde{t}_{\mathbf{Z}_p}\sum_{i=1}^{p}\Delta_{\mathbf{G}}W_i - \frac{1}{2}\|\varphi_{\mathbf{g}}\|^2_{\mathscr{H}_{\mathbf{Z}_p}}\right\}$$
$$\mathscr{U}_{\mathbf{Z}_p} := \exp\left\{-2\tilde{t}_{\mathbf{Z}_p}\sum_{i=1}^{p}\Delta_{\mathbf{G}}W_i - \frac{1}{2}\|\varphi_{\mathbf{g}}\|^2_{\mathscr{H}_{\mathbf{Z}_p}}\right\}, \quad (31)$$

where $\Delta_{\mathbf{G}}W_i := W_i(b_1,b_2) - W_i(b_1,a_2) - W_i(a_1,b_2) + W_i(a_1,a_2)$ *and* W_i *is the ith component of* $pr_{\mathbf{W}^{p\perp}}\Sigma^{-1/2}\mathbf{g}$*, which is given by*

$$W_i := \sum_{k=1}^{p}\sigma^*_{ik}g_k - \sum_{j=1}^{m}\left(\left\langle f_j, \sum_{k=1}^{p}\sigma^*_{ik}g_k\right\rangle_{L_2(P_0,\mathbf{G})}\right)_{i=1}^{p} f_j \in BV_H(\mathbf{G}) \quad (32)$$

with σ^*_{ik} *denoting the* (i,k)*th element of* $\Sigma^{-1/2}$*, say, for* $i,k = 1,\ldots,p$*. Furthermore, for the* $\mathbf{Z}_p^0$ *model, we have*

$$1 - (1-\alpha)\mathscr{L}_{\mathbf{Z}_p^0} \le \Psi_{\mathbf{Z}_p^0}\left(\tilde{t}_{\mathbf{Z}_p^0}, \varphi_{\mathbf{g}}\right) \le 1 - \left(1 - \Psi_{\mathbf{Z}_p^0}\left(\frac{1}{2}\tilde{t}_{\mathbf{Z}_p^0}, \mathbf{0}\right)\right)\mathscr{U}_{\mathbf{Z}_p^0}, \quad (33)$$

where

$$\mathscr{L}_{\mathbf{Z}_p^0} := \exp\left\{-\frac{1}{2}\|\varphi_{\mathbf{g}}\|^2_{\mathscr{H}_{\mathbf{Z}_p^0}}\right\} =: \mathscr{U}_{\mathbf{Z}_p^0}. \quad (34)$$

Proof. By using a rule for the probability of complement, we get for the $\mathbf{Z}_p$ model

$$\Psi_{\mathbf{Z}_p}\left(\tilde{t}_{\mathbf{Z}_p}, \varphi_{\mathbf{g}}\right) = 1 - \mathbf{P}\left\{\mathbf{x} \in \mathscr{C}^p(\mathscr{B}(\mathbf{G})) : \sup_{A\in\mathscr{B}(\mathbf{G})}\|\varphi_{\mathbf{g}}(A) + \mathbf{x}(A)\|_{\mathbb{R}^p} < \tilde{t}_{\mathbf{Z}_p}\right\} = 1 - \mathbf{P}\left\{\mathbf{x} \in \mathscr{C}^p(\mathscr{B}(\mathbf{G})) : \|\varphi_{\mathbf{g}}(A) + \mathbf{x}(A)\|_{\mathbb{R}^p} < \tilde{t}_{\mathbf{Z}_p}, \forall A \in \mathscr{B}(\mathbf{G})\right\}, \quad (35)$$

where by using transformation of variables, it can be further expressed as

$$\begin{aligned}\mathbf{P}\Big\{\mathbf{x}\in\mathscr{C}^p(\mathscr{B}(\mathbf{G})):\left\|\varphi_{\mathbf{g}}(A)+\mathbf{x}(A)\right\|_{\mathbb{R}^p}<\tilde{t}_{\mathbf{Z}_p},\quad\forall A\\ \in\mathscr{B}(\mathbf{G})\Big\} &= \int_{\mathscr{C}^p(\mathscr{B}(\mathbf{G}))}\mathbf{1}\Big\{\left\|\varphi_{\mathbf{g}}(A)+\mathbf{x}(A)\right\|_{\mathbb{R}^p}\\ &<\tilde{t}_{\mathbf{Z}_p},\forall A\in\mathscr{B}(\mathbf{G})\Big\}\,\mathbf{P}(d\mathbf{x})\\ &= \int_{\mathscr{C}^p(\mathscr{B}(\mathbf{G}))}\mathbf{1}\Big\{\|\mathbf{y}(A)\|_{\mathbb{R}^p}<\tilde{t}_{\mathbf{Z}_p},\forall A\in\mathscr{B}(\mathbf{G})\Big\}\\ &\quad\cdot\mathbf{P}_{\varphi_{\mathbf{g}}}(d\mathbf{y}).\end{aligned} \tag{36}$$

Next, Cameron-Martin formula (12) for the p-dimensional set-indexed Brownian sheet implicates

$$\begin{aligned}&\int_{\mathscr{C}^p(\mathscr{B}(\mathbf{G}))}\mathbf{1}\Big\{\|\mathbf{y}(A)\|_{\mathbb{R}^p}<\tilde{t}_{\mathbf{Z}_p},\ \forall A\in\mathscr{B}(\mathbf{G})\Big\}\,\mathbf{P}_{\varphi_{\mathbf{g}}}(d\mathbf{y})\\ &= \int_{\mathscr{C}^p(\mathscr{B}(\mathbf{G}))}\mathbf{1}\Big\{\|\mathbf{y}(A)\|_{\mathbb{R}^p}<\tilde{t}_{\mathbf{Z}_p},\ \forall A\in\mathscr{B}(\mathbf{G})\Big\}\\ &\times\exp\left\{\int_{\mathbf{G}}\left(pr_{\mathbf{W}^{p\perp}}\mathbf{\Sigma}^{-1/2}\mathbf{g}\right)^{\top}(\mathbf{t})\,d\mathbf{y}(\mathbf{t})-\frac{1}{2}\|\varphi_{\mathbf{g}}\|^2_{\mathbf{Z}_p}\right\}\\ &\cdot\mathbf{P}(d\mathbf{y})\\ &= \int_{\mathscr{C}^p(\mathscr{B}(\mathbf{G}))}\mathbf{1}\Big\{\|\mathbf{y}(A)\|_{\mathbb{R}^p}<\tilde{t}_{\mathbf{Z}_p},\ \forall A\in\mathscr{B}(\mathbf{G})\Big\}\\ &\times\exp\left\{\sum_{i=1}^{p}\int_{\mathbf{G}}W_i(\mathbf{t})\,dy_i(\mathbf{t})-\frac{1}{2}\|\varphi_{\mathbf{g}}\|^2_{\mathbf{Z}_p}\right\}\mathbf{P}(d\mathbf{y}).\end{aligned} \tag{37}$$

Since $\|\mathbf{y}(A)\|_{\mathbb{R}^p}<\tilde{t}_{\mathbf{Z}_p}$ means $-\tilde{t}_{\mathbf{Z}_p}<y_i(A)<\tilde{t}_{\mathbf{Z}_p}$, then under the indicator $\mathbf{1}\{\|\mathbf{y}(A)\|_{\mathbb{R}^p}<\tilde{t}_{\mathbf{Z}^0_p},\ \forall A\in\mathscr{B}(\mathbf{G})\}$ we get by recalling integration by parts formula on $\mathbf{G}$, cf. [24, 25] and the assumption that W_i is constant throughout the boundary of $\mathbf{G}$; for the $\mathbf{Z}_p$ model we get

$$\begin{aligned}&\int_{\mathscr{C}^p(\mathscr{B}(\mathbf{G}))}\mathbf{1}\Big\{\|\mathbf{y}(A)\|_{\mathbb{R}^p}<\tilde{t}_{\mathbf{Z}_p},\ \forall A\in\mathscr{B}(\mathbf{G})\Big\}\\ &\times\exp\left\{\sum_{i=1}^{p}\int_{\mathbf{G}}W_i(\mathbf{t})\,dy_i(\mathbf{t})-\frac{1}{2}\|\varphi_{\mathbf{g}}\|^2_{\mathscr{H}_{\mathbf{Z}_p}}\right\}\mathbf{P}(d\mathbf{y})\\ &\le\exp\left\{2\tilde{t}_{\mathbf{Z}_p}\sum_{i=1}^{p}\Delta_{\mathbf{G}}W_i-\frac{1}{2}\|\varphi_{\mathbf{g}}\|^2_{\mathscr{H}_{\mathbf{Z}_p}}\right\}\\ &\cdot\mathbb{P}\left\{\sup_{A\in\mathscr{B}(\mathbf{G})}\|\mathbf{Z}_p\|_{\mathscr{H}_{\mathbb{R}^p}}<\tilde{t}_{\mathbf{Z}_p}\right\}\\ &=\exp\left\{2\tilde{t}_{\mathbf{Z}_p}\sum_{i=1}^{p}\Delta_{\mathbf{G}}W_i-\frac{1}{2}\|\varphi_{\mathbf{g}}\|^2_{\mathscr{H}_{\mathbf{Z}_p}}\right\}\\ &\cdot\left(1-\Psi_{\mathbf{Z}_p}\left(\tilde{t}_{\mathbf{Z}_p},\mathbf{0}\right)\right).\end{aligned} \tag{38}$$

Thus, the lower bound in (30) is established. To prove the upper bound, we start with the following inequality:

$$\begin{aligned}\Psi_{\mathbf{Z}_p}\left(\tilde{t}_{\mathbf{Z}_p},\varphi_{\mathbf{g}}\right)&\le 1\\ &-\mathbb{P}\left\{\left\|\varphi_{\mathbf{g}}(A)+\mathbf{Z}_p(A)\right\|_{\mathbb{R}^p}\le\frac{1}{2}\tilde{t}_{\mathbf{Z}_p},\ \forall A\in\mathscr{B}(\mathbf{G})\right\}.\end{aligned} \tag{39}$$

By applying the similar technique as that used in deriving the preceding result and the implication

$$\|\mathbf{y}(A)\|_{\mathbb{R}^p}\le\frac{1}{2}\tilde{t}_{\mathbf{Z}_p}\Longrightarrow-\tilde{t}_{\mathbf{Z}_p}\le y_i(A)\le\tilde{t}_{\mathbf{Z}_p}\qquad\forall i=1,\dots,p, \tag{40}$$

under the indicator $\mathbf{1}\{\|\mathbf{y}(A)\|\le(1/2)\tilde{t}_{\mathbf{Z}_p},\ \forall A\in\mathscr{B}(\mathbf{G})\}$ we have, by the integration by parts, the following inequality:

$$\begin{aligned}&\mathbb{P}\left\{\left\|\varphi_{\mathbf{g}}(A)+\mathbf{Z}_p(A)\right\|_{\mathbb{R}^p}\le\frac{1}{2}\tilde{t}_{\mathbf{Z}_p},\ \forall A\in\mathscr{B}(\mathbf{G})\right\}\\ &=\int_{\mathscr{C}^p(\mathscr{B}(\mathbf{G}))}\mathbf{1}\left\{\|\mathbf{y}(A)\|\le\frac{1}{2}\tilde{t}_{\mathbf{Z}_p},\ \forall A\in\mathscr{B}(\mathbf{G})\right\}\\ &\times\exp\left\{\sum_{i=1}^{p}\int_{\mathbf{G}}W_i(\mathbf{t})\,dy_i(\mathbf{t})-\frac{1}{2}\|\varphi_{\mathbf{g}}\|^2_{\mathscr{H}_{\mathbf{Z}_p}}\right\}\mathbf{P}(d\mathbf{y})\\ &\ge\int_{\mathscr{C}^p(\mathscr{B}(\mathbf{G}))}\mathbf{1}\left\{\|\mathbf{y}(A)\|\le\frac{1}{2}\tilde{t}_{\mathbf{Z}_p},\ \forall A\in\mathscr{B}(\mathbf{G})\right\}\\ &\times\exp\left\{-2\tilde{t}_{\mathbf{Z}_p}\sum_{i=1}^{p}\Delta_{\mathbf{G}}W_i-\frac{1}{2}\|\varphi_{\mathbf{g}}\|^2_{\mathscr{H}_{\mathbf{Z}_p}}\right\}\mathbf{P}(d\mathbf{y})\\ &=\exp\left\{-2\tilde{t}_{\mathbf{Z}_p}\sum_{i=1}^{p}W_i(b_1,b_2)-\frac{1}{2}\|\varphi_{\mathbf{g}}\|^2_{\mathscr{H}_{\mathbf{Z}_p}}\right\}\\ &\cdot\left(1-\Psi_{\mathbf{Z}_p}\left(\frac{1}{2}\tilde{t}_{\mathbf{Z}_p},\mathbf{0}\right)\right),\end{aligned} \tag{41}$$

completing the proof for the $\mathbf{Z}_p$ model. To prove the lower and upper bounds (33) for the $\mathbf{Z}^0_p$ model, we start with the equality

$$\begin{aligned}1-\Psi_{\mathbf{Z}^0_p}\left(\tilde{t}_{\mathbf{Z}^0_p},\varphi_{\mathbf{g}}\right)&=\int_{\mathscr{C}^p(\mathscr{B}(\mathbf{G}))}\mathbf{1}\Big\{\|\mathbf{y}(A)\|_{\mathbb{R}^p}<\tilde{t}_{\mathbf{Z}^0_p},\ \forall A\in\mathscr{B}(\mathbf{G})\Big\}\\ &\times\exp\left\{\sum_{i=1}^{p}\int_{\mathbf{G}}W_i(\mathbf{t})\,dy_i(\mathbf{t})-\frac{1}{2}\|\varphi_{\mathbf{g}}\|^2_{\mathscr{H}_{\mathbf{Z}^0_p}}\right\}\mathbf{P}(d\mathbf{y}).\end{aligned} \tag{42}$$

Next by the integration by parts and the assumption that $\varphi_{\mathbf{g}}\in\mathscr{H}_{\mathbf{Z}^0_p}$ and W_i are constant on the boundary of $\mathbf{G}$, we have under $\mathbf{1}\{\|\mathbf{y}(A)\|_{\mathbb{R}^p}<\tilde{t}_{\mathbf{Z}^0_p},\ \forall A\in\mathscr{B}(\mathbf{G})\}$ and the fact $y_i(b_1,b_2)=y_i(b_1,a_2)=y_i(a_1,b_2)=y_i(a_1,a_2)=0$ that

$$\int_{\mathbf{G}}W_i(\mathbf{t})\,dy_i(\mathbf{t})<\tilde{t}_{\mathbf{Z}^0_p}\int_{\mathbf{G}}dW_i(t,s)=0. \tag{43}$$

Hence, $1 - \Psi_{\mathbf{Z}_p^0}(\tilde{t}_{\mathbf{Z}_p^0}, \varphi_{\mathbf{g}}) \leq \exp\{-(1/2)\|\varphi_{\mathbf{g}}\|^2_{\mathcal{H}_{\mathbf{Z}_p^0}}\}(1-\alpha)$, establishing the lower bound in (33). The similar argument as that used in the case of $\mathbf{Z}_p$ model can be applied in deriving the upper bound of $\Psi_{\mathbf{Z}_p^0}(\tilde{t}_{\mathbf{Z}_p^0}, \varphi_{\mathbf{g}})$ as follows:

$$\begin{aligned} 1 - \Psi_{\mathbf{Z}_p^0}\left(\tilde{t}_{\mathbf{Z}_p^0}, \varphi_{\mathbf{g}}\right) &= \int_{\mathscr{C}^p(\mathscr{B}(\mathbf{G}))} \mathbf{1}\left\{\|\mathbf{y}(A)\|_{\mathbb{R}^p} < \tilde{t}_{\mathbf{Z}_p^0},\ \forall A \in \mathscr{B}(\mathbf{G})\right\} \\ &\quad \times \exp\left\{\sum_{i=1}^{p}\int_{\mathbf{G}} W_i(\mathbf{t})\, dy_i(\mathbf{t}) - \frac{1}{2}\|\varphi_{\mathbf{g}}\|^2_{\mathcal{H}_{\mathbf{Z}_p^0}}\right\}\mathbf{P}(d\mathbf{y}) \\ &\geq \int_{\mathscr{C}^p(\mathscr{B}(\mathbf{G}))} \mathbf{1}\left\{\|\mathbf{y}(A)\|_{\mathbb{R}^p} \leq \frac{1}{2}\tilde{t}_{\mathbf{Z}_p^0},\ \forall A \in \mathscr{B}(\mathbf{G})\right\} \\ &\quad \times \exp\left\{\sum_{i=1}^{p}\int_{\mathbf{G}} W_i(\mathbf{t})\, dy_i(\mathbf{t}) - \frac{1}{2}\|\varphi_{\mathbf{g}}\|^2_{\mathcal{H}_{\mathbf{Z}_p^0}}\right\}\mathbf{P}(d\mathbf{y}) \\ &\geq \exp\left\{-\frac{1}{2}\|\varphi_{\mathbf{g}}\|^2_{\mathcal{H}_{\mathbf{Z}_p^0}}\right\}\left(1 - \Psi_{\mathbf{Z}_p^0}\left(\frac{1}{2}\tilde{t}_{\mathbf{Z}_p^0}, \varphi_{\mathbf{g}}\right)\right), \end{aligned} \tag{44}$$

establishing the proof.

Now we can derive other formulas for the rate of decay of $\Psi_{\mathbf{Z}_p}(\tilde{t}_{\mathbf{Z}_p}, \varphi_{\mathbf{g}})$ and $\Psi_{\mathbf{Z}_p^0}(\tilde{t}_{\mathbf{Z}_p^0}, \varphi_{\mathbf{g}})$ to α by applying the similar method as that utilized in deriving the formula in Corollary 3. However by Theorem 4 we lead to computationally more complicated results.

Corollary 5. *Under the condition of Theorem 7, it holds true that*

$$\begin{aligned} &-e^{\rho}\left(2t_{\mathbf{Z}_p}\sum_{i=1}^{p}\Delta_{\mathbf{G}}W_i + \frac{1}{2}\|\varphi_{\mathbf{g}}\|^2_{\mathcal{H}_{\mathbf{Z}_p}}\right) \\ &\quad\leq \Psi_{\mathbf{Z}_p}\left(\tilde{t}_{\mathbf{Z}_p}, \varphi_{\mathbf{g}}\right) - \alpha \\ &\quad\leq e^{\zeta}\left(2t_{\mathbf{Z}_p}\sum_{i=1}^{p}\Delta_{\mathbf{G}}W_i + \frac{1}{2}\|\varphi_{\mathbf{g}}\|^2_{\mathcal{H}_{\mathbf{Z}_p}}\right), \end{aligned} \tag{45}$$

for some mean values

$$\begin{aligned} \zeta &\in \left(-2t_{\mathbf{Z}_p}\sum_{i=1}^{p}\Delta_{\mathbf{G}}W_i - \frac{1}{2}\|\varphi_{\mathbf{g}}\|^2_{\mathcal{H}_{\mathbf{Z}_p}} + \ln\left(1 - \Psi_{\mathbf{Z}_p}\left(\frac{1}{2}\tilde{t}_{\mathbf{Z}_p}, \mathbf{0}\right)\right); \ln(1-\alpha)\right), \\ \rho &\in \left(\ln(1-\alpha); 2t_{\mathbf{Z}_p}\sum_{i=1}^{p}\Delta_{\mathbf{G}}W_i - \frac{1}{2}\|\varphi_{\mathbf{g}}\|^2_{\mathcal{H}_{\mathbf{Z}_p^0}} + \ln(1-\alpha)\right). \end{aligned} \tag{46}$$

In particular, if the mean values ρ and ζ are taken to be the same, then

$$\begin{aligned} &\left|\Psi_{\mathbf{Z}_p}\left(\tilde{t}_{\mathbf{Z}_p}, \varphi_{\mathbf{g}}\right) - \alpha\right| \\ &\quad\leq e^{\zeta}\left(2t_{\mathbf{Z}_p}\sum_{i=1}^{p}\Delta_{\mathbf{G}}W_i + \frac{1}{2}\|\varphi_{\mathbf{g}}\|^2_{\mathcal{H}_{\mathbf{Z}_p}}\right). \end{aligned} \tag{47}$$

Proof. From Inequality (30), we have, by applying the mean value theorem,

$$\begin{aligned} \Psi_{\mathbf{Z}_p}\left(\tilde{t}_{\mathbf{Z}_p}, \varphi_{\mathbf{g}}\right) - \alpha &\leq \exp\{\ln(1-\alpha)\} \\ &\quad - \exp\left\{-2t_{\mathbf{Z}_p}\sum_{i=1}^{p}\Delta_{\mathbf{G}}W_i - \frac{1}{2}\|\varphi_{\mathbf{g}}\|^2_{\mathcal{H}_{\mathbf{Z}_p}} + \ln\left(1 - \Psi_{\mathbf{Z}_p}\left(\frac{1}{2}\tilde{t}_{\mathbf{Z}_p}, \mathbf{0}\right)\right)\right\} \\ &= e^{\zeta}\left(\ln\left(\frac{1-\alpha}{1 - \Psi_{\mathbf{Z}_p}\left((1/2)\tilde{t}_{\mathbf{Z}_p}, \mathbf{0}\right)}\right) + 2t_{\mathbf{Z}_p}\sum_{i=1}^{p}\Delta_{\mathbf{G}}W_i + \frac{1}{2}\|\varphi_{\mathbf{g}}\|^2_{\mathcal{H}_{\mathbf{Z}_p}}\right) \\ &\leq e^{\zeta}\left(2t_{\mathbf{Z}_p}\sum_{i=1}^{p}\Delta_{\mathbf{G}}W_i + \frac{1}{2}\|\varphi_{\mathbf{g}}\|^2_{\mathcal{H}_{\mathbf{Z}_p}}\right), \end{aligned} \tag{48}$$

for some mean value ζ laid in the interval

$$\left(-2t_{\mathbf{Z}_p}\sum_{i=1}^{p}\Delta_{\mathbf{G}}W_i - \frac{1}{2}\|\varphi_{\mathbf{g}}\|^2_{\mathcal{H}_{\mathbf{Z}_p}} + \ln\left(1 - \Psi_{\mathbf{Z}_p}\left(\frac{1}{2}\tilde{t}_{\mathbf{Z}_p}, \mathbf{0}\right)\right); \ln(1-\alpha)\right). \tag{49}$$

Conversely, based on the lower bound formula (30), we get

$$\begin{aligned} \Psi_{\mathbf{Z}_p}\left(\tilde{t}_{\mathbf{Z}_p}, \varphi_{\mathbf{g}}\right) - \alpha &\geq \exp\{\ln(1-\alpha)\} \\ &\quad - \exp\left\{2t_{\mathbf{Z}_p}\sum_{i=1}^{p}\Delta_{\mathbf{G}}W_i - \frac{1}{2}\|\varphi_{\mathbf{g}}\|^2_{\mathcal{H}_{\mathbf{Z}_p}} + \ln(1-\alpha)\right\} \\ &= -\left(\exp\left\{2t_{\mathbf{Z}_p}\sum_{i=1}^{p}\Delta_{\mathbf{G}}W_i - \frac{1}{2}\|\varphi_{\mathbf{g}}\|^2_{\mathcal{H}_{\mathbf{Z}_p}} + \ln(1-\alpha)\right\} - \exp\{\ln(1-\alpha)\}\right) \\ &= -e^{\rho}\left(2t_{\mathbf{Z}_p}\sum_{i=1}^{p}\Delta_{\mathbf{G}}W_i - \frac{1}{2}\|\varphi_{\mathbf{g}}\|^2_{\mathbf{Z}_p}\right) \\ &\geq -e^{\rho}\left(2t_{\mathbf{Z}_p}\sum_{i=1}^{p}\Delta_{\mathbf{G}}W_i + \frac{1}{2}\|\varphi_{\mathbf{g}}\|^2_{\mathcal{H}_{\mathbf{Z}_p}}\right) \end{aligned} \tag{50}$$

for some mean value ρ within the interval

$$\left(\ln(1-\alpha);2t_{\mathbf{Z}_p}\sum_{i=1}^{p}\Delta_{\mathbf{G}}W_i-\frac{1}{2}\|\varphi_{\mathbf{g}}\|^2_{\mathscr{H}_{\mathbf{Z}_p^0}}+\ln(1-\alpha)\right). \quad (51)$$

Thus it can be concluded that $\Psi_{\mathbf{Z}_p}(\tilde{t}_{\mathbf{Z}_p},\varphi_{\mathbf{g}})-\alpha$ is laid in the following closed interval:

$$\left[-e^{\rho}\left(2t_{\mathbf{Z}_p}\sum_{i=1}^{p}\Delta_{\mathbf{G}}W_i+\frac{1}{2}\|\varphi_{\mathbf{g}}\|^2_{\mathscr{H}_{\mathbf{Z}_p}}\right);\; e^{\zeta}\left(2t_{\mathbf{Z}_p}\sum_{i=1}^{p}\Delta_{\mathbf{G}}W_i+\frac{1}{2}\|\varphi_{\mathbf{g}}\|^2_{\mathscr{H}_{\mathbf{Z}_p}}\right)\right]. \quad (52)$$

In particular, if the mean values ρ and ζ are taken to be the same, then

$$\left|\Psi_{\mathbf{Z}_p}(\tilde{t}_{\mathbf{Z}_p},\varphi_{\mathbf{g}})-\alpha\right| \le e^{\zeta}\left(2t_{\mathbf{Z}_p}\sum_{i=1}^{p}\Delta_{\mathbf{G}}W_i+\frac{1}{2}\|\varphi_{\mathbf{g}}\|^2_{\mathscr{H}_{\mathbf{Z}_p}}\right), \quad (53)$$

establishing the proof.

Analogously, from (33), we get

$$\begin{aligned}\Psi_{\mathbf{Z}_p^0}(\tilde{t}_{\mathbf{Z}_p^0},\varphi_{\mathbf{g}})-\alpha &\le \exp\{\ln(1-\alpha)\} - \exp\left\{\ln\left(1-\Psi_{\mathbf{Z}_p^0}\left(\frac{1}{2}\tilde{t}_{\mathbf{Z}_p^0},\mathbf{0}\right)\right)-\frac{1}{2}\|\varphi_{\mathbf{g}}\|^2_{\mathscr{H}_{\mathbf{Z}_p^0}}\right\}\\ &= e^{\tau}\left[\ln\left(\frac{1-\alpha}{1-\Psi_{\mathbf{Z}_p^0}((1/2)\tilde{t}_{\mathbf{Z}_p^0},\mathbf{0})}\right)+\frac{1}{2}\|\varphi_{\mathbf{g}}\|^2_{\mathscr{H}_{\mathbf{Z}_p^0}}\right]\\ &\le e^{\tau}\exp\left\{\frac{1}{2}\|\varphi_{\mathbf{g}}\|^2_{\mathscr{H}_{\mathbf{Z}_p^0}}\right\},\end{aligned} \quad (54)$$

for some $\tau \in (\ln(1-\Psi_{\mathbf{Z}_p^0}((1/2)\tilde{t}_{\mathbf{Z}_p^0},\mathbf{0}))-(1/2)\|\varphi_{\mathbf{g}}\|^2_{\mathscr{H}_{\mathbf{Z}_p^0}};\ln(1-\alpha))$. Conversely,

$$\begin{aligned}\Psi_{\mathbf{Z}_p^0}(\tilde{t}_{\mathbf{Z}_p^0},\varphi_{\mathbf{g}})-\alpha &\ge \exp\{\ln(1-\alpha)\} - \exp\left\{\ln(1-\alpha)-\frac{1}{2}\|\varphi_{\mathbf{g}}\|^2_{\mathscr{H}_{\mathbf{Z}_p^0}}\right\}\\ &= e^{\iota}\exp\left\{\frac{1}{2}\|\varphi_{\mathbf{g}}\|^2_{\mathscr{H}_{\mathbf{Z}_p^0}}\right\},\end{aligned} \quad (55)$$

for some $\iota \in (\ln(1-\alpha)-(1/2)\|\varphi_{\mathbf{g}}\|^2_{\mathscr{H}_{\mathbf{Z}_p^0}},\ln(1-\alpha))$. Particularly, for $\iota=\tau$, we get

$$\left|\Psi_{\mathbf{Z}_p^0}(\tilde{t}_{\mathbf{Z}_p^0},\varphi_{\mathbf{g}})-\alpha\right| \le e^{\tau}\exp\left\{\frac{1}{2}\|\varphi_{\mathbf{g}}\|^2_{\mathscr{H}_{\mathbf{Z}_p^0}}\right\}. \quad (56)$$

Thus, we proved the following corollary.

Corollary 6. *Under the condition of Theorem 7, we have*

$$e^{\iota}\exp\left\{\frac{1}{2}\|\varphi_{\mathbf{g}}\|^2_{\mathscr{H}_{\mathbf{Z}_p^0}}\right\} \le \Psi_{\mathbf{Z}_p^0}(\tilde{t}_{\mathbf{Z}_p^0},\varphi_{\mathbf{g}})-\alpha \le e^{\tau}\exp\left\{\frac{1}{2}\|\varphi_{\mathbf{g}}\|^2_{\mathscr{H}_{\mathbf{Z}_p^0}}\right\}, \quad (57)$$

for some τ and ι specified above. If ι and τ are chosen to be the same, then

$$\left|\Psi_{\mathbf{Z}_p^0}(\tilde{t}_{\mathbf{Z}_p^0},\varphi_{\mathbf{g}})-\alpha\right| \le e^{\tau}\exp\left\{\frac{1}{2}\|\varphi_{\mathbf{g}}\|^2_{\mathscr{H}_{\mathbf{Z}_p^0}}\right\}. \quad (58)$$

4. Comparison to Neyman-Pearson Test

Our aim in this section is to establish nonrandomized Neyman-Pearson test for the hypothesis defined in the preceding section. It is well known in the literatures of test theory that Neyman-Pearson test constitutes a most powerful (MP) test for simple hypotheses; see, e.g., Theorem 3.2.1 in [15]. If some criterion is satisfied, the test can be extended to a uniformly most powerful (UMP) test for composite hypotheses. In this section the behavior of the power function including the rate of decay to α will be investigated and compared to those of KS and CvM type tests studied in the preceding section.

Let $\mathbf{V}$ be a linear subspace generated by a set of known and orthonormal regression functions $\{f_1,\ldots,f_m,f_{m+1},\ldots,f_q\}\subset L_2(P_0,\mathbf{G})\cap BV_H(\mathbf{G})$ including $\mathbf{W}=[f_1,\ldots,f_m]$. In this section we consider the hypothesis $H_0:\mathbf{g}\in\mathbf{W}^p$ against $H_1:\mathbf{g}\in\mathbf{V}^p$ instead of $H_0:\mathbf{g}\in\mathbf{W}^p$ against $H_1:\mathbf{g}\notin\mathbf{W}^p$. The former is actually the common frame work of model check for multivariate regression in which one is testing whether or not $\mathbf{g}\in\mathbf{W}^p$ while observing $\mathbf{g}\in\mathbf{V}^p$; see [26] for reference. Suppose there exist $\mathbf{g}_1\in\mathbf{W}^p$ and $\mathbf{g}_2\in\mathbf{V}^p\cap\mathbf{W}^{p\perp}$, such that $\mathbf{g}=\mathbf{g}_1\oplus\mathbf{g}_2$. It is enough to consider the simple hypotheses

$$H_0:\mathbf{g}_2\equiv\mathbf{0}\ \ vs.\ \ H_1:\mathbf{g}_2\equiv\mathbf{f}_0, \quad \text{for some } \mathbf{f}_0\in\mathbf{V}^p\cap\mathbf{W}^{p\perp}. \quad (59)$$

Hence the p-dimensional set-indexed partial sums process of the residuals is given by $\mathscr{Y}=\mathscr{W}_{\mathbf{f},\mathbf{P}_0}$, when H_0 is true; otherwise $\mathscr{Y}=\mathbf{\Sigma}^{-1/2}pr_{\mathbf{W}^{\perp}_{\mathscr{H}_{\mathbf{Z}_p}}}S_{\mathbf{f}_0}+\mathscr{W}_{\mathbf{f},\mathbf{P}_0}$.

The following theorem presents an MP test of size α for testing (59). Here we exhibit again the application of Cameron-Marin density formula of the shifted measure $\mathbf{P}_{\varphi_{\mathbf{f}_0}}$ with respect to $\mathbf{P}$, for $\varphi_{\mathbf{f}_0}=\mathbf{\Sigma}^{-1/2}pr_{\mathbf{W}^{\perp}_{\mathscr{H}_{\mathbf{Z}_p}}}S_{\mathbf{f}_0}$. Recently, [4] investigated the asymptotic optimality of a test for the mean vector in multivariate regression by means of Neyman-Pearson test.

Theorem 7. *Suppose $\varphi_{\mathbf{f}_0}\in\mathscr{H}_{\mathscr{W}_{\mathbf{f},\mathbf{P}_0}}$. Neyman-Pearson test of size α for testing (59) will reject H_0, if and only if*

$$\mathbf{L}(\varphi_{\mathbf{f}_0},\mathscr{Y})\ge\Phi^{-1}(1-\alpha)\|\varphi_{\mathbf{f}_0}\|_{\mathscr{H}_{\mathscr{W}_{\mathbf{f},\mathbf{P}_0}}}. \quad (60)$$

Furthermore, suppose $\Gamma_{\mathcal{W}_{\mathbf{f},\mathbf{P}_0}} : \mathbf{V}^p \longrightarrow (0,1)$ *is the corresponding power function of the test. Then the value of the power, evaluated at any* $\mathbf{f} \in \mathbf{V}^p$, *is given by*

$$\Gamma_{\mathcal{W}_{\mathbf{f},\mathbf{P}_0}}(\mathbf{f}) = \Phi\left(\frac{\mathbf{L}(\varphi_{\mathbf{f}_0}, \varphi_{\mathbf{f}})}{\|\varphi_{\mathbf{f}_0}\|_{\mathcal{H}_{\mathcal{W}_{\mathbf{f},\mathbf{P}_0}}}} - \Phi^{-1}(1-\alpha)\right), \qquad \mathbf{f} \in \mathbf{V}^p \cap \mathbf{W}^{p\perp}, \tag{61}$$

and otherwise, $\Gamma_{\mathcal{W}_{\mathbf{f},\mathbf{P}_0}}(\mathbf{f}) = \alpha$.

Proof. Let $\chi_0(\mathcal{Y})$ and $\chi_1(\mathcal{Y})$ be the density of $\mathbf{P}_{\varphi_{\mathbf{f}_0}}$ with respect to $\mathbf{P}$ under H_0 and H_1, respectively. By Theorem 3.2.1 in [15], an MP test of size α for testing (59) will reject H_0, if and only if $\chi_0(\mathcal{Y})/\chi_1(\mathcal{Y}) \le k$, for a constant k such that

$$\mathbb{P}\left\{\omega \in \Omega : \frac{\chi_0(\mathcal{Y}(\omega))}{\chi_1(\mathcal{Y}(\omega))} \le k \,|H_0\right\} = \alpha. \tag{62}$$

Since $\chi_0(\mathcal{Y}) = 1$ and $\chi_1(\mathcal{Y}) = \exp\{\mathbf{L}(\varphi_{\mathbf{f}_0}, \mathcal{Y}) - (1/2)\|\varphi_{\mathbf{f}_0}\|^2_{\mathcal{H}_{\mathcal{W}_{\mathbf{f},\mathbf{P}_0}}}\}$, then by recalling the fact $\mathbf{L}(\varphi_{\mathbf{f}_0}, \mathcal{W}_{\mathbf{f},\mathbf{P}_0}) \sim N(0, \|\varphi_{\mathbf{f}_0}\|^2_{\mathcal{H}_{\mathcal{W}_{\mathbf{f},\mathbf{P}_0}}})$, we get

$$\begin{aligned}
&\mathbb{P}\left\{\omega \in \Omega : \frac{\chi_0(\mathcal{Y}(\omega))}{\chi_1(\mathcal{Y}(\omega))} \le k\,|H_0\right\} = \alpha \Longleftrightarrow \\
&\mathbf{P}\left\{\exp\left\{-\mathbf{L}(\varphi_{\mathbf{f}_0}, \mathcal{Y}) + \frac{1}{2}\|\varphi_{\mathbf{f}_0}\|^2_{\mathcal{H}_{\mathcal{W}_{\mathbf{f},\mathbf{P}_0}}}\right\} \le k\,|H_0\right\} = \alpha \Longleftrightarrow \\
&\mathbf{P}\left\{\mathbf{L}(\varphi_{\mathbf{f}_0}, \mathcal{Y}) \ge -\ln(k) + \frac{1}{2}\|\varphi_{\mathbf{f}_0}\|^2_{\mathcal{H}_{\mathcal{W}_{\mathbf{f},\mathbf{P}_0}}}\,|H_0\right\} = \alpha \Longleftrightarrow \\
&\mathbf{P}\left\{\mathbf{L}(\varphi_{\mathbf{f}_0}, \mathcal{W}_{\mathbf{f},\mathbf{P}_0}) \ge -\ln(k) + \frac{1}{2}\|\varphi_{\mathbf{f}_0}\|^2_{\mathcal{H}_{\mathcal{W}_{\mathbf{f},\mathbf{P}_0}}}\right\} = \alpha \Longleftrightarrow \\
&\mathbf{P}\left\{N(0,1) \ge \frac{-\ln(k) + (1/2)\|\varphi_{\mathbf{f}_0}\|^2_{\mathcal{H}_{\mathcal{W}_{\mathbf{f},\mathbf{P}_0}}}}{\|\varphi_{\mathbf{f}_0}\|_{\mathcal{H}_{\mathcal{W}_{\mathbf{f},\mathbf{P}_0}}}}\right\} = \alpha \Longleftrightarrow \\
&-\ln(k) + \frac{1}{2}\|\varphi_{\mathbf{f}_0}\|^2_{\mathcal{H}_{\mathcal{W}_{\mathbf{f},\mathbf{P}_0}}} = \Phi^{-1}(1-\alpha)\|\varphi_{\mathbf{f}_0}\|_{\mathcal{H}_{\mathcal{W}_{\mathbf{f},\mathbf{P}_0}}},
\end{aligned} \tag{63}$$

establishing the rejection region of the test. Next, we compute the power function for any $\mathbf{f} \in \mathbf{V}^p \cap \mathbf{W}^{p\perp}$. By the definition of $\Gamma_{\mathcal{W}_{\mathbf{f},\mathbf{P}_0}}$ and by the symmetry of Φ, we have

$$\begin{aligned}
\Gamma_{\mathcal{W}_{\mathbf{f},\mathbf{P}_0}}(\mathbf{f}) &= \mathbf{P}\left\{\mathbf{L}(\varphi_{\mathbf{f}_0}, \mathcal{Y}) \ge \Phi^{-1}(1-\alpha)\|\varphi_{\mathbf{f}_0}\|_{\mathcal{H}_{\mathcal{W}_{\mathbf{f},\mathbf{P}_0}}}\,|\mathbf{g}_2 \equiv \mathbf{f}\right\} \\
&= \mathbf{P}\left\{\mathbf{L}(\varphi_{\mathbf{f}_0}, \varphi_{\mathbf{f}} + \mathcal{W}_{\mathbf{f},\mathbf{P}_0}) \ge \Phi^{-1}(1-\alpha)\|\varphi_{\mathbf{f}_0}\|_{\mathcal{H}_{\mathcal{W}_{\mathbf{f},\mathbf{P}_0}}}\right\} = \mathbf{P}\left\{\mathbf{L}(\varphi_{\mathbf{f}_0}, \mathcal{W}_{\mathbf{f},\mathbf{P}_0}) \ge \Phi^{-1}(1-\alpha)\|\varphi_{\mathbf{f}_0}\|_{\mathcal{H}_{\mathcal{W}_{\mathbf{f},\mathbf{P}_0}}} - \mathbf{L}(\varphi_{\mathbf{f}_0}, \varphi_{\mathbf{f}})\right\} = 1 \\
&\quad - \Phi\left(\Phi^{-1}(1-\alpha) - \frac{\mathbf{L}(\varphi_{\mathbf{f}_0}, \varphi_{\mathbf{f}})}{\|\varphi_{\mathbf{f}_0}\|_{\mathcal{H}_{\mathcal{W}_{\mathbf{f},\mathbf{P}_0}}}}\right) \\
&= \Phi\left(\frac{\mathbf{L}(\varphi_{\mathbf{f}_0}, \varphi_{\mathbf{f}})}{\|\varphi_{\mathbf{f}_0}\|_{\mathcal{H}_{\mathcal{W}_{\mathbf{f},\mathbf{P}_0}}}} - \Phi^{-1}(1-\alpha)\right).
\end{aligned} \tag{64}$$

The last formula results in $\Gamma_{\mathcal{W}_{\mathbf{f},\mathbf{P}_0}}(\mathbf{f}) = \alpha$, when $\mathbf{f}$ vanishes. The proof of the theorem is complete.

The test presented in Theorem 7 depends on the choice of $\mathbf{f}$ specified under H_1. For example, if we consider $H_1 : \mathbf{g}_2 \equiv \mathbf{f}_1$, for some $\mathbf{f}_1 \in \mathbf{V}^p \cap \mathbf{W}^{p\perp}$, then H_0 is rejected at level α, if

$$\mathbf{L}(\varphi_{\mathbf{f}_1}, \mathcal{Y}) \ge \Phi^{-1}(1-\alpha)\|\varphi_{\mathbf{f}_1}\|_{\mathcal{H}_{\mathcal{W}_{\mathbf{f},\mathbf{P}_0}}}. \tag{65}$$

This means that the test cannot be extended as a uniformly most powerful (UMP) test for the composite alternative $H_1 : \mathbf{g}_2 \in \mathbf{V}^p \cap \mathbf{W}^{p\perp}$. It is also not a UMP test for more specific one-sided alternatives $H_1 : \mathbf{g}_2 > \mathbf{0}$ or $H_1 : \mathbf{g}_2 < \mathbf{0}$.

As discussed in the preceding section, we are also interested in investigating the rate of decay of $\Gamma_{\mathcal{W}_{\mathbf{f},\mathbf{P}_0}}(\mathbf{f})$ to $\alpha = \Gamma_{\mathcal{W}_{\mathbf{f},\mathbf{P}_0}}(\mathbf{0})$. Toward this topic the result of Theorem 7 leads us to the following important corollary. The proof is left since it can be handled by using the similar technique as in the proof of Corollary 3.

Corollary 8. *Let* $\mathbf{f}$ *be an element of* $\mathbf{V}^p \cap \mathbf{W}^{p\perp}$, *such that* $\varphi_{\mathbf{f}} \in \mathcal{H}_{\mathcal{W}_{\mathbf{f},\mathbf{P}_0}}$ *and* $\mathbf{L}(\varphi_{\mathbf{f}_0}, \varphi_{\mathbf{f}}) > 0$. *Then for every presigned* $\alpha \in (0,1)$, *it holds that*

$$\left|\Gamma_{\mathcal{W}_{\mathbf{f},\mathbf{P}_0}}(\mathbf{f}) - \alpha\right| \le \frac{1}{\sqrt{2\pi}} \frac{\mathbf{L}(\varphi_{\mathbf{f}_0}, \varphi_{\mathbf{f}})}{\|\varphi_{\mathbf{f}_0}\|_{\mathcal{H}_{\mathcal{W}_{\mathbf{f},\mathbf{P}_0}}}}. \tag{66}$$

In the case $\mathbf{L}(\varphi_{\mathbf{f}_0}, \varphi_{\mathbf{f}}) < 0$, *we have*

$$\left|\Gamma_{\mathcal{W}_{\mathbf{f},\mathbf{P}_0}}(\mathbf{f}) - \alpha\right| \le \frac{1}{\sqrt{2\pi}} \frac{\mathbf{L}^-(\varphi_{\mathbf{f}_0}, \varphi_{\mathbf{f}})}{\|\varphi_{\mathbf{f}_0}\|_{\mathcal{H}_{\mathcal{W}_{\mathbf{f},\mathbf{P}_0}}}}, \tag{67}$$

where $\mathbf{L}^-(\varphi_{\mathbf{f}_0}, \varphi_{\mathbf{f}}) := -\mathbf{L}(\varphi_{\mathbf{f}_0}, \varphi_{\mathbf{f}}) > 0$.

Corollary 8 states that how fast the power function $\Gamma_{\mathcal{W}_{\mathbf{f},\mathbf{P}_0}}(\mathbf{f})$ decays to α depends on some value determined by $\mathbf{L}(\varphi_{\mathbf{f}_0}, \varphi_{\mathbf{f}})$ whose structure is influenced by the type of $\mathcal{W}_{\mathbf{f},\mathbf{P}_0}$. For comparison study suppose that the simple hypothesis (59) is tested using the KS or CvM type test. Then by virtue of Corollary 2, we have

$$\left|\Psi_{\mathscr{W}_{\mathbf{f},P_0}}\left(\tilde{t}_{\mathscr{W}_{\mathbf{f},P_0}},\varphi_{\mathbf{f}}\right)-\alpha\right|\le\frac{1}{\sqrt{2\pi}}\frac{\mathbf{L}\left(\varphi_{\mathbf{f}},\varphi_{\mathbf{f}}\right)}{\left\|\varphi_{\mathbf{f}}\right\|_{\mathscr{H}_{\mathscr{W}_{\mathbf{f},P_0}}}}$$
$$\left|\Upsilon_{\mathscr{W}_{\mathbf{f},P_0}}\left(\tilde{t}_{\mathscr{W}_{\mathbf{f},P_0}},\varphi_{\mathbf{f}}\right)-\alpha\right|\le\frac{1}{\sqrt{2\pi}}\frac{\mathbf{L}\left(\varphi_{\mathbf{f}},\varphi_{\mathbf{f}}\right)}{\left\|\varphi_{\mathbf{f}}\right\|_{\mathscr{H}_{\mathscr{W}_{\mathbf{f},P_0}}}}. \tag{68}$$

Thus, in contrast to Corollary 8, the rate of decay of the KS and CvM type tests does not depend on $\mathbf{f}_0$ at all. Consequently, compared to Neyman-Pearson test, KS and CvM test cannot detect whether to take $\mathbf{f}$ larger or less than $\mathbf{f}_0$ in order to have faster or slower decay.

The result presented throughout this section will become more visible when we look at the model involving p-dimensional set-indexed Brownian sheet or pillow. For example suppose we observe the model $\mathscr{Y}=\mathbf{\Sigma}^{-1/2}pr_{\mathscr{W}^{\perp}_{\mathscr{H}_{\mathbf{Z}_p}}}S_{\mathbf{g}_2}+\mathbf{Z}^0_p$, for testing (59). Then H_0 is rejected at level α, if

$$\mathbf{L}\left(\varphi_{\mathbf{f}_0},\mathscr{Y}\right)\ge\Phi^{-1}(1-\alpha)\left\|\mathbf{\Sigma}^{-1/2}pr_{\mathbf{W}^{p\perp}}\mathbf{f}_0\right\|_{L_2^p(P_0,\mathbf{G})}, \tag{69}$$

where for the p-dimensional set-indexed Brownian pillow; we have

$$\mathbf{L}\left(\varphi_{\mathbf{f}_0},\mathscr{Y}\right)=\left\|\mathbf{\Sigma}^{-1/2}pr_{\mathbf{W}^{p\perp}}\mathbf{f}_0\right\|^2_{L_2^p(P_0,\mathbf{G})}+\int_{\mathbf{G}}\left(\mathbf{\Sigma}^{-1/2}pr_{\mathbf{W}^{p\perp}}\mathbf{f}_0\right)^{\top}d\mathbf{Z}^0_p, \tag{70}$$

Furthermore, we get, for fixed $\mathbf{f}_0\in\mathbf{V}^p\cap\mathbf{W}^{p\perp}$,

$$\begin{aligned}\frac{\mathbf{L}\left(\varphi_{\mathbf{f}_0},\varphi_{\mathbf{f}}\right)}{\left\|\varphi_{\mathbf{f}_0}\right\|_{\mathscr{H}_{\mathbf{Z}^0_p}}}&=\frac{\left\langle\varphi_{\mathbf{f}_0},\varphi_{\mathbf{f}}\right\rangle_{\mathscr{H}_{\mathbf{Z}^0_p}}}{\left\|\varphi_{\mathbf{f}_0}\right\|_{\mathscr{H}_{\mathbf{Z}^0_p}}}\\&=\frac{\left\langle\mathbf{\Sigma}^{-1/2}pr_{\mathscr{W}^{\perp}_{\mathscr{H}_{\mathbf{Z}_p}}}S_{\mathbf{f}_0},\mathbf{\Sigma}^{-1/2}pr_{\mathscr{W}^{\perp}_{\mathscr{H}_{\mathbf{Z}_p}}}S_{\mathbf{f}}\right\rangle_{\mathscr{H}_{\mathbf{Z}^0_p}}}{\left\|\varphi_{\mathbf{f}_0}\right\|_{\mathscr{H}_{\mathbf{Z}^0_p}}}\\&=\frac{\left\langle pr_{\mathbf{W}^{p\perp}}\mathbf{\Sigma}^{-1/2}\mathbf{f}_0,pr_{\mathbf{W}^{p\perp}}\mathbf{\Sigma}^{-1/2}\mathbf{f}\right\rangle_{L_2^p(P_0,\mathbf{G})}}{\left\|pr_{\mathbf{W}^{p\perp}}\mathbf{\Sigma}^{-1/2}\mathbf{f}_0\right\|_{L_2(P_0,\mathbf{G})}}\\&\le\frac{\left\|pr_{\mathbf{W}^{p\perp}}\mathbf{\Sigma}^{-1/2}\mathbf{f}_0\right\|_{L_2^p(P_0,\mathbf{G})}\left\|pr_{\mathbf{W}^{p\perp}}\mathbf{\Sigma}^{-1/2}\mathbf{f}\right\|_{L_2^p(P_0,\mathbf{G})}}{\left\|pr_{\mathbf{W}^{p\perp}}\mathbf{\Sigma}^{-1/2}\mathbf{f}_0\right\|_{L_2^p(P_0,\mathbf{G})}}\\&\le\left\|\mathbf{\Sigma}^{-1/2}\right\|_{\mathbb{R}^p}\left\|pr_{\mathbf{W}^{p\perp}}\right\|_{\mathbb{R}^p}\left\|\mathbf{f}\right\|_{L_2^p(P_0,\mathbf{G})},\end{aligned} \tag{71}$$

where the first inequality appears by Holder's inequality, whereas the second follows by the fact that $\mathbf{\Sigma}^{-1/2}$ and $pr_{\mathbf{W}^{p\perp}}$ represent continuous linear transformations on $L_2^p(P_0,\mathbf{G})$; therefore they are uniformly bounded, cf. [27], p. 26-27. Since $\|\cdot\|_{L_2^p(P_0,\mathbf{G})}$ is continuous on the closed subset $\mathbf{V}^p\cap\mathbf{W}^{p\perp}$, then $\|\cdot\|_{L_2^p(P_0,\mathbf{G})}$ is bounded on $\mathbf{V}^p\cap\mathbf{W}^{p\perp}$. Thus there exists $M>0$, such that M is the uniform upper bound for $|\Gamma_{\mathbf{Z}^0_p}(\mathbf{f})-\alpha|$. It is clear that M is also the uniform upper bounds of $|\Psi_{\mathbf{Z}^0_p}(\tilde{t}_{\mathbf{Z}^0_p},\varphi_{\mathbf{f}})-\alpha|$ as well as $|\Upsilon_{\mathbf{Z}^0_p}(\tilde{t}_{\mathbf{Z}^0_p},\varphi_{\mathbf{f}})-\alpha|$.

5. Simulation Study

In this section we investigate the behavior of $\Psi_{\mathbf{Z}^0_p}(\tilde{t}_{\mathbf{Z}^0_p},\varphi_{\mathbf{f}})$ and $\Upsilon_{\mathbf{Z}^0_p}(\tilde{q}_{\mathbf{Z}^0_p},\varphi_{\mathbf{f}})$ with respect to their lower and upper bounds derived in Corollary 3. The simulated model is represented by the trend plus noise process

$$\varphi_{\mathbf{g}}(A)+\mathbf{Z}^0_2(A),\quad A\in\mathscr{B}([0,1]\times[0,1]) \tag{72}$$

where $\mathbf{Z}^0_2$ is the two-dimensional Brownian pillow and

$$\varphi_{\mathbf{g}}(A)=\mathbf{\Sigma}^{-1/2}\int_A\rho(t+s,0)^{\top}\lambda^2(dt,ds),\quad A\in\mathscr{B}([0,1]\times[0,1]),\ \rho\in\mathbb{R}. \tag{73}$$

Such a model appears as the limit process of the two-dimensional set-indexed partial sums processes of the residuals of two variate regression model

$$(Y_1,Y_2)^{\top}=(g_1,g_2)^{\top}+(\varepsilon_1,\varepsilon_2)^{\top} \tag{74}$$

for testing whether or not a constant model holds true. That is, we test the hypothesis that

$$H_0:(g_1,g_2)^{\top}\in\mathbf{W}^2\ vs.\ H_1:(g_1,g_2)^{\top}\notin\mathbf{W}^2, \tag{75}$$

where $\mathbf{W}=[f_1]$, with $f_1(t,s)=1$, for $(t,s)\in\mathbf{G}:=[0,1]\times[0,1]$. For fixed $n\ge1$, the $n\times n\times2$ arrays of observation are generated from the model

$$\begin{pmatrix}Y_{\ell k1}\\Y_{\ell k2}\end{pmatrix}=\begin{pmatrix}2+\dfrac{\rho\ell}{n}+\dfrac{\rho k}{n}\\3+\dfrac{0\ell}{n}+\dfrac{0k}{n}\end{pmatrix}+\begin{pmatrix}\varepsilon_{\ell k1}\\\varepsilon_{\ell k2}\end{pmatrix},\quad 1\le\ell,\ k\le n, \tag{76}$$

according to an experimental design given by a regular lattice with $n\times n$ points on $\mathbf{G}$. Let $f_2(t,s)=t$ and $f_3(t,s)=s$, for $(t,s)\in\mathbf{G}$; then we equivalently have

$$\begin{pmatrix}Y_{\ell k1}\\Y_{\ell k2}\end{pmatrix}=\begin{pmatrix}2f_1\left(\frac{\ell}{n},\frac{k}{n}\right)+\rho f_2\left(\frac{\ell}{n},\frac{k}{n}\right)+\rho f_3\left(\frac{\ell}{n},\frac{k}{n}\right)\\3f_1\left(\frac{\ell}{n},\frac{k}{n}\right)+0f_2\left(\frac{\ell}{n},\frac{k}{n}\right)+0f_3\left(\frac{\ell}{n},\frac{k}{n}\right)\end{pmatrix}+\begin{pmatrix}\varepsilon_{\ell k1}\\\varepsilon_{\ell k2}\end{pmatrix}, \tag{77}$$

for $1\le\ell,k\le n$. Hence, if $\rho=0$, then the observations are from the model assumed under H_0. Otherwise, they support

TABLE 1: The numerical upper and lower bounds for $\Psi_{\mathbf{Z}_p^0}(\tilde{t}_{\mathbf{Z}_p^0}, \varphi_{\mathbf{g}})$ and $\Upsilon_{\mathbf{Z}_p^0}(\tilde{q}_{\mathbf{Z}_p^0}, \varphi_{\mathbf{g}})$. The size of the test is $\alpha = 0.05$.

ρ	$\mathscr{L}$	$\Psi_{\mathbf{Z}_p^0}\left(\tilde{t}_{\mathbf{Z}_p^0}, \varphi_{\mathbf{g}}\right)$	$\Upsilon_{\mathbf{Z}_p^0}\left(\tilde{t}_{\mathbf{Z}_p^0}, \varphi_{\mathbf{g}}\right)$	$\mathscr{U}$
0	0.04990	0.06100	0.05500	0.04990
1	0.01886	0.04900	0.05100	0.11280
2	0.00602	0.05600	0.06500	0.21809
3	0.00162	0.05700	0.07100	0.36483
4	0.00037	0.07400	0.07600	0.53486
5	6.9e-05	0.07500	0.09700	0.69867
6	1.1e-05	0.10600	0.13000	0.82988
7	1.5e-06	0.12600	0.14600	0.91724
8	1.6e-07	0.15700	0.21000	0.96561
9	1.5e-08	0.14700	0.19600	0.98787
10	1.2e-09	0.20300	0.27200	0.99639
11	7.3e-11	0.21000	0.28300	0.99909
12	3.9e-12	0.27500	0.35000	0.99981
13	1.7e-13	0.29600	0.39900	0.99997
14	6.4e-15	0.32800	0.45500	0.99999

H_1. In this simulation, the random vector $(\varepsilon_{\ell k1}, \varepsilon_{\ell k2})^\top$ is generated independently from the two-dimensional centered normal distribution with the covariance matrix given by

$$\mathbf{\Sigma} = \begin{pmatrix} 6.26 & -0.50 \\ -0.50 & 6.26 \end{pmatrix} \tag{78}$$

so that we have after some computations

$$\mathbf{\Sigma}^{-1} = \begin{pmatrix} 0.16077 & 0.01284 \\ 0.01284 & 0.16077 \end{pmatrix}, \quad \mathbf{\Sigma}^{-1/2} = \begin{pmatrix} 0.40064 & 0.01603 \\ 0.01603 & 0.40064 \end{pmatrix}. \tag{79}$$

Now, the norm of $\varphi_{\mathbf{g}}$ for the matrix $\mathbf{\Sigma}$ can be computed concretely as

$$\begin{aligned} \left\|\varphi_{\mathbf{g}}\right\|_{\mathscr{H}_{\mathbf{Z}_2^0}} &= \sqrt{\int_{[0,1]\times[0,1]} \rho^2 (t+s,0)\, \mathbf{\Sigma}^{-1} (t+s,0)\, dtds} \\ &= \rho 0.16077 \sqrt{\int_{[0,1]\times[0,1]} (t^2 + 2ts + s^2)\, dtds} \\ &= 0.43309\rho. \end{aligned} \tag{80}$$

The simulation results using a sample of size 50×50 with 1000 runs are exhibited in Table 1 and Figure 1 for $\alpha = 0.05$. The figures presented in Table 1 represent the values of the power functions of the KS and CvM tests together with the associated values of the lower ($\mathscr{L}$) and upper ($\mathscr{U}$) bounds evaluated at each given value of ρ utilizing the formulas given in Corollary 3, where in this case

$$\begin{aligned} \mathscr{L} &= \Phi\left(\Phi^{-1}(\alpha) - 0.43309\rho\right), \\ \mathscr{U} &= \Phi\left(\Phi^{-1}(\alpha) + 0.43309\rho\right). \end{aligned} \tag{81}$$

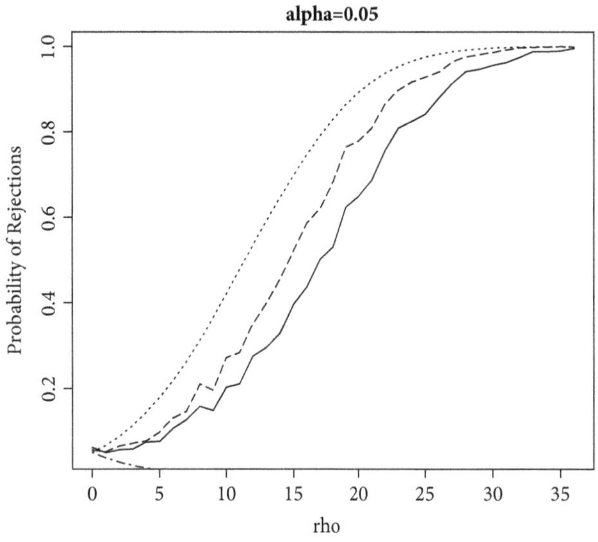

FIGURE 1: Upper (dotted line) and lower (dashed line) bounds for $\Psi_{\mathbf{Z}_p^0}(\tilde{t}_{\mathbf{Z}_p^0}, \varphi_{\mathrm{f}})$ (solid line), with $\alpha = 0.05$.

It is shown that the values of $\mathscr{L}$ are never exceeding the corresponding powers. Likewise, the values of $\mathscr{U}$ are also never preceding those of the corresponding power functions as suggested by the theory. Figure 1 presents the graphs of $\mathscr{L}$ (dotdash line), $\mathscr{U}$ (dotted line), $\Psi_{\mathbf{Z}_p^0}(\tilde{t}_{\mathbf{Z}_p^0}, \varphi_{\mathbf{g}})$ (smoothed line), and $\Upsilon_{\mathbf{Z}_p^0}(\tilde{q}_{\mathbf{Z}_p^0}, \varphi_{\mathbf{g}})$ (dashed line) scattered together in one panel. It can be seen that the curves of the power functions are laid within a band formed by the paired curve of $\mathscr{L}$ and $\mathscr{U}$ as they should be.

6. Concluding Remark

We have established the upper and lower bounds for the boundary crossing probability involving multivariate trend plus noise model. Our results give important contributions not only in the area of statistics, but also in other disciplines such as in finance mathematics and in statistical physics, where such probability model is also frequently encountered. It is important to note that the Cameron-Martin translation formula is valid if the trend function laid in the RKHS of the corresponding Gaussian process. In practice this is not always the case. Therefore further research must be conducted to be able to handle the problem appears in such situation.

Appendix

Proof of Theorem 1 . Let $\Xi := \{\mathbf{x} \in \mathscr{C}^p(\mathscr{B}(\mathbf{G})) : \mathbf{L}(\mathbf{h},\mathbf{x}) < r(\mathbf{E})\|\mathbf{h}\|_{\mathscr{H}_{\mathscr{W}_{\mathbf{f},P_0}}}\}$, for a fixed $\mathbf{h} \in \mathscr{H}_{\mathscr{W}_{\mathbf{f},P_0}}$. Then by recalling $\mathbf{L}(\mathbf{h},\mathscr{W}_{\mathbf{f},P_0}) \sim N(0,\|\mathbf{h}\|^2_{\mathscr{H}_{\mathscr{W}_{\mathbf{f},P_0}}})$, we have

$$\begin{aligned}
&\mathbf{P}(\Xi)\\
&= \mathbf{P}\left\{\mathbf{x} \in \mathscr{C}^p(\mathscr{B}(\mathbf{G})) : \mathbf{L}(\mathbf{h},\mathbf{x}) < r(\mathbf{E})\|\mathbf{h}\|_{\mathscr{H}_{\mathscr{W}_{\mathbf{f},P_0}}}\right\}\\
&= \mathbb{P}\left\{\omega \in \Omega : \mathbf{L}\left(\mathbf{h},\mathscr{W}_{\mathbf{f},P_0}(\omega)\right) < r(\mathbf{E})\|\mathbf{h}\|_{\mathscr{H}_{\mathscr{W}_{\mathbf{f},P_0}}}\right\}\\
&= \mathbb{P}\{\omega : N(0,1)(\omega) < r(\mathbf{E})\} = \Phi(r(\mathbf{E}))\\
&= \mathbf{P}(\mathbf{E}).
\end{aligned} \tag{A.1}$$

The last equality implicates

$$\begin{aligned}
\mathbf{P}(\mathbf{E}\setminus\Xi) &= \mathbf{P}(\mathbf{E}) - \mathbf{P}(\mathbf{E}\cap\Xi) = \mathbf{P}(\Xi) - \mathbf{P}(\mathbf{E}\cap\Xi)\\
&= \mathbf{P}(\Xi\setminus\mathbf{E})
\end{aligned} \tag{A.2}$$

We will show that $\mathbf{P}(\mathbf{E}-\mathbf{h}) \geq \mathbf{P}(\Xi-\mathbf{h})$. For this purpose we use the Cameron-Martin formula (12), the fact that $\mathbf{L}(\mathbf{h},\mathbf{x}) \geq r(\mathbf{E})\|\mathbf{h}\|_{\mathscr{H}_{\mathscr{W}_{\mathbf{f},P_0}}}$, whenever $\mathbf{x} \in \mathbf{E}\setminus\Xi$, and (A.2). Hence, we get

$$\begin{aligned}
\mathbf{P}(\mathbf{E}-\mathbf{h}) &= \mathbf{P}_{\mathbf{h}}(\mathbf{E}) = \mathbf{P}_{\mathbf{h}}(\mathbf{E}\setminus\Xi) + \mathbf{P}_{\mathbf{h}}(\mathbf{E}\cap\Xi)\\
&= \int_{\mathbf{E}\setminus\Xi} \exp\left\{\mathbf{L}(\mathbf{h},\mathbf{x}) - \frac{1}{2}\|\mathbf{h}\|^2_{\mathscr{H}_{\mathscr{W}_{\mathbf{f},P_0}}}\right\}\mathbf{P}(d\mathbf{x})\\
&\quad + \int_{\mathbf{E}\cap\Xi} \exp\left\{\mathbf{L}(\mathbf{h},\mathbf{x}) - \frac{1}{2}\|\mathbf{h}\|^2_{\mathscr{H}_{\mathscr{W}_{\mathbf{f},P_0}}}\right\}\mathbf{P}(d\mathbf{x})\\
&\geq \int_{\mathbf{E}\setminus\Xi} \exp\left\{r(\mathbf{E})\|\mathbf{h}\|_{\mathscr{H}_{\mathscr{W}_{\mathbf{f},P_0}}} - \frac{1}{2}\|\mathbf{h}\|^2_{\mathscr{H}_{\mathscr{W}_{\mathbf{f},P_0}}}\right\}\mathbf{P}(d\mathbf{x})\\
&\quad + \int_{\mathbf{E}\cap\Xi} \exp\left\{\mathbf{L}(\mathbf{h},\mathbf{x}) - \frac{1}{2}\|\mathbf{h}\|^2_{\mathscr{H}_{\mathscr{W}_{\mathbf{f},P_0}}}\right\}\mathbf{P}(d\mathbf{x})\\
&= \int_{\Xi\setminus\mathbf{E}} \exp\left\{r(\mathbf{E})\|\mathbf{h}\|_{\mathscr{H}_{\mathscr{W}_{\mathbf{f},P_0}}} - \frac{1}{2}\|\mathbf{h}\|^2_{\mathscr{H}_{\mathscr{W}_{\mathbf{f},P_0}}}\right\}\mathbf{P}(d\mathbf{x})\\
&\quad + \int_{\mathbf{E}\cap\Xi} \exp\left\{\mathbf{L}(\mathbf{h},\mathbf{x}) - \frac{1}{2}\|\mathbf{h}\|^2_{\mathscr{H}_{\mathscr{W}_{\mathbf{f},P_0}}}\right\}\mathbf{P}(d\mathbf{x}).
\end{aligned} \tag{A.3}$$

Next by the definition of Ξ, the term on the right-hand side of the last inequality is greater than the following one:

$$\begin{aligned}
&\int_{\Xi\setminus\mathbf{E}} \exp\left\{\mathbf{L}(\mathbf{h},\mathbf{x}) - \frac{1}{2}\|\mathbf{h}\|^2_{\mathscr{H}_{\mathscr{W}_{\mathbf{f},P_0}}}\right\}\mathbf{P}(d\mathbf{x})\\
&\quad + \int_{\mathbf{E}\cap\Xi} \exp\left\{\mathbf{L}(\mathbf{h},\mathbf{x}) - \frac{1}{2}\|\mathbf{h}\|^2_{\mathscr{H}_{\mathscr{W}_{\mathbf{f},P_0}}}\right\}\mathbf{P}(d\mathbf{x})
\end{aligned} \tag{A.4}$$

which is exactly the same with $\mathbf{P}(\Xi-\mathbf{h})$, establishing $\mathbf{P}(\mathbf{E}-\mathbf{h}) \geq \mathbf{P}(\Xi-\mathbf{h})$, where by the definition

$$\begin{aligned}
\mathbf{P}(\Xi-\mathbf{h}) &= \mathbf{P}\{\mathbf{x}-\mathbf{h} : \mathbf{x} \in \Xi\} = \mathbf{P}\Big\{\mathbf{x}\\
&\in \mathscr{C}^p(\mathscr{B}(\mathbf{G})) : \mathbf{L}(\mathbf{h},\mathbf{x}) < r(\mathbf{E})\|\mathbf{h}\|_{\mathscr{H}_{\mathscr{W}_{\mathbf{f},P_0}}}\\
&- \mathbf{L}(\mathbf{h},\mathbf{h})\Big\} = \mathbb{P}\left\{\omega \in \Omega : \mathbf{L}\left(\mathbf{h},\mathscr{W}_{\mathbf{f},P_0}(\omega)\right)\right.\\
&\left.< r(\mathbf{E})\|\mathbf{h}\|_{\mathscr{H}_{\mathscr{W}_{\mathbf{f},P_0}}} - \mathbf{L}(\mathbf{h},\mathbf{h})\right\} = \mathbb{P}\Bigg\{\omega\\
&\in \Omega : N(0,1)(\omega) < r(\mathbf{E}) - \frac{\mathbf{L}(\mathbf{h},\mathbf{h})}{\|\mathbf{h}\|_{\mathscr{H}_{\mathscr{W}_{\mathbf{f},P_0}}}}\Bigg\}\\
&= \Phi\left(r(\mathbf{E}) - \frac{\mathbf{L}(\mathbf{h},\mathbf{h})}{\|\mathbf{h}\|_{\mathscr{H}_{\mathscr{W}_{\mathbf{f},P_0}}}}\right).
\end{aligned} \tag{A.5}$$

We notice that the lower bound

$$\mathbf{P}(\mathbf{E}-\mathbf{h}) \geq \Phi\left(r(\mathbf{E}) - \frac{\mathbf{L}(\mathbf{h},\mathbf{h})}{\|\mathbf{h}\|_{\mathscr{H}_{\mathscr{W}_{\mathbf{f},P_0}}}}\right) \tag{A.6}$$

holds for any $\mathbf{E} \subset \mathscr{C}^p(\mathscr{B}(\mathbf{G}))$ and any constant $r(\mathbf{E}) = \Phi^{-1}(\mathbf{P}(\mathbf{E}))$. Hence it holds as well for the complement $\mathbf{E}^C$. That is,

$$\mathbf{P}\left(\mathbf{E}^C-\mathbf{h}\right) \geq \Phi\left(r\left(\mathbf{E}^C\right) - \frac{\mathbf{L}(\mathbf{h},\mathbf{h})}{\|\mathbf{h}\|_{\mathscr{H}_{\mathscr{W}_{\mathbf{f},P_0}}}}\right), \tag{A.7}$$

with $r(\mathbf{E}^C) = \Phi^{-1}(\mathbf{P}(\mathbf{E}^C))$. Since $\mathbf{P}(\mathbf{E}^C-\mathbf{h}) = 1-\mathbf{P}(\mathbf{E}-\mathbf{h})$ and $1-\Phi(t) = \Phi(-t)$, for any $t \in \mathbb{R}$ (by the symmetry of Φ), then we get the following:

$$\begin{aligned}
&\mathbf{P}\left(\mathbf{E}^C-\mathbf{h}\right) \geq \Phi\left(r\left(\mathbf{E}^C\right) - \frac{\mathbf{L}(\mathbf{h},\mathbf{h})}{\|\mathbf{h}\|_{\mathscr{H}_{\mathscr{W}_{\mathbf{f},P_0}}}}\right) \Longleftrightarrow\\
&\mathbf{P}(\mathbf{E}-\mathbf{h}) \leq \Phi\left(-r\left(\mathbf{E}^C\right) + \frac{\mathbf{L}(\mathbf{h},\mathbf{h})}{\|\mathbf{h}\|_{\mathscr{H}_{\mathscr{W}_{\mathbf{f},P_0}}}}\right).
\end{aligned} \tag{A.8}$$

On the other hand by the equality $r(\mathbf{E}^C) = \Phi^{-1}(\mathbf{P}(\mathbf{E}^C))$ and by the symmetry of Φ, it holds that

$$\begin{aligned} \Phi\left(r\left(\mathbf{E}^C\right)\right) &= 1 - \mathbf{P}(\mathbf{E}) \Longleftrightarrow \\ 1 - \Phi\left(r\left(\mathbf{E}^C\right)\right) &= \mathbf{P}(\mathbf{E}) \Longleftrightarrow \\ \Phi\left(-r\left(\mathbf{E}^C\right)\right) &= \mathbf{P}(\mathbf{E}) \Longleftrightarrow \\ -r\left(\mathbf{E}^C\right) &= \Phi^{-1}(\mathbf{P}(\mathbf{E})) = r(\mathbf{E}). \end{aligned} \tag{A.9}$$

Thus, we get the upper bound

$$\mathbf{P}(\mathbf{E} - \mathbf{h}) \leq \Phi\left(r(\mathbf{E}) + \frac{\mathbf{L}(\mathbf{h}, \mathbf{h})}{\|\mathbf{h}\|_{\mathscr{H}_{\mathscr{W}_{\mathbf{f},P_0}}}} \right), \tag{A.10}$$

which establishes the proof.

Conflicts of Interest

The author declares that they have no conflicts of interest.

Acknowledgments

This research was supported by the Ministry of Research, Technology, and Higher Education of the Republic of Indonesia through the KLN Research Award 2018. The author wishes to thank Professor Bischoff for valuable discussion during the preparation of the manuscript.

References

[1] W. Somayasa, G. N. A. Wibawa, and Y. B. Pasolon, "Multidimensional set-indexed partial sums method for checking the appropriateness of a multivariate spatial regression," *Mathematical Models and Methods in Applied Sciences*, vol. 9, pp. 700–713, 2015.

[2] W. Somayasa and G. N. Adhi Wibawa, "Asymptotic model-check for multivariate spatial regression with correlated responses," *Far East Journal of Mathematical Sciences*, vol. 98, no. 5, pp. 613–639, 2015.

[3] W. Somayasa, G.N.A. Wibawa, L. Hamimu, and L.O. Ngkoimani, "Asymptotic theory in model diagnostic for general multivariate spatial regression," *International Journal of Mathematics and Mathematical Sciences*, vol. 2016, Article ID 2601601, 16 pages, 2016.

[4] W. Somayasa and H. Budiman, "Testing the mean in multivariate regression using set-indexed Gaussian white noise," *Statistics and Its Interface*, vol. 11, no. 1, pp. 61–77, 2018.

[5] I. B. MacNeill, "Properties of sequences of partial sums of polynomial regression residuals with applications to tests for change of regression at unknown times," *The Annals of Statistics*, vol. 6, no. 2, pp. 422–433, 1978.

[6] I. B. MacNeill, "Limit processes for sequences of partial sums of regression residuals," *Annals of Probability*, vol. 6, no. 4, pp. 695–698, 1978.

[7] I. B. MacNeill and V. K. Jandhyala, "Change-point methods for spatial data," in *Multivariate environmental statistics*, G. P. Patil and C. R. Rao, Eds., pp. 288–306, Elsevier Science Publishers, 1993.

[8] L. Xie and I. B. MacNeill, "Spatial residual processes and boundary detection," *South African Statistical Journal*, vol. 40, no. 1, pp. 33–53, 2006.

[9] K. S. Alexander and R. Pyke, "A uniform central limit theorem for set-indexed partial-sum processes with finite variance," *Annals of Probability*, vol. 14, no. 2, pp. 582–597, 1986.

[10] R. Pyke, "A uniform central limit theorem for partial-sum processes indexed by sets," in *Probability, statistics and analysis*, vol. 79 of *London Math. Soc. Lecture Note Ser.*, pp. 219–240, Cambridge Univ. Press, Cambridge-New York, 1983.

[11] M. Lifshits, *Lectures on Gaussian processes*, SpringerBriefs in Mathematics, Springer, Berlin, Heidelberg, 2012.

[12] J. A. Clarkson and C. R. Adams, "On definitions of bounded variation for functions of two variables," *Transactions of the American Mathematical Society*, vol. 35, no. 4, pp. 824–854, 1933.

[13] W. Stute, "Nonparametric model checks for regression," *The Annals of Statistics*, vol. 25, no. 2, pp. 613–641, 1997.

[14] R. J. Serfling, *Approximation Theorems of Mathematical Statistics*, John Wiley & Sons, New York, NY, USA, 1980.

[15] E. L. Lehmann and J. P. Romano, *Testing statistical hypotheses*, Springer, New York, 3rd edition, 2005.

[16] P. Billingsley, *Convergence of Probability Measures*, John Wiley & Sons, New York, NY, USA, 2nd edition, 1999.

[17] A. Janssen and H. Ünlü, "Regions of alternatives with high and low power for goodness-of-fit tests," *Journal of Statistical Planning and Inference*, vol. 138, no. 8, pp. 2526–2543, 2008.

[18] E. Hashorva, "Boundary non-crossings of Brownian pillow," *Journal of Theoretical Probability*, vol. 23, no. 1, pp. 193–208, 2010.

[19] E. Hashorva and Y. Mishura, "Boundary noncrossings of additive Wiener fields," *Lithuanian Mathematical Journal*, vol. 54, no. 3, pp. 277–289, 2014.

[20] R. F. Bass, "Probability estimates for multiparameter Brownian processes," *Annals of Probability*, vol. 16, no. 1, pp. 251–264, 1988.

[21] V. L. Wenbo and J. Kuelbs, "Some shift inequalities for Gaussian measures," *Progress in Probability*, vol. 43, pp. 233–243, 1998.

[22] J. A. Wellner, "Gaussian white noise models: some results for monotone functions," in *Crossing boundaries: statistical essays in honor of Jack Hall*, vol. 43 of *IMS Lecture Notes Monogr. Ser.*, pp. 87–104, Inst. Math. Statist., Beachwood, OH, 2003.

[23] G. Samorodnitsky, "Probability tails of Gaussian extrema," *Stochastic Processes and Their Applications*, vol. 38, no. 1, pp. 55–84, 1991.

[24] F. Móricz, "Pointwise behavior of double Forier series of functions of bounded variation," *Monatshefte für Mathematik*, vol. 148, pp. 51–59, 2006.

[25] J. Yeh, "Cameron-Martin translation theorems in the Wiener space of functions of two variables," *Transactions of the American Mathematical Society*, vol. 107, no. 3, pp. 409–420, 1963.

[26] S. F. Arnold, *The theory of linear models and multivariate analysis*, John Wiley & Sons, Inc., New York, NY, USA, 1981.

[27] J. Conway, *A Course in Functional Analysis*, Graduate Texts in Mathematics, Springer, New York, NY, USA, 1990.

11

Solutions of First-Order Volterra Type Linear Integrodifferential Equations by Collocation Method

Olumuyiwa A. Agbolade[1,2] and Timothy A. Anake[2]

[1] *Department of Mathematics and Statistics, Federal Polytechnic, Ilaro, Ogun State, Nigeria*
[2] *Department of Mathematics, College of Science and Technology, Covenant University, Ota, Ogun State, Nigeria*

Correspondence should be addressed to Timothy A. Anake; timothy.anake@covenantuniversity.edu.ng

Academic Editor: Mehmet Sezer

The numerical solutions of linear integrodifferential equations of Volterra type have been considered. Power series is used as the basis polynomial to approximate the solution of the problem. Furthermore, standard and Chebyshev-Gauss-Lobatto collocation points were, respectively, chosen to collocate the approximate solution. Numerical experiments are performed on some sample problems already solved by homotopy analysis method and finite difference methods. Comparison of the absolute error is obtained from the present method and those from aforementioned methods. It is also observed that the absolute errors obtained are very low establishing convergence and computational efficiency.

1. Introduction

Integrodifferential equation is a hybrid of integral and differential equations which have found extensive applications in sciences and engineering since it was established by Volterra [1]. A special class of these equations are the Volterra type which have been used to model heat and mass diffusion processes, biological species coexisting together with increasing and decreasing rate of growth, electromagnetic theory, and ocean circulations, among others [2].

First-order integrodifferential equation (IDE) of the Volterra type is generally of the form

$$y' = f(t, y(t), z(t)) \quad y(t_0) = y_0, \tag{1}$$

where

$$z(t) = \int_{t_0}^{t} K(t, s, y(s))\, ds, \quad t \in I. \tag{2}$$

In solving (1), we seek the unknown function $y(t)$ given the kernel K, a nonsingular function defined on $S \times \mathbb{R}$ with $S := \{(t, s),\ t_0 \le s \le t \le T\}$. This kernel determines the nature of the solutions of integral equation (2) depending on its type [3]. In this paper, only separable or degenerate kernels have been considered.

The theory and application of integrodifferential equations are important subjects in applied mathematics. The existence and uniqueness of the solutions of integrodifferential equations, usually discussed in terms of their kernel, had been established already in Linz [1]. Generally, methods for solving integrodifferential equations combine methods of solving both integral and differential equations. Also, since closed form solutions may not be tractable for most applications, numerical methods are employed to obtain approximations to the exact solutions.

Some numerical approaches in literature include iterative methods [4], successive approximation methods [5], and standard integral collocation approximation methods [6]. Other methods such as power series methods, where Chebyshev and Legendre's polynomials are used as basis functions, have been applied to obtain solutions of some higher order IDE of linear type. Akyaz and Sezer [7], for instance, presented Chebyshev collocation method for solving linear integrodifferential equations by truncated Chebyshev series. Recently, Gegele et al. [8] used power and Chebyshev series approximation methods to find numerical solution to higher order linear Fredholm integrodifferential equations using collocation methods. The result presented showed that the methods can give accurate results when compared with the exact solution. These methods proved efficient in the respective applications from the results provided but they seem yet to be applied to integrodifferential equations of Volterra type.

It is our aim here to extend the approach in Gegele et al. [8] to obtain approximate solutions for integrodifferential equations of Volterra type.

In the next section, we shall discuss the derivation of our methods; then the implementation using some sample problems is presented in Section 3. Finally, in Section 4 we shall present the results and draw our conclusions.

2. Methodology

In the sequel, the combination of the power series approximation and collocation method is employed for the solution of IDE of Volterra type.

To proceed, (1) is reduced to the form

$$y'(x) = F(x) + \int_a^x K(x,t)\, y(t)\, dt, \tag{3a}$$

$$y^{(j)}(0) = a_j, \quad j = 0, 1, \tag{3b}$$

where $y^{(j)} = d^j y/dx^j$ and $y^{(0)} = y$. The initial conditions (3b) are required in order to find particular solutions of (3a).

Now, let the solution $y(x)$ of Volterra type IDE, (3a) and (3b), be analytic and therefore possess the power series

$$y(x) = \sum_{i=0}^{N} a_i x^i, \quad i \geq 0, \tag{4}$$

where x^is are monomial bases and a_is are real coefficients to be determined.

Substituting equation (4) into both sides of (3a) gives

$$\sum_{i=0}^{N} i a_i x^{(i-1)} = F(x) + \sum_{i=0}^{N} a_i \int_a^x t^i K(x,t)\, dt. \tag{5}$$

Hence,

$$F(x) = \sum_{i=0}^{N} a_i \left(i x^{(i-1)} - \int_a^x t^i K(x,t)\, dt \right), \tag{6}$$

where $F(x)$ and $K(x,t)$ are known functions.

For an arbitrary choice of N, (6) is obtained as a linear algebraic equation in $N+1$ unknowns as follows:

$$\begin{aligned} a_0 + a_1\tau_1(x) + a_2\tau_2(x) + \cdots + a_{N-1}\tau_{N-1}(x) \\ + a_N\tau_N(x_1) = F(x). \end{aligned} \tag{7}$$

We note that a_0 is given by the initial condition (3b) while the remaining a_i, $i = 1, \ldots, N$, are to be determined by collocation method.

To generate the collocation points, we shall consider two methods, namely, the standard and Chebyshev-Gauss-Lobatto Collocation Methods, respectively.

2.1. Standard Collocation Method (SCM). This method is used to determine the desired collocation points within an interval, say, $[\vartheta, \sigma]$, and is given by

$$x_p = \vartheta + \frac{(\sigma - \vartheta)}{N} p, \quad p = 1, 2, 3, \ldots, N. \tag{8}$$

2.2. Chebyshev-Gauss-Lobatto Collocation Method (CGLCM). The collocation points are obtained as follows:

$$x_p = \cos\left(\frac{\pi p}{N}\right) \quad p = 1, 2, 3, \ldots, N. \tag{9}$$

Interestingly, Chebyshev-Gauss-Lobatto points have also been used as collocation and interpolation points in the solutions of optimal control problems governed by Volterra integrodifferential equations [9, 10].

Using either of the two collocation points to collocate (7) together with the initial conditions given in (3b) will result in a system of $N+1$ linear algebraic equations in $N+1$ unknowns. Hence, the resultant matrix problem is as follows:

$$\begin{bmatrix} 1 & 0 & 0 & \cdots & 0 & 0 \\ 0 & \tau_1(x_1) & \tau_2(x_1) & \cdots & \tau_{N-1}(x_1) & \tau_N(x_1) \\ 0 & \tau_1(x_2) & \tau_2(x_2) & \cdots & \tau_{N-1}(x_2) & \tau_N(x_2) \\ \vdots & & \vdots & & & \vdots \\ 0 & \tau_1(x_{N-1}) & \tau_2(x_{N-1}) & \cdots & \tau_{N-1}(x_{N-1}) & \tau_N(x_{N-1}) \\ 0 & \tau_1(x_N) & \tau_2(x_N) & \cdots & \tau_{N-1}(x_N) & \tau_N(x_N) \end{bmatrix} \begin{bmatrix} a_0 \\ a_1 \\ a_2 \\ \vdots \\ a_{N-1} \\ a_N \end{bmatrix} = \begin{bmatrix} F(x_0) \\ F(x_1) \\ F(x_2) \\ \vdots \\ F(x_{N-1}) \\ F(x_N) \end{bmatrix}, \tag{10}$$

where $\tau_j(x_p)$, $j = 1, 2, 3, \ldots, N$, are polynomials evaluated at each collocation point x_p. The values of the unknowns can be obtained using any convenient method of solving matrix equations of the form $AX = B$, where A is invertible.

Substituting the values of the a_i, $i = 0, 1, 2, \ldots, N$, obtained from (4) yields the approximate solution. We note that the accuracy level desired for the approximate solution is determined by the degree of the approximating polynomial.

3. Results

In this section, standard and Chebyshev-Gauss-Lobatto collocation points have been employed, respectively, to solve sample problems as described in Section 2. The numerical solutions obtained using the present method had been compared with the exact solutions of the sample problems. Similarly, absolute errors of results from this present method have been compared with those obtained in Behrouz [11] by homotopy analysis method (HAM) and finite difference method (FDM) for the same problems.

The absolute error of computation is defined in all cases as follows:

$$|y(x_i) - Y(x_i)| \quad \vartheta \le x_i \le \sigma,\ i = 1, 2, 3, \ldots. \tag{11}$$

Problem 1.

$$y'(x) + y(x) = \left(x^2 + 2x + 1\right) e^{-x} + 5x^2 + 8 - \int_0^x t y(t)\, dt, \quad y(0) = 10. \tag{12}$$

Exact solution: $y(x) = 10 - xe^{-x}$.

Using SCM, we obtained the following approximate solutions:

$$y(x) = 10 - 0.999955x + 0.999606x^2 - 0.498312x^3 + 0.162656x^4 - 0.0362999x^5 + 0.00442505x^6. \tag{13}$$

Similarly using CGLCM, we obtained the approximate solution as follows:

$$y(x) = 10 - x + 1.00115x^2 - 0.499641x^3 + 0.163495x^4 - 0.0434569x^5 + 0.0107202x^6. \tag{14}$$

The solutions obtained from the implementation of the method for Problem 1 using SCM and CGLCM are compared with the exact solution in Table 1. Also absolute errors obtained are compared with absolute errors obtained from HAM and FDM in Table 2.

Problem 2.

$$y'(x) + y(x) = \int_0^x e^{t-x} y(t)\, dt, \quad y(0) = 1. \tag{15}$$

Exact solution: $y(x) = e^{-x} \cosh x$.

Using SCM, we obtained the following approximate solutions:

$$y(x) = 1 - 0.999759x + 0.997930x^2 - 0.658238x^3 + 0.314202x^4 - 0.107608x^5 + 0.0235161x^6 - 0.00237066x^7. \tag{16}$$

TABLE 1: Comparison of exact solution with numerical solutions for Problem 1.

x_i	Exact	SCM	CGLCM
0.0000	0.0000	0.0000	0.0000
0.0714	9.933495516	9.933497240	9.933501428
0.1429	9.876160300	9.876162226	9.876183411
0.2143	9.827046197	9.827047975	9.827095298
0.2857	9.785292202	9.785293839	9.785371414
0.3571	9.750116951	9.750118464	9.750223918
0.4286	9.720811832	9.720813173	9.720936698
0.5000	9.696734670	9.696735770	9.696860288
0.5714	9.677303930	9.677304767	9.677407809
0.6429	9.661993413	9.661994038	9.662051938
0.7143	9.650327386	9.650327889	9.650322899
0.7857	9.641876128	9.641876560	9.641807476
0.8571	9.636251847	9.636252142	9.636149063
0.9286	9.633104936	9.633104922	9.633048727
1.0000	9.632120559	9.632120150	9.632267300

Similarly, using CGLCM we obtained the approximate solution as follows:

$$y(x) = 1 - 0.99941x + 0.999958x^2 - 0.671295x^3 + 0.331381x^4 - 0.123419x^5 + 0.0496174x^6 - 0.0193272x^7. \tag{17}$$

The solutions obtained from the implementation of the method for Problem 2 using SCM and CGLCM are compared with the exact solution in Table 3. Also absolute errors obtained are compared with absolute errors obtained from HAM and FDM in Table 4.

4. Conclusion

In this paper, numerical solution of Volterra type integrodifferential equation of first order with degenerate kernels is obtained by power series collocation method based on two collocating points methods, namely, Standard Collocation Method (SCM) and Chebyshev-Gauss-Lobatto Collocation Method (CGLCM), presented.

The two methods for selecting collocation points yielded different schemes from which approximate solutions were obtained, respectively, and compared with the exact solutions as shown in Tables 1 and 3. From the results presented, the two methods gave good results for first-order integrodifferential equations of Volterra type.

The comparison of absolute errors of the results obtained by the present method with those by homotopy analysis method and finite difference method for the same problems revealed that the method is efficient and cheap for the numerical solutions of first-order integrodifferential equation of Volterra type as illustrated in Tables 2 and 4. The performance of the present method against homotopy analysis method is expected as the latter is a semianalytic method.

Table 2: Comparison of absolute errors for Problem 1

x_i	SCM	CGLCM	FDM	HPM
0.0000	0.0000	0.0000	0.0000	0.0000
0.0714	$1.72431E-06$	$5.91262E-06$	$2.85397E-04$	$5.15735E-07$
0.1429	$1.92637E-06$	$2.31105E-05$	$2.98284E-04$	$3.00036E-07$
0.2143	$1.77825E-06$	$4.91013E-05$	$5.43393E-04$	$2.80293E-06$
0.2857	$1.63695E-06$	$7.92123E-05$	$5.11413E-04$	$1.47980E-05$
0.3571	$1.51288E-06$	$1.06967E-04$	$7.15638E-04$	$4.60491E-05$
0.4286	$1.34028E-06$	$1.24865E-04$	$6.54200E-04$	$1.11168E-04$
0.5000	$1.09939E-06$	$1.25617E-04$	$8.18261E-04$	$2.40330E-04$
0.5714	$8.36590E-07$	$1.03879E-04$	$7.38321E-04$	$4.73070E-04$
0.6429	$6.24770E-07$	$5.85256E-05$	$8.64022E-04$	$8.52587E-04$
0.7143	$5.03018E-07$	$4.48751E-06$	$7.73248E-04$	$1.45361E-03$
0.7857	$4.31615E-07$	$6.86526E-05$	$8.63249E-04$	$2.36487E-03$
0.8571	$2.95402E-07$	$1.02783E-04$	$7.66939E-04$	$3.71115E-03$
0.9286	$1.39661E-08$	$5.62088E-05$	$8.24573E-04$	$5.63206E-04$
1.0000	$4.08829E-07$	$1.46741E-04$	$7.26353E-04$	$8.32344E-04$

Table 3: Comparison of exact solution with numerical solution for Problem 2.

x_i	Exact	SCM	CGLCM
0.0000	0.0000	0.0000	0.0000
0.0833	0.9232408624	0.9232506225	0.923287006
0.1667	0.8582656553	0.8582754765	0.858341248
0.2500	0.8032653299	0.8032737684	0.803340755
0.3333	0.7567085595	0.7567163057	0.756749451
0.4167	0.7172991043	0.7173066673	0.717278486
0.5000	0.6839397206	0.6839470433	0.683851372
0.5833	0.6557016120	0.6557084119	0.655570212
0.6667	0.6317985691	0.6318047191	0.631680300
0.7500	0.6115650801	0.6115707282	0.611530371
0.8333	0.5944378014	0.5944432050	0.594525790
0.9167	0.5799398730	0.5799451069	0.580071957
1.0000	0.5676676416	0.5676724400	0.567505200

Table 4: Comparison of absolute errors for Problem 2.

x_i	SCM	CGLC	FDM	HPM
0.0000	0.0000	0.0000	0.0000	0.0000
0.0833	$9.76008E-06$	$4.61438E-05$	$1.77203E-02$	$1.85469E-09$
0.1667	$9.82124E-06$	$7.55931E-05$	$2.16887E-03$	$3.13105E-10$
0.2500	$8.43856E-06$	$7.54254E-05$	$1.89273E-03$	$1.14368E-09$
0.3333	$7.74622E-06$	$4.08918E-05$	$4.52374E-03$	$8.37039E-11$
0.4167	$7.56304E-06$	$2.06182E-05$	$2.06181E-02$	$2.65354E-09$
0.5000	$7.32270E-06$	$8.83487E-05$	$7.13624E-03$	$3.14279E-10$
0.5833	$6.79994E-06$	$1.31400E-05$	$1.10585E-02$	$1.24270E-09$
0.6667	$6.15006E-06$	$1.18269E-05$	$8.20866E-03$	$5.57863E-10$
0.7500	$5.64809E-06$	$3.47095E-05$	$3.41335E-03$	$1.32579E-09$
0.8333	$5.40361E-06$	$8.79889E-05$	$8.16328E-03$	$6.81219E-10$
0.9167	$5.23390E-06$	$1.32084E-05$	$2.89396E-03$	$5.16015E-09$
1.0000	$4.79838E-06$	$1.62442E-05$	$3.27168E-03$	$9.48169E-09$

Conflicts of Interest

The authors declare that there are no conflicts of interest regarding the publication of this paper.

Acknowledgments

The authors are grateful to the Covenant University Centre for Research, Innovation and Development (CUCRID) for sponsoring this publication.

References

[1] P. Linz, *Analytic and Numerical Methods for Volterra Equations*, SIAM, Philadelphia, Pa, USA, 1985.

[2] M. Rahman, *Integral Equations and Their Applications*, WIT Press, Southampton, UK, 2007.

[3] G. Tiwari, "Solving Integral Equations: The Interactive Way!," 2014, https://gauravtiwari.org/.

[4] A. J. Jerri, *Introduction to Integral Equations with Applications*, Marcel Dekken, Inc, New York, NY, USA, 1985.

[5] R. K. Saeed, *Computational methods for solving system of linear Volterra integral and integro-differential equation [Ph.D. thesis]*, University of Salahaddin Hawler, College of Science, 2006.

[6] A. Abubakar and O. A. Taiwo, "Integral collocation approximation methods for the numerical solution of high-orders linear Fredholm-Volterra integro-differential equations," *American Journal of Computational and Applied Mathematics*, vol. 4, no. 4, pp. 111–117, 2014.

[7] A. Akyaz and M. Sezer, "A Chebychev collocation method for the solution of linear integro-differential equation," *International Journal of Computer Mathematics*, vol. 72, no. 4, pp. 491–507, 1999.

[8] O. A. Gegele, O. P. Evans, and D. Akoh, "Numerical solution of higher order linear Fredholm-integro-differential equations," *American Journal of Engineering Research*, vol. 8, no. 3, pp. 243–247, 2014.

[9] M. El-Kady and H. Moussa, "Monic Chebyshev approximations for solving optimal control problem with Volterra integro differential equations," *General Mathematics Notes*, vol. 14, no. 2, pp. 23–36, 2013.

[10] W. Zhang and H. Ma, "The Chebyshev-Legendre collocation method for a class of optimal control problems," *International Journal of Computer Mathematics*, vol. 85, no. 2, pp. 225–240, 2008.

[11] R. Behrouz, "Numerical solutions of the linear Volterra integro-differential equations: homotopy perturbation method and finite difference method," *World Applied Sciences Journal*, vol. 9, pp. 7–12, 2010.

12

On Minimizing the Ultimate Ruin Probability of an Insurer by Reinsurance

Christian Kasumo,[1] **Juma Kasozi**,[2] **and Dmitry Kuznetsov**[1]

[1]*Department of Applied Mathematics and Computational Science, Nelson Mandela African Institution of Science and Technology, P.O. Box 447, Arusha, Tanzania*
[2]*Department of Mathematics, Makerere University, P.O. Box 7062, Kampala, Uganda*

Correspondence should be addressed to Christian Kasumo; kasumoc@nm-aist.ac.tz

Academic Editor: Saeid Abbasbandy

We consider an insurance company whose reserves dynamics follow a diffusion-perturbed risk model. To reduce its risk, the company chooses to reinsure using proportional or excess-of-loss reinsurance. Using the Hamilton-Jacobi-Bellman (HJB) approach, we derive a second-order Volterra integrodifferential equation (VIDE) which we transform into a linear Volterra integral equation (VIE) of the second kind. We then proceed to solve this linear VIE numerically using the block-by-block method for the optimal reinsurance policy that minimizes the ultimate ruin probability for the chosen parameters. Numerical examples with both light- and heavy-tailed distributions are given. The results show that proportional reinsurance increases the survival of the company in both light- and heavy-tailed distributions for the Cramér-Lundberg and diffusion-perturbed models.

1. Introduction

When the surplus process of an insurance company falls below zero, the company is said to have experienced ruin. Insurance companies customarily take precautions to avoid ruin. These precautions are referred to as *control variables* and include investments, capital injections or refinancing, portfolio selection, and reinsurance arrangements, to mention but a few. This study focuses on reinsurance as a control measure. Reinsurance, sometimes referred to as "insurance for insurers," is the transfer of risk from a direct insurer (the cedent) to a second insurance carrier (the reinsurer). With reinsurance, the cedent passes on some of its premium income to a reinsurer who, in turn, covers a certain proportion of the claims that occur. It has been argued in the literature that reinsurance plays an important role in risk reduction for cedents in that it offers additional underwriting capacity for them and reduces the probability of a direct insurer's ruin. Apart from helping the cedent to manage financial risk, increase capacity, and achieve marketing goals, reinsurance also benefits policyholders by ensuring availability and affordability of necessary coverage.

Of interest in this paper are those studies which investigate more directly the effect of reinsurance on the ultimate ruin probability. The minimization of the probability of ruin for a company whose claim process evolves according to a Brownian motion with drift and is allowed to invest in a risky asset and to purchase quota-share reinsurance was considered in [1]. In this study, an analytical expression for the minimum ruin probability and the corresponding optimal controls were obtained. Kasozi et al. [2] studied the problem of controlling ultimate ruin probability by quota-share (QS) reinsurance arrangements. Under the assumption that the insurer could invest part of the surplus in a risk-free and risky asset, [2] found that quota-share reinsurance does reduce the probability of ruin and that for chosen parameter values the optimal QS retention $b^* \in (0.2, 0.4)$. This study also concluded that investment helps insurance companies to reduce their ruin probabilities but that the ruin probabilities increase when stock prices become more volatile. However, while Kasozi et al. [2] considered only quota-share reinsurance, this paper seeks to combine quota-share and excess-of-loss (XL) reinsurance for one and the same insurance portfolio, but in the absence of investment.

Liu and Yang [3] reconsidered the model in [4] and incorporated a risk-free interest rate. Since closed-form solutions could not be obtained in this case, they provided numerical results for optimal strategies for maximizing the survival probability under different claim-size distribution assumptions. Also using the results in [4], the problem of choosing a combination of investments and optimal dynamic proportional reinsurance to minimize ruin probabilities for an insurance company was investigated in [5] based on a controlled surplus process satisfying the stochastic differential equation $dX_t^{Ab} = (c - c(b_t) + \mu A_t)dt + \sigma A_t dW_t - b_t dS_t$, where $b_t \in [0,1]$ is a proportional reinsurance retention at time t, $c(b_t)$ is the dynamic reinsurance premium rate, $\{A_t\}$ is the amount invested in a risky asset at time t, and S_t is the aggregate claims process. But while [5] uses proportional reinsurance in minimizing ruin probabilities in the Cramér-Lundberg model, this paper considers proportional and excess-of-loss reinsurance in the diffusion-perturbed model.

More recently, taking ruin probability as a risk measure for the insurer, [6] investigated a dynamic optimal reinsurance problem with both fixed and proportional transaction costs for an insurer whose surplus process is modelled by a Brownian motion with positive drift. Under the assumption that the insurer takes noncheap proportional reinsurance, they formulated the problem as a mixed regular control and optimal stopping problem and established that the optimal reinsurance strategy was to never take reinsurance if proportional costs were high and to wait to take the reinsurance when the surplus hits a level. Additionally, they obtained an explicit expression for the survival probability under the optimal reinsurance strategy and found it to be larger than that with the aforementioned strategies. Hu and Zhang [7] introduced a general risk model involving dependence structure with common Poisson shocks. Under a combined quota-share and excess-of-loss reinsurance arrangements, they studied the optimal reinsurance strategy for maximizing the insurer's adjustment coefficient and established that excess-of-loss reinsurance was optimal from the insurer's point of view. Zhang and Liang [8] studied the optimal retentions for an insurance company that intends to transfer risk by means of a layer reinsurance treaty. Under the criterion of maximizing the adjustment coefficient, they obtained the closed-form expressions of the optimal results for the Brownian motion as well as the compound Poisson risk models and concluded that under the expected value principle excess-of-loss reinsurance is better than any other layer reinsurance strategies while under the variance premium principle pure excess-of-loss reinsurance is no longer the optimal layer reinsurance strategy. Both of these studies, however, used the criterion of maximizing the adjustment coefficient rather than minimizing the insurer's ruin probability.

This paper aims at combining proportional and excess-of-loss reinsurance for one and the same insurance portfolio. In proportional or "pro rata" reinsurance, the reinsurer indemnifies the cedent for a predetermined portion of the claims or losses, while in excess-of-loss (XL) reinsurance, which is nonproportional, the reinsurer indemnifies the cedent for all claims or losses or for a specified portion of them, but only if the claim sizes fall within a prespecified band. Excess-of-loss reinsurance has been defined in [9] as "a form of nonproportional reinsurance contract in which an insurer pays insurance claims up to a prefixed *retention level* and the rest are paid by a reinsurer." Mathematically, given retention level a, a claim of size X is divided into the cedent's payment $X \wedge a$ and the reinsurer's payment $X - X \wedge a$. The combination of proportional and excess-of-loss reinsurance has been in fact widely used in the construction of reinsurance models (see, e.g., [10]).

The models in this paper result in Volterra integral equations (VIEs) of the second kind which are solved using the block-by-block method, generally considered as the best of the higher order methods for solving Volterra integral equations of the second kind. The block-by-block methods are essentially extrapolation procedures which produce a block of values at a time. These methods can be of high order and still be self-starting. They do not require special starting procedures, are simple to use, and allow for easy switching of step-size [11].

The rest of the paper is organized as follows. Section 2 presents the formulation of the model and assumptions, followed, in Section 3, by a derivation of the HJB, integrodifferential, and integral equations. In Section 4, we present numerical results for some ruin probability models with reinsurance, using the exponential distribution for small˜claims and the Pareto distribution for large ones. Some conclusions and possible extensions of this study are given in Section 5.

2. Model Formulation

Let $(\Omega, \mathcal{F}, \{\mathcal{F}_t\}_{t\in\mathbb{R}^+}, \mathbb{P})$ be a filtered probability space containing all stochastic objects encountered in this paper and satisfying the usual conditions; that is, $\{\mathcal{F}_t\}_{t\in\mathbb{R}^+}$ is right-continuous and $\mathbb{P}$-complete. In the absence of reinsurance, the surplus of an insurance company is governed by the diffusion-perturbed classical risk process:

$$U_t = u + ct + \sigma W_t - \sum_{i=1}^{N_t} X_i, \quad t \geq 0, \tag{1}$$

where $u = U_0 \geq 0$ is the initial reserve, $c = (1+\theta)\lambda\mu > 0$ is the premium rate, θ is the safety loading, $\{N_t\}$ is a homogeneous Poisson process with intensity $\lambda > 0$, and $\{X_i\}$ is an i.i.d. sequence of strictly positive random variables with distribution function F. $S_t = \sum_{i=1}^{N_t} X_i$ is a compound Poisson process representing the cumulative amount of claims paid in the time interval $[0,t]$. The claim arrival process $\{N_t\}$ and claim sizes $\{X_i\}$ are assumed to be independent. Here $\{W_t\}$ is a standard one-dimensional Brownian motion independent of the compound Poisson process S_t. We assume that $\mathbb{E}[X_i] = \mu < \infty$ and $F(0) = 0$. The diffusion term σW_t denotes the fluctuations associated with the surplus of the insurance company at time t. Without volatility in the surplus and claim amounts, (1) becomes the well-known Cramér-Lundberg model or the classical risk process.

We proceed as in [12] where the insurer took a combination of quota-share and excess-of-loss reinsurance arrangements. Most of the actuarial literature dealing with reinsurance as a risk control mechanism only considers pure quota-share or excess-of-loss reinsurance. However, in reality the insurer has the choice of a combination of the two and hence the use of a combination of quota-share and XL reinsurance in this paper. We assume that the reinsurance is *cheap*, meaning that the reinsurer uses the same safety loading as the insurer. Let the quota-share retention level be $k \in [0,1]$. Then the insurer's aggregate claims, net of quota-share reinsurance, are kX. If the company also buys excess-of-loss reinsurance with a retention level $a \in [0,\infty)$, then the insurer's aggregate claims, net of quota-share and excess-of-loss reinsurance, are given by $kX \wedge a$. Given that $\overline{R}$ is a reinsurance strategy combining quota-share and excess-of-loss reinsurance, the insurer's controlled surplus process becomes

$$U_t^{\overline{R}} = u + c^{\overline{R}}t + \sigma W_t - \sum_{i=1}^{N_t} kX_i \wedge a, \tag{2}$$

where the insurance premium $c^{\overline{R}} = c - (1+\theta)\lambda\mathbb{E}[(kX_i - a)^+]$. The controlled surplus process (2) has dynamics

$$dU_t^{\overline{R}} = c^{\overline{R}}dt + \sigma dW_t - d\left(\sum_{i=1}^{N_t} kX_i \wedge a\right). \tag{3}$$

The time of ruin is defined as $\tau^{\overline{R}} = \inf\{t \geq 0 \mid U_t^{\overline{R}} < 0\}$ and the probability of ultimate ruin is defined as $\psi^{\overline{R}} = \mathbb{P}(U_t^{\overline{R}} < 0$ for some $t > 0)$. A reinsurance strategy $\overline{R}$ is said to be *admissible* if $k \in [0,1]$ and $a \in [0,\infty)$. The objective is to find the quota-share level k and the excess-of-loss retention limit a to minimize the insurer's risk or to maximize the insurer's survival probability. It should be noted that when the retention limit a of the excess-of-loss reinsurance is infinite, then the treaty becomes a *pure quota-share* reinsurance, while when the quota-share level $k = 1$, it becomes a *pure excess-of-loss* reinsurance treaty. The premium income of the insurance company is nonnegative if $c \geq (1+\theta)\lambda\mathbb{E}[(kX-a)^+]$. Therefore, we will let $\underline{a}$ be the XL retention level at which equality $c = (1+\theta)\lambda\mathbb{E}[(kX-\underline{a})^+]$ holds.

Define the value function of this problem as

$$\begin{aligned} \psi^{\overline{R}}(u) &= \mathbb{P}\left(U_t \leq 0 \text{ for some } t \geq 0 \mid U_0^{\overline{R}} = u\right) \\ &= \mathbb{P}\left(\tau^{\overline{R}} < \infty \mid U_0^{\overline{R}} = u\right), \end{aligned} \tag{4}$$

where $\psi^{\overline{R}}(u)$ is the probability of ultimate ruin under the policy $\overline{R}$ when the initial surplus is u. Then the objective is to find the optimal value function, that is, the minimal ruin probability

$$\psi(u) = \inf_{(k,a)\in\mathscr{R}} \psi^{\overline{R}}(u) \tag{5}$$

and optimal policy $(\overline{R})^* = (k^*, a^*)$ s.t. $\psi^{\overline{R}^*}(u) = \psi(u)$. Alternatively, we can find the values of k^* and a^* which maximize the probability of ultimate survival $\phi(u) = 1 - \psi(u)$, so that the optimal value function becomes

$$\phi(u) = \sup_{(k,a)\in\mathscr{R}} \phi^{\overline{R}}(u), \tag{6}$$

where $\mathscr{R}$ is the set of all reinsurance policies.

3. HJB, Integrodifferential, and Integral Equations

Lemma 1. *Assume that the survival probability $\phi(u)$ defined by (6) is twice continuously differentiable on $(0,\infty)$. Then $\phi(u)$ satisfies the HJB equation*

$$\begin{aligned} \sup_{(k,a)\in\mathscr{R}} \Bigg\{ &\frac{1}{2}\sigma^2\phi''(u) + c^{\overline{R}}\phi'(u) \\ &+ \lambda\int_0^u [\phi(u - kx \wedge a) - \phi(u)]\, dF(x) \Bigg\} = 0, \\ &\qquad u > 0, \end{aligned} \tag{7}$$

where $\mathscr{R}$ is the set of all reinsurance policies.

Proof. See [13].

We now present the verification theorem which is essential for solving the associated stochastic control problem.

Theorem 2. *Suppose $\Phi \in C^2$ is an increasing strictly concave function satisfying HJB equation (7) subject to the boundary conditions*

$$\begin{aligned} \Phi(u) &= 0 \quad \text{on } u < 0 \\ \Phi(0) &= 0 \quad \text{if } \sigma^2 > 0 \\ \lim_{u\to\infty} \Phi(u) &= 1 \end{aligned} \tag{8}$$

for $0 \leq u < \infty$. Then the maximal survival probability $\phi(u)$ given by (6) coincides with Φ. Furthermore, if $(\overline{R})^* = (k^*, a^*)$ satisfies

$$\begin{aligned} &\frac{1}{2}\sigma^2\Phi''(u) + c^{\overline{R}^*}\Phi'(u) \\ &\quad + \lambda\int_0^u [\Phi(u - k^*x \wedge a^*) - \Phi(u)]\, dF(x) = 0 \\ &\qquad \text{when } 0 \leq u < \infty \end{aligned} \tag{9}$$

then the policy $(\overline{R})^*$ is an optimal policy; that is, $\Phi(u) = \phi(u) = \phi^{\overline{R}^*}(u)$.

Proof. Let $\overline{R}$ be an arbitrary reinsurance strategy and let U^* be the surplus process when $\overline{R} = \overline{R}^*$. Choose $n > u$ and define

$T = \mathcal{T}_n = \inf\{t \mid U_t \notin [0,n]\}$. Note that $U_{T\wedge t} \in (-\infty, n]$ because the jumps are downwards. The process

$$M_t^1 = \sum_{i=1}^{N_{T\wedge t}} \left[\Phi\left(U_{T_i}\right) - \Phi\left(U_{T_i^-}\right)\right] - \lambda \int_0^{T\wedge t} \left[\int_0^{U_s} \Phi\left(U_s - kx \wedge a\right) dF(x) - \Phi\left(U_s\right)\right] ds \quad (10)$$

is a martingale. We write

$$\Phi\left(U_{T\wedge t}\right) = \Phi(u) + \Phi\left(U_{T\wedge t}\right) - \Phi\left(U_{T_{N_{T\wedge t}}}\right) + \sum_{i=1}^{N_{T\wedge t}} \left[\Phi\left(U_{T_i^-}\right) - \Phi\left(U_{T_{i-1}}\right)\right] + M_t^1 + \lambda \int_0^{T\wedge t} \left[\int_0^{U_s} \Phi\left(U_s - kx \wedge a\right) dF(x) - \Phi\left(U_s\right)\right] ds. \quad (11)$$

By Itô's formula,

$$\Phi\left(U_{T_i^-}\right) - \Phi\left(U_{T_{i-1}}\right) = \int_{T_{i-1}}^{T_i} \left[\frac{1}{2}\sigma^2 \Phi''\left(U_s\right) + c^{\overline{R}} \Phi'\left(U_s\right)\right] ds + \int_{T_{i-1}}^{T_i} \sigma \Phi'\left(U_s\right) dW_s. \quad (12)$$

The corresponding result holds for $\Phi(U_{T\wedge t}) - \Phi(U_{T_{N_{T\wedge t}}})$. Thus,

$$\Phi\left(U_{T\wedge t}\right) = \Phi(u) + \int_0^{T\wedge t} \left[\frac{1}{2}\sigma^2 \Phi''\left(U_s\right) + c^{\overline{R}} \Phi'\left(U_s\right) + \lambda \left(\int_0^{U_s} \Phi\left(U_s - kx \wedge a\right) dF(x) - \Phi\left(U_s\right)\right)\right] ds + \int_0^{T\wedge t} \sigma \Phi'\left(U_s\right) dW_s + M_t^1. \quad (13)$$

Using HJB equation (7), we find that

$$\Phi\left(U_{T\wedge t}\right) \le \Phi(u) + \int_0^{T\wedge t} \sigma \Phi'\left(U_s\right) dW_s + M_t^1 \quad (14)$$

and equality holds for U^*. Let $\{\mathcal{S}_m\}$ be a localization sequence of the stochastic integral, and set $\mathcal{T}_n^m = \mathcal{T}_n \wedge \mathcal{S}_m$. Taking expectations yields

$$\mathbb{E}\left[\Phi\left(U_{\mathcal{T}_n^m \wedge t}\right)\right] \le \Phi(u). \quad (15)$$

By bounded convergence, letting $m \to \infty$ and then $t \to \infty$, we have $\mathbb{E}[\Phi(U_{\mathcal{T}_n})] \le \Phi(u)$. It turns out that, for $\Phi(0) = 0$,

$$\mathbb{P}\left(\tau < \mathcal{T}_n, U_\tau = 0\right) + \Phi(n)\, \mathbb{P}\left(\mathcal{T}_n < \tau\right) = \mathbb{E}\left[\Phi\left(U_{\mathcal{T}_n}\right)\right] \le \Phi(u). \quad (16)$$

Note that $\mathbb{P}(\mathcal{T}_n < \tau) \ge \phi^{\overline{R}}(u)$. Because there is a strategy with $\phi^{\overline{R}}(u) > 0$, it follows that $\Phi(u)$ is bounded. We therefore let $n \to \infty$, yielding $\mathbb{E}[\Phi(U_\tau)] \le \Phi(u)$. In particular, we obtain

$$\phi^{\overline{R}}(u)\, \Phi(\infty) \le \phi^{\overline{R}}(u)\, \Phi(\infty) + \mathbb{P}\left(\tau < \infty, U_\tau = 0\right) \le \Phi(u) \quad (17)$$

which simplifies to

$$\phi^{\overline{R}}(u) \le \phi^{\overline{R}}(u) + \mathbb{P}\left(\tau < \infty, U_\tau = 0\right) \le \Phi(u) \quad (18)$$

since $\Phi(\infty) = 1$. For U^* we obtain an equality. In particular, $\{\Phi(U_t^*)\}$ is a martingale. It remains to show that $\mathbb{P}(U_\tau^* \ne 0) = 1$. Note first from HJB equation (7) that $F(x)$ must be continuous; if not, the integral in (7) is not continuous. Choose $\varepsilon > 0$ and consider the strategy $\overline{R} = \overline{R}^* \mathbf{1}_{u \ge \varepsilon}$. Let $T_\varepsilon = \inf\{t \mid U_t^* < \varepsilon\}$. By the martingale property, $\Phi(u) = \Phi(\infty)\mathbb{P}(T_\varepsilon = \infty) + \mathbb{E}[\Phi(T_\varepsilon), T_\varepsilon < \tau < \infty]$ which reduces to

$$\Phi(u) = \mathbb{P}\left(T_\varepsilon = \infty\right) + \mathbb{E}\left[\Phi\left(T_\varepsilon\right), T_\varepsilon < \tau < \infty\right] \quad (19)$$

the last term of which is bounded by $\Phi(\varepsilon)\mathbb{P}(T_\varepsilon < \tau < \infty)$. Since $F(x)$ is continuous, it must converge to zero as $\varepsilon \to 0$. Because $\mathbb{P}(T_\varepsilon = \infty) \to \phi^*(u)$, it follows that $\Phi(u) = \phi^*(u)\Phi(\infty)$ or $\Phi(u) = \phi^*(u) = \phi(u)$. That is, $\Phi(u)$ is the optimal value function and $\overline{R}^* = (\overline{R})^*$ is an optimal policy.

The integrodifferential equation corresponding to optimization problem (6) immediately follows from Theorem 2 as

$$\frac{1}{2}\sigma^2 \phi''(u) + c^{\overline{R}} \phi'(u) + \lambda \int_0^u \left[\phi\left(u - kx \wedge a\right) - \phi(u)\right] dF(x) = 0 \quad \text{for } 0 \le u < \infty. \quad (20)$$

This is an integrodifferential equation of Volterra type (VIDE). Solution of this equation will require that it is transformed into a Volterra integral equation (VIE) of the second kind using successive integration by parts. Hence the following theorem is obtained.

Theorem 3. *Integrodifferential equation (20) can be represented as a Volterra integral equation of the second kind:*

$$\phi(u) + \int_0^u K(u,x)\, \phi(x)\, dx = h(u), \quad (21)$$

where

(1) *If $u \le \underline{a} < a$, one has*

$$K(u,x) = -\frac{\lambda \overline{F}(u - kx)}{c^{\overline{R}}}$$
$$h(u) = \phi(0) \quad (22)$$

with $\overline{F}(x) = 1 - F(x)$, when there is no diffusion (i.e., when $\sigma^2 = 0$), and

$$K(u,x) = \frac{2\left(c^{\overline{R}} + \lambda G(u-kx) - \lambda(u-kx)\right)}{\sigma^2} \tag{23}$$
$$h(u) = u\phi'(0) \quad \text{if } \sigma^2 > 0$$

when there is diffusion.

(2) *If $\underline{a} < a < u$, one has*

$$K(u,x) = -\frac{\lambda H_1(x,u)}{c^{\overline{R}}} \tag{24}$$
$$h(u) = \phi(0)$$

with

$$H_1(x,u) = \begin{cases} \overline{F}(u-kx) & kx < a \\ 1 - (F(kx+a) - F(a)) & kx \geq a \end{cases} \tag{25}$$

when there is no diffusion, and

$$K(u,x) = \frac{2\left(c^{\overline{R}} + \lambda H_2(x,u) - \lambda(u-kx)\right)}{\sigma^2} \tag{26}$$
$$h(u) = u\phi'(0) \quad \text{if } \sigma^2 > 0$$

with

$$H_2(x,u) = \begin{cases} G(u-kx) & kx < a \\ (F(kx+a) - F(a))(u-kx) & kx \geq a \end{cases} \tag{27}$$

and $G(x) = \int_0^x F(v)dv$ when there is diffusion.

Proof. The proof for the case $u \leq \underline{a} < a$ is similar to the proof of Theorem 2.2 in [14] but with $r = \sigma_R^2 = 0$, $k = 1$, and $p = c^{\overline{R}}$. Here, we present the proof for the case $\underline{a} < a < u$.

Integrating (20) on $[0,z]$ with respect to u gives

$$\begin{aligned} 0 &= \frac{1}{2}\sigma^2\left[\phi'(z) - \phi'(0)\right] + c^{\overline{R}}\left[\phi(z) - \phi(0)\right] \\ &\quad - \lambda\int_0^z \phi(u)\,du \\ &\quad + \lambda\int_0^z\int_0^u \phi(u - kx \wedge a) f(x)\,dx\,du. \end{aligned} \tag{28}$$

To simplify the double integral in (28), we again use integration by parts and Fubini's Theorem (see [13]) to switch the order of integration and change the properties of the convolution integral. Thus,

$$\begin{aligned} &\int_0^z\int_0^u \phi(u - kx \wedge a) f(x)\,dx\,du \\ &\quad = \int_0^a F(z-kx)\phi(x)\,dx \\ &\quad\quad + \int_a^z \phi(v)\left[F(v+a) - F(a)\right]dv, \end{aligned} \tag{29}$$

where $v = u - kx$. Substituting into (28) gives

$$\begin{aligned} &\frac{1}{2}\sigma^2\phi'(z) - \frac{1}{2}\sigma^2\phi'(0) + c^{\overline{R}}\phi(z) - c^{\overline{R}}\phi(0) \\ &\quad - \lambda\int_0^z \phi(u)\,du + \lambda\left[\int_0^a F(z-kx)\phi(x)\,dx\right. \\ &\quad \left. + \int_a^z \phi(v)\left[F(v+a) - F(a)\right]dv\right] = 0. \end{aligned} \tag{30}$$

Replacing z with u, v and u with x, and $F(v+a)$ with $F(kx+a)$ gives

$$\begin{aligned} &\frac{1}{2}\sigma^2\phi'(u) - \frac{1}{2}\sigma^2\phi'(0) + c^{\overline{R}}\phi(u) - c^{\overline{R}}\phi(0) \\ &\quad - \lambda\int_0^u \phi(x)\,dx + \lambda\int_0^a F(u-kx)\phi(x)\,dx \\ &\quad + \lambda\int_a^u \left[F(kx+a) - F(a)\right]\phi(x)\,dx = 0. \end{aligned} \tag{31}$$

Setting $\sigma^2 = 0$ in (31) yields the case without diffusion

$$\begin{aligned} &\phi(u) - \frac{\lambda}{c^{\overline{R}}}\int_0^a \overline{F}(u-kx)\phi(x)\,dx \\ &\quad - \frac{\lambda}{c^{\overline{R}}}\int_a^u \left[1 - (F(kx+a) - F(a))\right]\phi(x)\,dx \\ &= \phi(0) \end{aligned} \tag{32}$$

from which the kernel is clearly $K(u,x) = -\lambda H_1(x,u)/c^{\overline{R}}$ with

$$H_1(x,u) = \begin{cases} \overline{F}(u-kx) & kx < a \\ 1 - (F(kx+a) - F(a)) & kx \geq a \end{cases} \tag{33}$$

and the forcing function is $h(u) = \phi(0)$ as given by (24).

For the case with diffusion, repeated integration by parts of (30) on $[0,u]$ with respect to z yields the desired result.

$$\begin{aligned} &\phi(u) + \frac{2}{\sigma^2}\int_0^u \left(c^{\overline{R}} - \lambda(u-kx)\right)\phi(x)\,dx \\ &\quad + \frac{2\lambda}{\sigma^2}\left[\int_0^a G(u-kx)\phi(x)\,dx\right. \\ &\quad \left. + \int_a^u \left[F(kx+a) - F(a)\right](u-kx)\phi(x)\,dx\right] \\ &= \frac{\sigma^2\left(\phi(0) + u\phi'(0)\right) + 2c^{\overline{R}}u\phi(0)}{\sigma^2} \end{aligned} \tag{34}$$

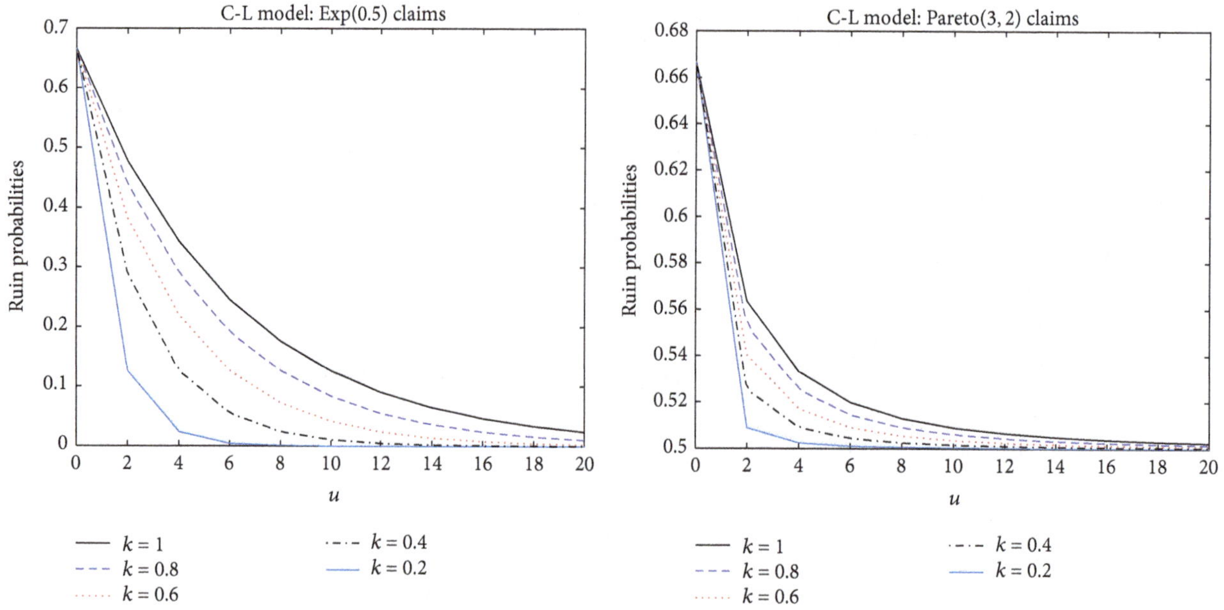

(a) Exp(0.5) claims

(b) Pareto(3, 2) claims

FIGURE 1: Ultimate ruin probabilities at different proportional retention levels in the Cramér-Lundberg model: $\lambda = 2$, $c = 6$.

which is a linear VIE of the second kind with $K(u, x)$ and $h(u)$ as given in (26).

4. Numerical Results

We solved (21) using the fourth-order block-by-block method, a full description of which can be found in [11, 14, 15]. Exp(β) refers to the exponential density $f(x) = \beta e^{-\beta x}$, so that the distribution function for the exponential distribution is $F(x) = 1 - e^{-\beta x}$ and its tail distribution is $\overline{F}(x) = 1 - F(x) = e^{-\beta x}$. The mean excess function for the exponential distribution is $e_F(x) = 1/\beta$ and $G(x) = x - (1/\beta)F(x)$. The Pareto(α, κ) distribution, which is a special case of the three-parameter Burr(α, κ, τ) distribution, has density $f(x) = \alpha\kappa^{\alpha}/(\kappa+x)^{\alpha+1}$ for $\alpha > 0$ and $\kappa = \alpha - 1 > 0$, and its distribution function is $F(x) = 1-(\kappa/(\kappa+x))^{\alpha}$. The tail distribution of the Pareto distribution is $\overline{F}(x) = (\kappa/(\kappa+x))^{\alpha}$ and its mean excess function is $e_F(x) = 1 + x/\kappa$, so that $G(x) = x - (1 + x/\kappa)F(x)$. A grid size of $h = 0.01$ was used throughout. The data simulations were performed using a Samsung Series 3 PC with an Intel Celeron 847 processor at 1.10 GHz and 6.0 GB RAM. To reduce computing time, the numerical method was implemented using the FORTRAN programming language, taking advantage of its DOUBLE PRECISION feature which gives a high degree of accuracy. The figures were constructed using MATLAB R2016a.

4.1. Ultimate Ruin Probability in the Cramér-Lundberg Model Compounded by Proportional Reinsurance. Here, the surplus process takes the form

$$U_t^{\overline{R}} = u + kct - \sum_{i=1}^{N_t} kX_i. \tag{35}$$

So, the survival probability $\phi(u)$ satisfies (21) and (22) with $a = \infty$ and $c^{\overline{R}} = kc$; that is, it satisfies a VIE of the second kind with kernel and forcing function given by

$$K(u, x) = -\frac{\lambda\overline{F}(u - kx)}{kc}$$
$$h(u) = \phi(0). \tag{36}$$

Figure 1 shows the ultimate ruin probabilities in the Cramér-Lundberg model for different proportional reinsurance retention levels k and provides validity for the assertion that reinsurance does in fact reduce the ruin probability, thus increasing the insurance company's chances of survival. The results for the case $k = 1$ (no reinsurance) are the same as those obtained in [14].

4.2. Ultimate Ruin Probability in the Cramér-Lundberg Model Compounded by Excess-of-Loss Reinsurance. This is the case of $k = 1$ and $\sigma = 0$, so the surplus process is

$$U_t^{\overline{R}} = u + c^{\overline{R}}t - \sum_{i=1}^{N_t} X_i \wedge a, \tag{37}$$

where $c^{\overline{R}} = c - (1 + \theta)\lambda\mathbb{E}[(X_i - a)^+]$. Here, for the case $\underline{a} < a < u$, the kernel and forcing function are given by

$$K(u, x) = -\frac{\lambda H(x, u)}{c^{\overline{R}}}$$
$$h(u) = \phi(0) \tag{38}$$

Table 1: Ruin probabilities for XL reins. in CLM: Exp(0.5) claims ($\lambda = 2, c = 6$).

u	$\psi_{\infty}(u)$	$\psi_{35}(u)$	$\psi_{30}(u)$	$\psi_{25}(u)$	$\psi_{20}(u)$
0	0.6667	0.6667	0.6667	0.6667	0.6667
2	0.4777	0.4777	0.4777	0.4777	0.4777
4	0.3423	0.3423	0.3423	0.3423	0.3422
6	0.2453	0.2453	0.2453	0.2453	0.2453
8	0.1757	0.1757	0.1757	0.1757	0.1757
10	0.1259	0.1259	0.1259	0.1259	0.1258
12	0.0902	0.0902	0.0902	0.0902	0.0901
14	0.0646	0.0646	0.0646	0.0646	0.0646
16	0.0463	0.0463	0.0463	0.0463	0.0462
18	0.0332	0.0332	0.0332	0.0332	0.0331
20	0.0238	0.0238	0.0238	0.0238	0.0237

Table 2: Ruin probabilities for XL reins. in CLM: Par(3, 2) claims ($\lambda = 2, c = 6$).

u	$\psi_{\infty}(u)$	$\psi_{35}(u)$	$\psi_{30}(u)$	$\psi_{25}(u)$	$\psi_{20}(u)$
0	0.6667	0.6667	0.6667	0.6667	0.6667
2	0.5634	0.5636	0.5637	0.5639	0.5641
4	0.5331	0.5335	0.5336	0.5338	0.5341
6	0.5198	0.5202	0.5204	0.5206	0.5210
8	0.5130	0.5134	0.5135	0.5138	0.5142
10	0.5090	0.5095	0.5096	0.5099	0.5103
12	0.5066	0.5070	0.5072	0.5074	0.5079
14	0.5050	0.5054	0.5056	0.5058	0.5063
16	0.5039	0.5043	0.5045	0.5048	0.5052
18	0.5031	0.5036	0.5037	0.5040	0.5044
20	0.5025	0.5030	0.5032	0.5034	0.5039

with

$$H(x,u) = \begin{cases} \overline{F}(u-x) & x < a \\ 1 - (F(x+a) - F(a)) & x \geq a. \end{cases} \tag{39}$$

This is simply (22) and (24) with $k = 1$ and $c^{\overline{R}} = c - (1 + \theta)\lambda\mathbb{E}[(X_i - a)^+]$.

Ruin probabilities for the Cramér-Lundberg model compounded by excess-of-loss (XL) reinsurance are given in Table 1 for different values of the XL retention level a ranging from 20 to infinity. Clearly, for Exp(0.5) claims, the ruin probabilities for the different retention levels reduce only very slightly as the retention level reduces. For Pareto(3, 2) claims, the ruin probabilities increase slightly as the retention level reduces (as shown in Table 2), meaning that it is optimal not to reinsure. But comparing these probabilities with Figure 1 leads to the conclusion that proportional reinsurance results in much lower ruin probabilities for the CLM as well as the perturbed model.

4.3. Ultimate Ruin Probability in the Perturbed Classical Risk Process Compounded by Proportional Reinsurance. The survival probability $\phi(u)$ satisfies (21) and (26) with $a = \infty$; that is,

$$\begin{aligned} &\phi(u) \\ &\quad + \frac{2}{\sigma^2}\int_0^u [kc - \lambda(u-kx) + \lambda G(u-kx)]\,\phi(x)\,dx \\ &= \frac{\sigma^2\left(\phi(0) + u\phi'(0)\right) - 2kcu\phi(0)}{\sigma^2} \end{aligned} \tag{40}$$

which is a VIE of the second kind with kernel and forcing function given, respectively, by

$$\begin{aligned} K(u,x) &= \frac{2[kc - \lambda(u-kx) + \lambda G(u-kx)]}{\sigma^2} \\ h(u) &= u\phi'(0) \quad \text{if } \sigma^2 > 0. \end{aligned} \tag{41}$$

Figure 2 depicts the ruin probabilities for the diffusion-perturbed model compounded by proportional reinsurance for different retention levels ranging from $k = 1$ (no reinsurance) to $k = 0.2$ (80% reinsurance). In the case of both Exp(0.5) claims and Pareto(3, 2) claims, applying proportional reinsurance significantly reduces the ultimate ruin probability of an insurance company.

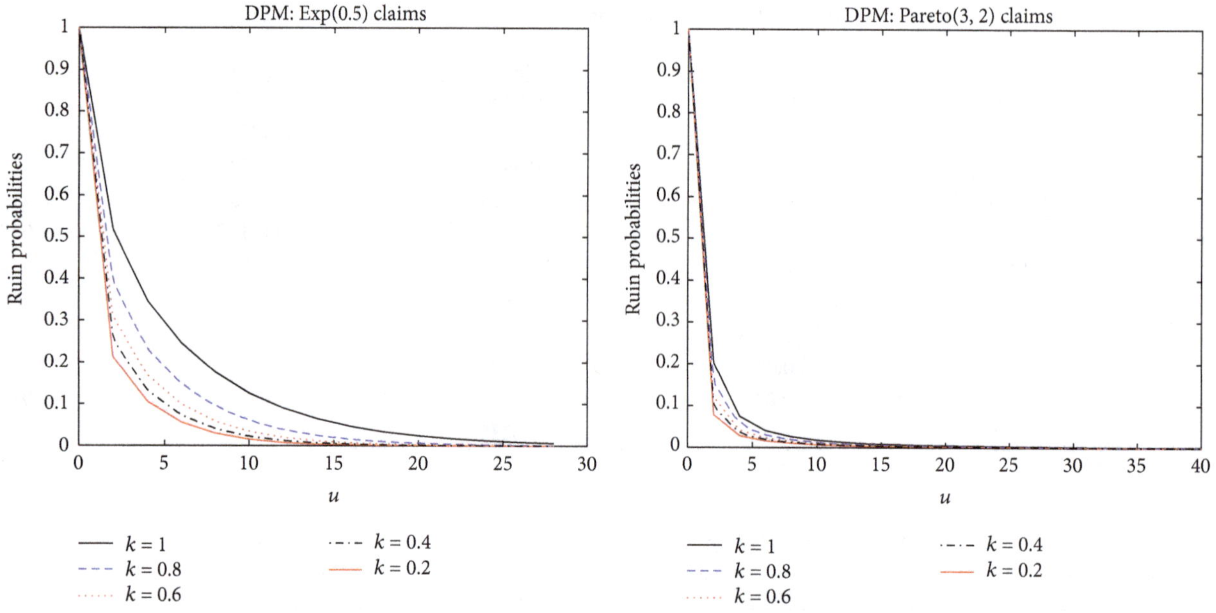

(a) Exp(0.5) claims

(b) Pareto(3, 2) claims

FIGURE 2: Ultimate ruin probabilities at different proportional retention levels in the diffusion-perturbed model: $\lambda = 2, c = 6, \sigma = 0.02$.

TABLE 3: Ruin probabilities for XL reins. in DPM: Exp(0.5) claims ($\lambda = 2, c = 6, \sigma = 0.02$).

u	$\psi_{\infty}(u)$	$\psi_{35}(u)$	$\psi_{30}(u)$	$\psi_{25}(u)$	$\psi_{20}(u)$
0	1.0000	1.0000	1.0000	1.0000	1.0000
2	0.5159	0.5159	0.5159	0.5155	0.5109
4	0.3467	0.3467	0.3466	0.3461	0.3399
6	0.2458	0.2458	0.2457	0.2451	0.2380
8	0.1759	0.1759	0.1758	0.1752	0.1674
10	0.1257	0.1258	0.1257	0.1250	0.1167
12	0.0901	0.0901	0.0901	0.0893	0.0807
14	0.0646	0.0646	0.0645	0.0638	0.0550
16	0.0463	0.0463	0.0463	0.0455	0.0365
18	0.0333	0.0333	0.0332	0.0324	0.0233
20	0.0240	0.0241	0.0240	0.0232	0.0140

4.4. Ultimate Ruin Probability in the Perturbed Classical Risk Process Compounded by Excess-of-Loss Reinsurance. The survival probability $\phi(u)$ satisfies a VIE of the second kind with kernel $K(u,x)$ as given in (23) (for the case $u \le \underline{a} < a$) and (26) (for the case $\underline{a} < a < u$), with $k = 1$, and forcing function $h(u) = u\phi'(0)$ in both cases. That is,

$$\text{for } u \le \underline{a} < a, K(u,x) = 2[c^{\overline{R}} + \lambda G(u-x) - \lambda(u-x)]/\sigma^2;$$

$$\text{for } \underline{a} < a < u, K(u,x) = 2[c^{\overline{R}} + \lambda H_2(x,u) - \lambda(u-x)]/\sigma^2$$

with

$$H_2(x,u) = \begin{cases} G(u-x) & x < a \\ (F(x+a) - F(a))(u-x) & x \ge a. \end{cases} \quad (42)$$

The impact of XL reinsurance on the ruin probabilities in a diffusion-perturbed model is evident from Table 3 which shows a reduction in the ruin probabilities for XL retentions not exceeding $a = 30$ for small claims. However, as can be seen from Table 4, the ruin probabilities for large claims are higher for values of a exceeding 150 but reduce significantly for values of a below 150. But again, if we compare these results with Figure 2 we see that the ruin probabilities are much lower for proportional reinsurance.

4.5. Optimal Reinsurance Strategy: Asymptotic Ruin Probabilities. It is known that the optimal quota-share retention k^* tends to the asymptotically optimal k^ρ that maximizes the adjustment coefficient ρ [13]. Therefore, since it was not possible to determine the optimal retention k^* from the results discussed in Sections 4.1–4.4, we will use asymptotic ruin probabilities. For illustrative purposes, we will now find

TABLE 4: Ruin probabilities for XL reins. in DPM: Par(3, 2) claims ($\lambda = 2, c = 6, \sigma = 0.02$).

u	$\psi_{\infty}(u)$	$\psi_{200}(u)$	$\psi_{150}(u)$	$\psi_{100}(u)$	$\psi_{50}(u)$
0	1.0000	1.0000	1.0000	1.0000	1.0000
2	0.2026	0.2029	0.2027	0.2022	0.1973
4	0.0744	0.0747	0.0745	0.0740	0.0683
6	0.0401	0.0405	0.0403	0.0397	0.0338
8	0.0257	0.0260	0.0258	0.0252	0.0192
10	0.0171	0.0174	0.0172	0.0167	0.0106
12	0.0124	0.0127	0.0125	0.0119	0.0058
14	0.0093	0.0096	0.0094	0.0088	0.0027
16	0.0072	0.0075	0.0073	0.0067	0.0006
18	0.0058	0.0061	0.0059	0.0054	0.0008
20	0.0050	0.0054	0.0052	0.0042	0.0015

TABLE 5: Asympt. ruin prob. for CLM with proportional reins. (Pareto claims) ($c = 6, \lambda = 2, \theta = \eta = 1$).

u	$\psi_1(u)$	$\psi_{0.6}(u)$	$\psi_{0.2}(u)$	$\psi_{0.05}(u)$	$\psi_{0.0125}(u)$	$\psi_{0.003125}(u)$
0	1.0000	1.0000	1.0000	1.0000	1.0000	1.0000
2	0.3333	0.2308	0.0909	0.0244	0.0062	0.0016
4	0.2000	0.1304	0.0476	0.0123	0.0031	0.0008
6	0.1429	0.0909	0.0323	0.0083	0.0021	0.0005
8	0.1111	0.0698	0.0244	0.0062	0.0016	0.0004
10	0.0909	0.0566	0.0196	0.0050	0.0012	0.0003
12	0.0769	0.0476	0.0164	0.0041	0.0010	0.0003
14	0.0667	0.0411	0.0141	0.0036	0.0009	0.0002
16	0.0588	0.0361	0.0123	0.0031	0.0008	0.0002
18	0.0526	0.0323	0.0110	0.0028	0.0007	0.0002
20	0.0476	0.0291	0.0099	0.0025	0.0006	0.0002

the optimal strategies only in the CLM for both the small and large claim cases.

4.5.1. Exponential Claims. We note, as in [13], that for exponential claims the optimal choice of the quota-share retention k that maximizes the adjustment coefficient $\rho(k)$ is given by

$$k^{\rho} = \min\left\{\left(1 - \frac{\eta}{\theta}\right)\left(1 + \frac{1}{\sqrt{1+\theta}}\right), 1\right\}, \qquad (43)$$

where θ and η are, respectively, the safety loadings of the reinsurer and insurer. Because maximizing the adjustment coefficient yields the asymptotically best strategy, we expect that the optimal retention k^* will tend to k^{ρ}. Since this study assumes cheap reinsurance (i.e., $\theta = \eta$), we have the fact that $k^{\rho} = 0$. That is, it is optimal for the insurance company to reinsure the entire portfolio or to take full proportional reinsurance.

4.5.2. Pareto Claims. For a given initial surplus u and a retention level $k \in [0, 1]$, let the calculated ruin probability be given by $\psi_k(u)$. Then, for large claims, the asymptotic values of the ruin probability are given by

$$\psi_k(u) = \frac{1}{k\theta - (\theta - \eta)} \frac{k}{1 + u/k}. \qquad (44)$$

This ruin probability is minimized when $k^{\rho} = 2(\theta-\eta)u/(\theta u - (\theta - \eta))$. Thus, for Pareto-distributed claims, assuming $\theta = \eta = 1$, we find that $\psi_k(u) = k/(k+u)$ and that $k^{\rho} = 0$ as well. The insurance company should reinsure the entire portfolio of risks. The results for different values of k are summarized in Table 5 and shown in Figure 3.

It is clear from Figure 3 that the ruin probabilities become smaller as $k \to 0$, meaning that the asymptotically optimal retention must be $k^{\rho} = 0$. This confirms the results shown in Figure 1. And since the optimal retention k^* tends to the asymptotically optimal k^{ρ} that maximizes the adjustment coefficient, it follows that $k^* = 0$. This means that the insurance company must cede the entire portfolio of risks to a reinsurer. We can therefore conclude that the optimal combinational quota-share and XL reinsurance strategy is $(k^*, a^*) = (0, \infty)$.

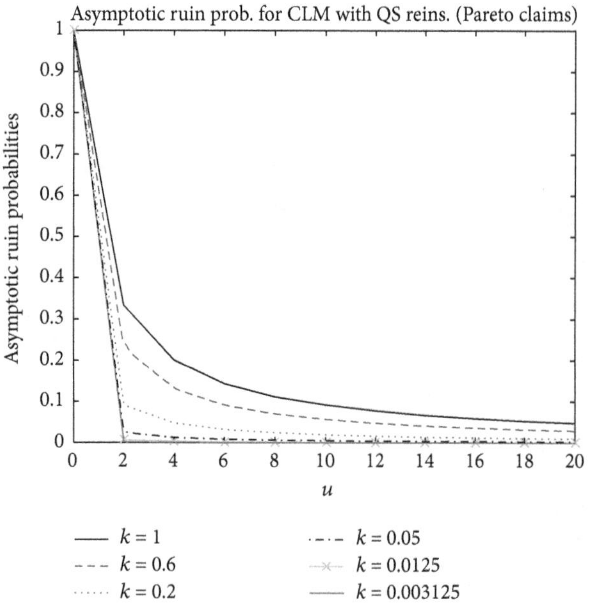

FIGURE 3: Asymptotic ruin probabilities for large claims in the CLM with proportional reinsurance ($c = 6, \lambda = 2, \theta = \eta = 1$).

5. Conclusion

While the results presented in the previous section show that proportional and XL reinsurance both result in a reduction in the ruin probabilities, the reduction is more drastic for Pareto than for exponential claims in both the Cramér-Lundberg and diffusion-perturbed models. On the one hand, a comparison of the figures presented in the foregoing shows that proportional reinsurance results in lower ruin probabilities than XL reinsurance and is therefore optimal. The optimal quota-share retention was found as $k^* = 0$, meaning that in both the small and large claim cases in the Cramér-Lundberg model, it is optimal for the insurance company to reinsure the whole portfolio using proportional reinsurance. Going by the results in Figure 3, the same conclusion can be drawn about the diffusion-perturbed model. Thus, the optimal combinational quota-share and XL reinsurance strategy is a pure quota-share reinsurance with $k^* = 0$; that is, $(k^*, a^*) = (0, \infty)$. It should be noted that full reinsurance is not ideal from the reinsurer's standpoint and this provides a strong argument for the use of noncheap reinsurance.

On the other hand, the literature shows that the optimal reinsurance strategy is a pure XL, that is, $(1, a^*)$ (see, e.g., [7, 8, 16]). Possible extensions to the work are the inclusion of investments and dividend payouts as well as considering noncheap reinsurance, whereby, for a given risk, the reinsurer requires more premium and therefore uses a higher safety loading, than the insurer.

Conflicts of Interest

The authors declare that there are no conflicts of interest regarding the publication of this paper.

Acknowledgments

This work was supported by Mulungushi University, the Nelson Mandela African Institution of Science and Technology, and the Zambian Ministry of Higher Education through the Support to Science and Technology Education Project (SSTEP) funded by the African Development Bank Group. The authors wish to also thank Christian Kasumo whose M.S. thesis [15] provided some theoretical background on proportional reinsurance, ultimate ruin probabilities, and the numerical solution of Volterra integral equations.

References

[1] S. D. Promislow and V. R. Young, "Minimizing the probability of ruin when claims follow Brownian motion with drift," *North American Actuarial Journal*, vol. 9, no. 3, pp. 109–128, 2005.

[2] J. Kasozi, C. W. Mahera, and F. Mayambala, "Controlling ultimate ruin probability by quota-share reinsurance arrangements," *International Journal of Applied Mathematics and Statistics*, vol. 49, no. 19, pp. 1-15, 2013.

[3] C. S. Liu and H. Yang, "Optimal investment for an insurer to minimize its probability of ruin," *North American Actuarial Journal*, vol. 8, no. 2, pp. 11–31, 2004.

[4] C. Hipp and M. Plum, "Optimal investment for insurers," *Insurance: Mathematics & Economics*, vol. 27, no. 2, pp. 215–228, 2000.

[5] H. Schmidli, "On minimizing the ruin probability by investment and reinsurance," *The Annals of Applied Probability*, vol. 12, no. 3, pp. 890–907, 2002.

[6] P. Li, M. Zhou, and C. Yin, "Optimal reinsurance with both proportional and fixed costs," *Statistics & Probability Letters*, vol. 106, pp. 134–141, 2015.

[7] X. Hu and L. Zhang, "Ruin probability in a correlated aggregate claims model with common Poisson shocks: application to reinsurance," *Methodology and Computing in Applied Probability*, vol. 18, no. 3, pp. 675–689, 2016.

[8] X. Zhang and Z. Liang, "Optimal layer reinsurance on the maximization of the adjustment coefficient," *Numerical Algebra, Control and Optimization*, vol. 6, no. 1, pp. 21–34, 2016.

[9] B.-G. Jang and K. T. Kim, "Optimal reinsurance and asset allocation under regime switching," *Journal of Banking & Finance*, vol. 56, pp. 37–47, 2015.

[10] R. Verlaak and J. Beirlant, "Optimal reinsurance programs. An optimal combination of several reinsurance protections on a heterogeneous insurance portfolio," *Insurance: Mathematics & Economics*, vol. 33, no. 2, pp. 381–403, 2003.

[11] P. Linz, *Analytical and Numerical Methods for Volterra Equations*, vol. 7 of *SIAM Studies in Applied Mathematics*, Society for Industrial and Applied Mathematics (SIAM), Philadelphia, Pa, USA, 1985.

[12] L. Centeno, "On combining quota-share and excess of loss," *ASTIN Bulletin*, vol. 15, no. 1, pp. 49–63, 1985.

[13] H. Schmidli, *Stochastic Control in Insurance*, Probability and Its Applications (New York), Springer, London, UK, 2008.

[14] J. Paulsen, J. Kasozi, and A. Steigen, "A numerical method to find the probability of ultimate ruin in the classical risk model with stochastic return on investments," *Insurance: Mathematics & Economics*, vol. 36, no. 3, pp. 399–420, 2005.

13

Nonlinear Waves in Rods and Beams of Power-Law Materials

Dongming Wei,[1] Piotr Skrzypacz,[1] and Xijun Yu[2]

[1]*Department of Mathematics, School of Science and Technology, Nazarbayev University, Astana 010000, Kazakhstan*
[2]*Institute of Applied Physics and Computational Mathematics, P.O. Box 8009, Beijing 100088, China*

Correspondence should be addressed to Dongming Wei; dongming.wei@nu.edu.kz

Academic Editor: Xin-Lin Gao

Some novel traveling waves and special solutions to the 1D nonlinear dynamic equations of rod and beam of power-law materials are found in closed forms. The traveling solutions represent waves of high elevation that propagates without change of forms in time. These waves resemble the usual kink waves except that they do not possess bounded elevations. The special solutions satisfying certain boundary and initial conditions are presented to demonstrate the nonlinear behavior of the materials. This note demonstrates the apparent distinctions between linear elastic and nonlinear plastic waves.

1. Introduction

Free vibrations of rods and beams of power-law materials are considered. Analytic traveling wave solutions to the wave equations for power-law materials (see [1, 2]) are obtained which represent kink waves of single elevation that propagates without change of forms in time. It is shown that, unlike the wave equations for linear materials, the nonlinear wave equations do not allow arbitrary traveling wave forms in an infinite rod or beam. The results demonstrate that the traveling fronts of the waves may sharpen or flatten as the wave speeds increase depending upon the power-law index n and the bulk modulus. For $n > 1$, the wave fronts sharpen, whereas for $0 < n < 1$, the fronts flatten as the wave speeds increase. The solutions also demonstrate that the speeds of the nonlinear traveling waves depend not only on the material properties but also on the initial energy-level. It is well known that the speeds of waves for the linear elastic materials ($n = 1$, Hooke's law) depend only on the material properties in contrast to that of the waves in nonlinear materials. As far as we know these solutions are not available in literature, even though there are numerous research papers and books devoted to the discovery and study of traveling waves in elastic and plastic solids (see [3–7] for details). In the case of rods and beams of finite length, we also present some special solutions satisfying certain boundary and initial conditions. The closed formula solutions are expressed in terms of non-Euclidean sine functions (cf. [8]), which differ from the Euclidean sine functions corresponding to the waves in rods and beams of linear elastic materials.

The note is organized as follows. In Section 2, the power-law constitutive stress-strain equation is introduced. In Section 3, the potential energy and derivations of the wave equations of power-law materials are outlined. In Sections 4 and 5, closed-form solutions are derived. And, finally the results are summarized in Section 6.

2. Hollomon's Equation

It is well known that, in uniaxial state, the following power-law stress and strain relation is used for certain elastoplastic materials:

$$\sigma = K\,|\varepsilon|^{n-1}\,\varepsilon, \quad 0 < n < \infty, \tag{1}$$

where σ is the axial stress, ε is the axial strain, and K and n are engineering constants with values depending on the specific material. The materials satisfying (1) sometimes are also referred to as Ludwick or as Hollomon's materials in literature (cf. [1, 2]). Many heat-treated metals are well-known power-law materials. For a given annealed metal or alloy, K and n depend on the heat treatment received by the metal or alloy. The values of n are typically between 0 and 1 for such metals. For a comprehensive list of experimental values of K and n of common annealed industrial metals, see, for example, [9]. For some geological materials, such as certain rocks or ice, however, the values of n are greater than 1. In some biological tissues, experiments also indicate that the power-law index n satisfies $0 < n < 1$ for bones such as tibia and femur, while $n > 1$ for cartilages such as common carotid artery and abdominal aorta (see, e.g., [10, 11]). For a given value of $0 < n < 1$, the stress-strain curve defined by (1) can result in a rapid increase in the yield stress for small strains or strain hardening. However, it can be the opposite for values of $n > 1$, for which large strains produce small stress or softening. For these reasons, n is called the strain-hardening or strain-softening exponent. Study of the mechanical properties of these heat-treated metals is very important in industries (see, e.g., [12], for stress analysis of beam columns made of Ludwick materials). If we allow $n = 1$, then (1) reduces to Hooke's law for linear elastic material and the constant K, also called the bulk modulus, equals the corresponding Young's modulus E. Power-law materials are a special case of a more general class of materials called Hencky plastics [13]. Physically, the constitutive equation (1) describes the hardening or softening of materials showing an elastic-plastic transition. In the following, bold letters are used to denote vectors or matrices. A vector is considered as a single row matrix. The transpose of a matrix $\mathbf{A}$ is denoted by $\mathbf{A}^T$, and the inner product of two vectors $\mathbf{u}$ and $\mathbf{v}$ by $\mathbf{uv}^T$. The time derivative $\partial\mathbf{u}/\partial t$ is denoted by $\dot{\mathbf{u}}$. Let $\mathbf{u}(x, y, z, t) = (u(x, y, z, t), v(x, y, z, t), w(x, y, z, t))$ denote the displacement vector,

$$\begin{aligned}
\varepsilon_x &= \frac{\partial u}{\partial x}, \\
\varepsilon_y &= \frac{\partial v}{\partial y}, \\
\varepsilon_z &= \frac{\partial w}{\partial z}, \\
\gamma_{xy} &= \frac{1}{2}\left(\frac{\partial u}{\partial y} + \frac{\partial v}{\partial x}\right), \\
\gamma_{yz} &= \frac{1}{2}\left(\frac{\partial v}{\partial z} + \frac{\partial w}{\partial y}\right), \\
\gamma_{zx} &= \frac{1}{2}\left(\frac{\partial w}{\partial x} + \frac{\partial u}{\partial z}\right)
\end{aligned} \tag{2}$$

the strain components, and σ_x, σ_y, σ_z, τ_{xy}, τ_{yz}, and τ_{zx} the corresponding stress components. The following generalized power law can be derived from the Hencky total deformation theory [13]:

$$\begin{Bmatrix} \sigma_x \\ \sigma_y \\ \sigma_z \\ \tau_{xy} \\ \tau_{yz} \\ \tau_{zx} \end{Bmatrix} = \frac{K\,\|D(\mathbf{u})\|^{n-1}}{(1+\nu)(1-2\nu)} \cdot \begin{pmatrix} 1-\nu & \nu & \nu & 0 & 0 & 0 \\ \nu & 1-\nu & \nu & 0 & 0 & 0 \\ \nu & \nu & 1-\nu & 0 & 0 & 0 \\ 0 & 0 & 0 & \frac{1-2\nu}{2} & 0 & 0 \\ 0 & 0 & 0 & 0 & \frac{1-2\nu}{2} & 0 \\ 0 & 0 & 0 & 0 & 0 & \frac{1-2\nu}{2} \end{pmatrix} \begin{Bmatrix} \varepsilon_x \\ \varepsilon_y \\ \varepsilon_z \\ \gamma_{xy} \\ \gamma_{yz} \\ \gamma_{zx} \end{Bmatrix}, \tag{3}$$

where $\|D(\mathbf{u})\| = \sqrt{\varepsilon_x^2 + \varepsilon_y^2 + +\varepsilon_z^2 + 2\gamma_{xy}^2 + 2\gamma_{yz}^2 + 2\gamma_{zx}^2}$, where n, K, and ν are the material constants; see also Wei [14]. Note that (3) is the three-dimensional version of (1). In the following two sections, wave equations of bars and beams made of the power-law elastoplastic materials are derived by (3) and the assumption of the Euler-Bernoulli beam theory. There are similar versions of generalized power-law stress-strain relations for strain-hardening or strain-softening material in the literature and similar wave equations can be derived (see, e.g., [15–20]).

3. The Nonlinear Wave Equations

The potential energy for a power-law elastoplastic body occupying a three-dimension body V can by defined by

$$U = \frac{1}{n+1}\int_V \sigma\varepsilon^T dV, \tag{4}$$

where $\boldsymbol{\varepsilon} = (\varepsilon_x, \varepsilon_y, \varepsilon_z, \gamma_{xy}, \gamma_{xz}, \gamma_{yz})$ and $\boldsymbol{\sigma} = (\sigma_\mathbf{x}, \sigma_\mathbf{y}, \sigma_\mathbf{z}, \tau_\mathbf{xy}, \tau_\mathbf{xz}, \tau_\mathbf{yz})$. The Lagrangian energy functional $I(\mathbf{u})$ equals the kinetic energy T minus the elastoplastic potential energy U plus the work W done by external force. It can be written as

$$\begin{aligned} I(\mathbf{u}) = {} & \frac{1}{2}\int_V \rho\dot{\mathbf{u}}\dot{\mathbf{u}}^T dV - \frac{1}{n+1}\int_V \sigma\varepsilon^T dV + \int_V \mathbf{f}\mathbf{u}^T dV \\ & + \int_{\partial V} \mathbf{t}\mathbf{u}^T dS, \end{aligned} \tag{5}$$

where ρ is the density, $\dot{\mathbf{u}} = (\dot{u}, \dot{v}, \dot{w})$ the velocity, $\mathbf{f} = (f_x, f_y, f_z)$ the body force, and $\mathbf{t} = (t_x, t_y, t_z)$ the surface force. See, for example, [21], for a standard definition of $I(\mathbf{u})$. For a uniaxial bar of infinite length with cross-sectional area $A(x)$, subject to axial force and zero surface force, we have $\boldsymbol{\sigma} = (\sigma_x, 0, 0, 0, 0, 0)$, $\mathbf{u} = (u(x,t), 0, 0)$, $\sigma_x = K|\varepsilon_x|^{n-1}\varepsilon_x$, $\mathbf{f} = (f(x,t), 0, 0)$, and $\mathbf{t} = (0, 0, 0)$. For an Euler beam of infinite length, it is assumed that the components of the displacement satisfy $u(x, y, t) = -y(\partial v/\partial x)$, $v = v(x,t)$, $w = 0$, $\mathbf{f} = (0, r(x,t), 0)$, and $\mathbf{t} = (0, 0, 0)$. Therefore $\varepsilon_x = \partial u/\partial x = -y(\partial^2 v/\partial x^2)$, $\varepsilon_{xy} = (1/2)(\partial u/\partial y + \partial v/\partial x) = 0$, and $\varepsilon_y = \varepsilon_{xz} = \varepsilon_{yz} = \varepsilon_z = 0$. The potential energies for the bar and the beam are given by

$$U = \frac{1}{n+1}\int_{-\infty}^{+\infty} KA\left|\frac{\partial u}{\partial x}\right|^{n+1} dx, \tag{6}$$

$$U = \frac{1}{n+1}\int_{-\infty}^{+\infty} KI_n\left|\frac{\partial^2 v}{\partial x^2}\right|^{n+1} dx, \tag{7}$$

respectively, where $I_n = \int_A |y|^{n+1} dy\,dz$ is the generalized second moment of inertia of the beam. The x-axis is taken to be the axial direction of the bar and the beam. For rods and beams of finite length L, the corresponding Lagrangian functions are given by replacing $-\infty$ and $+\infty$ in (6) and (7) by 0 and L, respectively. Note that the assumptions made in this section on the elastoplastic bars and the beams are standard assumptions frequently made for elastic bars and beams (see, e.g., [22, 23], for details). The corresponding linear wave equations of elastic bars and beams corresponding to $n = 1$ have been studied extensively.

For completeness, the derivation of the wave equations of the power-law materials given in [14] is outlined here. It is well known that Hamilton's principle seeks an equilibrium state in time dependent mechanical systems (see, e.g., [21]).

Specifically, Hamilton's principle requires that we seek a displacement $\mathbf{u}$ so that, for any time interval $[t_1, t_2]$, $\mathbf{u}(t_1) = \mathbf{u}(t_2)$ and $\dot{\mathbf{u}}(t_1) = \dot{\mathbf{u}}(t_2)$, and for all displacement of the form $\mathbf{u} + \tau\mathbf{v}$, where τ is any real number, the first variation of the energy functional I satisfies

$$\delta I = \int_{t_1}^{t_2} \frac{d}{d\tau} \left[I \left(\mathbf{u}(t) + \tau \mathbf{v}(t) \right) \right] \bigg|_{\tau=0} dt = 0 \qquad (8)$$

for all $\mathbf{v}$ satisfying $\mathbf{v}(t_1) = \mathbf{v}(t_2) = \mathbf{0}$ and $\dot{\mathbf{v}}(t_1) = \dot{\mathbf{v}}(t_2) = \mathbf{0}$. The combination $\mathbf{u}(t) + \tau\mathbf{v}(t)$ is referred to as an admissible displacement for the mechanical system since it is required to satisfy some boundary conditions. It can be shown that if the displacement $\mathbf{u}$ satisfies (8) of Hamilton's principle, then it must also satisfy a differential wave equation under certain conditions. In particular, suppose that the cross-sectional area, denoted by A, is a nonzero constant, and then for the rod, we have

$$\rho \frac{\partial^2 u}{\partial t^2} = K \frac{\partial}{\partial x} \left(\left| \frac{\partial u}{\partial x} \right|^{n-1} \frac{\partial u}{\partial x} \right) + f, \quad x \in \mathbb{R},\ t \in \mathbb{R}^+ \qquad (9)$$

and for the corresponding Euler beam

$$\rho A \frac{\partial^2 v}{\partial t^2} = -\frac{\partial^2}{\partial x^2} \left(K I_n \left| \frac{\partial^2 v}{\partial x^2} \right|^{n-1} \frac{\partial^2 v}{\partial x^2} \right) + Ar, \quad x \in \mathbb{R},\ t \in \mathbb{R}^+. \qquad (10)$$

When $n = 1$, (9) reduces to the standard wave equation for the elastic bar

$$\rho \frac{\partial^2 u}{\partial t^2} = K \frac{\partial^2 u}{\partial x^2} + f, \quad x \in \mathbb{R},\ t \in \mathbb{R}^+ \qquad (11)$$

and (10) to the standard wave equation for the elastic Euler beam

$$\rho A \frac{\partial^2 v}{\partial t^2} = -\frac{\partial^2}{\partial x^2} \left(K I \frac{\partial^2 v}{\partial x^2} \right) + Ar, \quad x \in \mathbb{R},\ t \in \mathbb{R}^+. \qquad (12)$$

The quantity I_n reduces to the second moment of inertia, $I_n = \int_A |y|^{n+1} dA$ reduces to I when $n = 1$ in the elastic beam theory, and the material constant K becomes Young's modulus E for linear elastic materials. In deriving the wave equations (9) and (10), we have made the assumption that the solutions u and v are continuously differentiable and their appropriate lower order derivatives are bounded or vanishing when $|x| \to \infty$. By (8), we get

$$\int_{t_1}^{t_2} \int_{-\infty}^{+\infty} \left(\rho A \dot{u} \dot{v} - KA \left| \frac{\partial u}{\partial x} \right|^{n-1} \frac{\partial u}{\partial x} \frac{\partial v}{\partial x} + A f v \right) dx\, dt = 0. \qquad (13)$$

Using integration by parts and interchange of the order of integration, with $v(t_1) = v(t_2) = 0$, and assuming that $\lim_{x\to\pm\infty} |\partial u(x,t)/\partial x|^{n-1} (\partial u(x,t)/\partial x)$ is bounded by a constant independent of t and $\lim_{x\to\pm\infty} v(x,t) = 0$ uniformly in t, we get the following:

$$\int_{t_1}^{t_2} \int_{-\infty}^{+\infty} \left(-\rho A \ddot{u} + \frac{\partial}{\partial x} \left(KA \left| \frac{\partial u}{\partial x} \right|^{n-1} \frac{\partial u}{\partial x} \right) + Af \right) \cdot v\, dx\, dt = 0 \qquad (14)$$

from (13). Since v, t_1, and t_2 are arbitrary and $A \neq 0$, we then get (9) from (14). The corresponding beam equation (10) can be derived similarly which was reported in [14].

4. Traveling Waves in Rods and Beams of Arbitrary Length

In the following we will derive some traveling wave solutions to (9) and (10) for $0 < n < \infty$ and $n \neq 1$. As far as we know, these solutions are not available in literature, even though there are numerous research papers and books devoted to the discovery and study of traveling waves in elastic and plastic solids. For the study of traveling waves in nonlinear beam equations based on Hooke's law ($n = 1$) for elastic materials, see, for example, [3–5]. Also, see [6, 7, 24, 25], for more results of traveling waves in solids. Assuming that $f = r = 0$ in (9) and (10), we have

$$\frac{\partial^2 u}{\partial t^2} = c^2 \frac{\partial}{\partial x} \left(\left| \frac{\partial u}{\partial x} \right|^{n-1} \frac{\partial u}{\partial x} \right), \quad x \in \mathbb{R},\ t \in \mathbb{R}^+ \qquad (15)$$

for the bar and

$$\frac{\partial^2 v}{\partial t^2} = \tilde{c}^2 \frac{\partial^2}{\partial x^2} \left(\left| \frac{\partial^2 v}{\partial x^2} \right|^{n-1} \frac{\partial^2 v}{\partial x^2} \right), \quad x \in \mathbb{R},\ t \in \mathbb{R}^+ \qquad (16)$$

for the beam, where $c^2 = K/\rho$ and $\tilde{c}^2 = -KI_n/\rho A$, respectively. We look for traveling wave solutions of the form $g(x - \lambda t)$ for both (15) and (16), where λ denotes a constant and g is a function to be determined. Let $\phi(t) = |t|^{n-1} t$, where n is the index in power-law (3). The inverse of ϕ is $\phi^{-1}(t) = |t|^{(1-n)/n} t$ since

$$\begin{aligned} \left(\phi \circ \phi^{-1} \right)(t) &= \left| |t|^{(1-n)/n} t \right|^{n-1} |t|^{(1-n)/n} t \\ &= |t|^{(n-1)/n} |t|^{(1-n)/n} t = t. \end{aligned} \qquad (17)$$

First, let $\xi = x - \lambda t$ and substitute $u(x,t) = g(\xi)$ into (15), so $\lambda^2 g'' = c^2 (\phi(g'))'$. After integration we get $\lambda^2 g' = c^2 \phi(g') + c_1$, where c_1 is an arbitrary constant. Suppose that $\lim_{x\to\infty} (\partial u/\partial x)(x,0) = A$. Since $u(x,0) = g(x)$ and $g'(x) = (\partial u/\partial x)(x,0)$, we have $c_1 = \lambda^2 A - c^2 \phi(A)$. Looking for nontrivial solutions for $n \neq 1$ and assuming that $A = (c/\lambda)^{2/(1-n)}$, we get $c_1 = 0$ and $g' = \pm (c/\lambda)^{2/(1-n)}$ which gives the following traveling wave solutions:

$$u(x,t) = \pm \left(\frac{c}{\lambda} \right)^{2/(1-n)} (x - \lambda t) + c_2 \qquad (18)$$

for the bar equation (15). Note that solution (18) includes some physically meaningful solutions. For example, let us consider a semi-infinite bar with initial displacement

$$u(x,0)=\begin{cases}\left(\dfrac{c}{\lambda}\right)^{2/(1-n)}x & \text{if } 0<x<+\infty\\ 0 & \text{if } -\infty<x\le 0,\end{cases} \tag{19}$$

and initial velocity

$$\dot{u}(x,0)=\begin{cases}-\lambda\left(\dfrac{c}{\lambda}\right)^{2/(1-n)} & \text{if } 0<x<+\infty\\ 0 & \text{if } -\infty<x\le 0\end{cases} \tag{20}$$

and boundary condition $\lim_{x\to+\infty}(\partial u/\partial x)(x,t)=(c/\lambda)^{2/(1-n)}$. A particular solution satisfying these conditions is given by

$$u(x,t)=\begin{cases}\left(\dfrac{c}{\lambda}\right)^{2/(1-n)}(x-\lambda t) & \text{if } 0<x-\lambda t<+\infty\\ 0 & \text{if } -\infty<x-\lambda t\le 0\end{cases} \tag{21}$$

which is obtained from (18). The physical interpretation of the initial condition (19) is that half of the bar is initially subject to a constant stress and the other half is free of stress and fixed in position, and the second initial condition (20) means that the bar is initially moving at a constant speed and half of it is instantaneously stopped. Solution (21) explains that if a prestressed semi-infinite axial power-law rod subject to initial conditions $\sigma(x,0) = K(c/\lambda)^{2n/(1-n)}$ and $\dot{u}(x,0) > 0$ and boundary conditions $u(0,t) = 0$ for $x = 0$ and lower order derivatives are bounded or vanishing when $|x| \to \infty$, then the displacement in the interval $[0,x]$ will be zero at time $t = [x^{1+n}|\dot{u}(x,0)|^{1-n}/c^2]^{1/(1+n)} = x/\lambda$ and the restoration of the deformed bar in interval $[0,x(t)]$ to its undeformed configuration has a moving boundary $x(t)$ which is expanding like a kink wave at a velocity $\lambda = c(K/\sigma(x,0))^{(1-n)/2n}$.

In the more general situation, for any value of c_1, the equation $P(t) = \lambda^2 t - c^2\phi(t) - c_1$ has at least one solution since it is continuous, $\lim_{t\to+\infty}P(t) = +\infty$ and $\lim_{t\to-\infty}P(t) = -\infty$. Let $P(\varepsilon_0) = 0$, and then $g' = \varepsilon_0$ satisfies $\lambda^2 g' = c^2\phi(g') + c_1$. We have the following similar solutions:

$$u(x,t)=\begin{cases}\varepsilon_0(x-\lambda t) & \text{if } 0<x-\lambda t<+\infty\\ 0 & \text{if } -\infty<x-\lambda t\le 0\end{cases} \tag{22}$$

satisfying the initial and boundary conditions

$$u(x,0)=\begin{cases}\varepsilon_0 x & \text{if } 0<x<+\infty\\ 0 & \text{if } -\infty<x\le 0,\end{cases}$$
$$\dot{u}(x,0)=\begin{cases}-\lambda\varepsilon_0 & \text{if } 0<x<+\infty\\ 0 & \text{if } -\infty<x\le 0\end{cases} \tag{23}$$

for $\lim_{x\to+\infty}(\partial u/\partial x)(x,t) = \varepsilon_0$ and $\lambda\varepsilon_0^2 = c^2\phi(\varepsilon_0) + c_1$.

For the linear elastic bar, $n = 1$, $\phi(g') = g'$, and if $c_1 = 0$, the equation $\lambda^2 g' = c^2\phi(g')$ is satisfied for any g' and also makes $\lambda = c$. This shows that the linear elastic bar equation allows arbitrary traveling wave forms g in $u(x,t) = g(x-\lambda t)$, and however the wave can travel only at a fixed velocity $\lambda = C$. If $c_1 \neq 0$, then equation $\lambda^2 g' = c^2\phi(g') + c_1$ gives $g' = \varepsilon_0 = (\lambda^2 - c^2)/c_1$ and the corresponding solution is (22), which is similar to the solutions for $n \neq 1$.

The above shows that the difference between the nonlinear solution ($n \neq 1$) and the linear case ($n = 1$) is that all the nonlinear traveling waves have the same shape and the traveling velocity depends not only on the material property but also on the initial stress-level while the linear traveling waves can take any form while keeping a fixed traveling velocity c that depends only on the material property.

Similarly, by substituting $v(x,t) = g(x-\lambda t)$ into the beam equation (16), we get $\lambda^2 g'' = \bar{c}^2(\phi(g''))''$. After integration twice, we get $\lambda^2 g = \bar{c}^2\phi(g'') + c_1\xi + c_2$, which gives $\lambda^2 g = \bar{c}^2\phi(g'')$ by setting $c_1 = c_2 = 0$. Let $w = g'$; we get $w(dw/dg) = g''$ and $\lambda^2 g = \bar{c}^2\phi(w(dw/dg))$. From the last equation, we get

$$w\,dw = -\phi^{-1}\left(\frac{\lambda^2 g}{|\bar{c}|^2}\right)dg = -\left|\frac{\lambda}{\bar{c}}\right|^{2/n}|g|^{(1-n)/n}g\,dg \tag{24}$$

which gives $(g')^2 = C - (2n/(1+n))|\lambda/\bar{c}|^{2/n}|g|^{(1+n)/n}$, where C is the integration constant.

We assume $g'(0) = 0$ and $g(0) > 0$. So, the traveling wave solutions for the corresponding elastoplastic Euler beam are given implicitly

$$x-\lambda t = \pm\int_0^{g(x-\lambda t)}\frac{ds}{\sqrt{2|\lambda/\bar{c}|^{2/n}(n/(n+1))\left(|g(0)|^{1/n}g(0)-|s|^{1/n}s\right)}} \tag{25}$$

which results in the following formula in terms of generalized trigonometric function defined in [26]

$$g(x-\lambda t) = g(0)\sin_{2,1+1/n}(B(x-\lambda t)), \tag{26}$$

where $B = |\bar{c}|^{1/n}/|g(0)|^{(n+1)/2n}g(0)|\lambda|^{1/n}\sqrt{2n/(n+1)}$. Notice that for $n = 1$ we obtain the well-known Euclidean sine traveling wave solution for the elastic Euler beam equation. We observe that the amplitude of the wave is determined by the initial condition $v(0,0)$. The traveling waves for elastoplastic beams can be applied to study piezoelectric robots; see [27].

5. Special Waves in Rods and Beams of Finite Length

Let us consider some special waves in rods and beams of finite length L with fixed ends. Equations (15) and (16) are solved for $x \in (0,L)$ and $t \in R^+$ with homogeneous Dirichlet boundary conditions $u(0,t) = u(L,t) = 0$ and special initial conditions. We present some special solutions by using the generalized trigonometric functions developed by Drábek and Manásevich [26]. By using the separation of variables

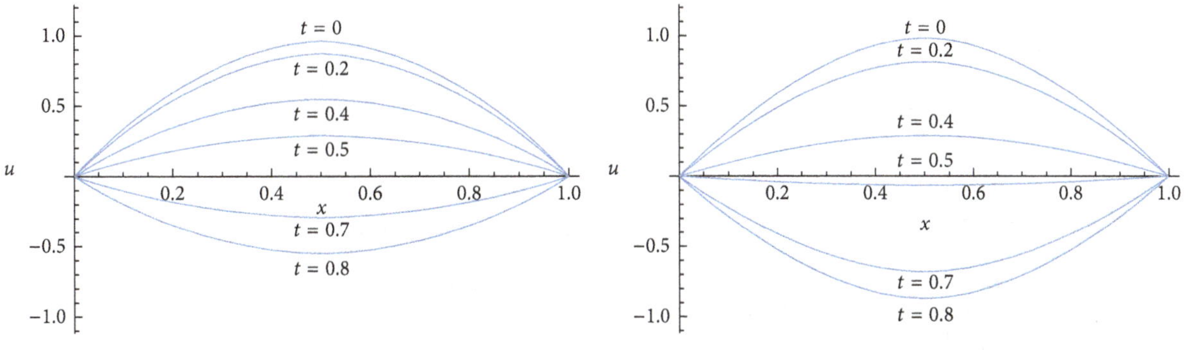

(a) (b)

FIGURE 1: Vibrating power-law strings: $n = 0.2$ (a) and $n = 0.5$ (b) at $t = 0, 0.2, 0.4, 0.5, 0.7, 0.8$.

$u(x,t) = X(x)T(t)$ in (15) and using boundary conditions $u(0,t) = u(L,t) = 0$, we have

$$\left(\left|X'\right|^{n-1} X'\right)' = \lambda X,$$
$$\ddot{T} - \lambda c^2 \left|T\right|^{n-1} T = 0. \tag{27}$$

From the first equation and the boundary conditions, we get

$$\left(\left|X'\right|^{n-1} X'\right)' = \lambda X,$$
$$X(0) = X(L) = 0. \tag{28}$$

By Thm 3.1 in [26], a sequence of solutions to the nonlinear eigenvalue problem (28) are given by $X_k = \sin_{n+1,2}(k(x\pi_{n+1,2}/L))$, where $\lambda_k = -((k+1)k/2)(\pi_{n+1,2}/L)^{n+1}$ and $\pi_{n+1,2} = \int_0^1 (1-t^2)^{-1/(n+1)} dt = B(1-1/(n+1), 1/2)$. Let us consider the initial conditions $u(x,0) = X_1(x)$ and $u_t(x,0) = 0$. In this case we have $T(0) = 1$ and we can solve the second equation in (27). A special solution of this initial value problem is given by

$$u(x,t) = \sin_{n+1,2}\left(\frac{x\pi_{n+1,2}}{L}\right) \cdot \sin_{2,n+1}\left(-\sqrt{2c}\left(\frac{\pi_{n+1,2}}{L}\right)^{(n+1)/2} t + \frac{\pi_{2,n+1}}{2}\right). \tag{29}$$

The time evolution of the special solutions for $c = 1$, $L = 1$, $n = 0.2$, and $n = 0.5$ is presented in Figure 1, respectively.

Similarly, by using the separation of variables $u(x,t) = X(x)T(t)$ in (16), we have

$$\left(\left|X''\right|^{n-1} X''\right)'' = \lambda X,$$
$$\ddot{T} - \lambda c^2 \left|T\right|^{n-1} T = 0. \tag{30}$$

From the first equation and the boundary condition, we get

$$\left(\left|X''\right|^{n-1} X''\right)'' = \lambda X,$$
$$X(0) = X'(0) = X(L) = X'(L) = 0. \tag{31}$$

An analytic solution to (31) is not available and is an open problem. This is a nonlinear and nonhomogeneous eigenvalue problem which belongs to an active area of research beyond the scope of this paper, and we post it here as an open problem. Since superposition principle can not be applied to nonlinear problems, the solutions to (15) and (16) with general initial and boundary conditions require further investigations.

6. Conclusions

Two nonlinear wave equations are derived: one is for the longitudinal vibrations of a power-law bar and the other is for vertical vibrations of the power-law Euler beam. Analytic traveling wave solutions are found for these two equations for free vibrations in terms of generalized sine functions of two parameters. We recovered the linear elastic waves as special cases of our solutions. The traditional ways of determining the vibrations of a structure made of untreated metals do not apply to the structures of heat-treated metals with hardening and softening mechanical properties. The obtained results can be useful in engineering applications of the power-law materials, such as heat-treated metals and polyimide plastics. Further study of wave propagation and vibrations in structures made of the power-law nonlinear bars and beams seems necessary.

Conflicts of Interest

The authors declare that there are no conflicts of interest regarding the publication of this paper.

References

[1] K. Lee, "Large deflections of cantilever beams of non-linear elastic material under a combined loading," *International Journal of Non-Linear Mechanics*, vol. 37, no. 3, pp. 439–443, 2001.

[2] G. Lewis and F. Monasa, "Large deflections of cantilever beams of non-linear materials of the Ludwick type subjected to an end moment," *International Journal of Non-Linear Mechanics*, vol. 17, no. 1, pp. 1–6, 1982.

[3] S. S. Antman, *Nonlinear Problems of Elasticity*, Springer-Verlag, New York, NY, USA, 1995.

[4] A. R. Champneys, P. J. McKenna, and P. A. Zegeling, "Solitary waves in nonlinear beam equations: stability, fission and fusion," *Nonlinear Dynamics*, vol. 21, no. 1, pp. 31–53, 2000.

[5] A. V. Porubov and M. G. Velarde, "Dispersive-dissipative solitons in nonlinear solids," *Wave Motion*, vol. 31, no. 3, pp. 197–207, 2000.

[6] W.-S. Duan and J.-B. Zhao, "Solitary waves in a quartic nonlinear elastic bar," *Chaos, Solitons and Fractals*, vol. 11, no. 8, pp. 1265–1267, 2000.

[7] V. I. Erofeev and N. V. Klyueva, "Solitons and nonlinear periodic strain waves in rods, plates, and shells (a review)," *Acoustical Physics*, vol. 48, no. 6, pp. 725–741, 2002.

[8] D. Wei, Y. Liu, and M. B. Elgindi, "Some generalized trigonometric sine functions and their applications," *Applied Mathematical Sciences*, vol. 6, no. 121-124, pp. 6053–6068, 2012.

[9] J. F. Shackelford, in *Introduction to Materials Science for Engineers*, Prentice-Hall, Englewood Cliffs, NJ, USA, 5th edition, 2000.

[10] Y. C. Fung, *Biomechanics: Mechanical Properties of Living Tissue*, Springer-Valag, New York, NY, USA, 2nd edition, 1993.

[11] A. J. Grodzinsky, R. D. Kamm, and D. A. Lauffenburger, "Quantitative aspects of tissue engineering: basic issues in kinetics, transport and mechanics," in *Principles of Tissue Engineering*, R. Lanza, R. Langer, and W. Chick, Eds., R.G. Landes Co, Austin, Tex, USA, 1997.

[12] X. Teng and T. Wierzbicki, "Crush response of an inclined beam-column," *Thin-Walled Structures*, vol. 41, no. 12, pp. 1129–1158, 2003.

[13] H. Hencky, "Zur Theorie plastischer Deformationen und der hierdurch im Material hervorgerufenen Nachspannungen," *Journal of Applied Mathematics and Mechanics*, vol. 4, no. 4, pp. 323–334, 1924.

[14] D. Wei, "Nonlinear wave equations arising in modeling of some power-law elasto-plastic structures," in *Proceedings of ICCP6 and CCP*, pp. 248–251, Beijing, China, 2005.

[15] C. Atkinson and C. Y. Chen, "On interface dislocations between dissimilar materials with non-linear stress-strain laws (plane strain)," *International Journal of Engineering Science*, vol. 37, no. 5, pp. 553–573, 1999.

[16] J. Y. Chen, Y. Huang, K. C. Hwang, and Z. C. Xia, "Plane-stress deformation in strain gradient plasticity," *Journal of Applied Mechanics, Transactions ASME*, vol. 67, no. 1, pp. 105–111, 2000.

[17] C. H. Chou, J. Pan, and S. C. Tang, "Nonproportional loading effects on elastic-plastic behavior based on stress resultants for thin plates of strain hardening materials," *International Journal of Plasticity*, vol. 10, no. 4, pp. 327–346, 1994.

[18] X.-L. Gao, "A mathematical analysis of the elasto-plastic plane stress problem of a power-law material," *IMA Journal of Applied Mathematics*, vol. 60, no. 2, pp. 139–149, 1998.

[19] A. E. Giannakopoulos, "Total deformation, plane-strain contact analysis of macroscopically homogeneous, compositionally graded materials with constant power-law strain hardening," *Journal of Applied Mechanics, Transactions ASME*, vol. 64, no. 4, pp. 853–860, 1997.

[20] Z. K. Wang, X. X. Wei, and X. L. Gao, "A general analytical solution of a strain-hardening elastoplastic plate containing a circular hole subject to biaxial loading-with applications in pressure vessels," *International Journal of Pressure Vessels and Piping*, vol. 47, no. 1, pp. 35–55, 1991.

[21] M. Petyt, *Introduction to Finite Element Vibration Analysis*, Cambridge University Press, Cambridge, UK, 1990.

[22] A. C. Ugural, *Stresses in Plates and Shells*, McGraw-Hill, New York, NY, USA, 1981.

[23] S. Timoshenko, D. H. Young, and W. Weaver Jr., *Vibration Problems in Engineering*, John Wiley & Sons, Hoboken, NJ, USA, 1974.

[24] J. Lamb, *Elements of Soliton Theory*, John Wiley & Sons, Inc, New York, NY, USA, 1982.

[25] M. Remoissenet, *Waves Called Solitons-Concepts and Experiments*, Springer, Berlin, Germany, 2nd edition, 1996.

[26] P. Drábek and R. Manásevich, "On the closed solution to some nonhomogeneous eigenvalue problems with p-Laplacian," *Differential and Integral Equations*, vol. 12, no. 6, pp. 773–788, 1999.

[27] H. Hariri, Y. Bernard, and A. Razek, "A traveling wave piezoelectric beam robot," *Smart Materials and Structures*, vol. 23, no. 2, Article ID 025013, 2014.

14

Assessing Heterogeneity for Factor Analysis Model with Continuous and Ordinal Outcomes

Ye-Mao Xia and Jian-Wei Gou

Department of Applied Mathematics, Nanjing Forestry University, Nanjing, Jiangsu 210037, China

Correspondence should be addressed to Ye-Mao Xia; ym_xia71@163.com

Academic Editor: Wei-Chiang Hong

Factor analysis models with continuous and ordinal responses are a useful tool for assessing relations between the latent variables and mixed observed responses. These models have been successfully applied to many different fields, including behavioral, educational, and social-psychological sciences. However, within the Bayesian analysis framework, most developments are constrained within parametric families, of which the particular distributions are specified for the parameters of interest. This leads to difficulty in dealing with outliers and/or distribution deviations. In this paper, we propose a Bayesian semiparametric modeling for factor analysis model with continuous and ordinal variables. A truncated stick-breaking prior is used to model the distributions of the intercept and/or covariance structural parameters. Bayesian posterior analysis is carried out through the simulation-based method. Blocked Gibbs sampler is implemented to draw observations from the complicated posterior. For model selection, the logarithm of pseudomarginal likelihood is developed to compare the competing models. Empirical results are presented to illustrate the application of the methodology.

1. Introduction

Owing to its wide applications in behavioral and social science researches, analysis of factor analysis models with mixed data structure has received a lot of attention; see [1–6]. However, most of these methods are mainly developed within particular parametric distribution families such as the exponential family or normal scale mixture family, which have a limited role in dealing with the distributional deviations, in particular heterogeneity or multimodality of the data. Though some robust methods are developed to downweight the influence of the outliers [7–12], most of them are still confined to dealing with unimodality and are less effective for the asymmetric and/or multimodal problems.

Recently, some authors focused on the Bayesian semiparametric modeling for latent variables model. For multivariate categorical data analysis, Kottas et al. [13] extended the traditional multivariate probit model [14–16] to a flexible underlying prior probability model. The usual single multivariate normal model for the latent variables is replaced by a mixture of normal priors with infinite number of components. And, for the latent variable model with fixed covariates and continuous responses, Lee et al. [17] established the semiparametric Bayesian hierarchal model for the structural equation models (SEMs) by relaxing the common normal distribution of exogenous factors to follow a finite-dimensional Dirichlet process [18]. Song et al. [19] developed a semiparametric Bayesian procedure for analyzing the latent variable model with unordered categorical data. For some recent advances in semiparametric analysis for factor analysis model, see [20–23] among others.

In this paper, we developed a Bayesian semiparametric approach for analyzing factor analysis model with mixed continuous and ordinal responses. The methods are twofold. Firstly, we extended Kottas, Müller, and Quintana's model to a more general multivariate model which contains factor variables. This extension aims to interpret the relationships between measurements and latent variables and explore correlations among the multiple manifest variables. Moreover, we treat the threshold parameters as unknown and estimate them simultaneously with other model parameters, thus providing a more flexible approach to fit the data. Secondly, we introduce the truncated Dirichlet process prior as the prior of the mean vector and variance-covariance parameters of unique errors and latent variables. This facilitates the interpretation of heterogeneity in the mean and/or covariance structure across the subjects.

This paper is organized as follows. We first introduce the Bayesian semiparametric modeling framework for factor analysis model with continuous and ordinal variables. We then present the Markov chain Monte Carlo procedure for parameters estimation and model selection. Simulation studies and a real example are provided to illustrate the performance of the proposed procedure. We close with some remarks and concluding comments.

2. Model Description

2.1. Factor Analysis Model with Continuous and Ordinal Responses. Suppose that a p-dimensional mixed observed vector $y_i = (x_i^T, z_i^T)^T$ contains r continuous variables $x_i = (x_{i1}, \ldots, x_{ir})^T$ and $s = p - r$ ordinal variables $z_i = (z_{i1}, \ldots, z_{is})^T$ with z_{ij} taking an integral value in $\mathbb{S}_j = \{0, 1, \ldots, b_j\}$ for $j = 1, \ldots, s$, $i = 1, \ldots, n$. We assume that the observed ordinal vector z_i is related to the unobserved continuous vector $u_i = (u_{i1}, \ldots, u_{is})^T$ through

$$z_{ij} = l \quad \text{if } \tau_{jl} < u_{ij} \le \tau_{j,l+1}, \; l \in \mathbb{S}_j, \tag{1}$$

where $\{\tau_{jl} : l = 0, \ldots, b_j, j = 1, \ldots, s\}$ is a set of unknown threshold parameters that define the categories: $-\infty = \tau_{j0} < \tau_{j1} < \cdots < \tau_{jb_j} < \tau_{j,b_j+1} = \infty$. Hence, for the jth variable z_{ij}, there are $b_j + 1$ categories.

Let $y_i^* = (x_i^T, u_i^T)^T$ denote the vector of continuous observed measurements and unobserved variables. For subject i, we formulate the dependence among y_{ij}^*'s through the following measurement model:

$$y_i^* = \mu + \Lambda\omega_i + \epsilon_i, \tag{2}$$

where μ is a $p \times 1$ intercept vector, Λ is a $p \times m$ factor loading matrix, ω_i is an $m \times 1$ vector of latent variables, and ϵ_i is a $p \times 1$ vector of measurement errors which is independent of ω_i. In many applications, ω_i may represent the hypothesized factors underlying manifest responses and/or unobserved heterogeneity not explained by covariates.

The latent variable model with mixed continuous and ordinal responses defined by (1) and (2) faces two sources of identification problems. The first one is associated with the determinacy of latent variables y^* in modeling of categorical variables, and the second one is related to the uniqueness of the factor loadings matrix. To solve the first problem, we use the common method [24] to fix endpoints τ_{j1} and τ_{jb_j} $(j = 1, \ldots, s)$ at preassigned values. For the second problem, we follow the usual practice in structural equation modeling to identify the covariance matrix of y_i^* by fixing appropriate elements in Λ at preassigned values.

Let θ be the parametric vector formed by the unknown parameters contained in $\{\mu, \Psi_\epsilon, \Phi\}$ and let ϑ denote the free parameters contained in factor loading matrices Λ and $\tau = (\tau_1^T, \ldots, \tau_s^T)^T$ with $\tau_j = (\tau_{j1}, \ldots, \tau_{jb_j})^T$. Based on the assumptions of (2), the conditional distribution of y_i^* given (θ, ϑ) is a normal distribution with mean vector μ and covariance matrix $\Sigma(\theta, \vartheta) = \Lambda\Phi\Lambda^T + \Psi_\epsilon$.

Note that the latent factors here play an important role in characterizing the associations between the observed variables. It can be seen clearly that z_i and x_i are dependent when ω_i is integrated out. The marginal density of y_i is given by

$$p(y_i \mid \vartheta, \theta) = \int p(x_i \mid \omega_i, \vartheta, \theta)\, p(z_i \mid \omega_i, \vartheta, \theta)\, p(\omega_i \mid \vartheta, \theta)\, d\omega_i \tag{3}$$

with

$$p(z_i \mid \omega_i, \vartheta, \theta) = \prod_{j=1}^{s} \left[\Phi_c\left(\frac{\tau_{jz_{ij}+1} - \mu_{r+j} - \Lambda_{r+j}^T \omega_i}{\psi_{\epsilon r+j}} \right) - \Phi_c\left(\frac{\tau_{jz_{ij}} - \mu_{r+j} - \Lambda_{r+j}^T \omega_i}{\psi_{\epsilon r+j}} \right) \right], \tag{4}$$

in which $\Phi_c(\cdot)$ is the standard normal cumulative distribution function.

2.2. Bayesian Semiparametric Hierarchical Modeling. Let $p(y_i^* \mid \theta, \vartheta)$ be the conditional density of y_i^* given (θ, ϑ) and denote by F a prior distribution function of θ. Suppose that F is proper; we define the following mixture density:

$$p(y_i^* \mid F, \vartheta) = \int p(y_i^* \mid \theta, \vartheta)\, F(d\theta), \tag{5}$$

in which $F(d\theta)$ is the conditional distribution of θ given F. By taking a prior for ϑ and restricting F to be a parametric family of distributions indexed by θ, we complete the Bayesian parametric model specification. However, this restriction severely constrains the estimation of θ and produces estimators that shrink data values toward the same points. A more flexible modeling for y_i^* is to treat F as random and assign a prior for it. For this end, we introduce a latent variable vector $\theta_i = \{\mu_i, \Psi_{\epsilon i}, \Phi_i\}$ and assume that, given θ_i, y_i^*'s are conditionally independent and drawn from $p(y_i^* \mid \theta_i, \vartheta)$. Furthermore, we suppose that θ_i's are independent and identically distributed (i.i.d.) according to F with a prior $\mathcal{P}$ on it. As a result, we break the mixture model $p(y_i^* \mid F, \vartheta)$ into

$$\begin{aligned} [y_i^* \mid \theta_i, \vartheta] &\overset{\text{ind}}{\sim} p(y_i^* \mid \theta_i, \vartheta), \\ [\theta_1, \ldots, \theta_n \mid F] &\overset{\text{iid}}{\sim} F, \quad F \sim \mathcal{P}, \end{aligned} \tag{6}$$

where "ind" means "independent" and $\mathcal{P}$ is a prior of F.

We consider the following truncated version of Dirichlet process for F:

$$\mathcal{P}(\cdot) = \mathcal{P}_G(\cdot) = \sum_{k=1}^{G} \pi_k \delta_{\theta_k^*}(\cdot), \tag{7}$$

in which $\delta_{\theta_k^*}(\cdot)$ denotes a discrete probability measure concentrated on atom θ_k^* and π_k ($k = 1, \ldots, G$), independent of θ_k^*, are random weights constructed through the following stick-breaking procedure:

$$\begin{aligned}\pi_1 &= V_1,\\ \pi_k &= (1 - V_1) \cdots (1 - V_{k-1}) V_k, \quad k = 2, \ldots, G - 1, \\ \pi_G &= (1 - V_1) \cdots (1 - V_{G-1}),\end{aligned} \tag{8}$$

with $V_k \overset{\text{i.i.d.}}{\sim} \text{Beta}(1, \alpha)$; θ_k^*'s are i.i.d with common distribution F_0.

Truncated Dirichlet process prior (7) can be considered as a truncation version of Dirichlet process [25–30] in the nonparametric Bayesian analysis. It can be shown that, under (7) and (8), for any Borel set A in $\mathbb{R}^p$,

$$\begin{aligned}\mathbb{E}F(A) &= F_0(A),\\ \text{Var}(F(A)) &= \frac{F_0(A)(1 - F_0(A))}{\alpha + 1}\left(1 + \frac{\alpha}{G}\right).\end{aligned} \tag{9}$$

This indicates that F_0 can be served as the starting point or guess of F and α determines the concentration of the prior around F_0. In practice, the value of G is either set to a large, predetermined value (e.g., $G \geq 100$) or chosen empirically. For instance, Ishwaran and Zarepour [31] suggested that the adequacy of the truncation level, G, can be assessed by evaluating moments of the tail probability. Our simulation results have shown that $G = 100$ is more than adequate for the model considered in the present context.

Now, we specify the distribution F_0. Recalling that by convention θ_k^* is the collection of $\{\mu_k^*, \Psi_{\epsilon k}^*, \Phi_k^*\}$, hence, we assume that

$$\begin{aligned}F_0 &(\mu_k^*, \Psi_{\epsilon k}^*, \Phi_k^* \mid \nu, \Sigma_\nu, R)\\ &= N(\mu_k^* \mid \nu, \Sigma_\nu) \cdot \prod_{j=1}^{m} \text{Gamma}^{-1}(\psi_{\epsilon kj}^* \mid \alpha_{\epsilon 0j}, \beta_{\epsilon 0j}) \\ &\quad \cdot \text{Wishart}^{-1}(\Phi_k^* \mid \rho_0, R^{-1}),\end{aligned} \tag{10}$$

where ν, Σ_ν, and R are hyperparameters, $\Sigma_\nu = \text{diag}\{\sigma_{\nu 1}, \ldots, \sigma_{\nu p}\}$ is a diagonal matrix with the kth diagonal element $\sigma_{\nu k}$, and R is an $m_2 \times m_2$ positive definite matrix; $\text{Gamma}^{-1}(\alpha_{\epsilon 0j}, \beta_{\epsilon 0j})$ refers to the inverse gamma distribution with shaper parameters $\alpha_{\epsilon 0j}$ and scale parameters $\beta_{\epsilon 0j}$, respectively, and Wishart^{-1} denotes the inverse Wishart distribution [32].

Modeling F in (7) into the random probability measure and incorporating the latent variable ω_i into (5) generate the following hierarchical model: for $i = 1, \ldots, n$,

$$\begin{aligned}(y_i^* \mid \omega_i, \theta_i, \vartheta) &\overset{\text{ind}}{\sim} N(\mu_i + \Lambda\omega_i, \Psi_{\epsilon i}),\\ (\omega_i \mid \theta_i) &\overset{\text{ind}}{\sim} N(0, \Phi_i),\\ (\theta_i \mid F) &\overset{\text{iid}}{\sim} F, \quad F \sim \mathscr{P}_G(\cdot),\end{aligned} \tag{11}$$

where $\mathscr{P}_G$ is given by (7) and (8).

3. Parameters Estimation and Model Selection

3.1. Prior Specifications and Estimation via Blocked Gibbs Sampler. Let $\Theta^* = \{\theta_k^* : k = 1, \ldots, G\}$. To implement Bayesian analysis, blocked Gibbs sampler is used to simulate observations from the posterior. The key for blocked Gibbs sampler is to recast model (11) completely by introducing the cluster variables $L = (L_1, \ldots, L_n)^T$ such that $\theta_i = \theta_{L_i}^*$. Consequently, the semiparametric hierarchical model (11) can be reformulated as the following framework:

$$\begin{aligned}(y_i^* \mid \omega_i, \theta_i, \vartheta) &\overset{\text{ind}}{\sim} N(\mu_i + \Lambda\omega_i, \Psi_{\epsilon i}),\\ (\omega_i \mid \theta_i) &\overset{\text{ind}}{\sim} N(0, \Phi_i),\\ (L_i = \cdot \mid \pi) &\overset{\text{iid}}{\sim} \sum_{k=1}^{G} \pi_k \delta_k(\cdot),\\ (\pi, \Theta^*) &\sim p(\pi)\, p(\Theta^*),\\ \vartheta &\sim p(\vartheta),\\ \tau &\sim p(\tau),\end{aligned} \tag{12}$$

where $p(\vartheta)$ is a prior of ϑ, $p(\pi)$ is the stick-breaking prior given by (8) with $[V_i \mid \alpha] \overset{\text{iid}}{\sim} \text{Beta}(1, \alpha)$, and $p(\Theta^*)$ is the joint distribution of Θ^* given by

$$\begin{aligned}p(\Theta^* \mid \nu, \Sigma_\nu, R) &= \prod_{k=1}^{G} p(\theta_k^* \mid \nu, \Sigma_\nu, R)\\ &= \prod_{k=1}^{G} p(\mu_k^*, \Psi_{\epsilon k}^*, \Phi_k^* \mid \nu, \Sigma_\nu, R),\\ [\mu_k^*, \Psi_{\epsilon k}^*, \Phi_k^* &\mid \nu, \Sigma_\nu, R] \overset{\text{iid}}{\sim} F_0\end{aligned} \tag{13}$$

in which $F_0(\cdot \mid \nu, \Sigma_\nu, R)$ is given in (10).

For the Bayesian analysis, we need to specify priors for the parameters involved in the model. The whole parameters can be divided into two parts: parametric component part $\{\vartheta, \tau\}$ and nonparametric component part $\{\nu, \Sigma_\nu, R, \alpha\}$. For the parametric components, we assume that $p(\vartheta, \tau) = p(\vartheta) p(\tau)$ with

$$\begin{aligned}p(\Lambda_k) &\overset{D}{=} N(\Lambda_{0k}, H_{\epsilon 0k}),\\ p(\tau) &= \prod_{j=1}^{s} p(\tau_j) = \prod_{j=1}^{s} p(\tau_{j,2}, \ldots, \tau_{j,b_j-1})\\ &\propto \prod_{j=1}^{s} I\{\tau_{j,2} < \cdots < \tau_{j,b_j-1}\},\end{aligned} \tag{14}$$

where Λ_k is a $p \times 1$ column vector that contains unknown parameters in the kth row of Λ.

For the hyperparameter $\beta = \{\nu, \Sigma_\nu, R, \alpha\}$, we consider the following conjugate priors:

$$\begin{aligned}\nu &\sim N(\mu_0, \Sigma_0),\\ R &\sim \text{Wishart}^{-1}(\rho_0^\phi, R_0^\phi),\\ \sigma_{\nu k} &\sim \text{Gamma}^{-1}(\kappa_1, \kappa_2),\\ \alpha &\sim \text{Gamma}(\tau_1, \tau_2).\end{aligned} \tag{15}$$

The hyperparameters $\mu_0, \Sigma_0, \Lambda_{0k}, H_{\epsilon 0k}, R_0^\phi, \alpha_{\epsilon 0j}, \beta_{\epsilon 0j}, \rho_0, \rho_0^\phi, \kappa_1, \kappa_2, \tau_1$, and τ_2 in (10), (14), and (15) are treated as known.

Let $Y(n \times p) = (y_1, \ldots, y_n)^T$, $\Omega = (\omega_1, \ldots, \omega_n)^T$, and $Y^* = (y_1^*, \ldots, y_n^*)^T$. Posterior analysis in relation to the complex $p(\vartheta, \beta \mid Y)$ is carried out through the data augmentation technique [33]. Specifically, we treat the latent quantities $\{\Omega, Y^*, \pi, \Theta^*, L\}$ as missing data and augment them with the observed data. A sequence of random observations is generated from the joint posterior distribution $p(\Omega, Y^*, \vartheta, \pi, \Theta^*, L, \beta \mid Y)$ by the blocked Gibbs sampler [31, 34], coupled with the Metropolis-Hastings algorithm [35, 36]: given $\{\Omega^{(l)}, Y^{*(l)}, \pi^{(l)}, \Theta^{*(l)}, L^{(l)}\}$ at the lth iteration

draw $\Omega^{(l+1)}$ from $p(\Omega \mid Y^{*(l)}, \vartheta^{(l)}, \pi^{(l)}, \Theta^{*(l)}, L^{(l)}, \beta^{(l)}, Y)$,

draw $(\vartheta^{(l+1)}, Y^{*(l+1)})$ from $p(\vartheta, Y^* \mid \Omega^{(l+1)}, \pi^{(l)}, \Theta^{*(l)}, L^{(l)}, \beta^{(l)}, Y)$,

draw $(\pi^{(l+1)}, \Theta^{*(l+1)})$ from $p(\pi, \Theta^* \mid \Omega^{(l+1)}, Y^{*(l+1)}, \vartheta^{(l+1)}, L^{(l)}, \beta^{(l)}, Y)$,

draw $L^{(l+1)}$ from $p(L \mid \Omega^{(l+1)}, Y^{*(l+1)}, \vartheta^{(l+1)}, \pi^{(l+1)}, \Theta^{*(l+1)}, \beta^{(l)}, Y)$,

draw $\beta^{(l+1)}$ from $p(\beta \mid \Omega^{(l+1)}, Y^{*(l+1)}, \vartheta^{(l+1)}, \pi^{(l+1)}, \Theta^{*(l+1)}, L^{(l+1)}, Y)$,

and form $\{\Omega^{(l+1)}, Y^{*(l+1)}, \pi^{(l+1)}, \Theta^{*(l+1)}, L^{(l+1)}\}$. It can be shown that as l tends to infinity, the empirical distribution of $\{\Omega^{(l)}, Y^{*(l)}, \pi^{(l)}, \Theta^{*(l)}, L^{(l)}\}$ converges to $p(\Omega, Y^*, \vartheta, \pi, \Theta^*, L, \beta \mid Y)$ at any geometrical rate. The full conditional distributions and the implementation of the above algorithm are given in the Appendix.

3.2. Model Selection. Model selection is an important issue in Bayesian semiparametric modeling for latent variable model since it is of practical interest to compare different modelings for factor analytic models. Formal Bayesian model selection is accomplished by comparing the marginal predictive distribution of data across models. Consider the problem of comparing competing models M_1 and M_2. Let $p(Y \mid M_1)$ and $p(Y \mid M_2)$ denote the marginal density of data Y under M_1 and M_2, respectively. A popular choice for selecting models is achieved via Bayes factor (BF) (e.g., [37–39]). However, in view of the fact that computing BF involves the high-dimensional density which is hard to estimate well, we prefer comparing the following logarithm of pseudomarginal likelihood (LPML) [40, 41]:

$$\text{LPML}(Y) = \sum_{i=1}^n \log(\text{CPO}_i), \tag{16}$$

where CPO_i is known as the conditional predictive ordinate (CPO) defined as

$$\begin{aligned}\text{CPO}_i &= p(y_i \mid Y_{(i)}) = \left[\int \frac{1}{p(y_i \mid Y_{(i)}, \vartheta, \Theta^*, \pi)} p(\vartheta, \Theta^*, \pi \mid Y)\, d\vartheta\, d\Theta^*\, d\pi\right]^{-1}\\ &= \left[\int \frac{1}{p(y_i \mid \vartheta, \Theta^*, \pi)} p(\vartheta, \Theta^*, \pi \mid Y)\, d\vartheta\, d\Theta^*\, d\pi\right]^{-1}.\end{aligned} \tag{17}$$

Here, $Y_{(i)}$ is the data set Y with y_i removed. Obviously, from (17), we can see that CPO_i is the marginal posterior predictive density of y_i given $Y_{(i)}$ and can be interpreted as the height of this marginal density at y_i. Thus, small values of LPML imply that Y does not support the model.

Based on MCMC sample $\{(\Theta^{*(t)}, \pi^{(t)}, L^{(t)}, \vartheta^{(t)}, \beta^{(t)}) : t = 1, \ldots, T\}$ already available in the estimation, a consistent estimate for LPML can be obtained via ergodic average given by

$$\widehat{\text{LPML}}(Y) = -\sum_{i=1}^n \log\left[\frac{1}{T}\sum_{t=1}^T \frac{1}{p(y_i \mid \vartheta^{(t)}, \Theta^{*(t)}, \pi^{(t)})}\right]. \tag{18}$$

It is noted that, under our proposed model,

$$\begin{aligned}p(y_i \mid \vartheta, \Theta^*, \pi) &= \int p(y_i \mid \omega_i, L_i, \vartheta, \Theta^*)\\ &\cdot p(\omega_i \mid L_i, \vartheta, \Theta^*)\, p(L_i \mid \pi)\, d\omega_i dL_i\end{aligned} \tag{19}$$

which is complicated due to the existence of Ω and L. This can be solved by Monte Carlo approximation. Specifically, given the current values $\{\vartheta^{(l)}, \Theta^{*(l)}, \pi^{(l)}\}$ at the lth iteration, we draw (i) $L_i^{l,h}$ from $p(L_i \mid \pi^{(l)})$ and (ii) $\omega_i^{l,h}$ from $p(\omega_i \mid L_i^{l,h}, \vartheta^{(l)}, \Theta^{*(l)})$ for $h = 1, \ldots, H$ and then evaluate $p(y_i \mid \vartheta^{(l)}, \Theta^{*(l)}, \pi^{(l)})$ at the observation y_i through

$$\begin{aligned}&\hat{p}(y_i \mid \vartheta^{(l)}, \Theta^{*(l)}, \pi^{(l)})\\ &\quad \approx \frac{1}{H}\sum_{h=1}^H p(y_i \mid \omega_i^{l,h}, L_i^{l,h}, \vartheta^{(l)}, \Theta^{*(l)}).\end{aligned} \tag{20}$$

Obviously, the distributions involved in (i) and (ii) are standard and sampling is rather straightforward and fast.

4. A Simulation Study

In this section, a simulation study to evaluate the performance of the proposed procedure is conducted. The goal is to assess the accuracy of estimates under parametric, partly exchangeable, and semiparametric modelings when data take on the multimodality or heterogeneity. We consider the situation in which each observed vector consists of three-dimensional continuous vector and three-dimensional ordinal vector with threshold values $\tau_j = (-1.0^*, -0.6, 0.3, 1.0^*)^T$ ($j = 1, 2, 3$). We generate Y by first generating Y^* with $y_i^* = (x_i^T, u_i^T)^T$ from the mixture of two factor analytic models with weights 0.45 and 0.55 and then transforming u_i into z_i ($s = 3$) via (1) to create the ordinal observations, where x_i represents a 6×1 observed continuous random vector and u_i is a 3×1 latent continuous random vector. Each component in the mixture model is specified through the following measurement model: for $m = 1, 2$,

$$y_i^* = \mu^{(m)} + \Lambda \omega_i^{(m)} + \epsilon_i^{(m)}, \quad \epsilon_i^{(m)} \sim N\left(0, \Psi_\epsilon^{(m)}\right). \tag{21}$$

The parameters involved in the components of mixture model are taken as $\mu^{(1)} = -1.5 \times 1_6$, $\mu^{(2)} = 1.0 \times 1_6$, $\Psi_\epsilon^{(1)} = 0.36 I_6$, $\Psi_\epsilon^{(2)} = I_6$,

$$\Lambda^T = \begin{bmatrix} 1^* & 0.8 & 0.8 & 0^* & 0^* & 0^* \\ 0^* & 0^* & 0^* & 1^* & 0.8 & 0.8 \end{bmatrix},$$
$$\Phi^{(1)} = \begin{pmatrix} 1 & -0.3 \\ -0.3 & 1.0 \end{pmatrix}, \tag{22}$$
$$\Phi^{(2)} = \begin{pmatrix} 1 & 0.6 \\ 0.6 & 1.0 \end{pmatrix},$$

in which 1_6 is a 6×1 vector with all elements equal to one and I_6 is a 6×6 identity matrix. The elements with asterisks involved in loading matrix Λ and threshold parameters $\{\tau_j\}_{j=1}^3$ are treated as fixed for identifying model (see Section 2.1). Based on these settings, random sample with size 500 is generated and 100 replications are completed for each combination.

Prior inputs in the prior distributions involved in the parametric components (see (14)) are as follows: $H_{\epsilon 0k}$ and $H_{\zeta 0}$ are diagonal matrices with the diagonal elements 1.0, and elements in $\{\Lambda_{0k}, \Pi_0\}$ are equal to the true values, while prior inputs in the prior distribution of superparameter β (see (15)) are $\mu_0 = 0_9$, $\Sigma_0 = 100 I_9$, $\kappa_1 = \kappa_2 = 0.001$, $R_0^\phi = 0.01 I_2$, $\rho_0 = \rho_0^\phi = 10$, $\alpha_{\epsilon 0k} = \beta_{\epsilon 0k} = 2.0$, and $\tau_1 = \tau_2 = 2.0$. Note that these values ensure approaching noninformative priors.

A few test runs are conducted to explore the effect of truncated levels on the estimates of unknown parameters and the convergence of the blocked Gibbs sampler. We take G = 50, 60, 70, 80, 90, 100, 200, and 300 and calculate the total sum of the root mean square (RMS) of estimates (see below for details). The resulting values are 1.9830, 1.7382, 1.6582, 1.5548, 1.4194, 1.4128, 1.4108, and 1.4101, respectively. It can be seen that the total sum of the root mean square (RMS) becomes rather stable when $G \geq 80$. In the following analysis, we set $G = 100$ in our data analysis. For the threshold parameters $\{\tau_{jk} : j = 1, 2, 3,\ k = 2, 3\}$, we choose $\sigma^2_{\mathrm{MH}jk} = 0.002$ (see Appendix) in MH algorithm to produce

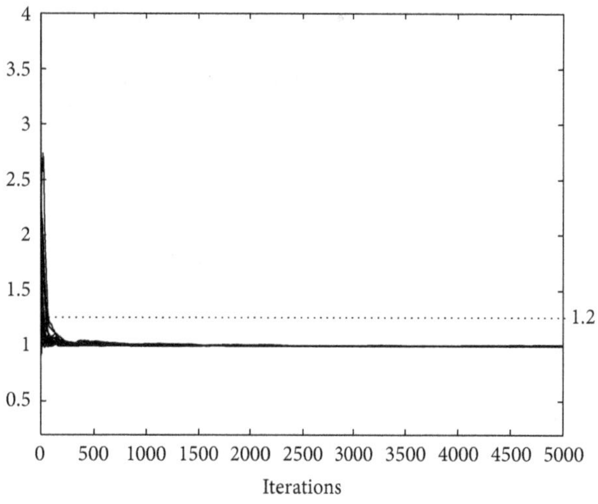

FIGURE 1: Plot of the values of EPSR of unknown parameters against the number of iterations under different starting values for the simulated data.

the acceptance rate about 0.40. Figure 1 gives the plots of EPSR (estimated potential scale reduction [42]) values of unknown free parameters in Λ, τ, and α against iterations for three groups of different starting values. It can be seen that the estimates converge in less than 1000 iterations. To be conservative, in the following analysis, we collect 3000 observations after 2000 "burn-in"s deleted to take posterior analysis. We first consider the performance of the proposed LMPL in model comparison. We compare the proposed model with the parametric model (denoted by PARA) and the partly exchangeable model (denoted by PAEX), which approximately correspond to $\alpha = +\infty$ and $\alpha = 0$ under our proposal, respectively. The parametric model is defined by

$$\begin{aligned} \left(y_i^* \mid \omega_i, \theta\right) &\overset{\text{ind}}{\sim} N\left(\mu + \Lambda \omega_i, \Psi_\epsilon\right), \\ \left(\omega_i \mid \Phi\right) &\overset{\text{iid}}{\sim} N\left(0, \Phi\right). \end{aligned} \tag{23}$$

The priors of the parameters are given by $p(\mu) \overset{D}{=} N(\mu_0, \Sigma_0)$, $p(\Lambda, \Psi_\epsilon) \overset{D}{=} \prod_{k=1}^9 N(\Lambda_{0k}, H_{0k}) \cdot \mathrm{Gamma}^{-1}(\alpha_{\epsilon 0k}, \beta_{\epsilon 0k})$, and $\Phi \sim \mathrm{Wishart}^{-1}(10, 7.0 I_2)$.

The partly exchangeable model is given by

$$\begin{aligned} \left(y_i^* \mid \omega_i, \theta_i, \vartheta\right) &\overset{\text{ind}}{\sim} N\left(\mu_i + \Lambda \omega_i, \Psi_{\epsilon i}\right), \\ \left(\omega_i \mid \theta_i\right) &\overset{\text{ind}}{\sim} N\left(0, \Phi_i\right), \end{aligned} \tag{24}$$

where $\theta_i = \{\mu_i, \Psi_{\epsilon i}, \Phi_i\}$ are i.i.d. with distribution $F_0(\cdot \mid \nu, \Sigma_\nu, R)$ given in (5); the priors for the unknown parameter vector Λ and hyperparametric vector $\{\nu, \Sigma_\nu, R\}$ are, respectively, given in (14) and (15).

Under the foregoing settings for the hyperparameters, observations obtained through the blocked Gibbs sampler are used to compute the values of LPML for each scenario across 100 replications. For the parametric and partly exchangeable model, computing values of LPML is very straightforward and standard. For the semiparametric model, we draw 50 observations for approximating $p(y_i \mid \vartheta, \Theta, \pi)$. The values of LPML under parametric model, semiparametric model, and partly exchangeable model are, respectively, −6684.740, −6255.553, and −8487.259 with standard deviations 62.509, 151.480, and 147.742. Based on the LPML criteria, semiparametric model is selected, which is consistent with the fact that the true model takes on the multimodes. Moreover, according to our empirical results, the correct rates of LPML selecting the true model across 100 replications are about 0.93.

Table 1 gives the biases (BIAS), root mean squares (RMS), and standard deviations (SD) of estimates of unknown parameters across 100 replications under semiparametric models and parametric and partly exchangeable model, respectively. The measures BIAS, RMS, and SD are given as

$$\mathrm{BIAS}\left(\hat{\vartheta}_j\right) = \left(\bar{\vartheta}_j - \vartheta_{j0}\right), \quad \bar{\vartheta}_j = \frac{1}{S}\sum_{r=1}^{S}\hat{\vartheta}_j^{(r)},$$

$$\mathrm{RMS}\left(\hat{\vartheta}_j\right) = \sqrt{\frac{1}{S}\sum_{r=1}^{S}\left(\hat{\vartheta}_j^{(r)} - \vartheta_{j0}\right)^2}, \tag{25}$$

$$\mathrm{SE}\left(\hat{\vartheta}_j\right) = \sqrt{\frac{1}{S}\sum_{r=1}^{S}\left(\hat{\vartheta}_j^{(r)} - \bar{\vartheta}_j\right)^2},$$

where S is the number of replications. It can be seen that estimates obtained through the proposed approach are reasonably accurate. The values of $\hat{\lambda}_{jk}$ under our approach are smaller than those under parametric and exchangeable modelings in terms of the absolute values of BIAS and RMS. The results show that ignoring heterogeneity among the data may lead to biased estimates and incorrect interpretation of the analyzed phenomena. This also reflects that the factor loadings λ_{jk} are not robust against the distributional deviations of inceptor, variance of unique errors, and covariances of latent factors.

Further simulation study is conducted to assess the performance of the proposed model and parametric model as well as the partly exchangeable model when data are generated from a single normal distribution. The population values of parameters are taken as $\mu = 0_6$, $\Psi_\epsilon = I_6$, and

$$\Phi = \begin{pmatrix} 1 & 0.3 \\ 0.3 & 1.0 \end{pmatrix}. \tag{26}$$

The values of factor loadings and threshold points are the same as those in previous mixture model. As usual, we take $G = 100$ for truncated levels. The sample size is set to 62 which is analogous to the real example. The inputs for superparameters involved in priors are set the same as that in mixture model. The results based on 100 replications are summarized in Table 2.

Based on Table 2, it can be found that the results obtained from our proposal are rather reasonable when compared to normal model, while partly exchangeable model gives serious biases. Moreover, we consider different inputs of superparameters in priors and find that the estimates are rather robust.

5. A Real Example

To illustrate the proposed procedure with a real example, a political-economic risk data set [43] was analyzed, which is adopted from Henisz's [44] political constraint index data set (POLCON), Marshall et al. [45] state failure problem sets (PITF), and Alvarez et al.'s [46] ACLP Political and Economic Database (ACLP). The data set is formed by the two economic indicators and three political variables from 62 countries. The first index is the log black market premium (BMP). This is a continuous variable which is usually used as a proxy for illegal economic activity. The second index is log real gross domestic product (GDP). It is used to measure the productivity of a country. The third variable is a measure of independence of the national judiciary. This is a binary variable: it takes 1 if the judiciary is judged to be independent and 0 otherwise. The next measurement, measuring the level of lack of expropriation risk threat (LE), is an ordered categorical variable coded with 0, 1, 2, 3, 4, and 5. The last variable is an expert judgment of measuring lack of corruption (LC). It is also an ordered categorical variable scaled with 0 to 5. The total sample size is 62 and the frequencies of each category occurring are equal to $\{34, 28\}$, $\{2, 6, 7, 19, 14, 14\}$, and $\{5, 11, 18, 11, 8, 9\}$, respectively. To unify scales of the continuous variables, the corresponding raw data were standardized.

Let $y^T = (\log \mathrm{BMP}, \log \mathrm{GDP}, \mathrm{IJ}, \mathrm{LE}, \mathrm{LC})$ be the vector of the observed variables. Based on the objective of this example, it is natural to group (i) {log BMP, log GDP} to an endogenous latent variable that can be interpreted as "economic factor, ξ" and (ii) {IJ, LE, LC} to an exogenous genotype latent variable that can be interpreted as "political factor, η." Hence, the following loading matrix Λ in the measurement equation with $\omega_i = (\eta_i, \xi_i)^T$ is considered:

$$\Lambda^T = \begin{bmatrix} 0^* & 0^* & 1^* & \lambda_{41} & \lambda_{51} \\ 1^* & \lambda_{22} & 0^* & 0^* & 0^* \end{bmatrix} \tag{27}$$

in which the ones and zeros are treated as known. Although other structures of Λ could be used, here we consider a nonoverlapped structure for clear interpretations of the latent variables: λ_{jk} measures the effect of ω_k on the observed variable y_j. Since the third variable is binary and the last two variables are measured on a six-point scale with each involving six thresholds, for model identification, we fix $\psi_{\epsilon 3} = 1$ and endpoints of thresholds $\tau_{31}, \tau_{35}, \tau_{41}$, and τ_{45} at −1.8486, 0.7527, −1.4007, and 1.0574, respectively. These fixed threshold values were chosen via $\tau_{jk} = \Phi^{-1}(\hat{p}_{jk})$, where $\hat{p}_{jk}$ are observed marginal proportions of the categories with $z_j < k$.

By primary data analysis, we find that the skewness and kurtosis of the first two variables are $\{-0.1340, -0.4319\}$ and $\{2.0892, 2.0958\}$, respectively. We also evaluate the predictive density function for continuous variables. Figure 2 gives the contours of posterior predictive density of pair (y_1, y_2) under parametric model and semiparametric model M^ϵ (see below) based on 60×60 grids. It can be seen that the data for pair (y_1, y_2) are heavy-tailed and the predictive density under semiparametric model captures the high frequency

TABLE 1: Summary of the estimates under the parametric, partly exchangeable, and semiparametric approaches in analyzing simulated data: mixture data.

Para.	PARA			PAEX			SEMI		
	BIAS	RMS	SD	BIAS	RMS	SD	BIAS	RMS	SD
λ_{21}	0.130	0.136	0.043	0.216	0.171	0.047	−0.022	0.066	0.070
λ_{31}	0.131	0.138	0.043	0.212	0.181	0.047	−0.026	0.080	0.070
λ_{52}	0.140	0.147	0.046	0.263	0.164	0.072	−0.015	0.093	0.092
λ_{62}	0.144	0.150	0.046	0.266	0.152	0.071	−0.012	0.098	0.092
τ_{12}	0.060	0.091	0.040	−0.110	0.014	0.052	0.007	0.062	0.038
τ_{13}	0.070	0.085	0.033	−0.216	0.050	0.060	0.007	0.067	0.040
τ_{22}	0.070	0.098	0.040	−0.113	0.016	0.051	0.008	0.056	0.038
τ_{23}	0.077	0.095	0.032	−0.186	0.038	0.059	0.013	0.062	0.042
τ_{32}	0.062	0.092	0.040	−0.118	0.016	0.051	0.013	0.069	0.038
τ_{33}	0.083	0.098	0.032	−0.184	0.037	0.059	0.018	0.070	0.042

TABLE 2: Summary of the estimates under the parametric, partly exchangeable, and semiparametric approaches in analyzing simulated data: normal data.

Para.	PARA			PAEX			SEMI		
	BIAS	RMS	SD	BIAS	RMS	SD	BIAS	RMS	SD
λ_{21}	0.109	0.103	0.068	0.216	0.201	0.147	0.152	0.125	0.130
λ_{31}	−0.114	0.108	0.066	0.212	0.201	0.207	0.138	0.137	0.120
λ_{52}	−0.122	0.116	0.101	0.253	0.134	0.172	0.146	0.139	0.141
λ_{62}	−0.123	0.114	0.102	0.366	0.136	0.271	0.144	0.118	0.176
τ_{12}	−0.012	0.004	0.052	−0.151	0.036	0.126	0.005	0.002	0.052
τ_{13}	−0.018	0.002	0.056	−0.153	0.041	0.126	−0.013	0.003	0.060
τ_{22}	−0.001	0.003	0.046	−0.110	0.014	0.052	−0.035	0.004	0.045
τ_{23}	0.016	0.003	0.051	−0.216	0.050	0.060	−0.022	0.004	0.055
τ_{32}	−0.001	0.003	0.046	−0.113	0.016	0.051	−0.001	0.002	0.048
τ_{33}	−0.018	0.002	0.051	−0.186	0.038	0.059	−0.012	0.003	0.058

region successfully while parametric model fails. For model comparison, we consider the following competing models:

$$\begin{aligned} M^{\mu}&: y_i^* = \mu_i + \Lambda\omega_i + \epsilon_i, \\ &\quad \epsilon_i \sim N(0, \Psi_\epsilon),\ \omega_i \sim N(0, \phi); \\ M^{\epsilon}&: y_i^* = \mu + \Lambda\omega_i + \epsilon_i, \\ &\quad \epsilon_i \sim N(0, \Psi_{\epsilon i}),\ \omega_i \sim N(0, \phi); \\ M^{\mu\epsilon}&: y_i^* = \mu_i + \Lambda\omega_i + \epsilon_i, \\ &\quad \epsilon_i \sim N(0, \Psi_{\epsilon i}),\ \omega_i \sim N(0, \phi); \\ M^{\mu\phi}&: y_i^* = \mu_i + \Lambda\omega_i + \epsilon_i, \\ &\quad \epsilon_i \sim N(0, \Psi_\epsilon),\ \omega_i \sim N(0, \phi_i); \\ M^{\epsilon\phi}&: y_i^* = \mu + \Lambda\omega_i + \epsilon_i, \\ &\quad \epsilon_i \sim N(0, \Psi_{\epsilon i}),\ \omega_i \sim N(0, \phi_i); \\ M^{\mu\epsilon\phi}&: y_i^* = \mu_i + \Lambda\omega_i + \epsilon_i, \\ &\quad \epsilon_i \sim N(0, \Psi_{\epsilon i}),\ \omega_i \sim N(0, \phi_i). \end{aligned} \tag{28}$$

The following two types of prior inputs are, respectively, used for the hyperparameters involved in the parametric components and semiparametric components: (I) $\Lambda_{\epsilon 0k} = \widetilde{\Lambda}_{\epsilon k}$, $H_{\epsilon 0k} = I_2$, $\gamma_0 = \widetilde{\gamma}$, $H_{\zeta 0} = 1$, $\mu_0 = \widetilde{\mu}$, $\Sigma_0 = \text{diag}\{S\}$, $\alpha_{\epsilon 0k} = 9.0$, $\beta_{\epsilon 0k} = (\alpha_{\epsilon 0k}-1)\widetilde{\psi}_{\epsilon k}$, $\rho_0 = \rho_0^\phi = 20$, $R_0^{\phi-1} = (\rho_0-2)\widetilde{\phi}$, $\kappa_1 = \kappa_2 = 8.0$, and $\tau_1 = \tau_2 = 8.0$, where $\widetilde{\theta}$ denotes the maximum likelihood estimates of θ under parametric model from analysis of a "control-group" sample and S is the polychoric correlation matrix obtained on the basis of single confirmatory factor analysis model; (II) $\lambda_{\epsilon 0jk} = 0$, $H_{\epsilon 0k} = 0.01I_2$, $\gamma_0 = 0$, $H_{\zeta 0} = 0.01$, $\mu_0 = 0_{5\times 1}$, $\Sigma_0 = 0.01I_5$, $\alpha_{\epsilon 0k} = \beta_{\epsilon 0k} = 2.0$, $\rho_0 = \rho_0^\phi = 10$, $R_0^{-1} = \rho_0 - 2$, $\kappa_1 = \kappa_2 = 0.01$, and $\tau_1 = \tau_2 = 0.01$. Note that prior (I) gives more information than prior (II) since it partly takes advantage of information from sample.

The proposed Bayesian semiparametric approach with $G = 50$ was applied to calculate the values of CPO and LPML. We draw 100,000 effective observations from the corresponding posteriors via the blocked Gibbs sampler and divide them into 100 batches equally. Table 3 gives the means and standard deviations of LPML under priors (I) and (II). The following facts can be found. (i) The values of LPML under prior (I) are larger than those under prior (II). This indicates that the LPML tends to choose the model with

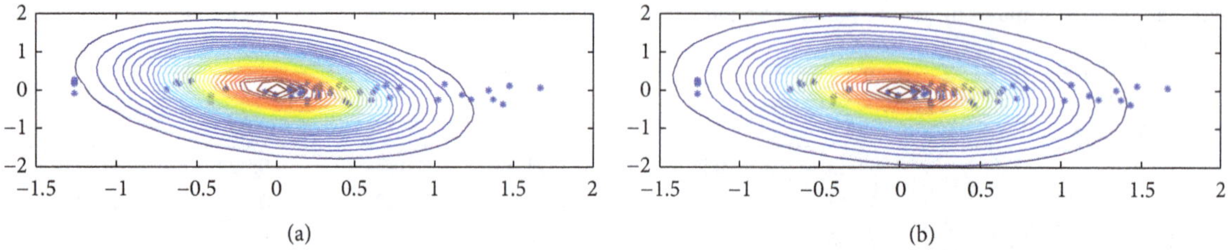

Figure 2: Plot of contours of posterior predictive density of pair (y_1, y_2) under parametric model and semiparametric model M^{ϵ}: (a) corresponds to parametric model and (b) corresponds to semiparametric model with $G = 50$.

Table 3: Mean and standard deviation (SD) of LPML in the political and economic risk data.

Model	LPML			
	BAY I		BAY II	
	Mean	SD	Mean	SD
M^{μ}	−253.0577	60.7693	−274.9704	14.7176
M^{ϵ}	−176.5289	110.0227	−188.8361	175.3741
$M^{\mu\epsilon}$	−193.2270	81.3209	−209.0063	77.0664
$M^{\mu\phi}$	−235.7465	15.3588	−235.7465	11.6911
$M^{\epsilon\phi}$	−267.4536	10.2084	−271.2875	10.4206
$M^{\mu\epsilon\phi}$	−255.2335	27.2208	−258.7124	15.8416

Table 4: Estimates and standard errors estimates of the parameters in analysis of political and economic risk data.

Parameter	Parametric model		M^{ϵ} model	
	Est.	SD	Est.	SD
λ_{22}	−0.155	0.083	−0.123	0.077
λ_{41}	0.418	0.104	0.846	0.088
λ_{51}	0.367	0.090	0.754	0.066
τ_{42}	−1.336	0.157	−1.340	0.055
τ_{43}	−0.905	0.167	−0.898	0.061
τ_{44}	−0.008	0.149	0.004	0.055
τ_{52}	−0.794	0.142	−0.787	0.046
τ_{53}	0.001	0.155	0.002	0.053
τ_{54}	0.566	0.136	0.567	0.035

informative prior. (ii) M^{ϵ} give the largest value. Among the posited models, M^{ϵ} is selected. We also compute the values of LPML for parametric model. They are −287.4262 and −288.6033 under priors (I) and (II) with standard deviations 6.287 and 5.065, respectively. Therefore, the data support the semiparametric model instead of parametric model.

Table 4 presents the estimates of factor loading λ_{jk} as well as their standard deviations with semiparametric and parametric model under prior (I). The factor loading estimates $\hat{\lambda}_{kj}$ in the measurement equation can be interpreted according to a standard confirmatory factor analysis model. The difference between the two approaches is obvious: the estimates of λ_{41} and λ_{51} under parametric model are only half of those under semiparametric model. Moreover, the standard deviations of estimates with parametric method are uniformly larger than that of semiparametric model. Since we identify illegal economic activity log PCR with economic factor ξ ($\lambda^*_{12} = 1$) and independent of judiciary with political factor η ($\lambda^*_{31} = 1$), respectively, the level of economic factor has a negative effect on real gross domestic product, while the level of political factor has positive effect on lack of expropriation risk threat and lack of corruption. The estimate $\hat{\lambda}_{22} = -0.123$ indicates that a one-unit increase in the level of economic factor leads to 0.123-unit decrease in the magnitude of gross domestic product. The interpretation of $\hat{\lambda}_{41}$ and $\hat{\lambda}_{51}$ is similar. The differences of estimates between parametric and semiparametric methods illustrate the effects of heavy tails of the data on the estimates.

6. Concluding Remarks

Parametric modeling for latent variable model with mixed data structure has long dominated Bayesian inference work, typically developed within the standard exponential family. Such modeling is often confused with handling the multimodal and unknown heterogeneous problems. In dealing with multimodality or increased heterogeneity in data, one naturally resorts to the finite mixture model [47, 48] which is more flexible and feasible to implement due to advances in simulation-based model fitting.

Rather than handling the large number of parameters resulting from the finite mixture models with a large number of components, we consider, in this paper, the finite-dimensional Dirichlet process mixture model for latent variable model with continuous and ordinal responses. The core of our proposal is to model the mean vector and/or variance-covariance parameters of unique errors and latent variables into the finite-dimensional stick-breaking priors. This will help to reveal the local dependence structure such as classification groups and clustering among the data. The blocked Gibbs sampler developed by Ishwaran and Zarepour [31], which takes advantage of the block updating and accelerates mixing in Gibbs sampling, is adapted here to cope with the posterior inference.

The proposed methodologies in this paper can be applied to more general latent variable models that include the multilevel SEMs [49] and longitudinal latent trait models [5] with discrete variables.

Appendix

Full Conditional Distributions

(1) Full Conditional Distribution $p(\vartheta, Y^* \mid \Omega, \pi, \Theta^*, L, \beta, Y)$. To draw (ϑ, Y^*) from $p(\vartheta, Y^* \mid \Omega, \pi, \Theta^*, L, \beta, Y)$,

we implement it by (i) drawing (τ, Y^*) from $p(\tau, Y^* \mid \Omega, \Lambda, \pi, \Theta^*, L, \beta, Y)$ and (ii) drawing Λ from $p(\Lambda \mid \Omega, \tau, Y^*\pi, \Theta^*, L, \beta, Y)$. The underlying reason is that drawing (τ, Y^*) from the joint conditional distribution as proposed here is more efficient than drawing τ and Y^* separately from the corresponding marginal conditional distribution (see Liu [50], Nandram and Chen [51], and Song and Lee [6]).

It can be shown that $p(\tau, Y^* \mid \Omega, \Lambda, \pi, \Theta^*, L, \beta, Y)$, not involving π and β, is given by

$$\begin{aligned} p\left(\tau, Y^* \mid \Omega, \Lambda, \Theta^*, L, Y\right) &= p\left(\tau \mid \Omega, \Lambda, \Theta^*, L, Y\right) \\ &\quad \cdot p\left(Y^* \mid \tau, \Omega, \Lambda, \Theta^*, L, Y\right) \\ &= \prod_{j=1}^{s} p\left(\tau_j \mid \Omega, \Lambda, \Theta_{(j)}, Y_{(j)}\right) \\ &\quad \cdot p\left(Y^*_{(j)} \mid \tau_j, \Omega, \Lambda, \Theta_{(j)}, Y_{(j)}\right), \end{aligned} \tag{A.1}$$

where $Y^*_{(j)} = \{y^*_{ij} : i = 1, \ldots, n\}$, $\Theta_{(j)} = \{\theta_{ij} : i = 1, \ldots, n\}$, and $Y_{(j)} = \{y_{ij} : i = 1, \ldots, n\}$. Further,

$$\begin{aligned} &p\left(\tau_j \mid \Omega, \Lambda, \Theta_{(j)}, Y_{(j)}\right) \\ &\quad \propto \prod_{k=1}^{b_j-1} \prod_{\{i: y_{ij}=k\}} \left\{\Phi\left(\psi_{\epsilon ij}^{-1/2}\left(\tau_{j,k+1} - \mu_{ij} - \Lambda_j^T\omega_i\right)\right)\right. \\ &\quad \left. - \Phi\left(\psi_{\epsilon ij}^{-1/2}\left(\tau_{j,k} - \mu_{ij} - \Lambda_j^T\omega_i\right)\right)\right\} I\left\{\tau_{j,k} < \tau_{j,k+1}\right\}, \\ &p\left(Y^*_{(j)} \mid \tau_j, \Omega, \Lambda, \Theta_{(j)}, Y_{(j)}\right) \propto \prod_{i=1}^{n} N\left(y^*_{ij} \mid \mu_{ij}\right. \\ &\quad \left. + \Lambda_j^T\omega_i, \psi_{\epsilon ij}\right) I\left\{\tau_{j,y_{ij}} < y^*_{ij} \le \tau_{j,y_{ij}+1}\right\}, \end{aligned} \tag{A.2}$$

where $\Phi(\cdot)$ is the cumulative distribution function of $N(0, 1)$. It is difficult to sample τ_j from $p(\tau_j \mid \Omega, \Lambda, \Theta_{(j)}, Y_{(j)})$ since this target distribution is nonstandard. We follow Cowles' routines [52] and use Metropolis-Hasting (MH) algorithm to sample observations from this complex conditional distribution. Specifically, given the current values $\tau_j^{(l)} = (\tau_{j,2}^{(l)}, \ldots, \tau_{j,b_j-1}^{(l)})^T$ at the lth iteration, generate a candidate vector $\tau_j^* = (\tau_{j,2}^*, \ldots, \tau_{j,b_j-1}^*)^T$ from the following truncated normal distribution:

$$\tau_{j,k}^* \sim N\left(\tau_{j,k}^{(l)}, \sigma_{\mathrm{MH}jk}^2\right) I\left\{\left(\tau_{j,k-1}^*, \tau_{j,k+1}^{(l)}\right]\right\}, \quad \text{for } k = 2, \ldots, b_j - 1. \tag{A.3}$$

Accept this candidate τ_j^* as $\tau_j^{(l+1)}$ with the probability $\min\{1, R_j\}$, where

$$\begin{aligned} R_j &= \prod_{k=2}^{b_j-1} \frac{\Phi\left(\sigma_{\mathrm{MH}jk}^{-1}\left[\tau_{j,k+1}^{(l)} - \tau_{j,k}^{(l)}\right]\right) - \Phi\left(\sigma_{\mathrm{MH}jk}^{-1}\left[\tau_{j,k-1}^{*} - \tau_{j,k}^{(l)}\right]\right)}{\Phi\left(\sigma_{\mathrm{MH}jk}^{-1}\left[\tau_{j,k+1}^{*} - \tau_{j,k}^{*}\right]\right) - \Phi\left(\sigma_{\mathrm{MH}jk}^{-1}\left[\tau_{j,k-1}^{(l)} - \tau_{j,k}^{*}\right]\right)} \\ &\quad \times \prod_{i=1}^{n} \frac{\Phi\left(\psi_{\epsilon ij}^{-1/2}\left(\tau_{j,y_{ij}+1}^{*} - \mu_{ij} - \Lambda_j^T\omega_i\right)\right) - \Phi\left(\psi_{\epsilon ij}^{-1/2}\left(\tau_{j,y_{ij}}^{*} - \mu_{ij} - \Lambda_j^T\omega_i\right)\right)}{\Phi\left(\psi_{\epsilon ij}^{-1/2}\left(\tau_{j,y_{ij}+1}^{(l)} - \mu_{ij} - \Lambda_j^T\omega_i\right)\right) - \Phi\left(\psi_{\epsilon ij}^{-1/2}\left(\tau_{j,y_{ij}}^{(l)} - \mu_{ij} - \Lambda_j^T\omega_i\right)\right)}. \end{aligned} \tag{A.4}$$

As pointed out by Cowles (1996) [52], the quantities $\sigma_{\mathrm{MH}jk}^2$ should be chosen carefully such that the average acceptance probability is about 0.30 or more.

For $p(\Lambda, \Pi \mid \Omega, \tau, Y^*, \pi, \Theta^*, L, \beta, Y)$, without loss of generality, we assume that the elements in Λ are all free. Let $y_{ij}^{**} = y_{ij}^* - \mu_j$. Under the prior distributions given in (14), we have

$$p\left(\Lambda \mid \Omega, Y^*, \Theta^*, L\right) \overset{D}{=} \prod_{k=1}^{p} N\left(m_{\epsilon k}, \Sigma_{\epsilon k}\right), \tag{A.5}$$

in which

$$\begin{aligned} m_{\epsilon k} &= \Sigma_{\epsilon k}\left(H_{\epsilon 0k}^{-1}\Lambda_{0k} + \sum_{i=1}^{n}\frac{\omega_i y_{ik}^{**}}{\psi_{\epsilon ik}}\right), \\ \Sigma_{\epsilon k} &= \left(\sum_{i=1}^{n}\psi_{\epsilon ik}^{-1}\omega_i\omega_i^T + H_{\epsilon 0k}^{-1}\right)^{-1}. \end{aligned} \tag{A.6}$$

(2) Full Conditional Distribution $p(\Omega \mid \tau, Y^*, \theta, \pi, \Theta^*, L, \beta, Y)$. It can be shown that the conditional distribution of Ω is given by

$$\begin{aligned} &p\left(\Omega \mid \tau, Y^*, \theta, \pi, \Theta^*, L, \beta, Y\right) \\ &\quad = \prod_{i=1}^{n} p\left(\omega_i \mid \theta, \theta_i, y_i^*\right), \end{aligned} \tag{A.7}$$

$$\left[\omega_i \mid \theta, \theta_i, y_i^*\right] \overset{D}{=} N\left(\Sigma_{\omega i}\Lambda^T\Psi_{\epsilon i}^{-1}\left(y_i^* - \mu_i\right), \Sigma_{\omega i}\right),$$

where $\Sigma_{\omega i} = (\Lambda^T\Psi_{\epsilon i}^{-1}\Lambda + \Phi_i^{-1})^{-1}$.

(3) The Full Conditional Distribution $p(\pi, \Theta^* \mid \Omega, Y^*, \vartheta, L, \beta, Y)$. It is clear that

$$\begin{aligned} &p\left(\pi, \Theta^* \mid \Omega, Y^*, \vartheta, L, \beta, Y\right) \\ &\quad = p\left(\pi \mid \Theta^*, \Omega, Y^*, \vartheta, L, \beta, Y\right) \\ &\quad \cdot p\left(\Theta^* \mid \pi, \Omega, Y^*, \vartheta, L, \beta, Y\right). \end{aligned} \tag{A.8}$$

Let $m_k = \#\{i : L_i = k\}$ be the number of L_i equal to k, for $k = 1, \ldots, G-1$. It can be shown that $p(\pi \mid \Theta^*, \Omega, Y^*, \vartheta, L, \beta, Y) = p(\pi \mid L, \alpha)$ is a generalized Dirichlet distribution, $\mathscr{GD}(a_1^*, b_1^*, \ldots, a_{G-1}^*, b_{G-1}^*)$ with $a_k^* = 1 + m_k$, $b_k^* = \alpha + \sum_{j=k+1}^{G} m_j$ $(k = 1, \ldots, G-1)$, which is constructed by

$$\pi_1 = V_1^*,$$
$$\pi_k = V_k^* \prod_{j=1}^{k-1} \left(1 - V_j^*\right) \quad (k = 2, \ldots, G-1), \tag{A.9}$$

where $V_j^* \overset{\text{ind}}{\sim} \text{Beta}(a_j^*, b_j^*)$.

For $p(\Theta^* \mid \pi, \Omega, Y^*, \vartheta, L, \beta, Y) = p(\Theta^* \mid \Omega, Y^*, \vartheta, L, \beta)$, let $L^* = \{L_1^*, \ldots, L_m^*\}$ be the unique set of L, $\Theta_{L^*}^* = \{\theta_{L_1^*}^*, \ldots, \theta_{L_m^*}^*\}$, and $\Theta_{(-L^*)}^*$ corresponding to those values in Θ^* with $\Theta_{L^*}^*$ excluded. Then,

$$p(\Theta^* \mid \Omega, Y^*, \vartheta, L, \beta) = p\left(\Theta_{(-L^*)}^* \mid \Omega, Y^*, \vartheta, L, \beta\right) \cdot p\left(\Theta_{L^*}^* \mid \Theta_{(-L^*)}^*, \Omega, Y^*, \vartheta, L, \beta\right) = p\left(\Theta_{(-L^*)}^* \mid \beta\right) p\left(\Theta_{L^*}^* \mid \Omega, Y^*, \vartheta, L, \beta\right). \tag{A.10}$$

Let $\mu^* = \{\mu_j^* : j = 1, \ldots, G\}$, $\Psi_\epsilon^* = \{\Psi_{\epsilon j}^* : j = 1, \ldots, G\}$, and $\Phi^* = \{\Phi_j^* : j = 1, \ldots, G\}$, and note that $\Theta^* = \{\mu^*, \Psi_\epsilon^*, \Phi^*\}$. The components of $\{\mu_{(-L^*)}^*, \Psi_{\epsilon(-L^*)}^*, \Phi_{(-L^*)}^*\}$ are easy to sample based on (14). Further,

$$p\left(\Theta_{L^*}^* \mid \Omega, Y^*, \vartheta, L, \beta\right) = \prod_{j=1}^{m} p\left(\theta_{L_j^*}^* \mid \Omega, Y^*, \vartheta, L, \beta\right) \tag{A.11}$$

which can be implemented by drawing: for $l \in L^*$

$$p\left(\mu_l^* \mid \Psi_{\epsilon l}^*, \Psi_{\zeta l}^*, \Phi_l^*, \Omega, Y^*, \vartheta, L, \beta\right) \sim N\left(\mu_l^*, \Sigma_l^*\right),$$
$$p\left(\Psi_{\epsilon l}^{*-1} \mid \mu_l^*, \Psi_{\zeta l}^*, \Phi_l^*, \Omega, Y^*, \vartheta, L, \beta\right) \sim \prod_{k=1}^{p} \text{Gamma}\left(\alpha_{\epsilon lk}^*, \beta_{\epsilon lk}^*\right), \tag{A.12}$$
$$p\left(\Phi_l^{*-1} \mid \mu_l^*, \Psi_{\epsilon l}^*, \Psi_{\zeta l}^*, \Omega, Y^*, \vartheta, L, \beta\right) \sim \text{Wishart}\left(\rho_{l0}^*, R_{l0}^*\right)$$

in which

$$\mu_l^* = \Sigma_l^* \left(\Sigma_\nu^{-1} \nu + \left(\Psi_{\epsilon l}^{*-1} \sum_{\{i:L_i=l\}} \left(y_i^* - \Lambda \omega_i\right)\right)\right),$$
$$\Sigma_l^* = \left[\Sigma_\nu^{-1} + m_l \Psi_{\epsilon l}^{*-1}\right]^{-1},$$
$$\alpha_{\epsilon lk}^* = \alpha_{\epsilon 0k} + m_l,$$
$$\beta_{\epsilon lk}^* = \beta_{\epsilon 0k} + 2^{-1} \sum_{\{i:L_i=l\}} \left(y_{ik}^* - \mu_{lk}^* - \Lambda_k^T \omega_i\right)^2,$$
$$\rho_{l0}^* = \rho_0^\phi + m_l,$$
$$R_{l0}^* = \left(R^{\phi-1} + \sum_{\{i:L_i=l\}} \omega_i \omega_i^T\right)^{-1}. \tag{A.13}$$

(4) Full Conditional Distribution $p(L \mid \Omega, \tau, Y^*, \vartheta, \pi, Z, Y)$. It can be shown that

$$p\left(L \mid \Omega, Y^*, \vartheta, \pi, \Theta^*, Y\right) = \prod_{i=1}^{n} p\left(L_i \mid \Omega, Y^*, \vartheta, \pi, \Theta^*, Y\right),$$
$$\left[L_i = \cdot \mid \Omega, \tau, Y^*, \vartheta, \pi, \Theta^*, Y\right] \overset{\text{iid}}{\sim} \sum_{k=1}^{G} \pi_{ik}^* \delta_k(\cdot), \tag{A.14}$$

where $\pi_{ik}^* = c_i \pi_k p(y_i^* \mid \omega_i, \theta_k^*, \vartheta) p(\omega_i \mid \theta_k^*)$ and c_i is a normalized constant such that $\sum_{k=1}^{G} \pi_{ik}^* = 1.0$.

(5) Full Conditional Distribution $p(\beta \mid \Omega, \tau, Y^*, \vartheta, \pi, Z, L, Y)$. Based on the priors given in (15), the full conditional distributions for components of hyperparameters β are given as follows:

$$[\nu \mid \Theta^*, \Sigma_\nu] \sim N(m_\nu, A_\nu),$$
$$[\Sigma_\nu \nu, \Theta^*] \sim \prod_{j=1}^{p} \text{Gamma}^{-1}\left(\kappa_1 + 0.5G, \kappa_2 + 0.5 \sum_{k=1}^{G} \left(\mu_{kj}^* - \mu_j\right)^2\right),$$
$$\left[R^{-1} \mid \Theta^*\right] \sim \text{Wishart}\left(G\rho_0 + \rho_0^\phi, \left(\sum_{k=1}^{G} \Phi_k^{*-1} + R_0^\phi\right)^{-1}\right),$$
$$[\alpha \mid \pi] \sim \text{Gamma}\left(\tau_1 + G - 1, \tau_2 - \log \pi_G\right), \tag{A.15}$$

where $m_\nu = A_\nu\{\Sigma_0^{-1}\mu_0 + \Sigma_\nu^{-1} \sum_{k=1}^{G} \mu_k^*\}$ and $A_\nu = (G\Sigma_\nu^{-1} + \Sigma_0^{-1})^{-1}$.

Competing Interests

The authors declare that they have no competing interests.

Acknowledgments

The work described in this paper was supported by the National Natural Science Fund (11471161), Nanjing Forestry University Foundation (163101004), and Technological Innovation Item of Personnel Division (013101001).

References

[1] B. Muthén, *LISCOMP: Analysis of Linear Statistical Equation with a Comprehensive Measurement Model*, Scientific Software, Mooresville, NC, USA, 1987.

[2] J.-Q. Shi and S.-Y. Lee, "Bayesian sampling-based approach for factor analysis models with continuous and polytomous data," *British Journal of Mathematical and Statistical Psychology*, vol. 51, no. 2, pp. 233–252, 1998.

[3] J.-Q. Shi and S.-Y. Lee, "Latent variable models with mixed continuous and polytomous data," *Journal of the Royal Statistical Society, Series B: Statistical Methodology*, vol. 62, no. 1, pp. 77–87, 2000.

[4] D. B. Dunson, "Bayesian latent variable models for clustered mixed outcomes," *Journal of the Royal Statistical Society—Series B: Statistical Methodology*, vol. 62, no. 2, pp. 355–366, 2000.

[5] D. B. Dunson, "Dynamic latent trait models for multidimensional longitudinal data," *Journal of the American Statistical Association*, vol. 98, no. 463, pp. 555–563, 2003.

[6] X.-Y. Song and S.-Y. Lee, "Bayesian analysis of two-level nonlinear structural equation models with continuous and polytomous data," *British Journal of Mathematical and Statistical Psychology*, vol. 57, no. 1, pp. 29–52, 2004.

[7] Y. Kano, M. Berkane, and P. M. Bentler, "Statistical inference based on pseudo-maximum likelihood estimators in elliptical populations," *Journal of the American Statistical Association*, vol. 88, no. 421, pp. 135–143, 1993.

[8] Y. Kano, "Consistency property of elliptical probability density functions," *Journal of Multivariate Analysis*, vol. 51, no. 1, pp. 139–147, 1994.

[9] S.-Y. Lee and Y.-M. Xia, "Maximum likelihood methods in treating outliers and symmetrically heavy-tailed distributions for nonlinear structural equation models with missing data," *Psychometrika*, vol. 71, no. 3, pp. 565–585, 2006.

[10] S.-Y. Lee and Y.-M. Xia, "A robust Bayesian approach for structural equation models with missing data," *Psychometrika*, vol. 73, no. 3, pp. 343–364, 2008.

[11] Y.-M. Xia, X.-Y. Song, and S.-Y. Lee, "Robust model fitting for the non linear structural equation model under normal theory," *British Journal of Mathematical and Statistical Psychology*, vol. 62, no. 3, pp. 529–568, 2009.

[12] Y.-M. Xia and Y.-A. Liu, "Robust Bayesian analysis and its applications for a factor analytic model with normal scale mixing," *Chinese Journal of Applied Probability and Statistics*, vol. 30, no. 4, pp. 423–438, 2014.

[13] A. Kottas, P. Müller, and F. Quintana, "Nonparametric Bayesian modeling for multivariate ordinal data," *Journal of Computational & Graphical Statistics*, vol. 14, no. 3, pp. 610–625, 2005.

[14] J. H. Albert and S. Chib, "Bayesian analysis of binary and polychotomous response data," *Journal of the American Statistical Association*, vol. 88, no. 422, pp. 669–679, 1993.

[15] M. K. Cowles, B. P. Carlin, and J. E. Connett, "Bayesian tobit modeling of longitudinal ordinal clinical trial compliance data with nonignorable missingness," *Journal of the American Statistical Association*, vol. 91, no. 433, pp. 86–98, 1996.

[16] M.-H. Chen and D. K. Dey, "Bayesian analysis for correlated ordinal data models," in *Generalized Linear Models: A Bayesian Perspectives*, D. K. Dey, S. Ghosh, and B. K. Mallick, Eds., pp. 133–157, Marcel Dekker, New York, NY, USA, 2000.

[17] S.-Y. Lee, B. Lu, and X.-Y. Song, "Semiparametric Bayesian analysis of structural equation models with fixed covariates," *Statistics in Medicine*, vol. 27, no. 13, pp. 2341–2360, 2008.

[18] H. Ishwaran and L. F. James, "Gibbs sampling methods for stick-breaking priors," *Journal of the American Statistical Association*, vol. 96, no. 453, pp. 161–173, 2001.

[19] X.-Y. Song, Y.-M. Xia, and S.-Y. Lee, "Bayesian semiparametric analysis of structural equation models with mixed continuous and unordered categorical variables," *Statistics in Medicine*, vol. 28, no. 17, pp. 2253–2276, 2009.

[20] X.-Y. Song, Y.-M. Xia, J.-H. Pan, and S.-Y. Lee, "Model comparison of bayesian semiparametric and parametric structural equation models," *Structural Equation Modeling-A Multidisciplinary Journal*, vol. 18, no. 1, pp. 55–72, 2011.

[21] Y.-M. Xia and A.-Y. Liu, "Bayesian semiparametric analysis for generalized linear latent variable model," in *The Proceedings of 2010 International Conference on Probability and Statistics of the International Institute for General Systems Studies*, Y. Jiang and G. Z. Wang, Eds., vol. 1 of *Advances on Probability and Statistics*, pp. 308–312, 2010.

[22] Y.-M. Xia and Y.-A. Liu, "Bayesian semiparametric analysis and model comparison for confirmatory factor model," *Chinese Journal of Applied Probability and Statistics*, In press.

[23] Y.-M. Xia and J.-W. Gou, "Semiparametric Bayesian analysis for factor analysis model mixed with hidden Markov model," *Applied Mathematics*, vol. 30, pp. 17–30, 2015.

[24] S.-Y. Lee and H.-T. Zhu, "Statistical analysis of nonlinear structural equation models with continuous and polytomous data," *British Journal of Mathematical and Statistical Psychology*, vol. 53, no. 2, pp. 209–232, 2000.

[25] T. S. Ferguson, "A Bayesian analysis of some nonparametric problems," *The Annals of Statistics*, vol. 1, pp. 209–230, 1973.

[26] C. E. Antoniak, "Mixtures of dirichlet processes with applications to bayesian nonparametric problems," *The Annals of Statistics*, vol. 2, no. 6, pp. 1152–1174, 1974.

[27] A. Y. Lo, "On a class of Bayesian nonparametric estimates. I. Density estimates," *The Annals of Statistics*, vol. 12, no. 1, pp. 351–357, 1984.

[28] M. D. Escobar, "Estimating normal means with a Dirichlet process prior," *Journal of the American Statistical Association*, vol. 89, no. 425, pp. 268–277, 1994.

[29] M. D. Escobar and M. West, "Bayesian density estimation and inference using mixtures," *Journal of the American Statistical Association*, vol. 90, no. 430, pp. 577–588, 1995.

[30] P. Müller and F. A. Quintana, "Nonparametric Bayesian data analysis," *Statistical Science*, vol. 19, no. 1, pp. 95–110, 2004.

[31] H. Ishwaran and M. Zarepour, "Markov chain Monte Carlo in approximate Dirichlet and beta two-parameter process hierarchical models," *Biometrika*, vol. 87, no. 2, pp. 371–390, 2000.

[32] A. Zellner, *An Introduction to Bayesian Inference in Econometrics*, John Wiley & Sons, New York, NY, USA, 1971.

[33] M. A. Tanner and W. H. Wong, "The calculation of posterior distributions by data augmentation," *Journal of the American Statistical Association*, vol. 82, no. 398, pp. 528–550, 1987.

[34] S. Geman and D. Geman, "Stochastic relaxation, gibbs distributions, and the bayesian restoration of images," *IEEE Transactions on Pattern Analysis and Machine Intelligence*, vol. 6, no. 6, pp. 721–741, 1984.

[35] N. Metropolis, A. W. Rosenbluth, M. N. Rosenbluth, A. H. Teller, and E. Teller, "Equation of state calculations by fast computing machines," *The Journal of Chemical Physics*, vol. 21, no. 6, pp. 1087–1092, 1953.

[36] W. K. Hastings, "Monte carlo sampling methods using Markov chains and their applications," *Biometrika*, vol. 57, no. 1, pp. 97–109, 1970.

[37] J. Albert and S. Chib, "Bayesian tests and model diagnostics in conditionally independent hierarchical models," *Journal of the American Statistical Association*, vol. 92, no. 439, pp. 916–925, 1997.

[38] R. E. Kass and A. E. Raftery, "Bayes factors," *Journal of the American Statistical Association*, vol. 90, no. 430, pp. 773–795, 1995.

[39] S. Basu and S. Chib, "Marginal likelihood and Bayes factors for Dirichlet process mixture models," *Journal of the American Statistical Association*, vol. 98, no. 461, pp. 224–235, 2003.

[40] S. Geisser and W. F. Eddy, "A predictive approach to model selection," *Journal of the American Statistical Association*, vol. 74, no. 365, pp. 153–160, 1979.

[41] S. Mukhopadhyay and A. E. Gelfand, "Dirichlet process mixed Generalized linear models," *Journal of the American Statistical Association*, vol. 92, no. 438, pp. 633–639, 1997.

[42] A. Gelman and D. B. Rubin, "Inference from iterative simulation using multiple sequences," *Statistical Science*, vol. 7, no. 4, pp. 457–472, 1992.

[43] K. M. Quinn, "Bayesian factor analysis for mixed ordinal and continuous responses," *Political Analysis*, vol. 12, no. 4, pp. 338–353, 2004.

[44] W. J. Henisz, *The Political Constraint Index (POLCON) Dataset*, 2002, http://www-management.wharton.upenn.edu/henisz/POLCON/ContactInfo.html.

[45] M. G. Marshall, R. G. Ted, and H. Barbara, *State Failure Task Force Problem Set*, 2002, http://www.cidcm.umd.edu/inscr/stfail/index.html.

[46] M. Alvarez, A. C. Jose, L. Fernando, and P. Adam, *ACLP Political and Economic Database*, 1999, http://www.ssc.upenn.edu/cheibub/data/.

[47] D. M. Titterington, A. F. M. Smith, and U. E. Makov, *Statistical Analysis of Finite Mixture Distributions*, John Wiley & Sons, Chichester, UK, 1985.

[48] H.-T. Zhu and S.-Y. Lee, "A Bayesian analysis of finite mixtures in the LISREL model," *Psychometrika*, vol. 66, no. 1, pp. 133–152, 2001.

[49] S.-Y. Lee and X.-Y. Song, "Maximum likelihood analysis of a general latent variable model with hierarchically mixed data," *Biometrics*, vol. 60, no. 3, pp. 624–636, 2004.

[50] J. S. Liu, "The collapsed Gibbs sampler in Bayesian computations with applications to a gene regulation problem," *Journal of the American Statistical Association*, vol. 89, no. 427, pp. 958–966, 1994.

[51] B. Nandram and M.-H. Chen, "Reparameterizing the generalized linear model to accelerate Gibbs sampler convergence," *Journal of Statistical Computation & Simulation*, vol. 54, no. 1–3, pp. 129–144, 1996.

[52] M. K. Cowles, "Accelerating Monte Carlo Markov chain convergence for cumulative-link generalized linear models," *Statistics and Computing*, vol. 6, no. 2, pp. 101–111, 1996.

15

Production Planning of a Failure-Prone Manufacturing System under Different Setup Scenarios

Guy-Richard Kibouka,[1] Donatien Nganga-Kouya,[1] Jean-Pierre Kenne,[2] Victor Songmene,[2] and Vladimir Polotski[2]

[1] *Mechanical Engineering Department, Omar Bongo University, École Normale Supérieure de l'Enseignement Technique, BP 3989, Libreville, Gabon*
[2] *Mechanical Engineering Department, University of Quebec, École de Technologie Supérieure, 1100 Notre Dame West, Montreal, QC, Canada H3C 1K3*

Correspondence should be addressed to Jean-Pierre Kenne; jean-pierre.kenne@etsmtl.ca

Academic Editor: Dar-Li Yang

This paper presents a control problem for the optimization of the production and setup activities of an industrial system operating in an uncertain environment. This system is subject to random disturbances (breakdowns and repairs). These disturbances can engender stock shortages. The considered industrial system represents a well-known production context in industry and consists of a machine producing two types of products. In order to switch production from one product type to another, a time factor and a reconfiguration cost for the machine are associated with the setup activities. The parts production rates and the setup strategies are the decision variables which influence the inventory and the capacity of the system. The objective of the study is to find the production and setup policies which minimize the setup and inventory costs, as well as those associated with shortages. A modeling approach based on stochastic optimal control theory and a numerical algorithm used to solve the obtained optimality conditions are presented. The contribution of the paper, for industrial systems not studied in the literature, is illustrated through a numerical example and a comparative study.

1. Introduction

The production and setup planning problem surfaces in manufacturing systems when significant cost and time are required to set up the production unit for the processing of multiple part types. The setup scheduling problem involves deciding which part type has to be processed next and when the production unit has to stop its current operations and make a setup change to begin the processing of that part type. The time required to switch from producing one part type to another and the associated cost are significant. Given that it is not realistic (or advantageous) to devote one machine to a single part type, different part types must share the same machine, and capacity is lost due to each setup change. In addition, the considered machine is subject to random breakdowns and repairs. It is therefore essential to jointly investigate setup scheduling and production policies in order to optimize the system performance measure of the failure-prone manufacturing system under study.

For the class of completely flexible machines (based on a crucial assumption that no setup time and cost are required when production is switched from one part type to another), an explicit formulation of the optimal control problem for an unreliable flexible machine which produces multiple part types is provided in [1]. In addition, Gharbi and Kenné [2] provided a suboptimal control policy for the multiple parts, multiple-machines problem. The considered planning problem falls under an important class of stochastic manufacturing systems involving nonflexible machines, given that the setup time and costs are considered when production is switched from one product type to another. This class of systems is a subset of manufacturing systems for which the problem of determining the optimal production policies has been considered by many authors. A significant portion of the research by the latter is based on a feedback formulation of the control problem in a dynamic manufacturing environment. It is shown in [3] that the optimal control policy has a special structure called the Hedging Point Policy (HPP) in the case of a single-machine, single product system. For such a policy, a nonnegative production surplus of parts, corresponding to optimal inventory levels, is maintained during times of excess capacity in order to hedge against future capacity shortages caused by machine failures for the case of a single-machine, two-product manufacturing system with setup (see [4]). Various researchers have considered the problems of setup scheduling in production using advanced optimization approaches in the context of multiple-product manufacturing systems. As recently stated in [5], the problems of sequence-dependent setup times have been attracting increasing interest [6]. Previous sequence-dependent setup times are studied using objective functions such as makespan, total completion time, and their combinations, with an emphasis on the learning aspects of the sorting algorithms. In the same context, Feng et al. [7] optimized various scheduling policies and then analyzed them from the point of view of their robustness to uncertainties and system parameter variations. The obtained setup policy had a cyclic policy structure resilient to parameter variations.

The stochastic optimal control problem of a manufacturing system with setup costs and time was formally presented in [4] following the series of papers published in the same domain by Sethi and Zhang [8], Yan and Zhang [9], Boukas and Kenné [10] and Hajji et al. [11]. The proposed models led to the optimality conditions described by the Hamilton Jacobi Bellman equations (HJB). Such equations are difficult to resolve analytically for more general cases. An explicit solution for such equations was obtained by Akella and Kumar [12] for a one-machine, one-product manufacturing system. Numerical methods based on the Kushner approach (see Kushner and Dupuis [13]) were used by Yan and Zhang [9] and Boukas and Kenné [10] for a one-machine, two-product manufacturing system. They were able to develop near-optimal control policies for production and setup scheduling in the case of a homogeneous and machine age-dependent Markovian process.

For the one-machine, two products' case, Yan and Zhang [9] provide a characterization of the optimal production and setup policy by four exclusive regions as a main result, while Bai and Elhafsi [14] focused their contribution on providing a suitable production and setup policy structure and obtained the so-called Hedging Corridor Policy (HCP). Following these studies, Gharbi et al. [4] developed a production and setup policy for unreliable multiple-machine, multiple part type manufacturing system, for which the production and setup policy are known across the sample space. They obtained a control policy called the Modified Hedging Corridor Policy (MHCP), qualified as more realistic and useful in the context of the production planning of manufacturing systems with setup.

The main contribution of this paper is to develop a production and setup policy for a more realistic unreliable one-machine, one-part type manufacturing system under appropriate assumptions in different industrial situations, called here *industrial scenarios*. The resultant control policy is more realistic and useful in the context of the production planning of manufacturing systems with setup. This paper's contribution is further illustrated through the fact that the proposed control policy guarantees a system performance for systems that have not yet been studied in the relevant literature. Our proposal is an extension of the works of Bai and Elhafsi [14], Boukas and Kenné [10], and Hajji et al. [11].

This paper is organized as follows: Section 2 presents the notations and main assumptions of the proposed model. Section 3 presents the statement of the optimal production and setup scheduling problem. The optimality conditions and numerical approach are presented in Section 4. Section 5 describes the numerical example with results analysis, and the paper is concluded in Section 6.

2. Model Assumptions and Hypotheses

This section presents the notations and assumptions used throughout this paper.

2.1. Notations

P_i: part type i ($i \in I = \{1, 2\}$),

θ_{ij}: setup time to go from P_i to P_j,

K_{ij}: setup cost to go from P_i to P_j,

d_i: rate of P_i product request,

$x(t)$: vector inventory levels/shortage, product type i,

p_i: product processing time, type i,

$u_i(t)$: production rate, product type i,

U_i^+: maximum production rate, product type i,

z_i: optimal inventory level, product type i,

$\alpha(t)$: stochastic process describing the dynamics of the machine,

S_{ij}: setup policy from product part type i to j,

$q_{\alpha\beta}$: transition rate, mode $\alpha\beta$,

c_i^-: shortage cost, product type i,

c_i^+: inventory cost, product type i,

ρ: cost discount rate,

$g(\cdot)$: cost function,

$R(\cdot)$: total cost function during setup,

$J(\cdot)$: total cost function,

$v(\cdot)$: value function.

2.2. Context and Assumptions. The following is a summary of the general context and main assumptions considered in this paper:

(1) The model is time-continuous.

(2) Raw materials for the production of each product part type are always available and unlimited.

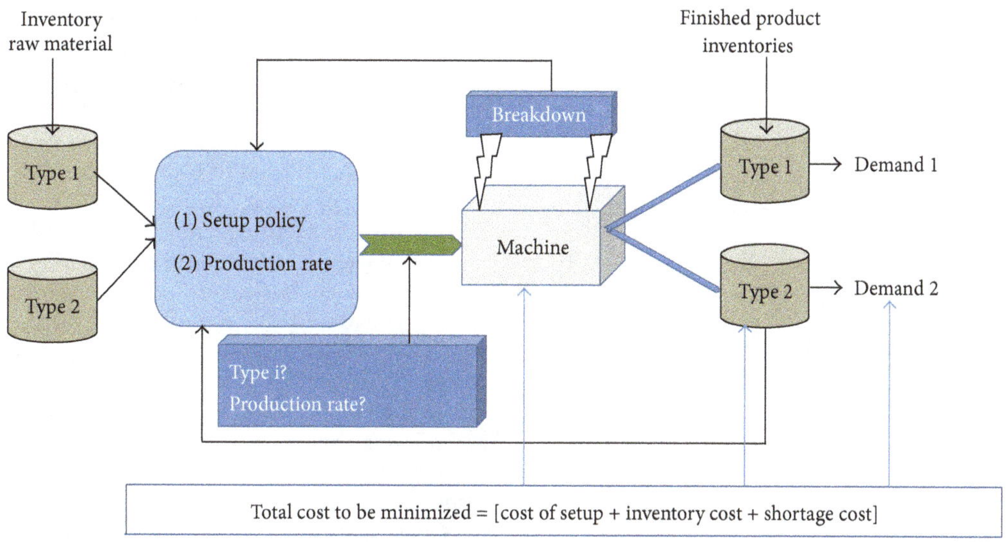

FIGURE 1: Manufacturing system studied.

(3) Customer demand of finished products for each part type is known and represented by a constant rate over time.

(4) The maximal production rate of each part type is known.

(5) All failures are instantly detected and repaired. A corrective maintenance action renews the production system to its initial state (as good as new condition).

(6) The machine shares the production of different product part types with significant setup time and cost.

(7) The shortage cost depends on the shortage quantity and time (average value ($/product/unit of time)).

(8) The holding cost depends on the mean inventory level (average value ($/product/unit of time)).

(9) For each product part type, once the production starts at a given rate, no adjustment of the rate will be allowed until either the machine is down (failure mode) or the current unit is completed.

We complete the assumptions by two hypotheses that help us to study different industrial contexts (or production scenarios) with setup.

Hypothesis 1. The setup operation is performed only when the machine is in operational mode and cannot be interrupted by any machine failure such that it has to be started all over again.

Hypothesis 2. The setup operation is only allowed if the machine is in operational mode, and the setup process is interrupted by failure such that it can be continued after a repair.

In this paper, we show how the hypotheses affect the optimality conditions of the associated stochastic optimal control problem. We then develop appropriate optimality conditions consisting of a modified form of the traditional HJB equations. We finally compare the results obtained for the two hypotheses (or contexts of production) in order to provide more realistic production and setup policies.

3. Problem Formulation

The production system presented in Figure 1 consists of one machine capable of producing two *different* part types. The machine is not completely flexible in the sense that the setup activities between the two part types involve both time and cost to switch from the production of one part type to another. The system under study is dynamic and the associated costs to be minimized are illustrated in Figure 1.

Let θ_{ij} and K_{ij} be the duration and the cost incurred for switching the production from P_i to P_j with $i \neq j$, respectively. Note that, for $i, j = 1, 2$ and $i \neq j$, $\theta_{ij} \geq 0$ and $K_{ij} \geq 0$.

The i-type product requires an average production time denoted as $p_i > 0$ $(i = 1, 2)$ and ordered with a constant demand rate d_i.

Let $x_i(t), u_i(t)$ *be* the stock level and the rate of production of two part types of products P_i, $i = 1, 2$, respectively.

Let **x**, **u**, and **d** denote the vectors $(x_1, x_2)^t$, $(u_1, u_2)^t$, and $(d_1, d_2)^t$, respectively, knowing that the notation A^t denotes the transpose of A.

At a given moment, we can describe the system by a hybrid state that consists of a continuous portion (stock dynamics) and a discrete portion (modes of the machine). A stochastic process $\xi(t)$ is used to describe the mode of the machine as follows:

$$\xi(t) = \begin{cases} 1 & \text{if the machine is operational} \\ 2 & \text{if the machine is under repair.} \end{cases} \quad (1)$$

The machine uptimes and downtimes are assumed to be exponentially distributed with rates p and q, respectively. Hence, the machine state evolves according to a continuous-time Markov process with modes $\xi(t) \in M = \{1, 2\}$. The states

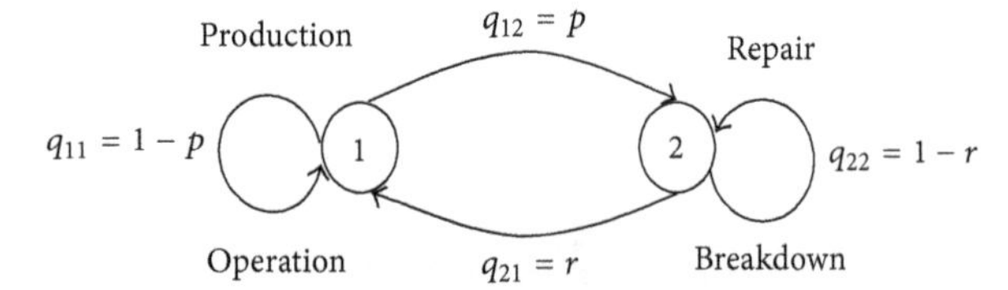

FIGURE 2: States transition diagram of the system studied.

transition diagram of the Markov chain associated with the machine dynamics is shown in Figure 2.

The evolution of machine states in the interval $(t, t + \delta t)$ can be expressed by

$$\begin{aligned}
&\text{prob}\,(\xi\,(t + \delta t) = 1 \mid \xi\,(t) = 2) = q_{12}\delta t + 0\,(\delta t)\,,\\
&\text{prob}\,(\xi\,(t + \delta t) = 1 \mid \xi\,(t) = 1) = 1 + q_{12}\delta t + 0\,(\delta t)\,,\\
&\text{prob}\,(\xi\,(t + \delta t) = 2 \mid \xi\,(t) = 1) = q_{21}\delta t + 0\,(\delta t)\,,\\
&\text{prob}\,(\xi\,(t + \delta t) = 2 \mid \xi\,(t) = 2) = 1 + q_{21}\delta t + 0\,(\delta t)\,.
\end{aligned} \tag{2}$$

The process $\xi(t)$ can be described by a transition rate matrix $Q = \{q_{\alpha\beta}\}$, defined by $q_{\alpha\beta} \geq 0$ si $\alpha \neq \beta$ and $q_{\alpha\alpha} = -\sum_{\alpha\neq\beta} q_{\alpha\beta}$, knowing that $\alpha, \beta \in M$.

The transition rate from a state $\alpha \in M$ to a state $\beta \in M$ at time t is defined by

$$\begin{aligned}
q_{\alpha\beta} &= \lim_{\delta t\to 0}\left[\text{prob}\frac{(\xi\,(t+\delta t) = \beta \mid \xi\,(t) = \alpha)}{\delta t}\right],\\
&\alpha \neq \beta \text{ knowing that } \lim_{\delta t\to 0}\frac{0\,(\delta t)}{\delta t} = 0.
\end{aligned} \tag{3}$$

The corresponding matrix of transition rates e is given in the following:

$$Q = \begin{pmatrix} -q_{12} & q_{12} \\ q_{21} & -q_{21} \end{pmatrix}. \tag{4}$$

The differential equation representing the dynamics of the finished products stocks is

$$\begin{aligned}
\dot{x}\,(t) &= \frac{dx\,(t)}{dt} = u\,(t) - d,\\
x\,(0) &= x_0,
\end{aligned} \tag{5}$$

where x_0 is the initial stock level.

The production rates satisfy the system capacity constraint given by the following equation:

$$0 \leq u_i\,(\cdot) \leq \grave{U}_i^+, \quad i = 1, 2, \tag{6}$$

where $\grave{U}_i^+$ denotes the maximal production rate of product i on the machine. The set of feasible production rates of the machine for a product i is given by

$$\Gamma\,(\alpha) = \left\{u : u = (u_1, u_2),\ 0 \leq u_i\,(\cdot) \leq \grave{U}_i^+,\ i = 1, 2\right\}. \tag{7}$$

The decision variables of the optimal control problem under study are production rates $u = (u_1, u_2)$ and a sequence of setups denoted by $\Omega = \{(\tau_0, i_0 i_1), (\tau_1, i_1 i_2), \ldots\}$. A setup (τ, ij) is defined by the time τ at which it begins and a pair ij, denoting that the system was already set up to produce part i, and is being switched to be able to produce part j. Let A denote the set of admissible decisions (Ω, u_1, u_2).

The instantaneous cost function depends on the state of the system (stock level, mode of the machine) and is given by

$$g\,(x, \alpha) = \left(c_1^+ x_1^+ + c_1^- x_1^- + c_2^+ x_2^+ + c_2^- x_2^-\right) + c^\alpha, \tag{8}$$

where c^α is the cost incurred at mode α of the machine (assuming that $c^1 = 0$ and $c^2 \neq 0$). Note that $x_i^+ = \max(0, x_i)$ and $x_i^- = \max(-x_i, 0)$; $c_i^+$and c_i^- are inventory and backlog costs for part type i per unit of product per unit of time, respectively.

Given that the setup cost is consumed at the beginning of the operation, the instantaneous cost as a function of the setups denoted as $R(\cdot)$ is therefore expressed by the following expression:

$$\begin{aligned}
R_{ij}\,(x, s) &= K_{ij}\,\text{Ind}\left\{s = \Theta_{ij}\right\} + \int_0^s e^{-\rho t} g\,(x - d)\,dt,\\
&s \in \left[0, \Theta_{ij}\right],\ i, j = 1, 2,\ i \neq j,
\end{aligned} \tag{9}$$

where ρ is the discount rate. The first part of (9) expresses the setup cost at the beginning of the operation. The second part evaluates the penalty incurred for an inventory during the setup, depending on the time remaining in the setup operation, denoted as s, with

$$\text{Ind}\left(s = \Theta_{ij}\right) = \begin{cases} 1 & \text{if } s = \Theta_{ij} \\ 0 & \text{otherwise.} \end{cases} \tag{10}$$

We can deduce the instantaneous cost function of the setup as follows:

$$\begin{aligned}
R_{12}\,(x, \Theta_{12}) &= K_{12}\\
&\quad + \int_0^{\Theta_{12}} e^{-\rho t} g\,(x_1 - d_1 t; x_2 - d_2 t)\,dt,\\
R_{21}\,(x, \Theta_{21}) &= K_{21}\\
&\quad + \int_0^{\Theta_{21}} e^{-\rho t} g\,(x_1 - d_1 t; x_2 - d_2 t)\,dt.
\end{aligned} \tag{11}$$

The total discounted cost over an infinite horizon can then be defined by the following expression:

$$\begin{aligned}
&J\,(i, x, \alpha, s, \Omega, u\,(\cdot))\\
&= \int_0^s e^{-\rho t} g\,(x\,(t))\,dt\\
&\quad + E_{i,x-ds,\alpha_s}\left[\int_s^\infty e^{-\rho t} g\,(x\,(t))\,dt + \sum_{i=0}^{\infty} e^{-\rho\tau_i} K_{i_1 i_{1+1}}\right].
\end{aligned} \tag{12}$$

The production planning problem is to find an admissible decision or control policy $(\Omega, u(\cdot))$ that minimizes $J(\cdot)$, given by (12). For the production of part type i, the value function can be given by the following:

$$v_i(x,\alpha,s) = \min_{(\Omega,u)\in A} J(i,x,\alpha,s,\Omega,u) \quad \forall x \in R^n,\ \alpha \in M. \tag{13}$$

In the next section, we present the optimality conditions and the corresponding discrete form obtained by the application of the numerical methods inspired from the Kushner approach (see [11] for more details).

4. Optimality Conditions and Numerical Approach

In this section, we present the modified HJB equations related to Hypotheses 1 and 2. We then compare the results obtained for those hypotheses to the results given by the application of the traditional form used in Yan and Zhang [9], Bourkas and Kenné [10], and Hajji et al. [11].

The value function $v_i(x,\alpha)$ that satisfies the HJB equations in mode 1 for Hypothesis 1 is

$$\min\left\{\min_{u\in\Gamma_i(\alpha)}\left[(u-d)(v_i)_x(x,\alpha)+g(x)+Qv_i(x,\cdot)(\alpha)\right] - \rho v_i(x,\alpha);\min_{j\neq i}\left[R_{ij}(x,\Theta_{ij})+(P_{ij})e^{-\rho\Theta_{ij}}\cdot v_j(x-d\Theta_{ij},1)+(1-P_{ij})e^{-\rho\Theta_{ij}}\cdot v_j(x-d\tau_{ij},1)\right]-v_i(x,\alpha)\right\}=0. \tag{14}$$

The value function $v_i(x,\alpha)$ that satisfies HJB equations in mode 1 for Hypothesis 2 is

$$\min\left\{\min_{u\in\Gamma_i(\alpha)}\left[(u-d)(v_i)_x(x,\alpha)+g(x)+Qv_i(x,\cdot)(\alpha)\right] - \rho v_i(x,\alpha); \min_{j\neq i}\left[R_{ij}(x,\widehat{\Theta_{ij}})+e^{-\rho\widehat{\Theta_{ij}}}\cdot v_j(x-d\widehat{\Theta_{ij}},1)\right]\right\}=0, \tag{15}$$

where $(v_i)_x(\cdot)$ is the gradient *of* $v_i(\cdot)$ related to x. Moreover,

$$Qv_i(x,\cdot)(\alpha)=\sum_{\alpha\neq\beta}q_{\alpha\beta}(v_i(x,\beta)-v_i(x,\alpha)). \tag{16}$$

Let $S_i(\alpha)$ be the machine configuration changes (setup) defined by the following:

(a) for Hypothesis 1,

$$S_i(\alpha)=\left\{x:\min_{j\neq i}\left[R_{ij}(x,\Theta_{ij})+(P_{ij})e^{-\rho\Theta_{ij}}\cdot v_j(x-d\Theta_{ij},1)+(1-P_{ij})e^{-\rho\Theta_{ij}}\cdot v_j(x-d\tau_{ij},1)\right]=v_i(x,\alpha)\right\}; \tag{17}$$

(b) for Hypothesis 2,

$$S_i(\alpha)=\left\{x:\min_{j\neq i}\left[R_{ij}(x,\widehat{\Theta_{ij}})+e^{-\rho\widehat{\Theta_{ij}}}\cdot v_j(x-d\widehat{\Theta_{ij}},1)\right]=v_i(x,\alpha)\right\}. \tag{18}$$

Let us use the Kushner approach method (Kushner and Dupuis [13]), as in [16], to develop the numerical form of HJB equations. The basic idea behind it consists in using an approximation scheme for the gradient of the value function. Let h_j, $j=1,2$, denote the length of the finite difference interval of the variable x_j. Using the finite difference approximation, $v(x,\alpha)$ could be given by $v_i^h(x,\alpha)$ and $(v_i)_{xj}(x,\alpha)$ by

$$(v_i)_{xj}(x,\alpha)=\begin{cases}\dfrac{1}{h_j}\left(v_i^h(x_1,\ldots,x_j+h_j,\ldots,x_2)-v_i^h(x_1,\ldots,x_j,\ldots,x_2)\right) & \text{if } u_j-d_j\geq 0\\ \dfrac{1}{h_j}\left(v_i^h(x_1,\ldots,x_j,\ldots,x_2)-v_i^h(x_1,\ldots,x_j-h_j,\ldots,x_2)\right) & \text{if } u_j-d_j<0.\end{cases} \tag{19}$$

The following expression can be deduced:

$$
\begin{aligned}
(u_j - d_j)(v_i)_{xj}(x,\alpha) &= \frac{|u_j - d_j|}{h_j} \\
&\cdot v_i^h(x_1,\ldots,x_j + h_j,\ldots,x_2)\,\mathrm{Ind}\{u_j - d_j \geq 0\} \\
&+ \frac{|u_j - d_j|}{h_j} v_i^h(x_1,\ldots,x_j - h_j,\ldots,x_2) \\
&\cdot \mathrm{Ind}\{u_j - d_j < 0\} - \frac{|u_j - d_j|}{h_j} \\
&\cdot v_i^h(x_1,\ldots,x_j,\ldots,x_2).
\end{aligned} \tag{20}
$$

Following the previous approximation, we can express (14) and (15) in terms of $v_i^h(x,\alpha)$ as follows:

(a) for Hypothesis 1,

$$
\begin{aligned}
v_i^h(x,\alpha) &= \min\left\{\min_{u\in\Gamma_i(\alpha)}\left\{\left(\rho + |q_{\alpha\alpha}| + \sum_{j=1}^{2}\frac{|u_j - d_j|}{h_j}\right)^{-1}\left[\sum_{j=1}^{2}\frac{|u_j - d_j|}{h_j} v_i^h(x(h_j,+)\,\mathrm{Ind}(u_j - d_j \geq 0) + v_i^h(x(h_j,-))\,\mathrm{Ind}(u_j - d_j < 0)) + g(x) + \sum_{\beta\neq\alpha} q_{\alpha\beta} * v_i^h(x,\beta)\right]\right\};\right. \\
&\left.\min_{j\neq i}\left[R_{ij}(x,\Theta_{ij}) + (P_{ij})e^{-\rho\Theta_{ij}}\cdot v_j(x - d\Theta_{ij},1) + (1 - P_{ij})e^{-\rho\Theta_{ij}}\cdot v_j(x - d\tau_{ij},1)\right]\right\};
\end{aligned} \tag{21}
$$

(b) for Hypothesis 2,

$$
\begin{aligned}
v_i^h(x,\alpha) &= \min\left\{\min_{u\in\Gamma_i(\alpha)}\left\{\left(\rho + |q_{\alpha\alpha}| + \sum_{j=1}^{2}\frac{|u_j - d_j|}{h_j}\right)^{-1}\left[\sum_{j=1}^{2}\frac{|u_j - d_j|}{h_j} v_i^h(x(h_j,+)\,\mathrm{Ind}(u_j - d_j \geq 0) + v_i^h(x(h_j,-))\,\mathrm{Ind}(u_j - d_j < 0)) + g(x) + \sum_{\beta\neq\alpha} q_{\alpha\beta} * v_i^h(x,\beta)\right]\right\};\right. \\
&\left.\min_{j\neq i}\left[R_{ij}(x,\widehat{\Theta_{ij}}) + e^{-\rho\widehat{\Theta_{ij}}}\cdot v_j(x - d\widehat{\Theta_{ij}},1)\right]\right\}
\end{aligned} \tag{22}
$$

with

$$
\begin{aligned}
v_i^h(x(h_j,+)) &= v_i^h(x_1,\ldots,x_j + h,\ldots,x_2), \\
v_i^h(x(h_j,-)) &= v_i^h(x_1,\ldots,x_j - h,\ldots,x_2).
\end{aligned} \tag{23}
$$

For all $\alpha \in M$, let us define the following expressions:

$$
\begin{aligned}
Q_h^{\alpha}(x,u) &= |q_{\alpha\alpha}| + \sum_{j=1}^{2}\frac{|u_j - d_j|}{h_j}, \\
P_h^{\alpha}(x,x \pm h_j,u) &= \frac{|u_j - d_j|}{h_j Q_h^{\alpha}(x,u)}, \\
\widetilde{P}_h^{\alpha}(x,\alpha,\beta,u) &= \frac{q_{\alpha\beta}}{Q_h^{\alpha}(x,u)}.
\end{aligned} \tag{24}
$$

Substituting these expressions in (21) and (22), we obtain the following two equations:

(a) for Hypothesis 1,

$$
\begin{aligned}
v_i^h(x,\alpha) &= \min\left\{\min_{u\in\Gamma_i(\alpha)}\left\{\frac{Q_h^{\alpha}(x,u)}{\rho + Q_h^{\alpha}(x,u)}\left\langle\sum_{j=1}^{2} P_h^{\alpha}(x,x \pm h_j,u)\,v_i^h(x,\alpha) + \sum_{\beta\neq\alpha}\widetilde{P}_h^{\alpha}(x,\alpha,\beta,u)\cdot v_i^h(x,\beta)\right\rangle + \frac{g(x)}{\rho + Q_h^{\alpha}(x,u)}\right\}\right. \\
&\left.\cdot\min_{j\neq i}\left[R_{ij}(x,\Theta_{ij}) + (P_{ij})e^{-\rho\Theta_{ij}}\cdot v_j(x - d\Theta_{ij},1) + (1 - P_{ij})e^{-\rho\Theta_{ij}}\cdot v_j(x - d\tau_{ij},1)\right]\right\};
\end{aligned} \tag{25}
$$

(b) for Hypothesis 2,

$$v_i^h(x,\alpha) = \min\left\{\min_{u\in\Gamma_i(\alpha)}\left\{\frac{Q_h^{\alpha}(x,u)}{\rho+Q_h^{\alpha}(x,u)}\left\langle\sum_{j=1}^{2}P_h^{\alpha}(x,x\pm h_j,u)\,v_i^h(x,\alpha)+\sum_{\beta\neq\alpha}\widetilde{P}_h^{\alpha}(x,\alpha,\beta,u)\cdot v_i^h(x,\beta)\right\rangle+\frac{g(x)}{\rho+Q_h^{\alpha}(x,u)}\right\}\right. \tag{26}$$
$$\left.\cdot\min_{j\neq i}\left[R_{ij}(x,\widehat{\Theta_{ij}})+e^{-\rho\widehat{\Theta_{ij}}}\cdot v_j(x-d\widehat{\Theta_{ij}},1)\right]\right\}.$$

Let us specify the terms of the previous equations, the discretization domain, and the limit conditions and illustrate the algorithm used to solve the modified numerical version of the HJB equations obtained. In addition, (25) and (26) correspond to four equations, expressing the optimality conditions concerning the production system under study, involving two products and a machine with two modes.

(i) Hypothesis 1. Let us denote by P_{12} the probability that the machine is in failure mode at the end of a setup from P_1 to P_2, if it was operational when the setup started, and by τ_{12} the corresponding failure time. Similarly, P_{21} is the probability that the machine is in failure mode at the end of a setup from P_2 to P_1, if it was operational when the setup started, with τ_{21} denoting the corresponding failure time. According to the random machine failure process,

$$P_{12} = e^{-q_{12}\Theta_{12}}. \tag{27}$$

In order to calculate τ_{12} we evaluate the conditional expectation $E\{t \mid t < \Theta_{12}\}$ and obtain

$$\tau_{12} = \frac{1}{q_{12}}\left(1-e^{-q_{12}\Theta_{12}}(1+q_{12}\Theta_{12})\right). \tag{28}$$

For the setup from product 2 to product 1, we have $P_{12} = e^{-q_{12}\Theta_{21}}$ and

$$\tau_{21} = \frac{1}{q_{12}}\left(1-e^{-q_{12}\Theta_{21}}(1+q_{12}\Theta_{21})\right). \tag{29}$$

We then have

$$v_1^h(x,1) = \min\left\{\min_{u_1\in\Gamma_1(1)}\left\{\frac{Q_h^{1}(x,u)}{\rho+Q_h^{1}(x,u)}\left\langle\sum_{j=1}^{2}P_h^{1}(x,x\pm h_j,u)\,v_1^h(x,1)+\widetilde{P}_h^{1}(x,1,2,u)\cdot v_1^h(x,2)\right\rangle+\frac{g(x)}{\rho+Q_h^{1}(x,u)}\right\}\right.$$
$$\left.\cdot\left[R_{12}(x,\Theta_{12})+(P_{12})e^{-\rho\Theta_{12}}\cdot v_2^h(x_1-d_1\Theta_{12},1,x_2-d_2\Theta_{12},1)+(1-P_{12})\left[e^{-\rho\tau_{12}}\cdot v_1^h(x_1-d_1\tau_{12},1,x_2-d_2\tau_{12},1)\right]\right]\right\},$$

$$v_2^h(x,1) = \min\left\{\min_{u_2\in\Gamma_2(1)}\left\{\frac{Q_h^{1}(x,u)}{\rho+Q_h^{1}(x,u)}\left\langle\sum_{j=1}^{2}P_h^{1}(x,x\pm h_j,u)\,v_2^h(x,1)+\widetilde{P}_h^{1}(x,1,2,u)\cdot v_2^h(x,2)\right\rangle+\frac{g(x)}{\rho+Q_h^{1}(x,u)}\right\}\right.$$
$$\left.\cdot\left[R_{21}(x,\Theta_{21})+(P_{12})e^{-\rho\Theta_{21}}\cdot v_1^h(x_1-d_1\Theta_{21},1,x_2-d_2\Theta_{21},1)+(1-P_{12})\left[e^{-\rho\tau_{21}}\cdot v_2^h(x_1-d_1\tau_{21},1,x_2-d_2\tau_{21},1)\right]\right]\right\}, \tag{30}$$

$$v_1^h(x,2) = \min\left\{\frac{Q_h^{2}(x,0)}{\rho+Q_h^{2}(x,0)}\left\langle\sum_{j=1}^{2}P_h^{2}(x,x\pm h_j,0)\,v_1^h(x,2)+\widetilde{P}_h^{1}(x,2,1,0)\cdot v_1^h(x,1)\right\rangle+\frac{g(x)}{\rho+Q_h^{2}(x,0)}\right\},$$

$$v_2^h(x,2) = \min\left\{\frac{Q_h^{2}(x,0)}{\rho+Q_h^{2}(x,0)}\left\langle\sum_{j=1}^{2}P_h^{2}(x,x\pm h_j,0)\,v_2^h(x,2)+\widetilde{P}_h^{1}(x,2,1,0)\cdot v_2^h(x,1)\right\rangle+\frac{g(x)}{\rho+Q_h^{2}(x,0)}\right\}.$$

We notice that the setup expression no longer appears in both failure mode equations and that the first two operational mode equations are different from the equivalent equations for the reference hypothesis of the literature.

(ii) Hypothesis 2. Using a known formula = $\Theta_{12}/\mathrm{MTTF}$, we get the terms of the corrections due to repairs $r_{12} = (\Theta_{12} * \mathrm{MTTR})/\mathrm{MTTF}$ and $r_{21} = (\Theta_{21} * \mathrm{MTTR})/\mathrm{MTTF}$ (here $\mathrm{MTTF} = 1/q_{12}$). Finally, the modified setup time is calculated:

$$\widehat{\Theta_{12}} = \Theta_{12} + \frac{\Theta_{12} * \mathrm{MTTR}}{\mathrm{MTTF}},$$

$$\widehat{\Theta_{21}} = \Theta_{21} + \frac{\Theta_{21} * \mathrm{MTTR}}{\mathrm{MTTF}},$$

$$v_1{}^h(x,1) = \min\left\{\min_{u_1\in\Gamma_1(1)}\left\{\frac{Q_h{}^1(x,u)}{\rho+Q_h{}^1(x,u)}\left\langle\sum_{j=1}^{2}P_h{}^1(x,x\pm h_j,u)\,v_1{}^h(x,1)+\widetilde{P}_h{}^1(x,1,2,u)\cdot v_1{}^h(x,2)\right\rangle+\frac{g(x)}{\rho+Q_h{}^1(x,u)}\right\}\cdot\left[R_{12}\left(x,\widehat{\Theta_{12}}\right)+e^{-\rho\widehat{\Theta_{12}}}\cdot v_2{}^h\left(x_1-d_1\widehat{\Theta_{12}},1,x_2-d_2\widehat{\Theta_{12}},1\right)\right]\right\},$$

$$v_2{}^h(x,1) = \min\left\{\min_{u_1\in\Gamma_1(1)}\left\{\frac{Q_h{}^1(x,u)}{\rho+Q_h{}^1(x,u)}\left\langle\sum_{j=1}^{2}P_h{}^1(x,x\pm h_j,u)\,v_2{}^h(x,1)+\widetilde{P}_h{}^1(x,1,2,u)\cdot v_2{}^h(x,2)\right\rangle+\frac{g(x)}{\rho+Q_h{}^1(x,u)}\right\}\cdot\left[R_{21}\left(x,\widehat{\Theta_{21}}\right)+e^{-\rho\widehat{\Theta_{21}}}\cdot v_1{}^h\left(x_1-d_1\widehat{\Theta_{21}},1,x_2-d_2\widehat{\Theta_{21}},1\right)\right]\right\},\quad (31)$$

$$v_1{}^h(x,2) = \min\left\{\frac{Q_h{}^2(x,0)}{\rho+Q_h{}^2(x,0)}\left\langle\sum_{j=1}^{2}P_h{}^2(x,x\pm h_j,0)\,v_1{}^h(x,2)+\widetilde{P}_h{}^1(x,2,1,0)\cdot v_1{}^h(x,1)\right\rangle+\frac{g(x)}{\rho+Q_h{}^2(x,0)}\right\},$$

$$v_2{}^h(x,2) = \min\left\{\frac{Q_h{}^2(x,0)}{\rho+Q_h{}^2(x,0)}\left\langle\sum_{j=1}^{2}P_h{}^2(x,x\pm h_j,0)\,v_2{}^h(x,2)+\widetilde{P}_h{}^1(x,2,1,0)\cdot v_2{}^h(x,1)\right\rangle+\frac{g(x)}{\rho+Q_h{}^2(x,0)}\right\}.$$

We also notice that the setup expression no longer appears in both of the failure mode equations and that the first two operational mode equations are similar to those of the reference hypothesis (with the setup times modified), but different from equations related to the case of Hypothesis 1. In the next section, we present a numerical example and analyze the results obtained in different situations.

5. Numerical Example and Results Analysis

In order to characterize the production and setup policies and to show the influence of the interactions between the setup procedure and the random machine failure process, for two new hypotheses in comparison with the reference hypothesis, we present three different cases in three data groups. The first two groups had identical economic parameters, and only technical parameters such as the setup duration were varied according to the groups. In the third group, both products had different economic parameters. Table 1 shows the constant parameters for the numerical examples considered, with

$$c_1^- = c_2^-;\quad c_1^+ = c_2^+;\quad K_{12} = K_{21};\quad d_1 = d_2;\quad U_1^+ = U_2^+ \quad (32)$$

for the two first groups.

In addition,

(i) group 1 represents the case $\Theta_{12} = \Theta_{21} = 1.25$;

(ii) group 2 represents the case $\Theta_{12} = \Theta_{21} = 1.75$;

(iii) group 3 is the case of two nonidentical products $\Theta_{12} = \Theta_{21} = 1.25$ and $C_1^+ = C_2^+ = 1$; $C_1^- = 20$; $C_2^- = 40$.

TABLE 1: Data parameters.

c_1^-	c_1^+	K_{12}	U_1^+	d_1	x_1^+	x_1^-	q_{12}	q_{21}	ρ	$h_{x_1^+}$
20	1	0.5	5	1.5	20	−5	0.1	0.8	0.1	0.2

TABLE 2: Setup times for machine tools (Boothroyd et al. [15]).

	Some nonproductive times for common machine tools		
Machine tool	Time to engage tool and so forth (s)	Basic setup time (h)	Additional setup per tool (h)
Horizontal band saw	—	0.17	—
Manual turret lathe	9	0.15	0.2
NC turret lathe	1.5	0.5	0.15
Milling machine	30	1.5	—
Drilling machine	9	1	—
Horizontal-boring machine	30	1.3	—
Broaching machine	13	0.6	—
Gear hobbling machine	39	0.9	—
Grinding machine	19	0.6	—
Internal grinding machine	24	0.6	—
Machining center	8	0.7	0.05

The setup times are given by the table of the machine tools setup times (see Table 2).

With these data, the system is capable of producing with setup at the request of both products if the following feasibility condition is satisfied:

$$U^i_{\max}\frac{q_{21}}{q_{12}+q_{21}}e^{-q_{12}\theta_{12}} > d_1, d_2 \quad (i=1,2). \qquad (33)$$

We present six figures for every case (three figures for each product). Each figure contains the production or setup policies for the two new hypotheses and for the reference hypothesis. For group 1, Figures 3 and 6, respectively, illustrate the production policy of P_1 and P_2 for three hypotheses. Then, also for group 1, Figures 4 and 7 illustrate the setup policies for P_1 to P_2 and for P_2 to P_1, respectively. Finally, for group 1, Figures 5 and 8, respectively, illustrate the association of both policies (production and setup) for P_1 and the association of both policies (production and setup) for P_2. In the next section, analyses of sensibility will be provided to study the effects of variation of the various costs on the control policies.

Group 1 (product type 1) $\Theta_{12} = \Theta_{21} = 1.25$ (see Figures 3–5).

Group 1 (product type 2) $\Theta_{12} = \Theta_{21} = 1.25$ (see Figures 6–8).

Group 2 (product type 1) $\Theta_{12} = \Theta_{21} = 1.75$ (see Figures 9–11).

Group 2 (product type 2) $\Theta_{12} = \Theta_{21} = 1.75$ (see Figures 12–14).

Group 3 (product type 1) $\Theta_{12} = \Theta_{21} = 1.25$ and $C_1^+ = 1$; $C_1^- = 20$ (see Figures 15–17).

Group 3 (product type 2) $\Theta_{12} = \Theta_{21} = 1.25$ and $C_1^+ = 1$; $C_1^- = 20$ (see Figures 18–20).

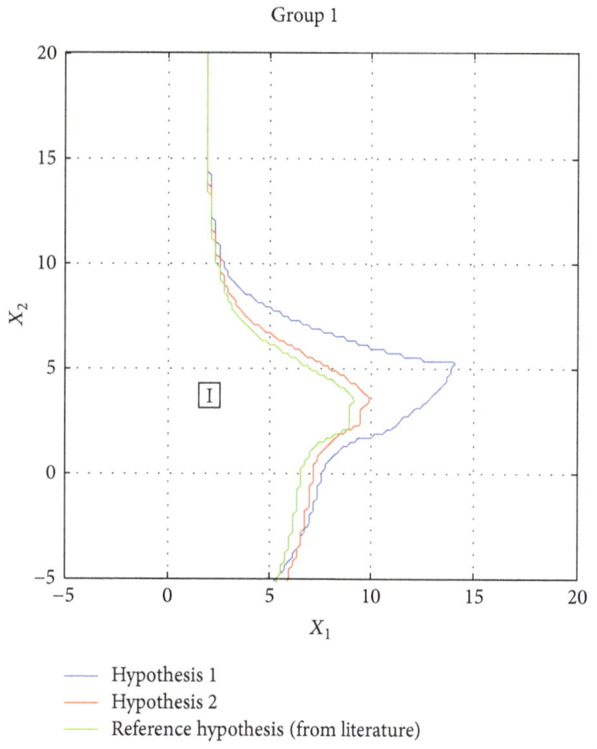

FIGURE 3: Production policy for P_1.

Let us now interpret the results obtained and illustrated by Figures 3–20. This analysis will allow us to present the structure of the production and setup policies for the two product part types and with regard to the formulated hypotheses.

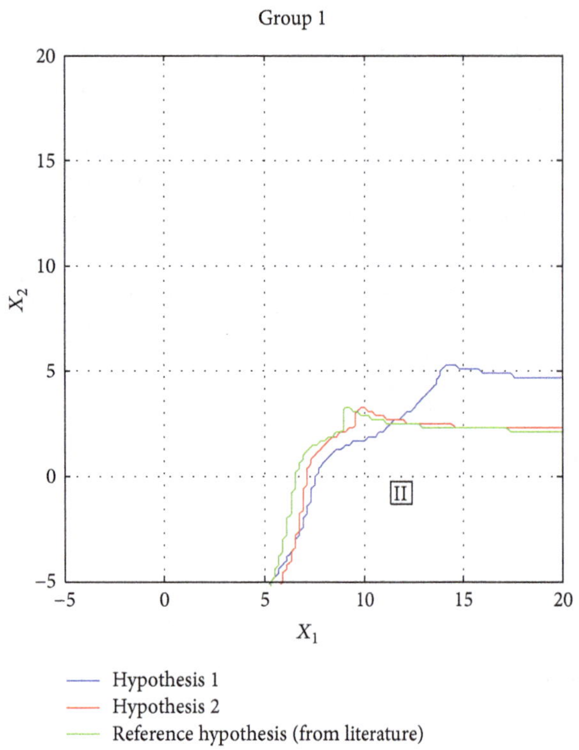

FIGURE 4: Setup policy (product 1 to product 2).

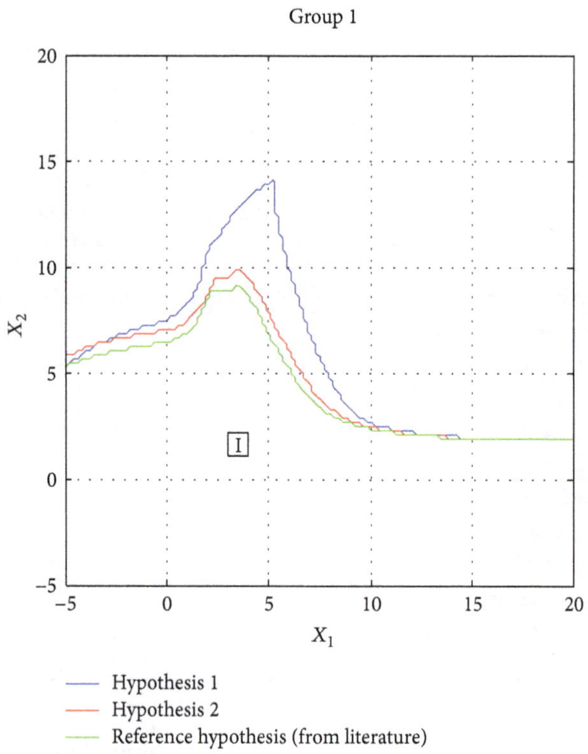

FIGURE 6: Production policy for P_2.

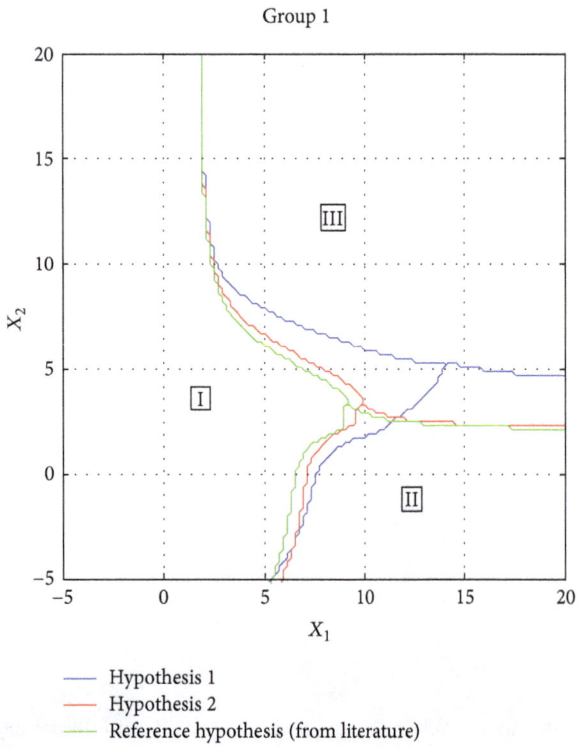

FIGURE 5: Production policy for P_1 and setup policy (P_1 to P_2).

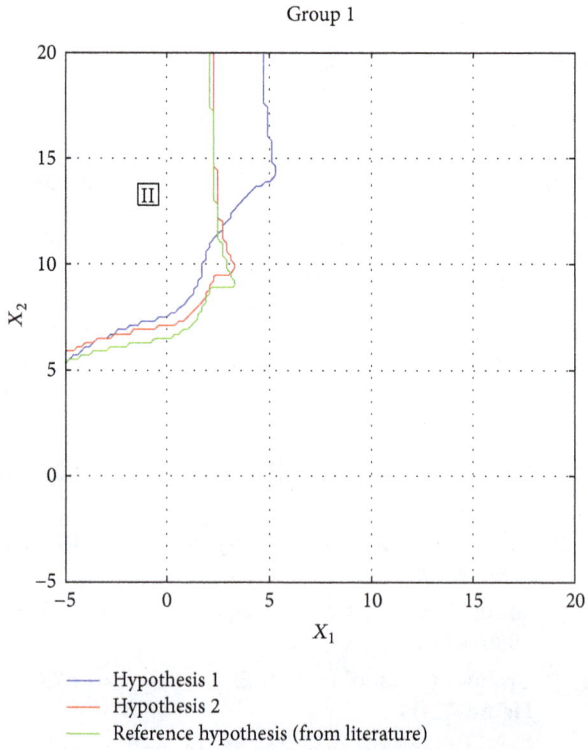

FIGURE 7: Setup policy (product 2 to product 1).

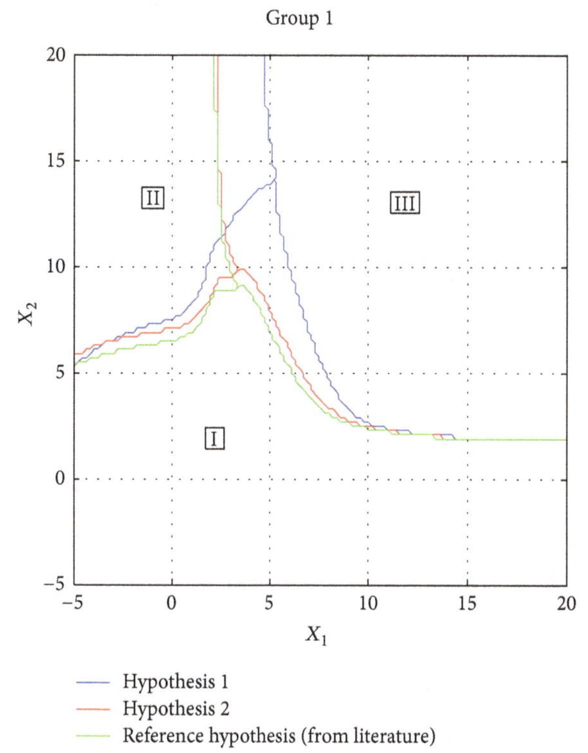

FIGURE 8: Production policy for P_2 and setup policy (P_2 to P_1).

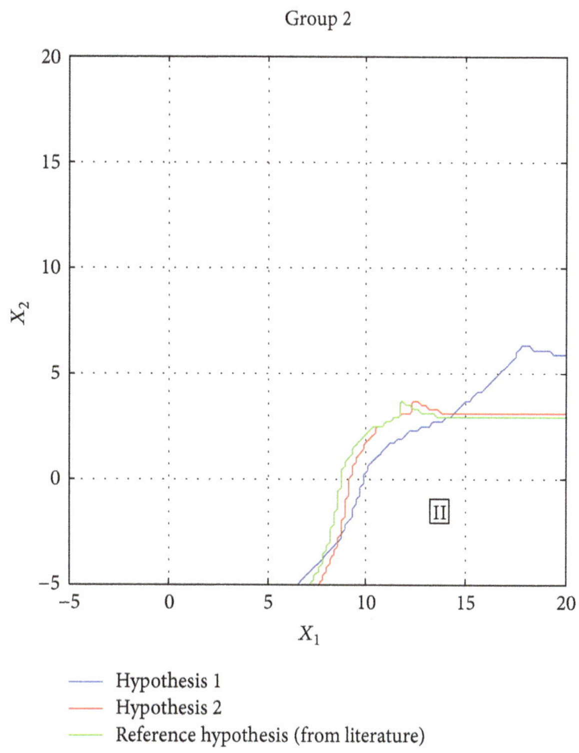

FIGURE 10: Setup policy (product 1 to product 2).

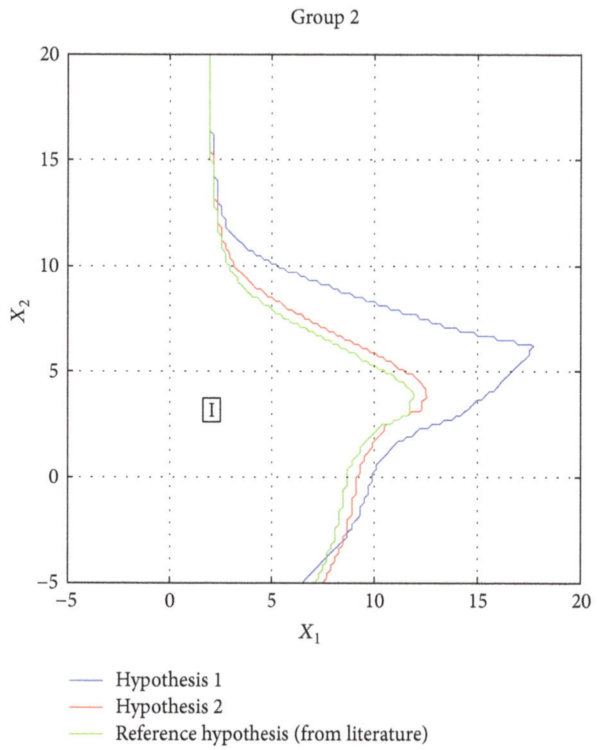

FIGURE 9: Production policy for P_1.

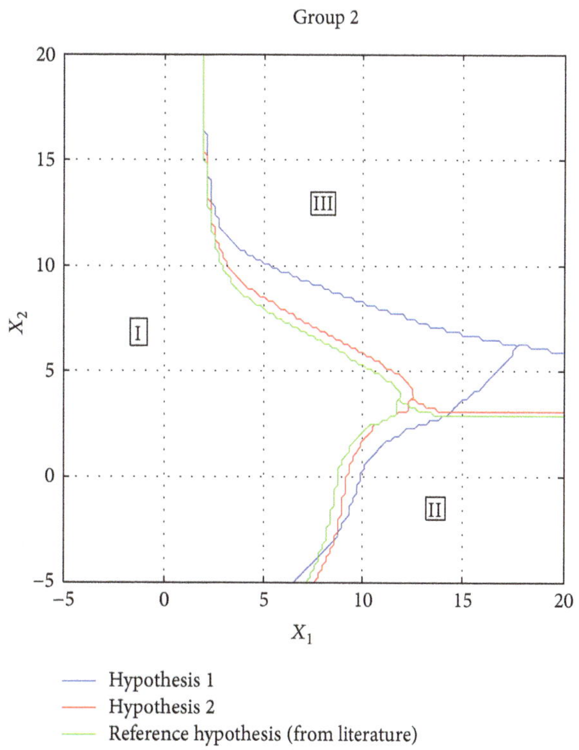

FIGURE 11: Production policy for P_1 and setup policy (product 1 to product 2).

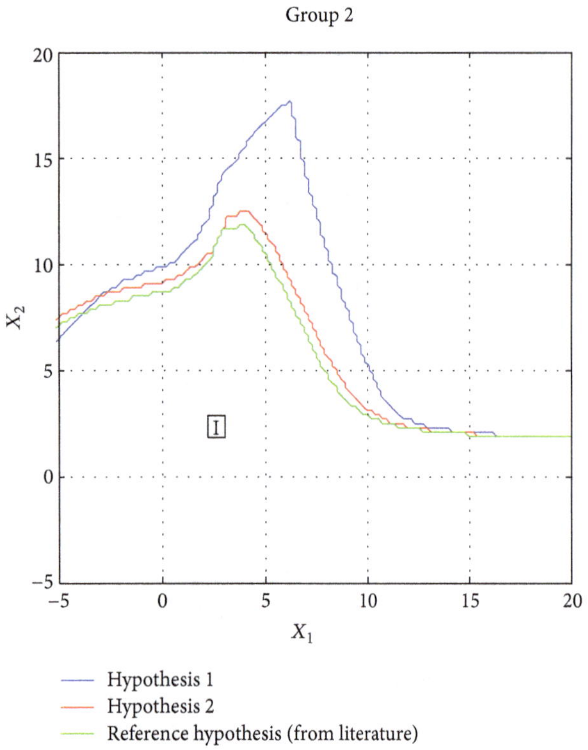

Figure 12: Production policy for P_2.

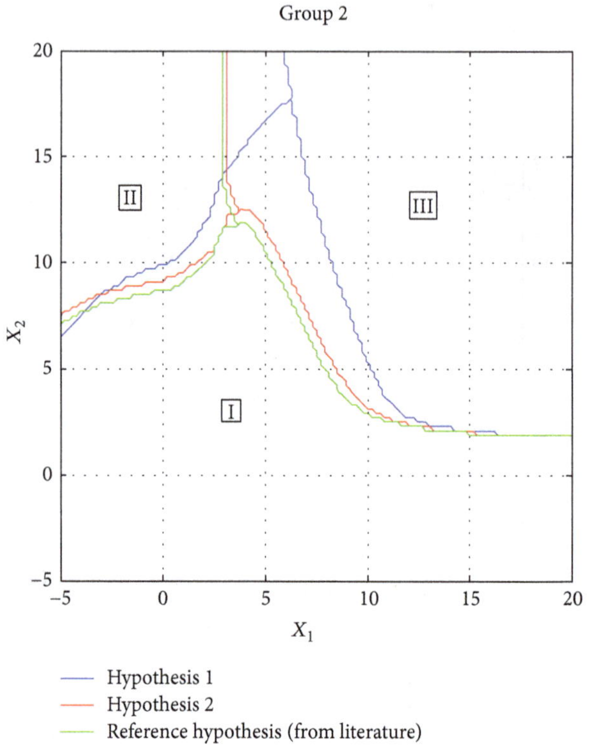

Figure 14: Production policy for P_2 and setup policy (product 2 to product 1).

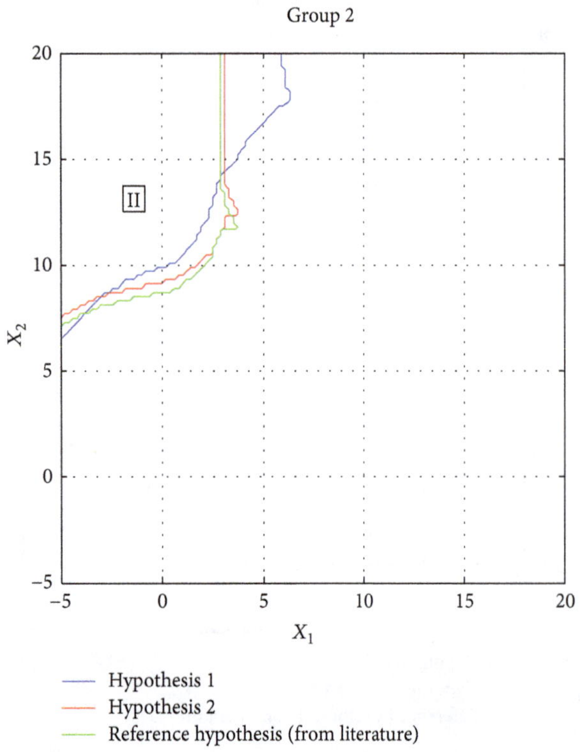

Figure 13: Setup policy (product 2 to product 1).

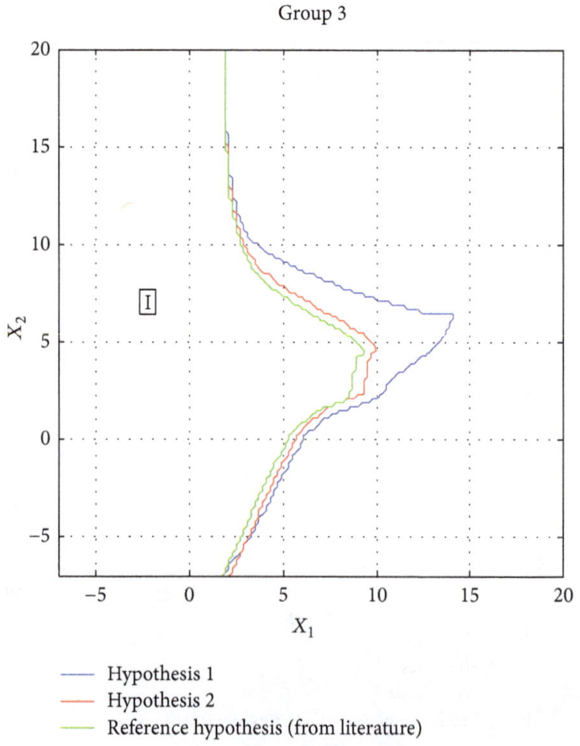

Figure 15: Production policy for P_1.

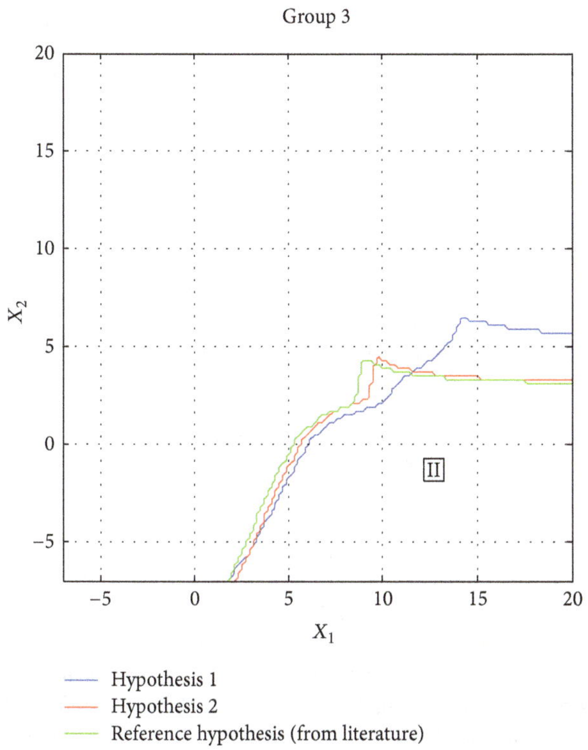

FIGURE 16: Setup policy (product 1 to product 2).

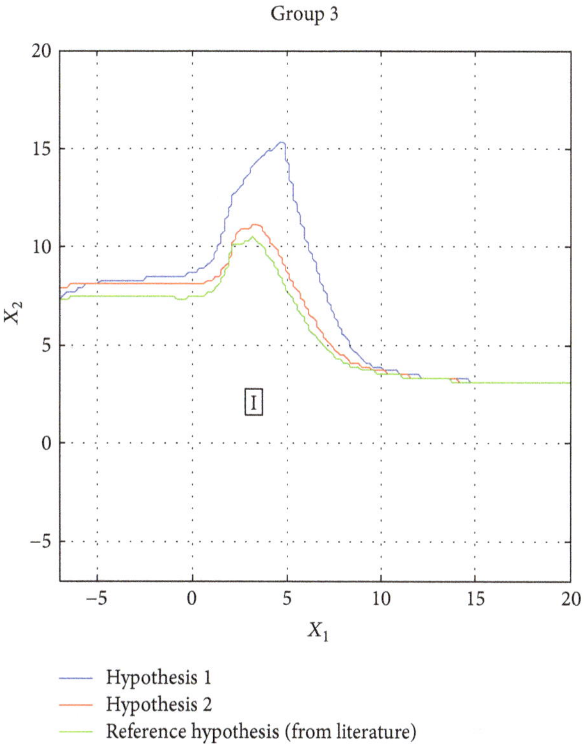

FIGURE 18: Production policy for P_2.

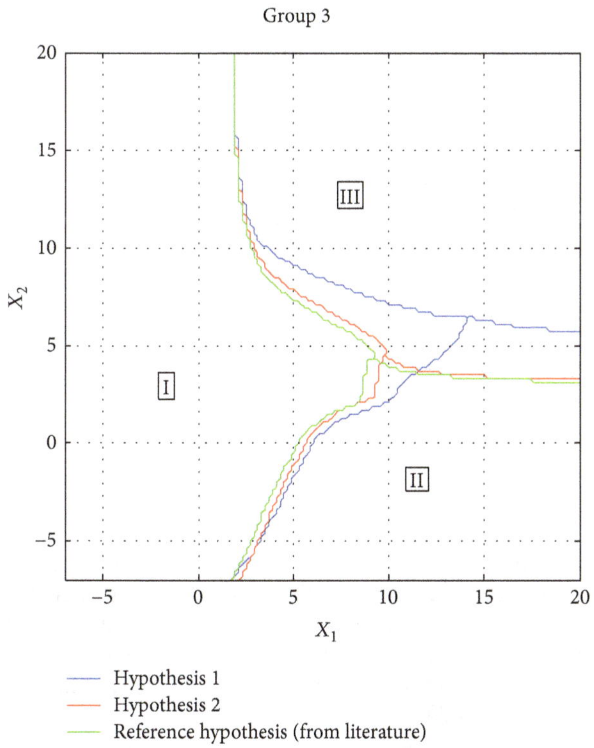

FIGURE 17: Production policy for P_1 and setup policy (product 1 to product 2).

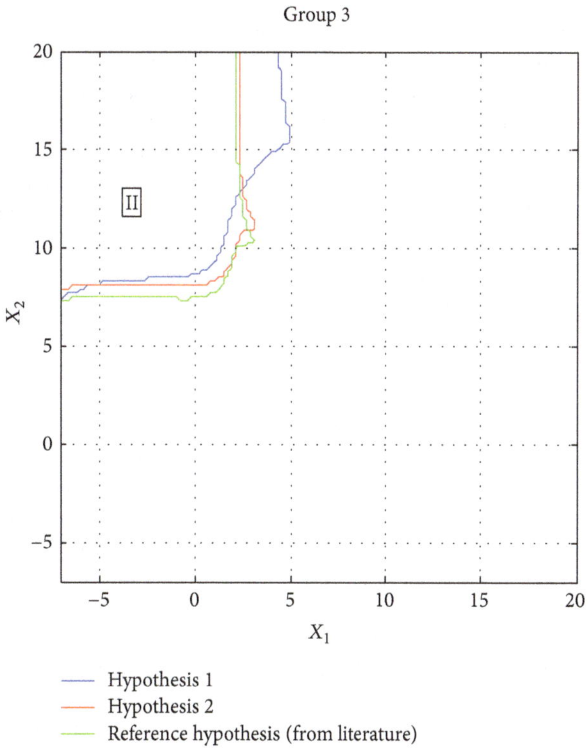

FIGURE 19: Setup policy (product 2 to product 1).

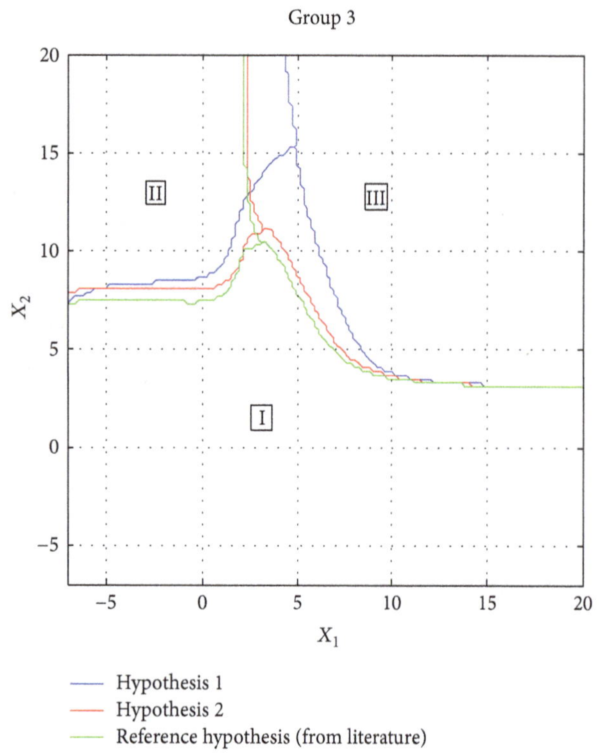

FIGURE 20: Production policy for P_2 and setup policy (product 2 to product 1).

5.1. Optimal Production Policy. An analysis of three cases shows that we have to produce at the maximum rate in region I when the machine is configured for the same product. In region II, the machine is configured (according to the setup policy) for the production of the other type of product. In region III, the production policy recommends stopping the machine and producing nothing by setting the production rate to zero. It is interesting to note that, in Hypothesis 1, the optimal inventory level is very big, contrary to Hypothesis 2 and to the reference hypothesis, as seen in Bai and Elhafsi [14], Boukas and Kenné [10], and Hajji et al. [11]. This increase in the inventory level is understandable given that if a breakdown occurs during the setup operation, we have to stop the setup until the machine repair is completed. This breakdown cancels all the data relative to the setup activities started before the failure. In this case, the operator has to resume the setup operation, which increases the overall setup time and leads to a loss of availability of the machine. According to Hypothesis 2, the optimal inventory level is slightly bigger than that of the reference hypothesis. This light increase in the inventory level is understandable, given that the setup operation must be stopped when a breakdown occurs. This breakdown does not cancel all the data relative to the setup activities (contrary to Hypothesis 1). The operator pursues the setup activities after the repair of the machine. The global setup time of Hypothesis 2 is then higher than that of the reference hypothesis. In fact, we have a loss of availability of the machine, but to a smaller extent than in the case of Hypothesis 1. The optimal production policy is

TABLE 3: Comparative study of optimal threshold inventory levels.

Group	Hypothesis 1	Hypothesis 2	Literature hypothesis
1	$Z_1 = 14$	$Z_1 = 9.8$	$Z_1 = 9$
	$Z_2 = 14$	$Z_2 = 9.8$	$Z_2 = 9$
2	$Z_1 = 17.6$	$Z_1 = 12.4$	$Z_1 = 11.8$
	$Z_2 = 17.6$	$Z_2 = 12.4$	$Z_2 = 11.8$
3	$Z_1 = 14$	$Z_1 = 9.8$	$Z_1 = 9$
	$Z_2 = 15.2$	$Z_2 = 11$	$Z_2 = 10.4$

then of hedging policy structure and can be expressed by the following two equations (for product 1 and product 2):

$$u_1(x_1, x_2) = \begin{cases} U_1^+ & \text{if } x_1 < Z_1(x_2) \\ 0 & \text{if } x_1 > Z_1(x_2), \end{cases}$$
$$u_2(x_1, x_2) = \begin{cases} U_2^+ & \text{if } x_2 < Z_1(x_1) \\ 0 & \text{if } x_2 > Z_1(x_1). \end{cases} \quad (34)$$

5.2. Optimal Setup Policy. By analyzing Figures 4 and 5 in region II (zone in which a setup from P_1 to P_2 or from P_2 to P_1 is allowed), we can observe that the setup policies of Hypotheses 1 and 2 give a margin bigger than that of the reference hypothesis. This trend reduces region III and so allows the system the possibility of performing the setup without any shortage risk for the other product. The setup policies are given in this case by the following expressions:

$$S_{12} = \begin{cases} 1 & \text{if } \begin{cases} x_1 \geq a_1, \\ x_2 \leq 0 \end{cases} \text{ or } \begin{cases} x_1 < 0, \\ x_2 \leq -b_1 \end{cases} \\ 0 & \text{otherwise,} \end{cases}$$
$$S_{21} = \begin{cases} 1 & \text{if } \begin{cases} x_1 \leq c_2 x_2 + b_2, \\ x_1 \geq 0,\ x_2 \geq 0 \end{cases} \text{ or } \begin{cases} x_2 \geq a_2, \\ x_1 \leq 0 \end{cases} \\ 0 & \text{otherwise} \end{cases} \quad (35)$$

with

$$\begin{aligned} &0 \leq a_1 \leq Z_1, \\ &b_1 \leq 0, \\ &0 \leq a_2 \leq Z_2, \\ &0 \leq c_2 a_2 + b_2 \leq Z_1. \end{aligned} \quad (36)$$

To conclude this section, we recapitulate the results obtained according to the critical thresholds Z_1 and Z_2, which characterize the production and setup policies presented by (34) and (35). The values of the optimal threshold inventory levels obtained numerically for the three groups of data are presented in Table 3.

These results show our contribution, given that all the previous works in the literature did not handle the case of industrial systems subjected to Hypotheses 1 and 2. In this paper, we determined the production structures and setup policies for these industrial systems.

6. Conclusion

This research clearly defines the production planning problem and the setup strategies for industrial systems subjected to specific hypotheses. In this paper, we considered two hypotheses that hold that a breakdown can occur during a setup activity. This breakdown can cancel (or may not cancel) the setup activities undertaken before the breakdown occurs. Hence, we propose new optimality conditions integrating the probability of breakdowns during the setup. A numerical approach is used to solve the optimality conditions obtained. A numerical example and a comparative analysis of the results for three groups of data allow us to determine the production structures and setup policies for manufacturing systems that have previously never been studied in the literature. This work can be extended to the cases of industrial systems allowing setup activities in all modes of the machine (operational or failure modes).

Conflict of Interests

The authors declare that there is no conflict of interests regarding the publication of this paper.

References

[1] S. P. Sethi and H. Zhang, "Average-cost optimal policies for an unreliable flexible multiproduct machine," *The International Journal of Flexible Manufacturing Systems*, vol. 11, no. 2, pp. 147–157, 1999.

[2] A. Gharbi and J. P. Kenné, "Optimal production control problem in stochastic multiple-product multiple-machine manufacturing systems," *IIE Transactions*, vol. 35, no. 10, pp. 941–952, 2003.

[3] K. A. Francie, K. Jean-Pierre, D. Pierre, S. Victor, and P. Vladimir, "Stochastic optimal control of manufacturing systems under production-dependent failure rates," *International Journal of Production Economics*, vol. 150, pp. 174–187, 2014.

[4] A. Gharbi, J.-P. Kenné, and A. Hajji, "Operational level-based policies in production rate control of unreliable manufacturing systems with set-ups," *International Journal of Production Research*, vol. 44, no. 3, pp. 545–567, 2006.

[5] V. Polotski, J.-P. Kenné, and A. Gharbi, "Failure-prone manufacturing systems with setups: feasibility and optimality under various hypotheses about perturbations and setup interplay," *International Journal of Mathematics in Operational Research*, vol. 7, no. 6, pp. 681–705, 2015.

[6] C. Koulamas and G. J. Kyparisis, "Single-machine scheduling problems with past-sequence-dependent setup times," *European Journal of Operational Research*, vol. 187, no. 3, pp. 1045–1049, 2008.

[7] W. Feng, L. Zheng, and J. Li, "The robustness of scheduling policies in multi-product manufacturing systems with sequence-dependent setup times and finite buffers," *Computers and Industrial Engineering*, vol. 63, no. 4, pp. 1145–1153, 2012.

[8] S. P. Sethi and Q. Zhang, *Hierarchical Decision Making in Stochastic Manufacturing Systems*, Birkhäuser, Boston, Mass, USA, 1994.

[9] H. Yan and Q. Zhang, "A numerical method in optimal production and setup scheduling of stochastic manufacturing systems," *IEEE Transactions on Automatic Control*, vol. 42, no. 10, pp. 1452–1455, 1997.

[10] E. K. Boukas and J. P. Kenné, *Maintenance and Production Control of Manufacturing Systems with Setups*, vol. 33 of *Lectures in Applied Mathematics*, American Mathematical Society, Providence, RI, USA, 1997.

[11] A. Hajji, A. Gharbi, and J. P. Kenne, "Production and set-up control of a failure-prone manufacturing system," *International Journal of Production Research*, vol. 42, no. 6, pp. 1107–1130, 2004.

[12] R. Akella and P. R. Kumar, "Optimal control of production rate in a failure prone manufacturing system," *IEEE Transactions on Automatic Control*, vol. 31, no. 2, pp. 116–126, 1986.

[13] H. J. Kushner and P. G. Dupuis, *Numerical Methods for Stochastic Control Problems in Continuous Time*, Springer, New York, NY, USA, 1992.

[14] S. X. Bai and M. Elhafsi, "Scheduling of an unreliable manufacturing system with nonresumable set-ups," *Computers and Industrial Engineering*, vol. 32, no. 4, pp. 909–925, 1997.

[15] G. Boothroyd, P. Dewhurst, and A. W. Knight, *Product Design for Manufacture and Assembly*, CRC Press, 3rd edition, 2010.

[16] A. F. Kouedeu, J. P. Kenné, P. Dejax, V. Songmene, and V. Polotski, "Production planning of a failure-prone manufacturing/remanufacturing system with production-dependent failure rates," *Applied Mathematics*, vol. 5, no. 10, pp. 1557–1572, 2014.

Permissions

All chapters in this book were first published in JAM, by Hindawi Publishing Corporation; hereby published with permission under the Creative Commons Attribution License or equivalent. Every chapter published in this book has been scrutinized by our experts. Their significance has been extensively debated. The topics covered herein carry significant findings which will fuel the growth of the discipline. They may even be implemented as practical applications or may be referred to as a beginning point for another development.

The contributors of this book come from diverse backgrounds, making this book a truly international effort. This book will bring forth new frontiers with its revolutionizing research information and detailed analysis of the nascent developments around the world.

We would like to thank all the contributing authors for lending their expertise to make the book truly unique. They have played a crucial role in the development of this book. Without their invaluable contributions this book wouldn't have been possible. They have made vital efforts to compile up to date information on the varied aspects of this subject to make this book a valuable addition to the collection of many professionals and students.

This book was conceptualized with the vision of imparting up-to-date information and advanced data in this field. To ensure the same, a matchless editorial board was set up. Every individual on the board went through rigorous rounds of assessment to prove their worth. After which they invested a large part of their time researching and compiling the most relevant data for our readers.

The editorial board has been involved in producing this book since its inception. They have spent rigorous hours researching and exploring the diverse topics which have resulted in the successful publishing of this book. They have passed on their knowledge of decades through this book. To expedite this challenging task, the publisher supported the team at every step. A small team of assistant editors was also appointed to further simplify the editing procedure and attain best results for the readers.

Apart from the editorial board, the designing team has also invested a significant amount of their time in understanding the subject and creating the most relevant covers. They scrutinized every image to scout for the most suitable representation of the subject and create an appropriate cover for the book.

The publishing team has been an ardent support to the editorial, designing and production team. Their endless efforts to recruit the best for this project, has resulted in the accomplishment of this book. They are a veteran in the field of academics and their pool of knowledge is as vast as their experience in printing. Their expertise and guidance has proved useful at every step. Their uncompromising quality standards have made this book an exceptional effort. Their encouragement from time to time has been an inspiration for everyone.

The publisher and the editorial board hope that this book will prove to be a valuable piece of knowledge for researchers, students, practitioners and scholars across the globe.

List of Contributors

Run Xu
Department of Mathematics, Qufu Normal University, Qufu, Shandong 273165, China

A. Kinfack Jeutsa
Higher Technical Teachers' Training College, The University of Buea, Buea Road, Kumba, Cameroon

A. Njifenjou
Faculty of Industrial Engineering, The University of Douala, Douala, Cameroon

J. Nganhou
National Advanced School of Engineering, The University of Yaounde I, Yaounde, Cameroon

Ahmed Farooq Qasim and Ekhlass S. AL-Rawi
College of Computer Sciences and Mathematics, University of Mosul, Iraq

Joshua Kiddy K. Asamoah and Francis T. Oduro
Department of Mathematics, Kwame Nkrumah University of Science and Technology, Kumasi, Ghana

Ebenezer Bonyah
Department of Statistics and Mathematics, Kumasi Technical University, Kumasi, Ghana

Baba Seidu
Department of Mathematics, University for Development Studies, Navrongo, Ghana

Moussa Kounta
The College of the Bahamas, School of Mathematics, Physics and Technology, Nassau, Bahamas

Silvia Martorano Raimundo
Faculdade de Medicina da Universidade de São Paulo, MLS and LIM01-HCFMUSP, São Paulo, SP, Brazil

Hyun Mo Yang
Universidade de Campinas, DMA-IMECC, Campinas, SP, Brazil

Eduardo Massad
Faculdade de Medicina da Universidade de São Paulo, MLS and LIM01-HCFMUSP, São Paulo, SP, Brazil
School of Applied Mathematics, Fundação Getúlio Vargas, Rio de Janeiro, RJ, Brazil

M. H. Daliri Birjandi, J. Saberi-Nadjafi and A. Ghorbani
Department of Applied Mathematics, School of Mathematical Sciences, Ferdowsi University of Mashhad, Mashhad, Iran

N. Ghawadri
Institute for Mathematical Research, Universiti Putra Malaysia (UPM), 43400 Serdang, Selangor, Malaysia

N. Senu, F. Ismail and Z. B. Ibrahim
Institute for Mathematical Research, Universiti Putra Malaysia (UPM), 43400 Serdang, Selangor, Malaysia
Department of Mathematics, Universiti Putra Malaysia (UPM), 43400 Serdang, Selangor, Malaysia

Srinivasa Rao Kola, Balakrishna Gudla and P. K. Niranjan
Department of Mathematical and Computational Sciences, National Institute of Technology Karnataka, Surathkal, India

Wayan Somayasa
Department of Mathematics, Halu Oleo University, Indonesia

Olumuyiwa A. Agbolade
Department of Mathematics and Statistics, Federal Polytechnic, Ilaro, Ogun State, Nigeria
Department of Mathematics, College of Science and Technology, Covenant University, Ota, Ogun State, Nigeria

Timothy A. Anake
Department of Mathematics, College of Science and Technology, Covenant University, Ota, Ogun State, Nigeria

Christian Kasumo and Dmitry Kuznetsov
Department of Applied Mathematics and Computational Science, Nelson Mandela African Institution of Science and Technology, Arusha, Tanzania

Juma Kasozi
Department of Mathematics, Makerere University, Kampala, Uganda

Dongming Wei and Piotr Skrzypacz
Department of Mathematics, School of Science and Technology, Nazarbayev University, Astana 010000, Kazakhstan

Xijun Yu
Institute of Applied Physics and Computational Mathematics, Beijing 100088, China

Ye-Mao Xia and Jian-Wei Gou
Department of Applied Mathematics, Nanjing Forestry University, Nanjing, Jiangsu 210037, China

Guy-Richard Kibouka and Donatien Nganga-Kouya
Mechanical Engineering Department, Omar Bongo University, École Normale Supérieure de l'Enseignement Technique, BP 3989, Libreville, Gabon

Jean-Pierre Kenne, Victor Songmene and Vladimir Polotski
Mechanical Engineering Department, University of Quebec, École de Technologie Supérieure, 1100 Notre Dame West, Montreal, QC, Canada H3C 1K3

Index

www.ingramcontent.com/pod-product-compliance
Lightning Source LLC
Chambersburg PA
CBHW050437200326
41458CB00014B/4976
* 9 7 8 1 6 3 2 3 8 7 3 1 8 *